RANGERS

Lead the Way

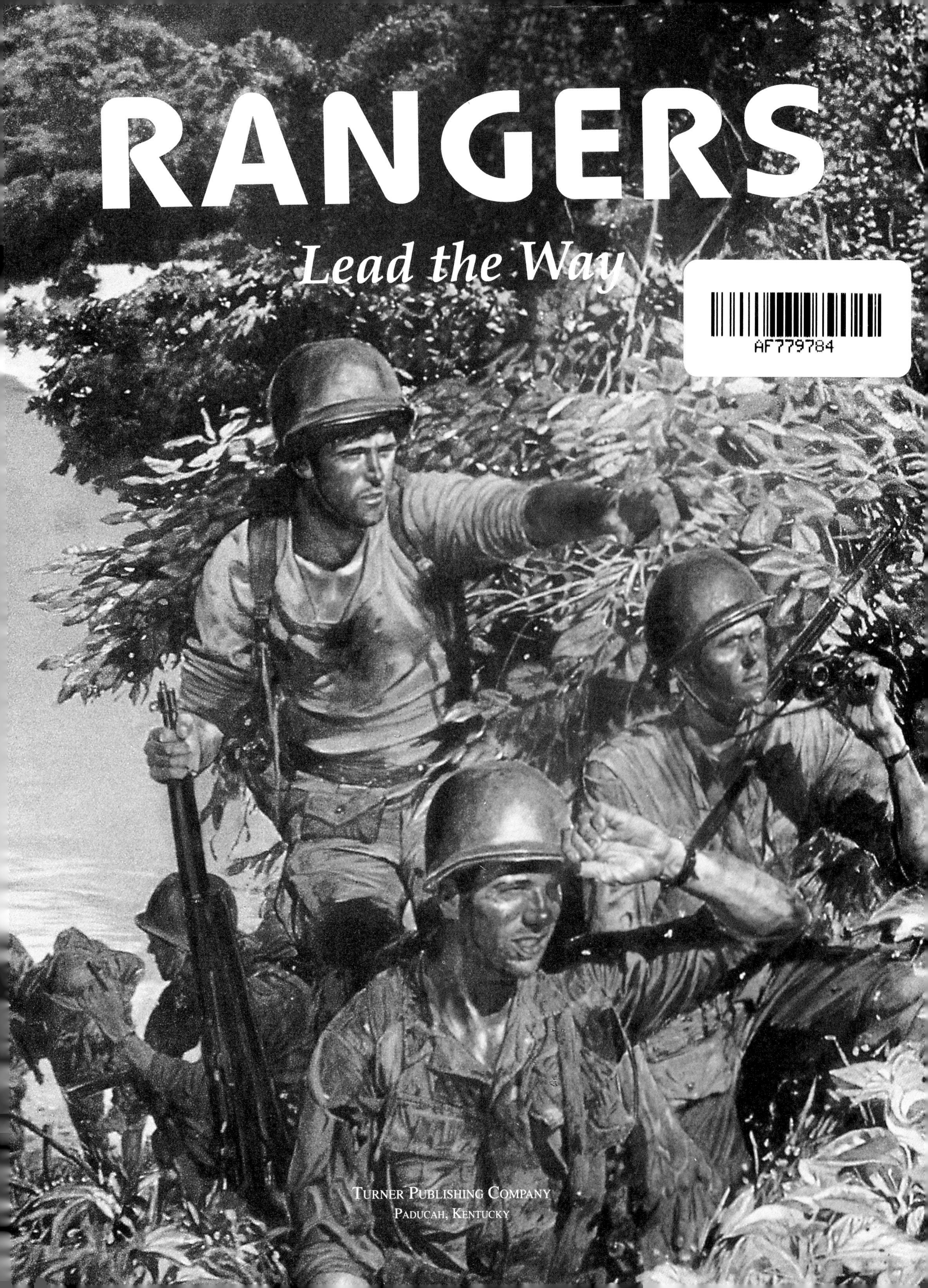

Turner Publishing Company
Paducah, Kentucky

Turner Publishing Company
412 Broadway • P.O. Box 3101
Paducah, Kentucky 42002-3101
(502) 443-0121

Author: COL Thomas H. Taylor, USA (Ret.)
Editor: Robert J. Martin

Library of Congress No. 95-60318
ISBN 1-56311-182-9

Limited Edition
Printed in the U.S.A.
Additional copies may be available directly from the Publisher.

Contents

Preface

Tribute to the Rangers of World War Two

The men of all six Ranger Battalions were from all walks of life and from all types of U.S. Army units. They came from the Regular Army, the National Guard, the Reserve Forces, and the Selective Service. They had one thing in common. THEY WERE ALL VOLUNTEERS. Every man knew that he, if he survived the harsh training, would become a member of an elite unit. He would be in top physical condition and in a unit that would be committed to the most vigorous combat, under the most difficult conditions and would be expected to be successful in each and every case. He could be injected into the combat arena by sea or air. He was expected to infiltrate enemy lines and to be able to fight in his rear areas and live off the land when necessary.

A Ranger Battalion was small when compared to a regular infantry battalion. A Ranger Battalion had only 370 enlisted men and 28 officers. Their accomplishments by far outweighed their small numbers. The Six Battalions fought in a total of 26 Campaigns, six Invasions, and who can count the innumerable battles and skirmishes? The Ranger Battalions received seven Presidential Unit Citations, and the 2nd and 5th Ranger Battalions each received a French Croix de Guerre with Silver-Gilt Star (Battalion Awards). The 6th Battalion received the Philippine Presidential Unit Citation. This Battalion would conduct a daring raid behind Japanese lines and rescue 511 American Prisoners of War at the infamous prison camp at Cabanatuan.

Rangers of the 1st Battalion fought at Dieppe where they suffered their first Americans to be killed in combat with their German foes and also they would be the first Americans to kill one of their German enemies. They were initially trained by the British Commandos and would fight with them at Dieppe. Their association with the 2nd Commando and the 41st Royal Marine Commando would be strengthened during the invasions of Italy at Salerno and again at Anzio and would develop into a close and lasting friendship that still exists today.

The Ranger Motto earned on that bloody beach called Omaha, "RANGERS LEAD THE WAY," is still the motto of today's Rangers.

Richard N. Hathaway, Jr., President
Ranger Battalions Association
of World War II

Dedication

One of the happiest days of my life was the day I presented the Merrill's Marauders colors at a ceremony in 1969 at Eglin Air Force Base activating the 75th Ranger Regiment. After a 20 year battle with the Department of Defense and the U.S. Congress by the Merrill's Marauders, Major General Kenneth D. Wilkins, Adjutant general, ordered that the 75th regiment, which traced its origin to the 5307th Composite Unit (Merrill's Marauders) would be the fitting unit to be activated as a parent organization, and authorized the cloth insignia of the Marauders to be incorporated into the design of their organizational flag. The regiment was to wear the metal insignia, as the distinctive insignia.

The modern day rangers performed so gallantly in campaigns in World war II. Darby's Rangers participated in battles on raids on Dieppe, D-Day landings, Africa, Sicily, Italy and the South Pacific who will always be remembered for performing so very well in 35 battles behind the enemy lines for a period of seven months, with a small force of 2,800, to recapture Burma, all without any relief. The Korean Rangers who distinguished themselves in battles against the North Koreans and Chinese under adverse conditions in an (unpopular war), but still persisted. The 75th Vietnam LRRP Companies which did such an outstanding job in Vietnam, without getting their true recognition, which they so rightfully deserved. The 75th ranger regiment that distinguished itself in Granada, Panama, Desert Storm and Somalia, always executing their missions to conduct special operations in support of the United States Policies and objective.

In my estimation, the Rangers of today are the best soldiers in the world, They are the best equipped, best trained, best disciplined, best motivated, and we old Rangers admire their spirit and abilities. We are proud that they have not only carried on the traditions and esprit de corps of the Rangers who came before them, but have even surpassed them in many ways. This book is dedicated to the memory of those Rangers who so gallantly sacrificed their lives in defense of their country and their comrades, as well as the Rangers who serve so ably today, and the future Rangers. These men have lived the last stanza of the ranger Creed, "Readily will I display the intestinal fortitude required to fight on to the ranger objective, and complete the mission, though I may be the lone survivor."

"Rangers Lead the Way"

Phillip B. Piazza, President
Merrill's Marauders Association

7th Ranger Inf. Co. attached to the U.S. 3rd Army.

10th Ranger Inf. Co. attached to the 45th Inf Div.

13th Ranger Inf. Co. attached to the 43rd Inf. Div.

8th Ranger Inf. Co. attached to the 24th Inf. Div.

11th Ranger Inf. Co. attached to the 40th Inf. Div.

14th Ranger Inf. Co. attached to the 4th Inf. Div.

8th Ranger Inf. Co. attached U. S. Eighth Army.

12th Ranger Inf. Co. attached to the 28th Inf. Div.

15th Ranger Inf. Co. attached to the 47th Inf. Div.

9th Ranger Inf. Co.

HQ Detachment and Ranger Co. A and B Trainees at Ft. Benning, GA, wore 3rd Army patch under the scroll.

Ranger Patch worn by most Rangers during the Korean War. It was later changed to individual companies.

Introduction

In this book, Tom Taylor takes up the story of Rangers in the 20th Century and has asked me to sketch their precedents. To do so requires revisiting the eastern seaboard in the 17th Century when virtually all British colonists in America were pioneers. Scattered, isolated and independent in a forest wilderness, they had to rely upon themselves for defense against hostile Indians. The best of the hunters and woodsmen 'ranged' out from settlements to provide early warning of Indian raids. Such men became known as 'rangers'.

When they banded together in small groups, militarily speaking they were the first long range reconnaissance patrols (lurps'). But rangers were not militarily organized, not even during the French and Indian Wars when they were employed as trackers and scouts for the British Army. Rogers Rangers in the Revolutionary War was the first ranger organization. The value of such units on the frontier is implied in the Second Amendment to the U.S. Constitution which recognizes the importance of a well armed militia, and that therefore the right of citizens to bar arms shall not be denied.

In the Civil War, ranger-type units fought mostly for the South, usually taking the names of their commanders like Ashby, Morgan, Mosby, and most notably Nathan Bedford Forrest, recognized as one of the most effective commanders on either side. It was a fitting coincidence that the 2nd and 5th Ranger Battalions in World War II first trained at Camp Forrest, Tennessee. In the Civil War, Rangers were usually mounted raiders; as light cavalry they also provided long and excellent service in the wars against the Indian nations of the Great Plains.

Ever independent, constantly seeking the freedom of initiative and elbow room for their operations, rangers made conventional army commanders uneasy and left them ungrateful. In 1891, at the close of the Indian Wars, rangers disappeared from the Army's roles.

They would not be heard of again until World War II, but then the name Ranger would ring with honor in famous battles, to echo in Korea and Vietnam; to resound thereafter in every US Army action for the rest of the century—and forever after, whenever the best are needed to lead the way.

Rangers: from World War II Forward tells about the most stirring and significant Ranger actions in this century. The story continues, to be written by Rangers who read this book. Your time has come to lead the way as you follow into history some of the greatest soldiers who ever lived and died for victory.

Robert W. Black
Carlisle, PA

Publisher's Message

It is indeed my privilege to publish the history of the Rangers. The Rangers have distinguished themselves as unique, brave, innovative, disciplined, and well-trained strike forces that categorizes them as "elite." Few attain this type of honor, but repeatedly throughout their proud lineage the tradition echoed in the Ranger Motto "Rangers Lead The Way" has been maintained.

This tradition is reflected in this first volume which accounts the heroic deeds of Darby's Rangers, Merrill's Marauders, the Rangers of World War II, and the Rangers of the Korean War. The Ranger history is so distinctive that in order to adequately and thoroughly cover the entire story a second volume is currently being prepared for release continuing the saga through Vietnam, Grenada, Panama, Desert Storm, and Somalia. This history is being written and we will publish it soon.

A special thank you to Ranger Philip B. Piazza, President of the Merrill's Marauders Association; to Ranger Richard N. Hathaway, Jr., President of the Ranger Battalions Association of World War II; to Ranger Rex G. Masters, President 1994-1997 of the Ranger Infantry Companies (Airborne), Association of the Korean War; to Ranger Robert F. Gilbert, President 75th Ranger Regiment, 1986-88, 1990-92; and to the 75th Ranger Regiment and Training Brigade. Their support and efforts helped make this book a reality. Also, I express my gratitude to Col. Thomas H. Taylor, USA (Ret) for his exhaustive research and writing of this volume.

To the Rangers who are members of the Ranger Hall of Fame and all Rangers past, present, and future, I salute you for your dedication and selfless service for the peace of our Nation.

Other Turner Publishing Company titles of interest are: *Merrill's Marauders; USA Airborne 50th Anniversary; 187th Airborne (RCT); Army Airborne Troop Carriers; 11th Airborne; 173rd Airborne; 101st Airborne Division; Korean War Veterans Memorial; Korean War Ex-Prisoners of War; Chosin Few, North Korea November-December, 1950; Korean War Veterans; The Forgotten War. . . Remembered; 1st Cavalry Division, Korea, June 1950 to January 1952; 1st Cavalry Division, Vietnam, August 1965 to December 1969; 2nd Infantry Division, The Legacy of the Purple Heart*, and many more. For a catalog listing approximately 300 military titles or to order please write or call: Turner Publishing Company, P.O. Box 3101, Paducah, KY, 42002-3101, 1-800-788-3350.

Dave Turner

Dave Turner, President
Turner Publishing Company

Author's Foreword

Ranger or ranger? From the start, a writer is bedeviled by the first letter of the subject. Should 'ranger' be capitalized or not? Perhaps the simplest answer is that of a grammarian who would say that Ranger is a proper noun while ranger is an adjective. Thus, using the adjective, Merrills Marauders could be considered a ranger-type force, and the seizure of Pointe du Hoc a famous ranger operation. Using the noun, members of the 75th Regiment are clearly Rangers as were members of Darby's command.

From these simplest examples, nomenclature becomes more complicated: Ranger is capitalized when used as an adjectival noun like the Ranger School, and a graduate of that school may be called a Ranger though he is not necessarily a member of a Ranger unit. Furthermore, there are probably clerks working in the Ranger School who are not Rangers themselves and therefore do not wear the Ranger tab.

Such semantic confusion is sometimes amusing and often annoying— but actually a reflection of ranger history and the fundamental question that may never be permanently answered: what are Rangers and what should they do?

To get to the origin of the question requires history preceding World War II; Bob Black has sketched these precedents in his excellent introduction that highlights the uniquely American and individualistic character of early rangers. Carrying forward Black's thoughts into the 20th Century, it should be noted that the United States had no ranger units—by any definition—in World War I. So the outbreak of World War II was a clear dividing line in ranger history, and really the start of a new story about Rangers as formally organized units rather than as informal groupings of individuals.

Before starting at this new beginning, I'd like to look at the end; i.e., the present-day status of Rangers. Though profound consideration evidently has been given to the fundamental question, the roles and mission of Rangers still overlap others. Organizationally, Ranger units at present are part of a four-star joint command that includes army Special Forces, certain aviation, PSYOP and civil affairs units, as well as navy SEALS and USAF Air Commandos. Together these sometimes odd bedfellows are designated Special Operations Forces, primarily oriented on unconventional warfare. But to just consider the army components, PSYOP and civil affairs have conventional applications—and so do the Rangers. Organizations change, but questions stay the same.

Now back to the beginning—to 1942, the year of turning points during World War II in which the United States has been a combatant for only a few months. The great pivotal battles of El Alamein, Stalingrad and Midway have yet to be fought. It is April, and the axis tide of conquest still floods the Eastern Hemisphere. America has yet to contribute much to the Allied cause except to gear up for production of war materials that will make it 'the arsenal of democracy'.

The Navy is still reeling from its losses at Pearl Harbor. The Air Force doesn't exist: ground based warplanes are the Air Corps (USAAC), a branch of the Army. Ground forces are vastly undermanned, poorly equipped, and virtually untrained in large scale maneuvers.

Transforming this sorry state of preparedness is the responsibility of Army Chief of Staff, George Marshall, undoubtedly the greatest figure ever to hold that position. High on his list of pressing priorities is training; not just live fire but also OJT—against the Nazis.

Marshall has already endorsed America's first great strategic decision: to fight a holding action against the Japanese, though they were the only Axis power to have actually attacked U.S forces, and instead to make the national main effort against Germany, the most powerful enemy. Europe has been at war for nearly three years. Though the British are most experienced in defeat and retreat, their forces nonetheless contain valuable veterans who could teach the American cherries a lot.

Especially the British Commandos. Except in the armored war raging back and forth across North African deserts, Commandos are the only western soldiers in contact with German forces, albeit through sporadic hit-and-run pin pricks along the coast of occupied Europe. So Marshall sends Col. Lucian Truscott over to London to see if the Commandos would take a couple of dozen American infantrymen along on raids to acquire a little combat experience.

No one is yet using the word 'ranger', capitalized or otherwise. All that Truscott hopes for is to develop training cadres to inject a few combat veterans into the new draftee army that is expanding by tens of thousands of recruits every month. Thus began a dichotomy that still exists in ranger employment: the direct action function and the training function. Though raids with the Commandos were what today is called direct action to take tactical objectives, for the Americans involved the purpose was primarily to gain some experience that could be used in training green troops.

During different periods of this history, one function alternately has predominated over the other. Generally, direct action is paramount in war time—the original Truscott period being the most notable exception—while training (i.e., spreading ranger craft throughout the army) has been the foremost function between wars.

But the mind-set of senior commanders has also had significant influence on which function predominated. Some generals you will read about liked to use Rangers as a unit while others preferred to scatter ranger expertise throughout line infantry. Some didn't like rangers at all, believing that they could do nothing that good infantry could not; and, moreover, rangers units weren't worth the special attention they required while sucking up some of the best soldiers who could be better employed as line infantry. British Field Marshal William Slim, with whom Merrills Marauders had to coordinate, scoffed at specialized units like Rangers, and once said, "Should all soldiers be able to climb trees, or must I designate units to be tree-climbers and have acorns sewn on their shoulders?" In the early 1960's, Gen. Bruce Clark, commanding all US forces in Germany, even resisted the wear of Ranger tabs on field uniforms.

Both officers had a point even if they carried it to extremes. Raids and ambushes, the most typical direct actions of Rangers, should also be the stock in trade for competent infantry. Critics of ranger units argue that because there is nothing special about these operations, why bother with the expense of having specialized units to perform them?

This argument is simplistic and fails to acknowledge how degrees of difficulty distinguish ranger from standard infantry operations. Yes, a raid is a raid no matter who performs it; but it is as accurate to say that a drive is a drive whether it be at 55 mph on a freeway or at blinding speed on hairpin turns of the Indianapolis 500. A Formula One racer is far more expensive and requires immensely more care than a station wagon, but experience has shown that both are worth the price for their respective purposes.

That analogy indicates the justification for Ranger units: they have been needed in the past and have proven their cost effectiveness. Those who would argue otherwise assume themselves to know better than the likes of MacArthur, Eisenhower, Bradley and Patton, who—along with the great majority of America's most distinguished commanders—supported and gratefully employed Ranger units.

Indeed field generals have always formed rather than dissolved Ranger formations. It has been the budgeteers and bureaucrats who have removed Ranger units from the rolls, only to see them reactivated in time of need; that is, whenever the shooting got serious.

But field generals have not always been godfathers for rangers. Merrills Marauders' regard for Gen. Stilwell and his chief of staff grew from wariness to hostility, and finally to outright hatred. Then, besides what I'd call 'the dispersion error' propounded by commanders like Slim, there has been the opposite error of occasionally 'conventionalizing' Ranger units as if they were a straight infantry outfits of equivalent size. This has meant 'throwing them [rangers] into the line', a most regretable practice, especially in Korea. Even allowing for the desperate tactical situation that faced Gen. Walker in 1951, the misuse of Rangers as straight infantry, was disastrous at the time and inexcusable even in hindsight.

Inexcusable because using rangers as foxhole-fillers not only deprives the field general of unique offensive capabilities but also taxes his other resources when Ranger units have no organic combat support. The debacle at Cisterna is the most tragic example (in ways, Mogadishu is another) of trying to conventionalize rangers while ignoring their direct support requirements in such a role. It is possible to convert that Formula One racer into a station wagon but never—this history shows—worth the effort, to say nothing of the lost lives. Add heavy bumpers, attach a trailer and luggage racks, retool the engine, change the compression ratio and tires: yet still there is no station wagon, but rather and extremely inefficient contraption that will soon break down, probably before it can haul its payload. It is more likely to go off the road, causing tow trucks and mechanics to fix what should never have been built in the first place, and detracting from vital maintenance for the owner's other vehicles.

Ranger veterans of WW II and Korea are all too familiar with this most grievous of misuses. Their bloody experience (the lessons learned) probably spared rangers of Vietnam and later periods from many similar ordeals. However there is a relatively minor misuse—maybe I should call it just poor use of time—which, to my surprise, was rarely criticized in unit histories: that of what rangers normally do when not conducting or preparing for missions. Over and over, without complaint, journal entries tell of securing the CP of higher headquarters and/or guarding enemy POW's. Sharpening training, it seems, would have been a much more productive activity, because it is the high level and specialization of training that really distinguishes rangers from line infantry.

The above discussion organizes the format of this book. Training and organizing are covered as after action functions . Though the original strap hanging with Commandos had a training purpose, from then on lessons learned from direct action informed training doctrine. As doctrine developed, it began to shape the conduct of direct action, and a 'chicken or egg' relationship arose from the two functions.

But it was direct action; i.e., ranger combat operations, that gained worldwide admiration and indeed immortal fame for America's Rangers and like units like Merrill's Marauders. Their history is already well told in unit histories such as those cited in the bibliography. I've not added original research to the war stories that comprise most of these pages, but rather rearranged them, first by wars and then by mission.

For my two mission categories I've used the simplest descriptors, 'good' and 'bad', because this is how I've most often heard soldiers describe a mission when looking back after an action. The two words do not necessarily summarize the outcome, though a good mission is more likely to be successful than a bad one. Nor do 'good' and 'bad' at all describe the conduct of a mission: indeed, the worse the mission, the more likely will Rangers perform superhuman feats in trying to accomplish it.

'Good' in this book means a mission in which Rangers were suitably employed and adequately supported.*[*FN: Suitability includes the duration of employment; i.e., a good mission that is carried on too long can become a bad mission.] 'Bad' means that one or both of these factors were missing.

Goodness and badness will be considered in my introduction to each operation. Most will fall in the category of good even though it is rare for a senior commander to have all the resources he could use to support his units in combat. Moreover, war is an exercise in expediency: a commander does what has to be done with the resources at hand. As this history shows, especially in Korea, often Rangers were the only resource left to desperately strapped generals.

They may have been strapped because of flawed planning or a number of other defects, personal or within their staffs. It is not my purpose to second guess, for instance, whether Gen. Lucas was sufficiently aggressive at Anzio. What matters in this book is that, for whatever reasons, his campaign plan resulted in a bad mission for three Ranger battalions.

I've been asked to justify why good and bad missions are the organizing feature of this book. A fellow historian noted that every type of unit could categorize their history in this way, but they don't because 'bad missions' sounds like bellyaching. For example, he said, the lst Infantry Division had a 'bad' mission at Omaha Beach because they were inadequately supported in taking out the fortifications that boxed them in.

That observation misses the point. The lst Infantry Division was organized and equipped to seize and hold major terrain—to seize it amphibiously if necessary. Somewhere in a field manual, that capability was spelled out. So Omaha Beach was a tough mission for The Big Red One, not a bad mission. V Corps suitably employed them, and they were supported by the unprecedented air and sea power of Operation Overlord.

Equally important support came in the form of intelligence: the pill boxes that gave them so much trouble had been located and analyzed. By contrast, inadequate intelligence support appears to be the most significant factor in bad ranger missions. Where an infantry division can use brute force to compensate for faulty intelligence, Rangers rarely have that luxury. Mass can be the overriding principle of war for a division, but surprise is the foremost principle for Rangers; so when intelligence does not provide the elements for surprise, Rangers have little to fall back on.

What they do fall back on is training, courage and resourcefulness— the three hallmarks of Rangers individually, and also of their operations.

Their training imbues them with self reliance and confidence in their buddies. It prepares them for the debilitating exhaustion of combat because they have exceeded normal limits of endurance in training. Ranger training also emphasises preparation for the unexpected, and rehearsals are more the norm than for any other army unit.

Their courage stands out because courage is most often required when backs are against the wall. That is a position Rangers regularly find themselves because of the high danger inherent in their operations. Almost by definition, ranger ops are more dangerous than conventional; otherwise conventional units would perform them.

Their resourcefulness is conspicuous because Rangers most often operate behind enemy lines or at least separated from adjacent friendly units. Lines of support and communication are typically thin and tenuous. Therefore Rangers habitually make do, bootstrap, and improvise. They have to.

Now to return to April l942, the hour of rangers' rebirth. The enemy situation is ominous; the friendly situation precarious. The United States—indeed the entire free world—is searching for a spark that will light hopes for victory.

A few American officers in Great Britain are gathering tinder so that a spark can become a flame. They envision small, attack-oriented detachments like Commandos, but they have received a caveat from Gen. Eisenhower to create forces distinctly American in character.

"I hope," Ike wrote Col. Truscott, "you will find some other name than 'Commandos', for the glamor of that name will always remain—and properly so—British."

For months to come, Truscott was too involved recruiting and organizing his testbed outfit (originally about 300 men) to coin a name for them. Years later, several claimants nominated themselves as originators. Lord Louis Mountbatten, commander of Allied combined operations, was one; Capt. Anthony Levorio, a former reporter for the *New York Times*, was another. But the honor should probably go to Truscott himself, who wrote in his postwar memoirs: "[From the] many names recommended, I selected Rangers."

He never explained why he chose that distinctive word, but clearly Truscott was harkening back to earliest American history—even prior to 1776—when men, far more than weapons, were the preeminent element in warfare. The opening phases of World War II already had indicated the ascendant importance of modern machines and science as factors in war. So by invoking a name from the pre-industrial days of the colonial frontier, Truscott was making a counter-point and highlighting an eternal truth: men matter most.

They did in l942 and forever after. The men he dubbed Rangers would prove his point for the next three years, and start a tradition of invincibility that will always endure.

Brig. Gen. Frank D. Merrill (left) and Lt. Gen. Joseph W. Stilwell talk during a meeting at Division headquarters of the American Gallahad Forces in Maubum, Burma, 5 May 1944.

Part One
World War Two

Number 2 gun Crew, Merrill's Marauders, at Mhpum Ga Hill, 1944.

Prologue

In the beginning training preceded direct action, and organization preceded training. Rangers originally were to be organized from existing US Army assets in the United Kingdom, and these were scant in 1942. Truscott (recently promoted to brigadier general) was to acquire volunteers from the 34th Infantry and lst Armored Divisions, the only ground combat formations in country. Not surprisingly, neither division commander was enthusiastic about a provisional, experimental outfit siphoning off his most aggressive and gung ho soldiers.

But Truscott had the backing of Marshall and Eisenhower, so the pipeline opened and volunteers trickled into a little town in Northern Ireland with the improbable name, Carrickfergus. The most important of the volunteers would be its commander, the aide de camp to MG Hartle, CG of the 34th Inf. Division.

Like all generals' aides, William Darby was among the first to learn what was going on in and around the division. That included Truscott's levy for rangers, whose concept and future excited Darby but put him in an awkward position with his boss who begrudged having to set up interviews with Truscott's recruiters. Luckily, Darby got an unexpected boost from the division chief of staff, Col. Edmond Leavey.

The three were riding a jeep to church in Belfast when the subject of rangers came up, and Hartle commented, rather contemptuously, that Truscott's infant outfit did not yet even have a commander. Leavey slyly suggested that if the new CO came from the 34th Division, this could produce some leverage on him to seek most of his volunteers from the lst Armored or elsewhere in V Corps. Nominate someone from the 34th for the command, in other words, and he might agree to go away.

Neither of the other two officers in the jeep could tell if Hartle's nod was one of assent or cynicism. Leavey knew that Darby didn't want to spend the war looking after a general. Boldly, Leavey popped a suggestion: "Why not put in Bill here for the job?"

Hartle thought a moment— the longest moment of Darby's life— then asked, "What about it, Bill?"

Though the jeep had not yet reached the church, Darby's prayer had already been answered.

William O. Darby's impact and influence on the fledgling rangers is such that he is universally recognized as their 'father'. Indeed, for the first year of ranger existence they were known as Darby's Rangers just as a similar force in Burma was to be called Merrill's Marauders. Darby was also a model of what came to be the ranger image. Handsome, even by Hollywood standards, Darby was a raw boned product of Middle America and its creed that success is to be earned the old fashioned way—by work. Darby had to work against the odds even to be appointed to West Point in 1929. He was not his congressman's first choice but was admitted only when those ahead of him failed their tests.

Darby was less than a star cadet, as neither academics nor (surprisingly) athletics, were easy for him. What he is remembered for at West Point was his leadership: a direct, forceful approach to problems, but most of all a natural ability to inspire men to join, rather than follow, him in the solution of those problems. Modern psychologists might describe Darby as a consensual leader but that misses the mark—he was not one to take a vote. But if, as was the practice sometimes in the Civil War, soldiers elected their leaders, then Darby would have won every vote. Appropriately, he was selected one of thirteen company commanders in the Corps of Cadets.

History ranks Darby as one of America's greatest infantrymen, but he chose artillery for his branch when he was commissioned in 1933. This may have been because West Point graduates selected among the combat arms according to class standing which was determined primarily by academic achievement. Generally the top scholars went into the engineers, the next highest into artillery. The bottom of a class was 'ranked' into cavalry or infantry. Perhaps Darby felt that artillery provided the best opportunity for him to be recognized in comparison with the pick of his classmates—we don't know, for sure, because he never expressed himself on the subject of his branch choice.

In the long run his choice was ideal because expertise in artillery support would be as scarce as it was vital for Rangers. Artillery also provided Darby's grounding in the functions of higher headquarters that would control ranger operations. Moreover, the army of the 1930s was just beginning to be motorized, so as a junior officer he had to learn a lot about pack howitzers, mortars and indirect fire in mountainous terrain. The prime movers for World War II rangers often would be four-legged, and Darby first became familiar with horses and mules in the artillery.

His branch choice is just the first of several seemingly chance events that in fact prepared Darby for his unique future as the founder and leader of rangers. In 1940, after a variety of normal battery and battalion-level assignments, he was fortunate enough to participate in the first corp exercises conducted since World War I, called the Louisiana Maneuvers. The several major headquarters involved showed scant proficiency in large unit movements, especially compared with the blitzkriegs that were swallowing Europe, but at least Captain Darby got some sense of the scale of problems at the level which was approaching the US Army, even though the United States was not yet at war.

Foremost among these daunting problems was the likelihood of American counter-invasions to liberate western Europe. Invasions meant many troops coming over many beaches—as indeed Rangers would in North Africa, Sicily, Italy and Normandy. In the prewar army, however, beach assaults were a sidelight (and a specialty of the Marines); but again, by good fortune, Darby became involved in amphibious exercises on the beaches of Puerto Rico where he also gained some appreciation for the capabilities and limitations of air as well as naval gunfire support.

Fate seemed to guide Darby once more when in 1942 he received orders to Hawaii—only to have them cancelled, and instead be ap-

Lt. Gen. George S. Patton congratulates William O. Darby after presenting him with a DSC. Patton then offered Darby the command of the combat team and a full colonelcy. Darby preferred to stay with the boys.

pointed MG Hartle's aide. It is interesting to speculate about Darby's future if he had followed his original orders to the Pacific where he would have first entered combat with the 24th or 25th Infantry Divisions (both were formed at that time from the Hawaiian Division). No doubt his epitaph would have been the same as that delivered by Gen. Truscott, eventually the Fifth Army Commander:

"Never have I known a more gallant, heroic officer."

A few—very few—other officers justifiably could be given a similar eulogy. None, however, could match Darby's dedication to a single organization: on three occasions he turned down promotion in order to stay with his rangers. Such selflessness has just never been demonstrated by anyone else, at anytime, in any army. Darby was one of a kind.

As the most successful commanders are, he was also lucky. His luck was demonstrated by the foregoing chain of events that prepared and positioned him to found the Rangers. He also survived three serious wounds. Luck, however, is an exhaustible quantity. Darby's ran out two days before the end of the war in Italy when the Germans were finally able to kill him, ironically, by artillery fire.

"He whom the gods love dies young", is a Greek proverb. Darby was 43. He followed hundreds of other rangers, most younger than he, who were KIA in Italy. But like them, he fell while leading the way.

In June, 1942, Darby was not leading but organizing a force to lead. Col. Leavey's stratagem did not work: Darby drew 50% of his original battalion from the 34th Infantry Division, 40% from the 1st Armored, the rest from V Corps. All were volunteers, and nearly 20% dropped or were washed out. Most remaining rangers had an infantry MOS.

Darby's mission (classified secret) directed him to form "a training and demonstration" battalion that would eventually "return as many men as possible to their original units... in order to provide battle-trained personnel in ALL units..." (emphasis in the original). Darby, of course, had other plans that definitely did NOT include sending his volunteers back to their old outfits as training cadre.

Instead, the 1st Ranger Battalion was organized for combat. It contained about as many officers as a line infantry battalion but less than half as many enlisted men. This ratio was designed for employment in small packets, as there were six rifle companies (three officers, 62 EM each), rather than the standard three, and was driven by the size of British landing craft used at that time for Commando raids. No consideration was afforded to aerial delivery, because the US Army did not yet have any airborne forces and was just beginning a jump school.

The oldest original ranger EM was 33, the youngest 18, and the average age about 25. By today's standards, physical requirements were not very strict: just 20/20 eyesight with no eyeglasses or night blindness. And no removable dentures! Neither were there height/weight limitations, and no initial physical fitness test, though ranger training (modeled on the Commandos) soon determined whether a candidate was in good enough shape. If Darby's screening and interviewing process does not seem very thorough, that was because he had but ten days to create his battalion from a standing start.

The composition of the 1st Battalion also reflected the haste pervading the entire US Army as it increased a hundredfold during its first full year at war. The 34th Infantry and 1st Armor, from whom the first rangers were drawn, were both regular army divisions but they were filled with draftees. So most rangers were also draftees who then became volunteers. Likewise, the majority of ranger officers were not regular army but rather reservists or national guardsmen. This composition is notable, in that the most elite ground force—America's shock troops— arose from elements not originally present in the regular army.

To scan the startup ranger TO&E is to marvel at what Darby could have hoped to accomplish with such primitive equipment which included many World War I artifacts like the 1903 Springfield bolt-action rifle, saucer helmets, semaphore flags, canvas leggings (worn over brogans)— and brass knuckles. Of course most of these items would be updated or replaced during the war, but two weapons are used by Rangers to this day: the 60 mm mortar and the Sykes knife, the latter a hand-down from the Commandos.

While basic ranger training got under way in Northern Ireland, and later progressed in Scotland, the S3 section fanned out to learn fieldcraft and specialized tactics from the Commandos. Relations between these two elite and similar forces began and remained on an excellent note. This harmony was somewhat a contrast to counterpart relations between the two countries' conventional forces.

The British Army felt that the newly arrived Americans were expecting to do too much too soon. This reflected the strategic argument between the high commands of the English-speaking allies: the U.S. pressed for a cross-channel invasion as soon as possible, but the U.K. preferred a scattered offensive against soft points around the periphery of Hitler's 'Fortress Europe' while the Soviet Army bled the cream of the *Wehrmacht*.

On the working level, senior allied staffs found themselves divided by national objectives and the attitudes that derived from them. Consequently, ranking British officers viewed 'the

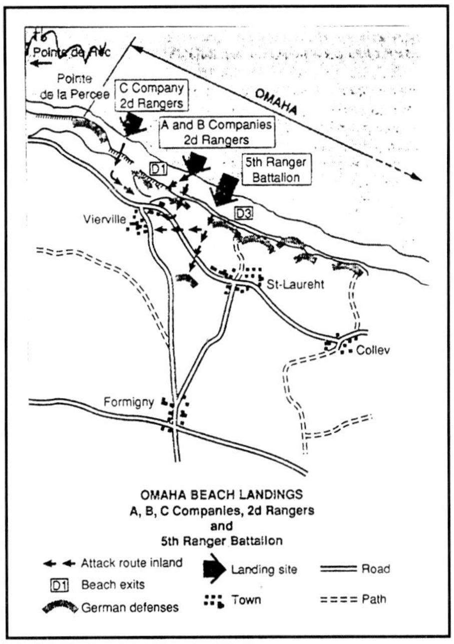

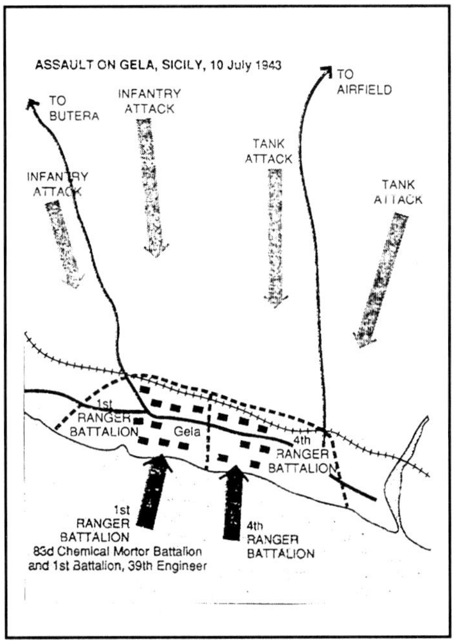

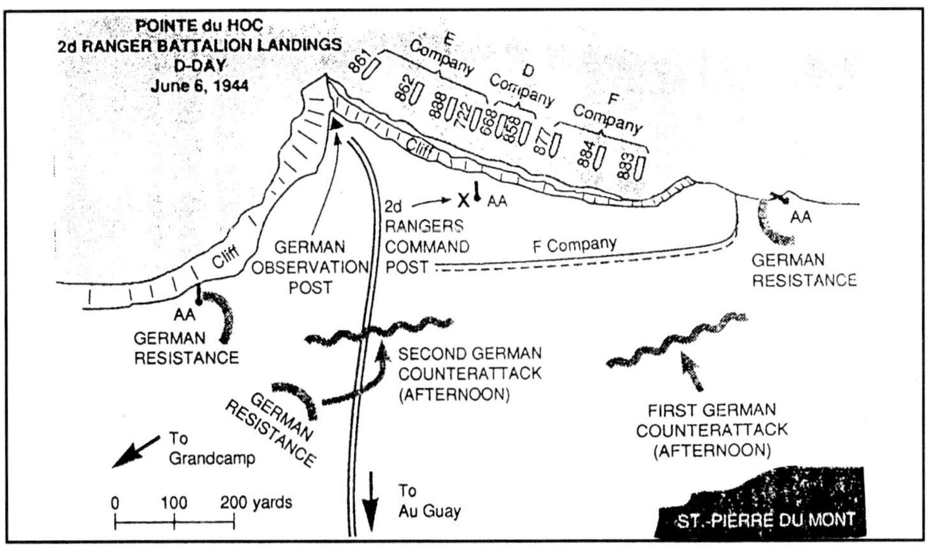

Yanks' to be naive and brash. American found their counterparts to be defeatist and condescending.

But where the rubber met the road—i.e., where Commandos met Rangers—there was selfless common purpose because the two would tackle tactical objectives in much the same way, no matter what strategy was overriding. Initially the strategy followed was that of the British: the peripheral strategy of raiding along the north European coast. Americans would sometimes go along for the ride, to gain experience, really to train and shake down their organization and develop their own doctrine.

The first and most significant of all combined Commando-Ranger operations was the mother of all raids, on the French coastal town of Dieppe, population at that time approximately 25,000. Six officers and forty-four EM (about a tenth of all the rangers in the world) accompanied the Commandos, not as US units but integrated right into the British organization. This would never happen again, not by reason of allied friction but because, after Dieppe, Darby felt his rangers could conduct their own operations. Coordination and mutual regard would continue throughout the war; at Salerno, Commandos and Rangers would even fight side by side, but not in the coach-and-pupil relationship that obtained up until Dieppe.

Before taking this history to the French coast, mention should be made of the ranger training that preceded and was concurrent with the Dieppe raid. In subsequent sections, training will be discussed after the direct actions (i.e., combat) covered in each section. However for this section alone, startup ranger training is most appropriately considered first.

The nature of this training will be familiar to rangers of any era: lots of endurance marches, land navigation, patrolling, obstacle courses, hand to hand combat, cross-training, insertion techniques, field first aid, etc.—the most strenuous of these activities conducted at night—and the worse the weather, the better the training. Great Britain lacked many facilities for training but had a surfeit of wretched weather!

What stands out in reviewing rangers' startup training is its foremost purpose and emphases. More than anything else, this was initially a weeding out and placement process. Darby needed to know, most of all, who in his new command could cut it and in what ways. This aspect of training would not be as important later when exacting prerequisites and finely tuned screening could determine pretty much who could cut it and in what capacities. Darby had a different problem. No one had much of an idea what anyone else could do. Training had to show individual capabilities before they were tested in the crucible of combat.

For example, Darby needed to be sure that every member of his battalion could function in the extreme rigors of the field. This need established his policy that each ranger, including himself and his staff, undergo identical training.

This requirement caused his S4 to ask, "I'll go on the march or I'll get your supplies. Which do you want?"

Such situations were a continuing conundrum. It might be that the S4 would fall out in the march but was irreplaceable in getting supplies. Different rangers had different talents. Darby, and all commanders after him, had to make trade offs.

The other prominent aspect of startup training was in the emphasis on getting to objectives. Infantry usually walks, armor rides. But for Rangers, the objective often requires a specialized delivery method. For Darby that meant boats. About a third of training time he designated for amphibious proficiency. As a minimum requirement, all Rangers had to swim 50 yards with boots and combat uniform.

Faking of satisfactory swimming ability was probably the most widespread falsification in all ranger training reports. Darby no doubt knew it was. Swimming is an Achilles heel for some of the bravest men. Darby didn't want to lose them. Several literally had to 'sink or swim' at Dieppe. Somehow none drowned.

Good Missions

I. Dieppe, France, August, 1942

Strategically and tactically the Allied raid on Dieppe in August 1942 was a bloody repulse. However, Dieppe is classified a 'good' mission because the Rangers participating (while attached to Commandos) were suitably employed and adequately supported by higher headquarters with good intelligence and abundant, though small-bore, indirect firepower. The fate of the conventional allied ground forces involved is a different story.

Dieppe was unique: no raid matching its size ever was or would again be conducted by either side in World War II. Its major strategic purpose was twofold: 1. Probe the German coastal defenses in order develop the best plans for a decisive invasion, hopefully scheduled for 1943. 2. Draw off German reserves from the Eastern Front where the Red Army was struggling for survival. Secondary purposes were to shake down combined staffs; test various aspects of amphibious/airborne operations (the original plan called for a company-size drop which was later scrubbed); and to 'blood' a Canadian division.

In just nine hours, that last purpose was a ghastly success. The Canadian 2d Division, with Royal Marines and the Essex Scottish Fusiliers, comprised the great bulk of the raiding force. Commando companies played a secondary though vital role on the flanks of the main attack.

The objective was to seize the city of Dieppe and await the German reaction, which was hoped to be relatively slow and uncoordinated. Relying upon the miraculous evacuation from Dunkirk as precedent, Lord Mountbatten's staff were confident that the raiding force could withdraw in good order and with an acceptable level of casualties. Tac air and small caliber naval gunfire were deemed to be sufficient offshore support. In consideration of the French, Dieppe was to be spared heavy bombardment.

The raid then, as all raids do, relied more upon surprise than firepower. Sometimes the most direct approach achieves the maximum surprise. This was the hope for the Canadians who would storm the town with a frontal assault, crossing the narrow shale shingle beaches into the protection of the sturdily built town before the Germans could launch a significant counterattack.

It was the mission of two Commando companies, supported by air strikes, to take down batteries on both flanks of the main landing. The distance between these well-prepared gun positions was about eleven miles, with Dieppe midway.

As strap hangars, the Rangers were not informed of tactical objectives until the force put to sea. Six Rangers accompanied the Canadians, and were pretty much ignored as observers. With the Commandos, 44 Rangers were paired with a counterpart and expected to function as members of a team. The British were particularly pleased that their new buddies had just been armed with the semiautomatic M-1 rifle, a state of the art weapon at the time. So Rangers were able to carry 176 rounds of 30 caliber ammo in eight-round clips, while the Commandos' combat load was only 100 rounds for their bolt-action rifles.

Canadian war correspondent, Ross Munro, remembered twilight on 18 August 1942 as one of the most beautiful of the summer. The sea was smooth, untroubled by a slight breeze, and the sky clear after a lavender sunset. The raiding ships had just cleared harbor, then made their way slowly as they approached minefields. The boats rode along, "gracefully curling white foam at their bows, and a phosphorescent wake that stood out like diamonds on black velvet."

This setting—a dramatic contrast to that of the Overlord fleet two years later— seemed a good omen for the Dieppe raiders, especially when they cleared the minefield without incident. They were within ten miles of their beaches when bad luck met them before the dawn.

German patrol boats, escorting a convoy along the littoral, happened upon the raiding flotilla, and a full fledged sea battle broke out. Landing craft, though unarmed and made of plywood, largely survived to reached their landfalls, but the Germans on shore were alerted and aroused. Surprise—the ingredient most needed for success—had been partially lost. That is, the Germans became aware that a force (but not its size) was coming in with the rising tide. Their 302nd Infantry Division, positioned to counterattack encroachments of static coastline defenses, consequently had a contingency plan to defend Dieppe and they began to execute it efficiently. They rushed into the town, occupying the very buildings which the Canadians had hoped to use for protection against the coastal batteries. Now it was all the more important for the Commandos to take out those guns.

Facing the Commando/Rangers were cliffs and bluffs which could not be climbed from the narrow beach, and penetrated in only a few places by water courses. The Germans had recognized these valleys as critical terrain and prepared positions to block them.

The original Commando plan was to come ashore just before BMNT, sneak up the valleys and be at the Germans' throats before daylight. But the unexpected sea battle delayed the schedule and scattered some landing craft (which the British called 'Higgins boats'). By the time the first wave came onto the steep beaches, German gunners had them under automatic fire.

On the eastern flank, the Commandos' objective were the coastal batteries near Berneval. Egress from the beach was blocked by barbed-wire barriers. Unfortunately the breaching charges had all been stowed on a single landing craft which did not make it to shore. Under heavy but sporadic and poorly aimed fire, the Berneval force had to snip their way through with wire cutters, a slow and sweaty process that a dozen Commandos did not survive.

The raiders had arrived on two beaches (Yellow I & II) about a mile apart. This split was an advantage because fire from the main German position could not pin down the entire force. With the enemy concentrating on Yellow I, the raiders on Yellow II moved off the beach smartly and mounted the bluff to take the batteries under surprise fire. The Germans tried to depress the tubes but it was not enough (a factor the Commandos' meticulous planning had considered), and their shells only blew big holes in far away countryside.

But infantry from inland soon rushed to the defense of the guns. This enemy reinforcement had also been a threat weighed in the Commandos' plans, and to be solved by aerial interdiction. However the *Luftwaffe* had reacted swiftly, scrambling fighters from all over northern France. In roiling, screaming dogfights overhead, the opposing air forces battled to a standoff. Without air superiority, the raiders on the ground were short of close air support, especially at the Yellow beaches.

For the Yellow II element, missing fighter bombers were not essential—the Commandos had climbed the most dangerous obstacles and, though outnumbered, were holding their own while investing the batteries which were their objective.

But until they had, the big German guns had focused on the Yellow I element and decimated them. Wriggling through barb-wired entanglements, the Commandos then faced grenades tossed down the vertical cliff and pre-registered fire. The German machine guns seemed to have an infinite supply of ammo. Great British soldiers went down, one after the other, over and over again. The best of the elite met their doom executing a plan that seemed to validate the British strategic foreboding that frontal assaults in the 20th Century were unconscionable, a fact proven in World War I, and its principal lesson that had not yet been learned if World War II was to be fought in the way that produced the carnage at Dieppe. Courage was no substitute when men charged against machine guns.

Halfway up a cliff, the last Commando officer was cut down. With no pause in the attack, Lt. Loustalot ot the Rangers grimly took command of British troops he had known for less than a day. His group reached the top before he himself was killed assaulting a machine gun nest. A fellow ranger spent a moment of prayer beside Loustalot's body before rejoining the attack.

Thus the rangers had been blooded for the first time but to what avail? Loustalot's newly acquired combat experience would never be of value to the cherry Rangers in Great Britain—but his memory would. He took command without being asked. He did what the situation required. He led the way.

Landing with the Canadians was Ranger Lt. Joe Randall. He barely made it ashore before he became the first American to be killed in ground combat in the European Theater of Operations (ETO)*.[*FN: The first American serviceman to lose his life in World War II was USAAC Captain Robert Losey. As military attache to Norway, he was killed by a German bomb on 21 April 1940 near Lillehammer, the site of the 1994 Winter Olympics.] On the western flank, where the beaches were designated Orange I and II, Ranger Cpl. Frank Koons was credited to be the first American to kill a German in World War II.

He did so while sniping after his Commando team maneuvered into concealed positions where they had a clear field of fire on a unsuspecting German battery. How Koons and his comrades accomplished their mission was told in this dispatch from British war correspondent, A. B. Austin:

We had grounded on the shingle at full tide, a few yards from the base of cold-looking, unscalable, hundred-foot, overhanging chalk-white cliffs. As we struggled to the cliff-foot, bent over by our loads, a German machine gun began to stutter from up above.

Heavy weapons from our support craft answered. Red hot tracers flashed past each other between cliff top and sea, looking vicious

Machine gun crew defends the perimeter.

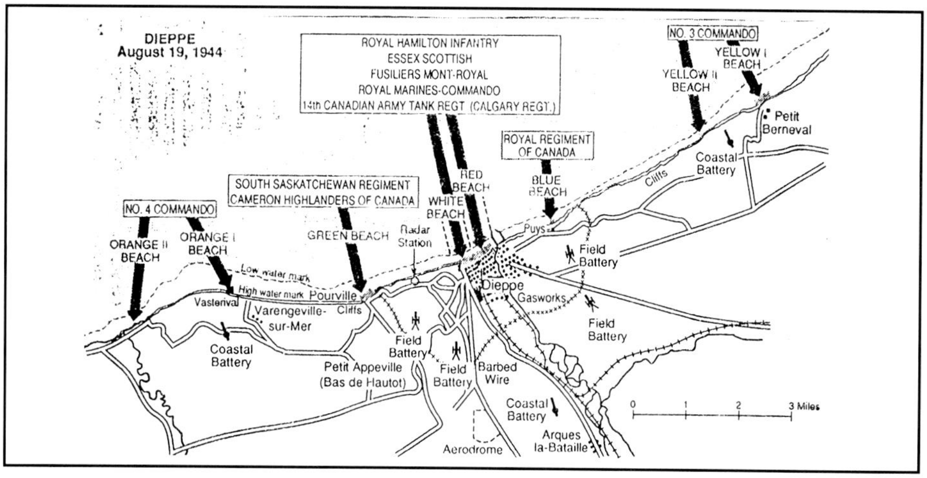

This assault boat loaded with Canadians streaks for the fire-swept shores of Dieppe. They fought valiantly. Their losses were heavy.

but surprisingly slow, like the lighted matches boys flick across a room.

At this same moment the other half of 4 Commando, led by Lord Lovat himself, had made their landing a little farther west [Orange II]. Facing no cliffs the size that were confronting us, Lovat's force began their dash across low ground, past cottages and farms to take the batteries from the rear, while our force made the frontal attack covered by mortar fire.

We found in the first few minutes that a crack in the cliffs to the left [east] was so crammed full of barbed wire that obviously we would not have time to breach it. A second crack, about twenty feet wide, ended at an almost vertical stone staircase built decades ago, apparently for the convenience of holiday bathers and beachcombers. At the top, the crack widened into a long gully which widened into woods and fields. That gully was to be the axis of our attack to the batteries.

A pioneer [engineer] detachment climbed to the top of the stairs to thrust their long, pipe-like bangalore torpedoes under the first two banks of barbed wire. They lit the fuse and dodged back under cover.

In a few seconds we heard the sharp roar, muffled by the cliffs but still loud enough to make your head ring.

"And now," I thought, "the Krauts will be really roused, and that gully will be impossible."

But no. The Commando spearhead, scrambling through the chalk rubble, the smashed concrete and flattened wire, crept cautiously up the gully and spread out like skirmishers. They were soon followed by the mortar platoon.

Then the enemy howitzers, as yet unseen, opened fire. Toward where I don't know—but it was a dull, hollow sound like detonations in a quarry. Their first volley depressed me for a moment. It seemed we didn't have complete surprise after all. The howitzers could still wreck our main forces coming into Dieppe who, from what I could hear, sounded desperately engaged.

The light had grown just enough for German observers to spot landing craft a fair way out to sea. I watched foaming columns rise as if large rocks were plunging into the water, and felt a little cheered. For I saw only spouts of water, not the flash and smoke from hits. Perhaps we'd be in time yet.

*As I crouched by the signallers' wireless set [radio], waiting for the spearhead troop to report what they'd come upon, I could see the angry, soaring exchange of fire across the sea approaches to Dieppe. These fireworks were more than matched by the aerial combat overhead, a greater spectacle than I'd ever before witnessed, even in the Battle of Britain. * [*FN: During the Dieppe raid, the total loss from both sides was more than 200 aircraft, the most ever shot down in a single day over Europe.] The sky was scored with the slow tracks of incendiary shells, while the entire Dieppe basin began to rumble and thud like the growling crater of a volcano.*

Four cannon-firing Hurricanes dived out of the aerial battle to lend us a hand. Their targets, I learned, were machine gun posts and anti-aircraft artillery to the left and right of Orange I. The gun flashes of the Hurricanes, coming in head on, made them look as if they were being pulled through the air by jets of flame instead of propellers.

Troopers beside me waved, and shouted "Give 'em hell!". They were excited and pleased by their first sight of air support.

Boston bombers, higher up, hurried across the coast on their way to hit German batteries behind Dieppe. Their 'eggs' were still in the air when a mighty explosion, louder and longer than any blast we'd heard that morning, caused us to kiss the ground. It seemed to be the mother and father of all explosions, more immense than the biggest bomb I had heard in the London blitz.

We brushed ourselves off. I asked what had happened—whether it be good or bad—then ducked when projectiles from a German mortar, somewhere beyond the woods, dropped close by. My answer came when Major Roberts, an Irish Guardsman, came back through the trees. When he reached our position, Roberts was grinning with pleasure.

"We've just knocked out their ammunition dump," he told us, as if his announcement called for a toast. "Put a mortar bang on top of it! Bloody fools—all they've got, stored in one lot. Now let's see the Huns fight with bayonets..."

I ran down the gully and the staircase with my first dispatch to England. It read, "West-most German battery demolished at 0650."

"Demolished" was a bit of an exaggeration, but the west-most battery had been truly neutralized. The accurate Commando mortar round, of which Maj. Roberts was so justifiably proud, certainly was the decisive blow in the battle at the Orange beaches and assured success of Lovat's mission. However, the German howitzers remained intact and could be resupplied with ammo by counterattacking infantry.

Two Rangers, Koons and Sgt. Alex Szima, were part of the 'spearhead' whose departure up the gully Austin described. They came within 150 yards of the first howitzer emplacement where the artillerymen did not yet seem to realize that they were an objective of the raid. From the concealment of a farm yard, Koons dropped a German loader, the first of thousands of soldiers in many wars whom Rangers would change in enemy morning reports from duty to KIA.

Equally accurate Commando fire removed more artillerymen, but after having the first hour of combat go against them, first-line German infantry arrived from reserve positions inland to begin searching fire for the raiders, beginning with mortars and then progressing into a direct firefight.

The Commandos with Koons and Szima were soon greatly outgunned and outnumbered, but they were spared further blows when Lovat's encircling force ambushed a platoon of Germans moving up to reinforce the counterattack. The Commandos killed every one of them.

With Lovat—who went into battle wearing a cashmere sweater and carrying a sporting rifle as if out for a grouse hunt—were Rangers Ken Stempson and Bill Brady who in the first hours had already exercised several techniques practiced in ranger startup training, just as Darby had hope they would: they had thrown netting over barbed wire for a demolition team to cross; they had performed field first aid; they had climbed a 30-foot cliff on a tubular ladder; with their M-ls they had shot machine gun ammo bearers who then "rolled like jackrabbits" (said Stempson).

It was Ranger Brady—who had been shot in the butt within minutes of landing— who then received the final test in what to this point had seemed a live fire graduation exercise in all the skills and thrills of typical ranger operations. Darby's S3 might have labeled this test, 'Interdicting enemy communications while under fire'.

Brady, with a Commando buddy named Finney, came upon the telephone line which provided commo between the German batteries and their spotters on the coast. The line, of course, had to be snipped; but it was under observation by enemy infantry a hundred yards away.

Brady could have imagined this sort of exchange with with his training 'grader':

Grader: "Ranger, how are you going to cut that line?"

Brady: "Sir, I'll get a bazooka and hit the point where the line crosses the telephone pole."

Grader: "No bazookas available."

Brady: "Then I'll mark the pole with smoke and call in tac air."

Grader: "Tac air is fully committed somewhere else."

Brady: "Then I'll adjust artillery on the pole."

Grader: "You don't have any direct support artillery."

Brady: "Mortars?"

Grader: "Only 60 mm. Do you think they're accurate enough?"

Brady: "No, sir."

Grader: "So what are your actions and order, Ranger Brady?"

Brady: "Sir, I'm going to have my buddy climb that pole and cut the wire."

Grader: "Your buddy? What's wrong with you?

Brady: "I"m taller than he is. He can stand on my shoulders."

Thus, equally exposed to German rifle fire that splintered the telephone pole, Finney climbed atop Brady like some intrepid metaphor of Allied solidarity. For the longest and most painful moments since their umbilical chords had been cut, Finney clamped his wire cutters till the phone lines dropped. Neither man was wounded. Both were decorated for their feat.

The Commandos had by now removed the western batteries as a factor in the battle for Dieppe, but to complete his mission Lovat felt it necessary to destroy the guns—otherwise they might hinder the withdrawal. With the Orange II force closing in from the west and Roberts' force from the north, the raiders fixed bayonets and launched a coordinated assault. It carried the gun emplacements within a half hour, but at great cost.

Some of the Commando dead had to be left among the German bodies, but with his mission accomplished Lovat could not burden his withdrawal that was already slowed by the wounded and prisoners. Some of the hardest fighting occurred during the pull back to the beaches, but this most difficult of infantry maneuvers—withdrawal under pressure and fire—the Commandos had planned for and well rehearsed, a proficiency that greatly impressed the Rangers with them.

Finally lodged in a tight perimeter around a single evacuation site, Lovat could call in the RAF and naval guns to hit anything moving toward the beach. As the landing craft approached, Spitfires put down a smoke screen. Wading out to neck depth, the raiders embarked for safety. Casualties had been high—perhaps 20 percent—but had been expected to be more than 60 percent. If the raid had gone as well elsewhere as it had with Lovat, Dieppe would have be remembered as a great victory.

But in the town itself, the Canadians could not gain a foothold despite unsurpassed courage and determination. Theirs is another and utterly tragic story, especially for an army as small as Canada's.

The raiders at the Yellow beaches nearly escaped tragedy, but it overtook them through no fault of theirs when landing craft did not arrive to evacuate some of them. Till then they had gone for their objective with a scheme of maneuver similar to Lovat's: a direct pressure force and an encircling force. Though outnumbered two to one, both were able to take the batteries under direct fire and out of action for a crucial period. This coordinated attack, though unsupported from the air or sea, had nearly the effect of destroying the guns, and is well deemed to have been a mission accomplished.

Breaking off their siege of the batteries, the raiders found that German infantry had infiltrated between them and the sea. It turned out then that lifting the siege was very wise because the Commandos needed all remaining ammo to fight their way back to the beach. They also needed any transportation they could get, so Ranger Edwin Furru ambushed a German staff car, got in and began a nervy drive toward the beach. Alas, the air support that would have been so valuable earlier suddenly appeared and Furru in the *Wehrmacht* car was strafed and wounded.

That was not the end of his bad luck. Shooting and falling back again and again—sometimes with captured weapons which confused and even caused the Germans closing in to cease fire—Furru's group, that included two other Rangers, found a cave in the cliffside from where they could hold out if their evacuation landing craft came in on schedule.

In spite of fervent prayers, the Higgins boats never appeared as they did for the rest of the other Yellow beach raiders. Down to single digits in ammo, with wounded men dying around him, the Commando lieutenant leading Furru's group wisely surrendered to the German company that had surrounded them.

If that lieutenant had capitulated two months later, he and his men could have been executed by their captors because of what became infamously known as the Germans' 'Commando Order'—applying, presumably, with equal force to Rangers:

For some time our enemies [i.e., the Allies] have been using in their warfare certain methods which are outside the International Geneva Conventions [utterly untrue]. Especially brutal and treacherous is the behavior of the so-called Commandos who are partially recruited from freed criminals in enemy countries [this canard refers to refugees and resistance fighters from occupied nations who sometimes, as at Dieppe, accompanied raids]. From captured orders it is divulged that Commandos are directed not only to shackle prisoners but also kill them on the spot if they represent a burden to the pursuit of Commando objectives, or could otherwise be a hindrance. Finally, orders have been found in which the killing of prisoners is required on principle [equally untrue].

It is therefore ordered:

From now on all enemies on so-called Commando missions, even if to all appearances they are soldiers in uniform, whether armed or unarmed, in battle or in flight, are to be slaughtered to the last man.

It does not make any difference whether they are landed from ships or aircraft or dropped by parachute. Even if these individuals, when found, should apparently be prepared to give themselves up, no pardon nor clemency is to be granted them by any field commander. Instead they are to be executed by any convenient means, and a full report rendered to O.K.W. for publication in the Report of the Military Forces...

Any imprisonment, even under military guard, in stockades is strictly prohibited, no matter if this is intended for but a short time. The only period such criminals are to be kept alive is for the most rigorous interrogation...

This order is intended for commanders only and must not, under any circumstances, fall into enemy hands.

This order was signed by Adolf Hitler himself.

Hitler's order did, of course, "fall into enemy hands" soon after it was published, and has given Rangers food for thought, not only during World War II but forever after. That's because ranger attacks, taking place as they often do in the enemy rear, resemble guerrilla warfare and the fuzzy rules of war that thus apply.

The bottom line for Rangers of all eras is that they can not, as POWs, expect the same protection as that afforded conventional prisoners. In the most practical terms, Rangers must do or die because if they don't do (including an escape withdrawal), there is a high probability they will die, even if captured. Those are the ground rules Rangers have to accept, and do.

Rangers Jacobson, Bresnahan and Furru were the first POWs. Luckily Hitler's order was not yet published; instead the three were taken straight to Berlin for interrogation about the new American force. They revealed little, and were shipped to a special high security *Stalag*. Later they were joined by dozens of ranger POWs captured in Italy. It is not surprising that percentage-wise, more Rangers escaped German captivity than any other group.

It must also be acknowledged that, to their credit, the regular German army most often ignored Hitler's 'Commando Order' and, in spite of the real peril of defying Der Fuehrer, treated Commandos and Rangers with decency equal to that provided other Anglo-American POWs.

This treatment suggests that soldiers—whatever the war, no matter how incendiary the mutual hatred—eventually come to appreciate the essential character of their enemy. Such essentials may not necessarily be appealing or evoke humane reactions. In the Pacific, for instance, Japanese troops were regarded as subhuman— for that's the way they had treated those whom they had conquered or captured.

Vietnam Rangers too had precious little regard for the humanity of their Asiatic foes. Yet the Japanese and Vietnamese examples are not demonstrations of racism but rather of reciprocity. That is, in war there is a corollary of the Golden Rule that goes: "We'll do unto you what you do unto us."

By this rule, war between armies in the ETO was at times almost chivalrous. Here, for example, is a message passed through the lines during the Battle of the Bulge, a communication between opposing commanders that would have been appropriate in the 19th Century:

North of Mont Rigi
17 December 1944, 16:00 hours

General M. D. Taylor
Commander of the 101st Airborne Division

My dear General:

Herewith I send you six American soldiers taken prisoner, as well as several of my regiment injured in the recent jump [a regimental drop to seize deep objectives].

I flatter myself to trust that you remember me: I opposed your division at Carentan [Normandy], where you demanded my surrender, and in Holland I fought your landings.

I now earnestly request that you take as good care of my wounded as my regiment did when, on the above occasions, wounded 'Screaming Eagles' fell into our hands.

With highest respect,

Baron von der Heydte
Commander, 6th Parachute Regiment

For the military failure that it was, Dieppe (like Doolittle's raid on Tokyo earlier in 1942) nevertheless was a boost for Allied morale which had lagged since the war began three years earlier. One could generalize that high profile raids like these often have an unpredictable affect on public and political opinion, often an affect that contrasts with the military results.

For instance, Dieppe, Doolittle, and Son Tay (the attempt to free POWs in North Vietnam) all produced negligible military results, and the first two at staggering cost to the raiders. Yet these outcomes were ignored by the public who instead were heartened by the effort. Desert One was aborted with no combat casualties, but the failure of its mission produced widespread gloom and pessimism out of proportion to its shortfall. Indeed, Desert One is acknowledged to have contributed to President Carter's defeat for re-election. Another perverse contrast between military result and public reaction is the example of Mogadishu where the raid's mission was accomplished (snatching a number of high-level aides to a warlord), and in the ensuing battle losses were inflicted in a ratio of about 15:1 in favor of the Rangers. But public and political reaction to the U.S. casualties began to dissolve the commitment to Somalia.

Thus do raids, especially those conducted

Lt. Col. Charles E. Beach, 3rd. Bn. Com., left, shakes hands with Admiral Lord Louis Mountbatten. Rear seats, left, Sgt. Doyer; right Gen. Frank Merrill.

while there is little widespread combat elsewhere, produce repercussions far beyond their military intent. To predict such outcomes is the most difficult task of the overall commander.

Casualty figures, of course, were not made public after Dieppe; otherwise public reaction could not have been nearly so favorable. Of the 5,000 Canadian who embarked, only 2,200 returned alive. Two-thousand were captured. Of the seven major Canadian formations engaged, only one returned with its commanding officer. In but nine hours Canada lost more men than the combined total of their casualties in both France and Italy during the rest of the war.

Commando losses were not nearly as severe Fewer than a hundred did not come home from the raid. The Rangers suffered three killed, three captured, one missing and about a dozen wounded.

The lessons so dearly learned from Dieppe were strategic as well as tactical. From the other side, the German high command viewed the raid as demonstrating the Allies' vital need to capture a major port during the first hours of an invasion. They consequently devoted tremendous resources to converting port cities into fortresses. The Allies fortunately foresaw what the Germans perceived. Dieppe convinced Mountbatten that beaches, not cities, were the way to enter the continent. It remained true that port facilities would be immediately needed, so he turned ingenious minds to the task of creating artificial ports—which was exactly what was done at Normandy.

At the tactical level, Ranger veterans of Dieppe had plenty of lessons learned to tell Darby:

1. Ships and landing craft had to be 'combat loaded', even if it was inefficient, so that vital cargo (like the wire breaching equipment at Yellow Beach) could not all be lost with the sinking of a single vessel.

2. "Don't send a boy to do a man's job" when it comes to off shore fire support. Cruisers and battleships from the Royal Navy were not employed at Dieppe because admirals felt the waters to be too constrictive. The small guns of destroyers were not heavy enough to knock out well prepared emplacements.

3. If you're going to count on air support, you must first be able to count on air superiority. Air parity isn't enough.

4. Bring along captured weapons, and use them so that the enemy thinks he may be in a firefight with himself.

5. Plan for everything, the way the Commandos do. This was a restatement of the motto of Rogers, the original ranger: "Don't forget nothing."

As smoke still drifted up from the allied disaster at Dieppe, the German commander flashed this message to Berlin:

19 August 1740 hours: No armed Englishman remains on the Continent.

This fact applied to armed Americans as well, and remained true for another year. The British strategy seemed to have been borne out by the Dieppe experience—Hitler's 'Fortress Europe' was too tough to crack. Better to nibble on the edges where his forces were weaker and scattered. That was in North Africa. The American high command agreed, for the moment, but men like Darby vowed to return— after he stood up his new force in desert sands. Though he had not yet heard enemy fire, Darby had already won a significant victory: his 'training and demonstration' mission was suspended (though it would crop up again).

Another significant step toward permanence for Rangers was the development and adaptation of a shoulder patch. Among the American Dieppe veterans, a sentiment prevailed that the insignia should honor the Commandos with whom World War II Rangers had first experienced combat, so the basic design was to be a scroll like that worn by their British counterparts.

Darby announced a contest for the design and color scheme of the scroll. First prize was a three-day pass, won by Sgt. Anthony Rada of Headquarters Company who came up with a bright blue background with red trim and white lettering, the same color scheme of the 82nd Infantry Division (not yet airborne). But wartime Britain did not have the desired blue dye, perhaps because their army was muting their shoulder patches the way the Americans would in Vietnam some 25 years later.

Black dye was plentiful, however, and thus became the background of the ranger scroll. Unintentionally this black, red, and white combination mimicked the colors of Hitler's swastika, an irony that has been explained by some original Rangers as symbolizing how they would terrorize the Nazis the way the Nazis had terrorized Europe.

So Darby had his shoulder patch but not authorization to wear it. Truscott once again came to his aid by endorsing the scroll, citing its influence on morale and noting that "rear area commandos" (the World War II equivalent of REMF's) were able to impersonate Rangers, for the purpose of impressing British women, because Rangers could not be identified by a distinctive patch.

Whether some general above Truscott empathized with this romantic handicap for authentic Rangers is not revealed in any written history; indeed, documented approval for Darby's request to wear the scroll has not been found despite research. Even Truscott approved it sewn only on the long wool overcoat of that time, and not to be worn into combat. Nevertheless, men of lst Battalion began to wear it everywhere, even on their 'short timer' stocking caps.

There is much more to be told about ranger insignia, and this is as good a point as any for digression to do so:

Darby, of course, had more important things on his mind than informing the Army's Heraldic Branch about his patch, so officially it did not exist. To complicate matters, the 2nd Ranger Battalion was activated around this time at Camp Forrest, Tennessee, and a year later were issued a bright blue, diamond-shaped shoulder patch with 'Rangers' superimposed in gold letters, and also edged in gold. The insignia did not catch on because of its similarity to the logo of Sun Oil Company (Sunoco). Fights with non-rangers occurred when they drove up to men of 2nd Battalion and said "Fill 'er up..."

Despite the unpopularity of the 'Sunoco' patch, 2nd Battalion was obliged to wear it through D Day, but as they fought deeper into Europe they dropped it in favor of the scroll of its sister battalions.

Across the world in the Pacific, the 6th Battalion adopted a trench knife insignia but later came around to the scroll.

5th Battalion led the cause to make the scroll legal by petitioning the War Department for official sanction, but their request was denied. After World War II, the Heraldic Branch abolished the 'Sunoco' patch but prescribed nothing in its stead. So Rangers in Korea wore the scroll without authorization. As did the lurps and ranger companies in Vietnam, even after the 75th Regiment was created. It was not until the early 1980s, after Rangers had distinguished themselves in Grenada, did the scroll pass muster with the Department of the Army.

Long before 1980, the black and gold ranger 'tab' was approved and worn above unit patches worldwide— but that's another story to be covered after we return to Darby and his original Rangers, proudly wearing their new, though unauthorized, insignia.

Good Missions

II. Arzew, Algeria, November, 1942

The opening of a front in Algeria and Morocco was an effort to seize the initiative from the Germans in North Africa, thereby distracting Hitler from his decisive campaign against the Russians. After the tide-changing battle of El Alamein in October 1942, the British Eighth Army had Rommel on the run in Libya; but he had retreated before, only to turn on his pursuers and rout them. A sudden amphibious attack on his rear could finish him off and take the vaunted Afrika Korps out of the war.

As an allied field commander for the first time, Eisenhower would direct Operation Torch, including the US 3rd Infantry Division which sailed directly from the United States to seize Casablanca. Together with regimental-size landings near two other ports on the Moroccan coast, MG Patton was to command the Western Task Force. Except for the 39th RCT, Eastern Task Force was all British; its mission was to seize the city of Algiers.

Central Task Force's objective was Oran, and the lst Ranger Battalion would be the spearhead for one half of the double envelopment of that vital navy base.

As with all amphibious invasions, early seizure of a major port was essential. The best tactics to do so were still being debated among the allies. Remembering the debacle of the Canadians at Dieppe, Patton eschewed a direct attack Casablanca harbor. The nearby beaches were wide and gentle, so he brought the entire 3rd ID over them, and encircled the port from inland.

Eastern and Central Task Forces tried both beach and harbor approaches. The latter failed at Algiers and Oran but the Rangers' harbor attack succeeded at Arzew. All beach landings in Operation Torch were successful.

Several factors made Arzew unique in Ranger history:

It was the first time the original battalion was committed into combat as a unit, the first time it was under American command, and the first time, of course, it had operated outside Europe. It was also the first and only time in World War II that Rangers would fight against forces other than those of an Axis power. In Algeria lst Battalion was opposed by the overseas army of Vichy France, troops who three years earlier were allied with the British against the Germans. The opposing force made quite a difference from Dieppe which Arzew resembled in other important respects.

Additionally, Arzew was unusual in that it was not a raid but an attack to seize and hold key terrain until relieved. This was vital because without securing Arzew, Oran could not be enveloped from inland. Arzew was also the beginning of a long and mutually rewarding association with the lst Infantry Division (ID), and introduction to its colorful commander, MG Terry Allen. The Rangers liked him from the start, and he came to love the Rangers.

As at Dieppe, two coastal batteries controlled the sea approaches to the port and were the Rangers' objectives. This time it was not enough to neutralize the batteries or hit them and run. They had to be taken and held until follow-on forces consolidated control of the port and continued operations inland.

One battery (Objective Darby) sat atop the highest ground in the area and dominated all adjacent landing sites; the other (Objective Dammer) was at the foot of that hill about a mile away, and its guns covered Arzew's harbor from point blank range. For maximum surprise, both objectives had to be hit simultaneously and at night, for at dawn a dozen allied ships would be anchored within range of the batteries. Also to maximize surprise, there would be no naval gunfire to 'prep' either objective. For its maiden operation, the Rangers had been handed a formidable mission that would have sobered veteran Commandos.

The Rangers were scheduled to go ashore in Africa on 8 November 1942, some three months after Dieppe. In the interim, they had continued training in Scotland where they were assigned to II Corps who attached them to the lst ID. From this new all-American arrangement, the principal satisfaction for Rangers was to get away from British chow. The arsenal of democracy also had a far better breadbasket, including a larder of coffee instead of the British staple, tea. Like all Rangers, its new mess team had to be cross trained, developing a 'feed and fire' capability that proved very worthwhile when, it later operations, Darby's cooks and KP's were his only reserve.

On 13 October, lst Battalion embarked for Africa on three British transports. The Rangers' memories of Scotland would be lifelong. Some marriages resulted from acquaintances developed with Scottish lasses, while the unique hospitality of wartime could never be forgotten. Whenever they were off duty, Rangers had been overwhelmed with invitations to homes and cottages where rare Scotch and friendliness flowed.

First Battalion would never return to such pleasant surroundings, for fate kept them in the Mediterranean Theater for the rest of the war. It took them nearly a month to sail into the Mediterranean as the great Torch flotilla assembled from around the Atlantic. The landings would be as far flung, scattered along a thousand miles of coastline.

With three weeks at sea, Darby had time to review his plan and critique backbriefs, but really the most important preparations had been made already. Arzew was out of range of allied tac air, and with no naval gunfire planned, he realized that 60mm mortars could not provide sufficient support if effective indirect fire were required. So in Scotland Darby had converted a company into an 81 mm mortar battery, and made it the sole exception to his cross training policy: D Company would be his pocket artillery and do nothing else.*[*FN: By comparison, Darby's KP's also doubled as his MP's.] This artillery would be almost self-propelled, thanks to the ingenuity of Sgt. Bob Ehalt who fabricated two-wheeled carts for the tubes, carts that could be dragged over almost any terrain that their crews could walk (or climb).

Of the two objectives the highest battery, consisting of four l05mm guns, was farthest from the landing sites and looked to be the tougher. Naturally Darby reserved command of this force for himself with four Ranger companies including D Co.. The lower (harbor) objective, three 75mm howitzers, would be the mission for Maj. Herman Dammer, the battalion XO with A and B companies. Objective Darby would probably be more strongly garrisoned by the French but the sea approach to Objective Dammer was much more ticklish.

Arzew's harbor was artificial, compromised of two long rock jetties almost perpendicular to each other with only a hundred-yard passage between them. There was also a boom that was lowered or raised to close or open the passage. It had to be open—or Dammer's force would have to break through it—for them to land close enough to the howitzers to suddenly overpower them. Despite otherwise excellent intelligence from aerial photos and French agents, there was no way of knowing whether the boom would be up or down at 0130 when Dammer's men headed toward it in Higgins Boats with motors muffled.

Darby's force had a safer but nonetheless daunting landing site, a tiny cove of sand protected by reef and hidden by rock outcroppings from view at sea; that is, no navigational aids but high peril for the helmsmen of the Higgins Boats. At first the British Navy rejected Darby's proposal to use the hidden cove, but the sailors did not want to appear less bold than the sol-

1st. Ranger Battalion, blackfaced and weary, disembark at a southern English port.

diers, so the site was approved with crossed fingers.

The Center Task Force ships put into Gibraltor for fuel and water on 5 November. No one was allowed to debark or even go above deck because the deception plan was for ships to appear as a supply convoy (some troop ships were disguised as tankers) bound for the eastern Mediterranean, hauling cargo to Montgomery or perhaps to the British garrison on the island of Malta. When abreast of Oran, the flotilla would change course to the south, the Rangers transferring to their Higgins Boats.

The deception was effective. The French along the Algerian coast did not even bother to blackout; indeed, the Germans had told them to leave coastal lights on because this provided silhouette illumination for a U boat attack they planned on the convoy. So harbor lights and navigational buoys were on at Arzew when Dammer's force approached in eight nearly silent boats.

Despite the favorable circumstances (the raised boom most of all), blood chilling problems loomed ahead. First, the harbor was crowded with moored ships and lighters, making maneuver room very tight and causing the lead boat to ram a jetty and recircle the little fleet for another try at landing. The helmsman became disoriented in the dark and another landing was aborted. Then some sleepy Frenchman, perhaps noticing unusual traffic in the small harbor, literally 'lowered the boom', thus trapping the British boats and making the seizure of Dammer's objective a life or death matter for everyone.

Nonplussed, the Rangers finally were able to scramble onto the jetty and moved out smartly to their tasks. The first was to block reinforcements to the fort. Unerringly, B Company located the street intersections (with which they were familiar from aerial photos and sand table models) to set up road blocks while A Company hurried through the slumbering town toward the old French fort that housed the battery.

As yet not a shot had been fired, and Dammer prayed things would stay this quiet. His scouts, Bruder and Lacolle, were first to approach the fort. Rather than continue to rely on stealth, they boldly approached the two sentries like French soldiers returning from late night social activities. While still in the shadows, the Americans were challenged. They didn't know the password, but instead just said in French, "We are friends", a bluff that allowed them to slip into the fort without further examination.

From a roadblock there was sound of firing—some French soldier had tried to flee from capture by B Company. This was an automatic signal for A Company to rush the fort. Aided by Bruder and Lacosse inside, they scaled the walls. Someone in the garrison set off a siren, triggering lst Platoon's charge on the guns. To announce their attack to 2nd Platoon, which was their fire support, lst Platoon called out "Hi ho Silver!", and 2nd Platoon called back, "Away!" The coordinated attack carried the howtizers without any more gunfire—with silver bullets or otherwise.

However, the French, aroused by the siren, began to rally for a counterattack. B Company had all the approaches to the fort covered, so A Company had only to deal with the garrison within the walls.

The barracks was the main threat, but Ranger Murray Katzen went at it with his Tommy gun and grenades. The high volume of supporting fire sounded even greater, echoing around the dark fort, and Katzen was able to get the drop on forty-two stunned Frenchman as they fell out to engage the attackers. None of the howitzer crews reached their pieces. The French commandant never even made it out of bed. Within twenty minutes of the time Bruder and Lacolle entered the fort, it was in Dammer's hands with no casualties except a French sentry killed when A Company came over the walls.

At 0215, Dammer fired the flare signalling that his objective was secure. This was a great relief for MG Allen, but he could hear and observe considerable fire in Arzew.

Along the waterfront this was B Company engaging French reinforcements who were headed toward the fort in a reaction drill set off by the siren. Many stumbled into B Company's ambushes and were bagged, but other French troops began sniping, some effectively, wounding two Rangers. B Company answered with considerable fire which attracted Allen's notice.

He had heard nothing from Darby's force, and this disturbed him. Neither had Dammer whose landing had been scheduled to precede Darby's by thirty minutes.

Delays for Darby's force began even before they debarked into their landing craft. One had dumped all its Rangers into the sea (which was fortunately calm) when a davit stuck, tipping the prow perpendicular into the water. Among the cargo that went to the bottom was all the communications equipment, including flares vital to signal the results of the mission most crucial for Allen's landings in the morning.

This loss was not immediately perceived by Darby as the other boats swarmed around in the darkness, picking up struggling combat loaded Rangers. It was miraculous that none drowned, but the survivors were mixed among their rescuers so that unit integrity was impossible to preserve.

Worse still was the confusion in the flotilla that its commander could not sort out before all boats headed for shore.

"Have you any idea where we are?" Darby asked him in the darkness.

"I haven't the foggiest," was the frustrated reply.

The captain of a British support ship was better oriented, and took commendable initiative by leading the scrambled landing craft toward their landfall. Enroute Darby was able to restore some order, and the excellent navigation of the British seamen brought the force into the hidden cove without further incident.

They were grateful to be ashore, but immediately confronted by the cliffs hemming in the strip of beach. As the Higgins Boats backed out of the cove, the Rangers were challenged by a lone sentry. His rifle could have alerted the artillerymen at Objective Darby, but he was quickly disarmed and captured.

Climbing the cliffs in the dark with hundred pound combat loads was slow work, and the Rangers were behind schedule; but once they reached the narrow, winding mountain road they knew led to their objective, the plan looked doable—if they could achieve surprise.

In a stealthy column of twos, the four companies listened for any signs that the enemy on their objective had been alerted by the sporadic fire from B Company's first contact in the town. Darby made it to a long, twisting ravine that led from the road to the rear of his objective before the fort's siren wailed and wakened the night. Now there was no hope that the French artillerymen would be asleep—but if things went right they could still be surprised.

Darby whispered for D Company to emplace their mortars at the foot of the ravine, and left G Company with them in reserve. The range to the guns would be only a hundred yards, but the three attacking companies would have to climb much farther because of how the ravine zig zagged toward the peak. The going was tediously slow, but in about an hour they came on line outside the 14-foot wide band of double apron barbed wire that protected their objective.

So far, so good—but as Rangers were cutting the wire, they received searching fire from the high ground. This was a signal for D Company to fire its first volley in anger. Their unregistered tubes nevertheless were right on target, firing for effect without further adjustment.

The French had not bothered to erect overhead cover, so the 81mm's blew the gun emplacements apart. The assaulting Rangers took this opportunity to finish cutting lanes in the wire, then rushed the guns while the mortar barrage fell right in front of them. It was a coordinated attack for the books, capturing three-hundred dazed enemy at the loss of a single Ranger WIA.

The objective secure, the guns wired for demolition in case of counterattack, Darby outposted the hill and ordered patrols to other high ground to locate any possible French forces that might affect the morning's landings. He also sent a patrol to link up with Dammer at the harbor fort. Dammer in turn had sent a contact patrol up to Darby. By 0400, approximately ninety minutes before the main landings, lst Battalion had, in Allen's words, "accomplished their initial mission with great dash and vigor."

But a crucial link was missing. Because of the flares lost at sea, Darby had no way to signal his 100% success to Allen. The main landing forces awaited the report through a prearranged series of multicolored flares. Now all Darby had was green. He fired these over and over, but his message went unperceived on the ships, so that the landings were delayed two hours till a small boat came in to learn the situation from Dammer.

Consequently the main landings had to come ashore in broad daylight, and were harassed by sniper fire. Nevertheless, Arzew was secure by noon, and the next day Rangers were leading the advance on Oran, where C Company met serious resistance outside the town of St. Cloud.

Before continuing with lst Battalion inland, the lessons learned for Arzew come to mind just as they did for Darby who filed them with mental notes for the next amphibious operations that he knew were coming, because soon MG Patton would be pulled out to begin planning the American contribution to the invasion of Sicily.

The first and primary lesson had already been raised after Dieppe when the loss of obstacle breaching equipment prevented the Commandos from moving off Yellow Beach. This equipment never reached shore because it was all stored on a single boat which sank. Now nearly the same problem had arisen, for the same sort of reason, when Darby's signal equipment plunged into the Mediterranean. The Arzew experience forever convinced him of the need for cross loading and redundancy in communications.

The second lesson was a positive one: if Rangers can be delivered, they can deliver. Despite the dangerous wanderings of Dammer's force inside Arzew harbor, his boats reached their destination pretty much on time. From there the land phase of their operation went off virtually without hitches. So too with Darby's force. Once they got ashore, they moved like clockwork in a time bomb.

So the shoreline was like the line of departure. For conventional forces, movement to the LD is routine, but for Rangers they must most often be delivered to the LD, usually by another service and sometimes even the service of another nation as the Royal Navy did so well at Arzew. With a bold but thoughtful plan, thoroughly rehearsed and understood by all its participants, Rangers could take it from there if delivered undetected to their LD. MG Allen became the first convert from the lesson of Arzew, and he would be joined by many other generals as the war moved east.

Relieved on their objectives by the lst ID, lst Battalion became Allen's troubleshooters for the next few days. The first trouble was at St. Cloud, a town about half way on the main road to Oran. Lead elements of the l8th RCT*[*RCT: a temporary combined arms task force built around an infantry regiment; i.e., regimental combat team.] rushed the town the day after landing but were repulsed by a tough battalion of Foreign Legionaires and a regiment of Tunisian troops who had weapons of all calibers and plenty of ammo. The buildings of St. Cloud were stone, and clear fields of fire covered the road from Arzew.

Allen decided to bypass St. Cloud and tasked Darby to reinforce the l8th RCT battalion that would contain the garrison. With possible amphibious operations in the offing around Oran, Darby convinced G3 that he should detach only one company for the St. Cloud mission. That was C Company, commanded by Lt. Gordon Klefman.

He reported into the infantry battalion commander outside the town and suggested that C Company march south that night to ambush the garrison's supply line. That mission was approved and the Rangers moved out. Before dawn they were rewarded by the sight of supply trucks halted on the road. Klefman immediately ordered a hasty attack with two platoons just at daybreak when the rising sun was at their backs and in the eyes of the surprised French column.

But surprise soon turned on the Rangers when they discovered that the column was not merely a supply convoy but reinforcements for St. Cloud, fully supported by mortars, machine guns and even 75mm artillery. C Company took up fire and maneuver, but Klefman quickly realized he had bitten off more than even his Rangers could chew. Scraping at the rocky ground with entrenching tools, they were pinned down in helmet defilade for the rest of the day. Moving around to emplace and encourage his men, Klefman was killed by mortar fragments. He, Eskola and Nystrom were the first Rangers KIA in Africa; C Company also lost nine others wounded, and in lst Battalion's after action report was described as "badly shot up".

The lesson learned at St. Cloud? That Ranger operations rely, more than conventional forces do, on precise intelligence. Klefman did not have time for even a hasty reconnaissance to determine the strength of the force in front of him, and he is in no way to be faulted for his bold decision to develop the situation by attacking. As he knew it would be, his decision was a gamble. It didn't pay off except to remind the Rangers that surprise is their best ally—and can be a fatal foe if possessed by a heavier enemy force. And Rangers usually face a heavier force.

First Battalion gained further combat experience when E Company supported Allen's drive along the coast toward Oran which fell to the allies on l0 November. Once again they impressed him, this time by appropriating a half-track to eliminate a distant enemy machine gun nest. Half-tracks were not in the Ranger TO&E (indeed there was no Ranger TO&E), but adept and experienced in cross training, they were able to take up whatever equipment at hand to do whatever job came up.

Ranger versatility became prominent again when lst Battalion went into II Corps reserve where they guarded POWs and important installations. Additionally, they became an ad hoc civil affairs unit when Darby was designated 'mayor' of Arzew. This was a ticklish position because, of course, he had not been elected, did not speak either Arabic or French, and neither was he a military governor because of intricate and delicate relations with the Algerian French which Eisenhower succeeded in partially splitting off from the Vichy regime.

With such distractions, Darby was uneasy that his battalion would lose their combat edge. By reminding higher headquarters of the extensive training that went into the Arzew missions, he was able to collect his companies and get them back into the field to hone skills that would otherwise atrophy. Especially assaults on difficult coastlines. One of his training exercises was to repeat the 'hidden cove' landings just as had been done in lst Battalion's 'cherry' combat mission.

Darby also knew how to sell Ranger capabilities, staging several beach-crossing and cliff-climbing demonstrations for conventional senior commanders. They were very interested because amphibious operations were the prospect for all II Corps units. In late l942 there was no one—including Marines in the Pacific—who had more expertise and experience in amphibious operations than the Rangers who had been in existence for less than a year.

Meanwhile the war in North Africa had moved eastward in what would be known as 'the race for Tunis', and within three months the Rangers joined it. Why they were not employed earlier was because of the herky-jerk progress of Eisenhower's campaign. Allied coordination had many teething problems, and the French remained a wild card, unreliable for both sides. By New Years, 1943, the front had moved into Tunisia but the race for its capital had bogged down. Later that month, lst Battalion received its first replacements, a hundred EM and six officers fresh from ranger training at Camp Forrest, TN.

They were warmly greeted by lst Battalion veterans—with rifle and automatic fire!—sprayed around the trucks that delivered them. The cherries must have wondered what kind of outfit they were joining as sand spurted from the nearby strike of bullets. Looking back fifty years, the live fire welcome seems an incredibly dangerous and undisciplined act that Darby could hardly have allowed. But the rough and tumble times must be recalled. Soldiers by the millions had already been killed or maimed, with millions more sure to be added before World War II ended. In other words death was likely, and safety therefore treated with some nonchalance.

Moreover, what better way to impress the cherries that Rangers must always be prepared for the unexpected than to shoot at them as they entered a base camp? They were kept together as a cohort, forming G Company. No doubt their baptism by friendly fire unified them still more. At the end of January, when lst Battalion flew out to Tunisia, the allied front had stabilized, much to the disappointment of the overall commander, British Gen. Anderson. Logistical difficulties, increased by the onset of winter rains, had stalled his advance but even more of a problem was the Luftwaffe *from whom the allies had not yet been able to wrest air superiority. This situation allowed Rommel to fall back and linkup with axis forces in Tunisia under Gen. von Arnim who was receiving reinforcements, including two panzer and one infantry division, by ship and air from Sicily. Rommel, meanwhile, had dug in on good defensive terrain called the Mareth line, and was prepared to face off with Montgomery again.*

So the main forces on both sides were concentrated near Mareth and Medjez El Bab (about 30 miles west of Tunis). Between them was a reconnaissance and surveillance line that ran for hundreds of miles along a mountain range known as the Eastern Dorsal. Axis and allied patrols probed at each others scattered outposts in and around these mountains. Both sides were economizing their forces in this no man's desert, hoping to cause the other to commit mobile reserves that would weaken his line along decisive areas of the front.

MG Allen saw such an opportunity at the southern end of the Eastern Dorsal. He requested the attachment of lst Battalion for a raid on a strong Italian outpost near Sened. The Rangers flew out to Gafsa, and their first crack at the Italians.

Good Missions:

III. Sened, Tunisia, February, 1943

The objective for the Rangers' first non-amphibious operation was the high ground overlooking a railroad town, the full name of which was Sened Station. The Germans and Italians were not using the railroad because of allied aerial interdiction, but the track bed was a high speed avenue to the sea that von Arnim could not leave uncovered. If the Italians manning Sened were seriously hit, he would have to race panzers to the rescue, and he had no panzers to spare.

To achieve the strategic purpose of convincing the Germans to commit some of their mobile reserve toward Sened, the Ranger raid had to give the appearance of at least a limited objective attack. That is, they would have to seize their objective, if only temporarily, rather than just hit and run so that the Italians would call for a reaction force from the Germans. Sizing up the intelligence about the Italian force on the high ground around Sened, Darby was pessimistic that his objective could be held by the three companies he decided to commit. But they would give it a try—and the Italians a bloody nose, for sure.

D Company had been reconverted back to infantry, but the 8lmm mortars remained available in HQ Company. As he had at Arzew, Darby planned to use his pocket artillery to cut off enemy reinforcements and retreat, rather than prep the objective prior to his infantry's assault. Again the infantry would rely on surprise rather than overwhelming fire support. There was none available anyway: Darby would be out of friendly artillery range, and since surprise meant a night attack, tac air support was virtually impossible.

Surprise depended most of all on precise intelligence, and this was not as good as it been for Arzew. In the fluctuating fighting of the previous month, American forces had taken Sened Station only to be run off by counterattacks. Nevertheless there were a few Americans who had been on the ground and their information was useful. Moreover, a Free French outpost had been observing the Italians from sixteen miles away, and were somewhat familiar with their habits. Some valuable intelligence Darby learned from the French was that the two enemy companies guarding Sened had recently arrived from Italy. They appeared to be a competent outfit but had not patrolled forward much, so might not be completely familiar with the best approaches for a raid. Most importantly, the French did not believe that their new Italian foes sent out listening posts at night. This assumption was a key in Darby's planning of the raid.

In contrast to Arzew, Darby did not have good aerial photos of his objective. Photo intelligence was too committed elsewhere in the theater to provide support for what was a small operation in the scale of things.

Second only to intelligence as the most important ingredient in the element of surprise was stealth. Stealth requires accurate land navigation, and for that the terrain around Sened was excellent. Practically without vegetation, the widely separated mountains provided good landmarks, and the clear desert sky allowed some visibility even at night. Sound carried farther in this clear air, but with careful footsteps a large force could move cross country quite silently.

A leaders' aerial recon was denied because of lack of aircraft, so on 10 February A, E and F Companies, along with the 81mm mortar battery, departed in trucks from their staging area outside Gafsa for a long range look at their objective from the French outpost. Even with binoculars the Rangers could not see much that would help them except for the approach route. They got under way, companies abreast, that night and crossed twelve miles of rugged terrain before concealing themselves with brush and shelter halves to wait out the next day in a saddle where they could overlook their objective and plan the final route.

But Darby felt he needed to know more about the actual Italian positions. With Dammer, he set off at noon for a personal recon that took him within two miles of the outposts, close enough where he could hear Italians talking and watch them sprawled outside their dugouts.

With this final intelligence input, Darby was able to brief his leaders, and compasses were set for the assault. As soon as the sun made its sudden plunge behind the mountains, the Rangers moved out. There was no H Hour scheduled so that they could move as slowly and stealthily as they needed to. Until they confirmed that the Italians had not deployed LP's, they had to be prepared for silent killing in case there were listening posts between them and their objective. The attack would begin when the Italians discovered them; hopefully that would be the moment before Sykes knives were at their throats.

Like most successful operations, the real thing closely resembled training exercises. For silence, the Rangers donned woolen caps instead of helmets, taped metal parts which could make noise, then jumped up and down to determine if anything could be heard. To blend with the night, they blackened skin with burnt cork, and moved out through ravines that kept them off the skyline. The moon set at 2300; prior to that time, Darby would halt the approach well short of the objective before the Rangers set out again.

Their control measures were hooded, red-filter flashlights with which signals could be passed to the rear. Crossing the last flat ground in company columns, guides flashed for platoons to move forward to about 500 yards from the Italians. From this position, Darby could see from the lights when his entire force was on line. It took more than an hour, but finally he got the signal that the Rangers were spread out on a half mile skirmish line. Then he passed up the word to fix bayonets and sneak forward.

Even before the first round was fired, Darby was breathing easier: his assumption that the raiders would encounter no listening posts had proven correct.

The attackers were within 200 yards of the unwary Italians before the extremely rocky terrain began to give them away. A stubbed boot, the slight roll of a rock—something informed a sentry that there was movement in front of the Italian positions. A machine gun on the left started a reconnaissance by fire.

Fortunately, it was wild high, and not returned by the Rangers who continued to torturously creep up the final grade. Even Darby's 8lmm's remained silent, waiting for the sound of American weapons before barraging the Italians.

That thunderclap was yet to come as the Rangers crept forward for a silent assault if that were possible. It wasn't. Green and white tracers began to shift their fire too close to be ignored. Rangers leaped from a crouch, launching a shower of grenades. Submachine gunners opened up first for maximum shock.

From where Darby observed the criss-crossing fire, it seemed a bloody battle was raging and the outcome very much in doubt. But for the attackers it was soon apparent they were on the way to a one-sided win. They had hit with too much sudden fire power for the defenders to react effectively. Explosions, flashes of light, stuttering American guns, screaming Italian wounded, supercharged the attackers and panicked the defenders.

Then 81's rumbled on the Italians' motor pool and command dugouts. All hell had broken out of the late night calm, too much for the Italians to withstand. The only action they were capable of now was escape, and even for that they had to dodge a rain of mortars.

Rangers on the march over Tunisian hills.

The shooting was over in twenty minutes. Seventy-five Italian dead lay strewn among their strong positions. Darby's orders were to take ten prisoners and eleven were led away. Five machine guns and an anti-tank cannon were destroyed after capture.

Ranger Elmer Garrison, decapitated by an Italian shell, was the only US KIA; and though there were also twenty WIA's, none of the wounds were serious enough to require litters. Moving onto the objective and hearing his subordinates reports, Darby had every reason to feel proud and grateful for the results. Yet he had a profound worry—the distant growling of tracked vehicles in the main valley to the east.

Optimum mission accomplishment was to cause a significant reaction by German armor. Darby had to decide if those tracks he heard was that reaction, and if so would it be too fast for the Rangers to withdraw unscathed. They had brought along only a few bazookas. The last thing he wanted to see by the dawn's early light was a platoon of panzers encircling his men on the objective. It would be nice to make a thorough search of the Italian dugouts, pick up some trophies, pretend that the lst ID was behind the Rangers and about to pour through the gap and down the railroad to split Rommel from von Arnim—raise hell and fears, in other words, as Rangers so love to do, but...

Darby must have been urged and tempted to stay longer, but his decision was to quit while he was ahead. He ordered immediate evacuation and dispersal, splitting the retreat into three groups so that enemy planes in the morning would have less to see. Also for speed, he ordered that Garrison's body be left with the Italian dead. This was a ground rule for early Rangers—it also applied to wounded, enemy or friendly—if a man couldn't keep up, or if a body was too much of a burden, he had to be abandoned. The eleventh Italian prisoner that night had been wounded. He couldn't keep up and was killed.

Force marching for the rest of the night, all raiders made it back to the French outpost from where, all during the next day, they could watch German armored patrols sweeping the evacuation route. Darby's instinct once again had been right: in raids like that on Sened, there's a time to hold and a time to fold. His decision this time was to cash in his chips without any further gambling. For that reason the raid was a brilliant success, hailed throughout II Corps.*[*FN: Sened resulted in twelve Silver Stars and two battlefield promotions for the Rangers.]

There were few lessons learned from Sened that had not already been repeatedly practiced in training: use of terrain in the approach march, a concealed assembly area as close to the objective as possible, precise land navigation to release points, reliable control measures, holding fire even when the enemy opened up, furious and unrelenting assault. Sened verified and validated these tactics the Rangers had learned from Commandos then modified to their own liking. Even with their new company of cherries, Darby's Rangers were a now a veteran force feared by their foes.

But to date those foes had been only junior varsity. The French had really not known which side they were supposed to be fighting on, and the Italians (whom, earlier, Hitler had to bail out against even the lowly Albanians) were regarded with contempt by both sides. Rangers looked forward meeting the first team of the Axis—the Germans' Afrika Korps— and the wait would not be long.

Line moves forward: Rangers and Tank Destroyer moving across plain in search of enemy.

Two Rangers watch artillery landing, Tunisian Campaign.

Bad Missions:

I. Dernia Pass, February, 1943

Smoking and joking in after action relaxation back at Gafsa, the raiders speculated how German armor had arrived so quickly on the scene. The answer was that Rommel was about to launch a huge spoiling attack to knock Anderson off balance and take advantage of Allied dispersion and poor coordination. As the most untested force opposing him, Rommel decided to hit II Corps, and hit them where they were thinnest. That was around Gafsa. With hardly time to celebrate the success at Sened, lst Battalion was trucked out of Gafsa the next day only hours ahead of Afrika Korps' arrival. The two elite forces would introduce themselves at a mountain pass near Dernia.

Rommel's bold reconnaissance in force threatened to buckle the entire II Corps front, and indeed it bowed back into Algeria. At the southern end of the bulge, lst Battalion was acting as rear guard for withdrawing elements of the lst ID. In open country, ideal for panzer envelopments, the Rangers were hardly suited for the job. With only bazookas for anti-tank protection, they had to rely on foot mobility to keep out of traps. They marched day and night, harried much of the time by dive bombers, and never long out of earshot of enemy tanks advancing parallel to the American retreat.

It was a maddening time for the Rangers whose wartime experience so far had been one of uninterrupted success. They wanted to turn and fight, but knew that would be fatal folly until they reached some defensible terrain. Till then, hunger, thirst, and sleepless fatigue would be challenge enough and more.

In four days they covered a hundred miles, a major success in a negative way, for avoiding battle with German armor. Chaos reigned throughout II Corps, and communications between headquarters was spotty; but on 17 February Darby received orders to move across Dernia Pass and block it. With their last reserves of energy, Rangers scrambled up the commanding heights and fell exhausted to await the onrushing Afrika Korps. This was a time for inspirational leadership, and Darby tried to instill it by moving among his men and announcing, "If they send tanks up at us, God help those tanks!"

The response from the Rangers was gallows guffaws. First Battalion would sell their lives dearly, but bullets and bazookas were not going to knock out Rommel's veteran panzers. Neither would artillery, but Darby set about to get all he could so that even if German tanks pushed through the pass, infantry and thin-skinned vehicles would not.

Luckily lst Division's artillery had regrouped and were accepting fire missions. Darby plotted concentrations all through the pass. Spread over a goose egg ten miles in circumference, his battalion was just a series of outposts and his men largely target acquisition observers for friendly artillery and tac air once the Allied air force joined the battle.

Just the same, there were plenty of infantry contacts as the wide spread fighting criss-crossed in attacks and counterattacks that more resembled naval than land warfare. Rangers influenced the fray by day and night patrols, adjusting artillery, and setting hasty ambushes of opportunity.

One such ten-man patrol illustrates the fluid situation around Dernia Pass. Into the mountains toward Kasserine went Capt. Paul Hermsen where he dispersed his men into an area recon. After four hours they were to reassemble at a predetermined rally point. Hermsen missed the rendezvous but met a counter-reconnaissance patrol of twenty Italians led by a German. They saw him just as he saw them; but Hermsen acted first, tossing his second grenade while his first was still in the air. They killed the German and two Italians, then Hermsen opened up with his M-fir, sending the others to the ground.

Still greatly outnumbered, but with the element of surprise, he started a get away. But as he rose to reload, he was pinned to the ground by a bayonet through his cartridge belt—an enemy soldier from another patrol had gotten behind him, and his next bayonet thrust would impale Hermsen. But the Italian could not extract the bayonet from the thick webbing of the cartridge belt. As Hermsen struggled vainly (the bayonet had also pinned his spine to the ground) salvation came in the form of US artillery, called in by members of his patrol who had been drawn by the sound of fire. Hermsen's mates were too few to save him from capture, but two years later he did escape, first from the Germans and then the Russians, an odyssey that ended after VE Day in Germany.

In the overall picture, confusion was nearly as high with the Germans as with the Americans in Tunisia. Rommel's thrust was to be coordinated with a supporting attack by Von Arnim, but the two generals had disagreements that proved costly. Their pincer on Tebessa assumed seizure of Kasserine Pass by Von Arnim, but the lst ID and lst Armored Division stiffened here and the penetration was blocked, a development of which Rommel remained unaware. Consequently he dispatched service support units toward Dernia Pass on the expectation that his panzers had cleared it. They had not. The Rangers simply could not stop them with artillery alone.

But they sure could deter some panzers and play hell with wheeled vehicles, especially after Darby scrounged some 37mm anti-tank guns from retreating US units. The Germans drove into his ambush on the morning of 20 February, when two sedans, a truck and motorcycle thought they were on an 'administrative' move through the pass—and never came out.

Maybe even more than raids, ambushes are Rangers' favorite operation. In raids, the enemy sets the place and rangers set the time; while in ambushes Rangers set the place and the enemy sets the time. Accordingly, Rangers have more control of an ambush than a raid. Their first ambush in World War II was not l00% successful because two vehicles escaped it, but the word did not get back to following convoys who blundered into the killing zone that afternoon. Two German troop carriers and a supply truck were destroyed and a number of MP's captured. They thought they were to proceed to Tebessa and direct traffic there as the German advance rolled on.

Finally word got back to the Afrika Korps that Dernia Pass was not a peaceful place to travel. German artillery and Stukas began to pummel Ranger positions, but Darby was not going to just sit and take it. Lt. George Sunshine went out in a jeep to be a very forward observer for artillery counterfire. Sunshine beamed as concentrations made life miserable for armor massing east of the pass. In time the Germans spotted him and two armored cars raced out to drive him off his OP. He hurried back to Ranger lines just before revenge could be exacted.

For the next two weeks Rangers and Afrika Korpsmen played cat and mouse this way. It was a new role for Darby's men, acting as eyes and ears for higher headquarters, a role that foreshadowed a principal activity of Rangers in Vietnam. Then the German spoiling attack sputtered out, though Rommel managed to withdraw largely unmolested.

Lessons learned from his offensive was that Rangers have important defensive functions even though they are designed for attack missions. Adept as they are in ambushes, Rangers proved to be effective as a rear guard for larger formations, even in poor defensive terrain, because they are accustomed to operating independently with little guidance from and reliance on higher headquarters.

Notwithstanding their effectiveness as MG Allen's troubleshooters, the Ranger action at Dernia Pas is classified as a bad mission according to the parameters set in this book: poor support for a conventional mission, principally intelligence and fire support. This, of course, was not the fault of Allen but rather an unexpected situation that forced Rangers into an unsuitable role.

Veterans of lst Battalion harbor no ill feelings about Dernia Pas the way they do about other bad missions. Rangers realize, more so than most soldiers, that war is fraught with unforeseen crises and stop gap measures. Allen would have thrown his aide de camp into a foxhole if that would have helped stop Rommel. First Battalion accepted that sometimes in emergencies the wrong people can be sent to the right places for the right reasons. So lst Battalion shrugged off the role they played-and played so well—at Dernia Pass.

The only long term harm to emerge from the mission was a tendency of some conventional staff officers to look at a Ranger symbol on their troop list—and seeing it to be that of an infantry battalion—therefore considered it to be interchangeable with an infantry battalion. Perhaps this notion first surfaced in the Dernia Pass operation, because it seemed to be reinforced by the commitment of lst Battalion right after the battle of El Guettar, when an infantry battalion was needed in the lst ID's line and the G3 plugged in Darby's Rangers.

Good Missions:

IV. El Guettar, March l943

Rommel's repulse at Kasserine Pass was the last effective Axis offensive of the North African war and the beginning of the last act in that theater. The Eastern Dorsal was the only defensible terrain to protect him from an Allied push to the sea, and the Germans held onto it grimly all through March.

Field Marshal Alexander was now the overall Allied ground commander; his most pressing problems were to redeploy his intermingled units into national sectors and create a mobile reserve. Then he set about to coordinate his moves with Montgomery's who was preparing to dislodge Afrika Korps from the Mareth Line. On 6 March, Rommel tried to beat him to the punch with an attack on Medenine, but it failed with the loss of a third of its tanks. Dispirited and ill, Rommel was recalled to Germany, never to see the desert again.

On that very day, Patton took command of II Corps. As was often the case throughout the war, he was not pleased with the mission he received. On the basis of American performance at the beginning of the Kasserine battles, Alexander had little confidence in their ability to close with and destroy first line German formations. Consequently, while Montgomery and Anderson hit Von Arnim's flanks full force, II Corps was merely to occupy the enemy along their front by conducting a limited attack toward Maknassy.

Patton saluted, but had other plans. He ordered the lst Infantry and lst Armored divisions to break through at El Guettar and Sened Station respectively, and prepare to continue the attack with the objective of Sfax on the coast.

Logically, lst Ranger Battalion should have been attached to the lst Armored and spearheaded their drive with a repeat performance at Sened, but Darby was happy to work again with MG Allen who had an equally important mission for the Rangers near El Guettar.

The town had little military importance, but the jagged ridges to the east, a massif called Djebel El Ank, overwatched a crucial pass from which to the east emerged the two main roads to the sea. He who controlled the pass controlled those roads. As lst ID's advance guard, lst Rangers Battalion moved out on 18 March to find a way to take that pass.

They found El Guettar unoccupied but the enemy firmly entrenched on the forward slopes of Djebel El Ank, flanking the intersection of the Gafsa-Sfax and Gafsa-Gabes roads. The defenders were ready and waiting to turn back any attempt to take this key terrain. Supersonic cracks from a few of the famous German 88s seemed to send greetings to the Rangers from the 10th Panzer Division which was positioned to counterattack any penetration of the pass that was held by 6,000 well equipped troops of the Italian Centauro Armored Division.

With the intelligence lst Battalion had developed, Allen formed his plan to force the pass which narrowed like a funnel from west to east. He would simultaneously launch an RCT up each of the roads into the mouth of the funnel. Darby was to tackle the Sfax road, moving out ahead of the 26th RCT along ridges to the north. Attached to the Rangers was a 4.2 mortar company. Darby had his doubts about how they could keep up in such difficult terrain, but his confidence resided in Lt. Walter Wojcik who, in the Sened raid, proved that if there was a route between hell and heaven, he could find it.

On 20 March, Wojcik took his first recon patrol out to find a way as lst ID watched and waited. Clambering up ravines and between giant boulders, his scouts could see how vital was their mission: across the neck of the pass were 88s protected by machine-gun nests, with plenty of mortars to hit every approach into the pass. If Wojcik couldn't find a way around the enemy positions, a frontal assault by the 1st ID would be suicidal.

'Impassable' is a subjective term. The British in Malaya considered the jungle impassable, but the Japanese learned to pass through it and flank their foes time after time. Rommel considered the deserts to his south impassable, but the British Long Range Desert Group found ways to operate in those sandy wastes with cross country vehicles to great effect. So too the Italians on Djebal El Ank considered the jumbled, jutting heights to the north impassable, and for ordinary troops indeed they were. But not for Wojicik's patrols.

The arid, nearly vertical slopes brought down small avalanches on Rangers clawing hand over hand to reach outcroppings where they could catch a moment's rest. The ridgelines were narrow roller coasters where a misstep meant a hundred foot tumble into a rocky moraine. Crags, precipices, sheer-faced walls would have defied mountain goats—but not Darby's Rangers. One by one they reached a bluff that overlooked the Italians' main defenses. Panting, scratched through torn clothing, and sored muscled, they could see that this was the place—but could anyone else reach it combat loaded? And in darkness?

Reporting back to Darby, Wojcik did not minimize the probably impossible ordeal if the battalion tried to duplicate the climb of his patrol. Rangers habitually move by night but this time Darby felt, from Wojcik's report, that the climb could only be accomplished in daylight. Moreover, the wild massif offered enough concealment so a night approach might not be necessary.

But time was of the essence. Montgomery was hammering on the Mareth Line, so Patton pushed Allen to get started. Allen, in turn, leaned on Darby. The CG would not tell his subordinate how to do the job, but Allen's implication was clear: 26th RCT should start down the Sfax road on 22 March so as to hit the pass simultaneously with the RCt attacking on the Gabes road. Despite Wojcik's misgivings, Darby felt he had no choice but to attempt the climb on the night of 21 March.

The battalion, with its attached 4.2 inch mortars, moved out at twilight, led by Wojcik's guides. Down gorges and up again, teetering on thin jagged ridges, the Rangers in single file clung to the ramparts while waiting for the point to make headway. Ropes helped to cross wide crevices and scale perpendicular traverses, and often seventy-pound packs were passed along hand to hand.

The main body had negotiated a thousand vertical feet in three miles when they hoped that the worst might be behind them. But they were only half way, and the going only got tougher. Men slipped and slid, some saved by those beneath them. Luckily for the hardest climbing, a quarter moon emerged to illuminate a two-mile snake slithering up ragged heaps of rock, dipping into rugged saddles and then up, up again—clawing, halting, groping, cursing silently, but climbing, always climbing till the moon set once more.

The file had been broken by wide gaps and fear of losing the way when a flashlight signalled from an elevation that seemed as high as the stars. Drenched with sweat in spite of the cold mountain air, Darby perceived the first good news of the night: the point had reached the top of the climb. It would be hours before the Rangers closed on the small plateau that would be the assembly area for their dawn attack. The 4.2 battery had lagged behind—their tubes and shells had to raised by human chains— and would not be in position to support the attack, but the rest of the force was perfectly positioned to take the Italians from the rear and from above.

At 0600, a bugle sounded 'charge'! The rock walls of the pass echoed back the notes but not before small arms fire crackled to the astonishment and horror of the Italians who were all facing the wrong way. Light machine guns poured out a base of fire as three hundred Rangers broke out of the hills, descending on the run, leaping between rocks like mountain goats, and carrying the camps of the Italians who were lighting their breakfast fires.

Few fought back against the ferocious surprise attack, and their positions were perched on such jutting terrain that escape was nearly impossible. Two hundred were bagged in the first hour and about thirty killed. The 4.2s were finally emplaced and began to punish the opposite side of the pass where machine gun nests still posed a threat. Darby ordered a company to descend into the pass and take out those machine guns, maintaining the momentum of the surprise attack. The Rangers did, climbing down then up once more, cleaning out the Italian positions till the south side of the pass was silent.

With all going so well, Darby signalled down for the 26th RCT to come ahead. They did, rounding up another thousand prisoners enroute. The most difficult obstacle on the road to Sfax was removed, at the loss of one Ranger wounded, capping what was probably the most remarkable American feat of military mountaineering since the battle for Missionary Ridge in the Civil War.

Allen was ecstatic as his attack pressed forward toward the narrow neck of the funnel. He pulled the Rangers off the mountain, and told them to rest up in reserve— but their break was short. A day later, the 10th Panzer Division counterattacked to bottle up American forces in the pass. Pressed into duty as a regular infantry battalion, the Rangers moved up under intense enemy artillery to man the line between the 16th Infantry Regiment. Trying to scrape out positions in the rubble of rock, Darby's men took casualties from air attacks. Even at this late stage the Allies did not own the skies.

Next morning at dawn, the 10th Panzer launched their do or die effort to retake the pass with sixty tanks and two battalions of panzer grenadiers, supported by two battalions of artillery. The lightly armed Rangers felt a full share of the attack, and did more firing than in any previous day of the war. Also their experience in adjusting artillery at Dernia Pass helped greatly to turn back the Germans once more; but the Rangers' equalizing advantage was the high ground (with excellent observation) they defended. With supporting artillery using direct fire, and bazooka teams to stop tanks that the artillery missed, lst Battalion held the heights against waves of attacks. This was not the sort of mission for which Rangers were designed or equipped but the same kind of coolness, and confidence in one another, that had gained them renowned as raiders and spearheads served them well when panzers climbed up toward them and Stukas strafed their shallow positions.

With the 10th Panzer Division repulsed, Allen sent the Rangers forward once again in the vanguard, this time toward commanding terrain a few miles east of the pass. The Germans were not about to retreat from this last chunk of high ground between them and the sea. First Battalion seized it with a rush, but the Germans came back with such heavy artillery and air strikes that the RCT on Darby's flank had to pull back. On 25 March, the Rangers experienced their first night attack when German paratroopers attempted to retake lst Battalion's high ground. It was a battle royal and a storm of fire but the Rangers held, withdrawing the next day as rear guard for the l8th Infantry Regiment.

First ID surged forward again, this time breaking into the coastal plain and linkup with the British Eighth Army. First Battalion had a few screening missions in the following weeks, but the hard fighting— and within a couple of months, the war— in North Africa was over, the first Allied victory in any theater.

Training and Organization

The North African section of this book has, obviously, been devoted to direct action by Rangers with only a glance at their co-equal mission of training. From the time lst Battalion set sail from Scotland for the Mediterranean, ranger training went on apace but with a different focus than originally intended by higher HQ.

It will be recalled that Darby's initial mission included—indeed emphasized—that hav-

Without a parachute, this Ranger appears to be running through the air as he leaps from a 24-foot obstacle during training.

ing become operational his 'training and demonstration' battalion would return its volunteers to the units from which they came. The pressing demands of World War II in 1942 did not eliminate that requirement but, to Darby's immense relief, it was effectively set aside so far as 1st Battalion was concerned.

To repeat the rationale for the training mission, it was that ranger capabilities should be instilled, at least in every infantry division. At the time this policy was approved by general officers in mid-1942, those capabilities were rather vague and might be paraphrased, "We want our own commandos at division level". Which meant that, on order, ranger-trained infantry should be able to land on tough beaches, climb cliffs, and skillfully conduct raids or ambushes.

That was the original standard, but the exploits of 1st Battalion raised the bar exponentially. Breaking into a harbor like Arzew, attacking swiftly in the darkness as at Sened, force marching as fast as cavalry to Dernia Pass, passing over impassable terrain as at El Guettar—these were feats of arms, not just top infantry capabilities. By comparison, any well conditioned soldier should be able to run a mile in eight minutes. What Darby's Rangers were doing was running it in five. Or to use a musical example, a high school band can play Pomp and Circumstances but they won't sound like a symphony orchestra.

By the end of the North African campaign, it was clear to everyone that no straight infantry battalion could be expected to do what 1st Battalion had done, no mater how good the infantry's cadre. Rangers, in other words, had earned a niche in combat which they would keep to themselves.

And with powerful backers like generals Patton and Allen, there was no longer any concern that Darby's Rangers would be broken up into cadres and returned to the 34th ID and the 1st Armored. No indeed! Yet follow-on ranger units would face that sort of predicament, and in the case of the short lived 29th Rangers, they would be reabsorbed by the conventional army.

To pick up the thread of the training story, on 30 September 1942, II Corps (then in UK preparing for Operation Torch) was directed to organize the 2nd Provisional Ranger Battalion identical to the 1st :

"It will be used," the directive read, "as a training unit for a maximum number of selected officers and enlisted men of combat units *to receive training experience in actual combat, after which they will return to their organizations.* (emphasis added)"

There is somewhat of a difference between the above wording and Darby's original directive. He was to train up then break up after "participating" in raids with the Commandos. 2nd Battalion, on the other hand, was to train, experience combat, then break up. In this second directive it is implied that the battalion would remain active by bringing in replacements, train and return them in cycle after cycle. But it was the "experience combat" proviso that offered 2nd Battalion the opportunity to break out of the training cycle and become a full time combat force as 1st Battalion had.

That was not to be. MG Mark Clark commanded II Corps at that time. He turned to his only infantry division, the 29th, for ranger volunteers. The 29th was a National Guard division from Virginia and Maryland. With the fame Rangers had gained at Dieppe, there was no shortage of volunteers.

Since they were all from the 29th, a confusion in names resulted: officially designated the 2nd Provisional Ranger Battalion, they were better known as the 29th Rangers.

It formed up on 20 December 1942. At that time the 29th ID had been reassigned from II Corps and remained in Scotland to prepare for the invasion of France. Training was with the Commandos. In contrast to 1st Battalion's leggings, 2nd Battalion wore paratrooper boots though they were not airborne. In February 1943 they put on a full dress demonstration of their capabilities, probably to convince the VIP observers that rangers would be wasted as 'school troops', as they derisively called that mission.

Through the first half of 1943, 29th Rangers a participated in parcels with Commando raids and generally distinguished themselves. But on 18 October they were officially disbanded. A staff report concluded,

..The life of the unit is coming to an unnatural end. It is to be returned to its basic units because the War Department felt they could bring over better trained ranger-type units from the States. What a lot of waste had gone into this. These men, while benefiting tremendously from their training as Rangers, are no longer a striking force. Their experience will probably not be put to use [in the 29th ID] if past conduct of the division is any criterion. The men themselves are terribly broken up over the fate of their unit. AWOL's have already started, a thing which is rare in Ranger or Commando units.

In closing down their last cantonment, some 29th Rangers shot up their Quonset huts, almost understandable considering their frustration and disappointment. Yet the other side of their subsequent history must also be considered. It is not at all a complete history because no official record was kept of their service with the 29th ID from D to VE Day. What is known is that by a count of valor decorations the ex-Rangers were way ahead of their comrades in the division. They never fought together as a unit within the 29th ID, but from all reports they provided a much needed leaven of aggressiveness and leadership far out of proportion to their numbers.

So the short story of the 29th Rangers does not settle the eternal question of which should be predominant, direct action or training. What can be proposed generally is that, in wartime, after sufficient direct action forces are available, training of conventional units in ranger craft is certainly valuable; and no doubt a valid training method is to bring Rangers into conventional units. There is also theoretical merit for those Rangers to have come from the units they will train, and something can be said for returning them there to serve again in an envronment with which they are familiar. But to organize and perform direct action as a ranger unit, then return to the originating units as cadre spawns morale problems that offset any advantages. In years to come, the best answer was to be an army-wide ranger school that filled out both Ranger and conventional units, but that satisfactory solution is far ahead of where this story stood in 1943.

Back in Tunisia and at Camp Forrest, TN, the army was seeing to it that enough ranger units did become available, even if they did not include the ill starred 29th Rangers. First Battalion's performance at Dieppe and Arzew were probably the stimulus for 2nd Army to form the 2nd Ranger Battalion*[*FN: Note this was not the 2nd Provisional Ranger Battalion, the official designation of the 29th Rangers] in March, 1943. With only two week to train up for activation, 2nd Battalion was somewhat 'shake and bake', to use an expression from Vietnam. But its manpower pool was nearly limitless: 2nd Battalion was authorized to recruit volunteers from throughout the stateside army.

Darby (now a lieutenant colonel) also obtained almost the same authorization, and it provided salvation from the quarrelsome problem of 'raiding' other combat units for volunteers, a practice that set conventional generals' teeth on edge. Instead he could draw on stateside replacement centers for the 52 additional officers and 1,000 EM he needed to fulfill Patton's requirement for two more ranger battalions—the 3rd and 4th— for the forthcoming invasion of Sicily. 2nd Battalion was reserved for the European Theater of Operations (ETO; i.e., the invasion of France). The Mediterranean was and remained a separate theater. Though most of its ground forces would eventually be transferred to the ETO by 1944, the Mediterranean Theater had to rely on its own assets, including a pool for ranger volunteers if insufficient numbers were obtained from stateside 'repo depots'.

With Patton's endorsement, Darby's requirements were approved by the War Department.*[*FN: The War Department was the name for the Pentagon until that colossal, all-service complex was built in 1942 and became operational in 1943.] However, an annoyingly familiar proviso accompanied the authorization: "After need for these battalions is passed [When would that be? After the war?], personnel therein might be returned to parent organizations so that the personnel might attain the highest rating commensurate with proven ability."

At least the first "might" was preferable wording to Darby's initial authorization in UK when he was directed to return his Rangers as cadre. The second "might" is interesting in its suggestion that men in Ranger units were likely to be held at grades lower than their proven ability, and there was some basis for such speculation. As an elite unit, Ranger squad members were no doubt capable of being squad leaders in conventional units, and so on up the chain. The War Department could have been hinting, slyly, that men who wanted fast promotion should not look at Rangers as a career field.*[*FN: A question of similar interest is whether valor awards are more difficult to win in elite units like Rangers and Special Forces where exposure to danger is more routine, and so expectations of bravery are higher. The author is unaware of any studies on the subject, and one would be difficult to pursue because of several variables involved— probably the most significant of which are the difference in appetite for publicity by commanders, and by the rhetorical ability of their respective adjutants in writing up citations. The latter is somewhat a function of the former because command emphasis is certainly an important factor.

So far as unit citations are concerned, Rangers come off extremely well: the six battalions in WW II garnered a total of six Presidential Unit Citations plus several others. Six battalions amount to two regiments. Except for the phenomenal Japanese-American 442nd RCT, combine the PUC's of any two conventional regiments, including Marines, and their total is four PUC's.]

There is little evidence that many Rangers moved out in order to move up. Instead they were grateful to be Rangers at all because the status of their battalions was still provisional, even after the exploits in North Africa.

With the activation of 3rd and 4th Battalions approved, Eisenhower—still the supreme commander in the Mediterranean—gave Darby written approval to go forth and recruit in "any and all replacement depots in this theater." Darby himself, along with two other officers and four sergeants, did just that, seeking not supermen but merely clean records and excellent physical condition. Except for technicians (there was a grade of 'tech sergeant' in those days), volunteers were to be PFC's so that none would outrank Ranger veterans.

Put three battalions together and you have a regiment, but that logic did not apply to the 'old' lst Battalion and its new sisters, the 3rd and 4th. No regimental organization or headquarters was authorized, and not without reason: the three battalions were not expected to fight as a regiment, but rather to be attached singly to divisions or possibly corps.

Nevertheless, while they were together for training, a ranger HQ of some kind was needed for control and coordination. Darby had to take the staff for this non-headquarters out of his hide, and assume dual command of it and lst Battalion.

Next he had to decide on the 'regimental' composition. When his first replacements reached him from Camp Forrest, Darby kept them together as a company. Should this 'cohort' policy apply as well to the two new battalions? No, Darby decided. Instead—and ironically—he cadred them with veteran leaders from lst Battalion, much the way conventional generals had been insisting be done with Rangers returning to their originating units.*[*FN: Some lst Battalion veterans were also sent to Camp Forrest to leaven the 2nd Battalion there.]

Darby's reasoning for his reorganization deserves detailed consideration because as Rangers expanded, not only in WW II but thereafter, the same factors came into play. Though he never committed his thoughts to writing on the subject, it is assumed that he had a number of alternatives, each with accompanying pros and cons:

l. Retain all the new personnel in 3rd and 4th Battalions, placing them under two seasoned commanders from lst Battalion.

The advantages would be a level line of experience in the new battalions which would rise steadily through training and combat. Coherence and familiarity would be optimized, and lst Battalion would keep its highest proficiency.

The principal disadvantage would be uneven proficiency between lst and the new battalions. This difference could translate into higher casualties for the latter, though this might be compensated for by initially giving then softer missions. But no one could guarantee if there would be any soft ranger missions coming up.

2. Put seasoned lst Battalion vets in command of the new battalions as in Option #l, but also sprinkle vets throughout the rest of the chain so that there would be experience at every link.

The obvious advantage here is that casualties among the cherries would be minimized because of experienced leaders in key positions.

The disadvantage is equally obvious, that of diminished proficiency in lst Battalion. There would also be the tendency of lst Battalion vets to stick together in their new battalions, crowding out emergent leaders among the cherries and, in the worst case, creating a sense of "them and us".

Option #2 would produce three battalions of roughly equal experience and proficiency, but this is neither a pro nor a con for this option because what was added to 3rd and 4th Battalions was subtracted from lst. That is, if lst Battalion is rated as l0 on some scale of combat proficiency, and the new battalions are rated 4, the sum is still l8 no matter what the mix. Option #2 just comes closest to making all three 6s; that's all.

3. Command and cadre the new battalions as in Option #2, but rather than spread vets evenly, concentrate them in a couple of companies where they would be a battalion strike force until the cherries came up to par.

The advantages here are that a third of each battalion would be combat ready immediately, and vets could stick together which was largely

First Anniversary of Rangers—MG Terry Allen, Lt. Col. William O. Darby, Capt. Roy Murray and Gen Terry Allen do the honors.

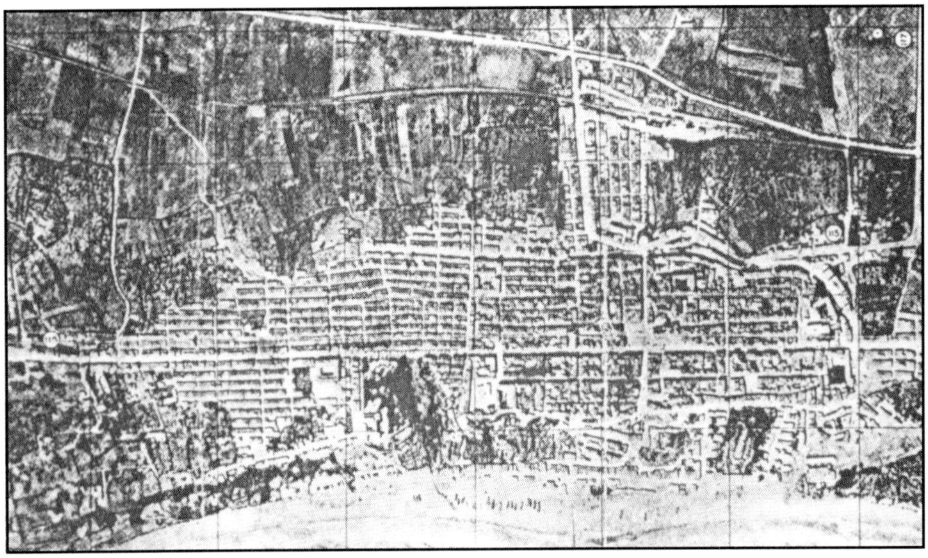

Gela, Sicily, where the 1st and 4th Battalions landed amid heavy opposition and seized the entire town in four hours.

Gen. Matthew Ridgway

their preference. Competition between 'old' and 'new' companies within a battalion would probably be salutary.

A disadvantage is that the new battalion CO would have to be very careful not to give the appearance of favoring veteran companies, or consistently giving them the toughest missions. Moreover, combat experience would seep more slowly into the new companies than under Option #3.

4. Create a permanent 'school' company out of lst Battalion, and have it train the cherries of the new battalions as well as lst Battalion replacements.

The advantage is uniformity of training throughout the Rangers under the best instructors. Additionally, limited duty Rangers (e.g., those who had been seriously wounded but not totally disabled) could stay with the Rangers and still contribute to their success.

However, school troops is what Darby never wanted Rangers to be—he had fought that concept from the beginning. What's more, his best instructors probably included some of his best leaders who would bridle at being removed from combat operations.

Darby's decision was closest to Option #2. He placed his XO, Maj. Dammer in command of 3rd Battalion, and one of his top captains, Roy Murray, Jr. in command of 4th. From A and B Companies of lst Battalion came the cadre for 3rd Battalion; from E and F Companies came 4th Battalion's cadre. Probably in order to even out experience, Darby kept his newest company, G, intact in lst Battalion. Of course personalities, the interaction of which a remote historian can only guess about, must have had great weight in Darby's decision. What reflects so well on him personally is how he did not keep the best to himself, for he remained CO of lst Battalion as well as overall Ranger commander. A lesser man might well have skimmed the cream and concentrated it in lst Battalion to enhance his own reputation in future action.

But in May, 1943, Darby's only desire was to let cream rise to the top, whether it be veterans or cherries. To do so he drove his men through six weeks of merciless training in the Atlas Mountains where both heat and cold added to the exhaustion that comes closest to resembling the rigors of combat. Though lst Battalion vets remember the Atlas as the hardest training they ever endured, the wash out rate among cherries was lower than it had been in Scotland.

Basically the vets taught them what the Commandos had taught the original Rangers. But one new feature was added: house to house combat. Though their upcoming mission was still top secret, the vets accurately guessed that Rangers would once more be asked to seize a port.

A break from the training ordeal occurred on the first anniversary of the Rangers. With a Sykes knife, Maj. Dammer cut a cake with one giant candle held by Darby and Allen. What a year it had been! What a year lay ahead . . .

Good Missions:

V. Gela and Licata, Sicily, July, l943

As long as Sicily was used by Axis air forces, Allied shipping could be choked midway in the Mediterranean. For this reason, the island at least had to be neutralized. If it were conquered, the domestic repercussions in Italy might bring down Mussolini and push the new government to surrender. Those were the two principal reasons Sicily, rather than Sardinia or Corsica, was chosen for Eisenhower's next strategic objective in a sea and air invasion called Operation Husky.

Approximately the size of Vermont, Sicily looks like a deformed football that deflated when kicked by the toe of Italy. Mountain ranges crowd the northeastern coast between the Simeto River and Messina, but the rest of the coast is flat, rolling into scattered hill masses inland. Topographically, Sicily has much in common with North Africa, so good training terrain for Operation Husky was readily available to the invasion forces.

The greatest, and most pleasant, contrast between Sicily and North Africa was the climate. Though only a hundred miles apart, there was a twenty degrees temperature differential in favor of the European island. Said MG Matthew Ridgway, CG of the 82nd Airborne Division in the invasion,

"No war was ever fought, I think, in more ideal weather than was the Sicilian campaign. The soft spring [though this was July] lay upon the land like a benediction; the days were warm, the nights cool, and there was no rain. After the frying pan of Africa, the balmy weather was like heaven to the troops who had suffered through Morocco and Tunisia."

Many Husky veterans would recall the 38-day offensive similarly, including Ridgway's DIVARTY commander:

"Most of the way it was as pleasant a campaign as one is likely to find in war. No one was very angry at anyone. The Italians had no desire to die for the King, much less for Hitler. In towns we passed through, the villagers greeted us as liberators and many rushed up to ask in broken English about relatives in America."

That's how the campaign would end, but it began with some of the most furious fighting of the war, and with Rangers at crucial points.

The invasion fleet assembled at ports from Algeria all the way to Syria. The 3rd ID, with the 3rd Ranger Battalion attached, would try the first large scale shore-to-shore amphibious operation of the war, boarding state of the art landing craft (LST, LCI, and LCT) in Africa and move directly to the beaches of Licata, obviating the need to load then debark from large ships.

The British Eighth Army would land near Syracuse; between them and Licata was II Corps, compromised of the green 45th ID and veteran lst ID (lst and 4th Ranger Battalions attached). Darby's objective was the fishing port of Gela.

The American landings would be commanded by Patton's newly formed Seventh Army. MG Dennis of Alexander's staff described the events leading up to H Hour:

On 9 July the convoys of Seventh and Eighth Armies were closing towards their rendezvous areas east and west of Malta. The sea was rough and many men in landing craft suffered from it. The swell threatened to make the beaching of assault craft hazardous. However, during the night the wind began to slacken and, as the risk of postponing the assault was greater than the hazard of continuing, it was decided to go forward with the plan.

The seaborne assault was an outstanding success, for as a result of the gale the enemy garrisons, already wearied by false alerts, had relaxed vigilance. Complete tactical surprise was achieved, and though some of the beaches came under fire from enemy guns, these were soon silenced. Naval supporting fire was admirable, while artillery FO's and liaison teams in bombarding ships worked together smoothly.

Patton described how things looked off of Gela:

The coastline as far as we could see was a mass of flames. In some places it looked like burning mountains; in other places, simply like a line of fire . . .These fires, we learned, were caused by our high explosives igniting the wheatfields. Fortunately for the Sicilians, the wheat had been harvested, but the dry stubble was still there and burned like hell.

At intervals along the flaming shore, searchlights would suddenly flare and sweep the water. This was very interesting, because while the light made it so bright on deck that we could read papers, the enemy was apparently unable to see us from the land. I think this had something to do with refraction and the haze caused by the fires.

Whenever one of these searchlights flicked on, innumerable tracers from our destroyers and patrol boats hurried toward it like bees returning to a hive. That searchlight immediately went out.

I knew H Hour was near when the davit for a landing boat gave way and the boat crashed into my porthole. For a moment we thought that a bomb had hit us. As we were sleeping in full pack, we immediately went up on deck to watch the start of the show.

What Patton would see was the outcome of a risky amphibious attack that been debated by Darby and Allen's staff. The risk, essentially, was in the similarities to Dieppe. Both ports were about the same size; though there were no bluffs at the waters edge for Rangers to scale at Gela, there were two vital coastal batteries to silence quite like Dieppe. And this time the town had to be taken, not just raided.

Rather than enveloping the town through landings on the flanks, Darby agreed to go at Gela directly much the way the Canadians had done so valiantly but disastrously at Dieppe. Darby's surprising decision resulted from study of aerial photos (agent intelligence was not as good as for Arzew) that convinced him that Gela was well prepared for defense from either land or sea. The batteries, for example, had 360 degree traverse rather than being pointed only to sea as they had been at Dieppe. Camouflaged pillboxes also covered inland approaches. But probably the most convincing factor weighing against envelopment was the time it would take: first to reduce the batteries, then fighting through the maze of pillboxes, all with little cover or concealment for the Rangers. Too costly and too slow. Patton had to start moving supplies through Gela fast or the invasion would lose momentum before it really kicked into gear.

The alternative to envelopment was hardly attractive. To take Gela head on meant sailing over a sand bar at the mouth of the port then, on the town beaches, cutting through mounds of barbed wire and across minefields. A 300-yard fishing pier divided the harbor. Darby made it the boundary for the lst Battalion on the west, 4th Battalion (Capt. Roy Murray, Jr. commanding) on the east.

This Ranger mini-regiment, supported by an engineer and 4.2 mortar battalion, was designated Force X, with the objective of clearing the town, seizing the high ground that controlled it, eliminating the threat of indirect fire on the port, and expanding the beachhead. A tall order, with many chances for mishap, all starting at midnight when a submarine was supposed to guide the first waves into the harbor.

First with sand tables, then with mock-ups, and finally with full dress rehearsals, Force X in Africa had prepared for the over the beach assault thoroughly, including participation of the very sailors who would put them ashore. The only element missing from rehearsals was the submarine which would guide them on D Day. At H Hour too it was missing.

To the music of Glenn Miller's *American Patrol* piped out from a ship's loudspeaker, lst and 4th Battalions scrambled down into their landing craft that soon began circling, waiting for the submarine guide that never appeared. Darby couldn't lose any more time. Recruiting a naval officer who felt confident he could find the way, Darby called out in the darkness for the other boats to follow him.

The searchlights that had so fascinated Patton converged their beams on the bow wakes of the Rangers' first wave. Tracers crossed in the night until naval gunfire shot out the lights. Still, preregistered mortar concentrations fell among the boats. From their boats the Rangers fought back, even with bazookas whose backblasts added pyrotechnics to the swelling fire fight.

The offshore situation seemed to be developing into a "Dieppe in the dark", as one veteran of both landings described Gela. And it got worse before it got better. As the first wave reached a quarter mile from the long pier, Italian engineers blew it up with a preset chain of mighty charges that launched huge chunks of rock high into the air. Their plunging splashes were a metaphor for Allen's hopes of landing heavy equipment over the pier. Darby had more immediate concerns when a boat in the first wave went aground on the sandbar and began to capsize.

Uncertain how close they were to shore, Rangers jumped into the shallow water but many slipped, heavily laden with equipment, off the sandbar into a deep tidal eddy. Despite other boats called to the rescue, seventeen Rangers drowned.

The rescue delay caused boats of the first wave to become mixed with those from the second. The assault was getting off to a bad start, causing some of Allen's staff at sea to wonder if enveloping landings outside town would not have been a better plan. But Rangers knew the old army adage that a poor plan well executed is better than a good plan poorly executed. "Rangers lead the way" became a motto, and leading was what they would continue to do despite accidents, bad luck and whatever the enemy threw at them.

They were throwing plenty. Their buildings were solid stone (there is little wood on Sicily), and stood up even to the naval bombardment, so that machine gun nests firing from ground floor windows swept the harbor. The Rangers just had to take their casualties in order to get at these machine guns fast. Barbed wire was hacked apart and mines ignored to rush the first row of buildings off the beach. The mines

Looking down the muzzle of a Tiger 88.

The Hermann Goering Panzer Division being aimed at US artillery as they attempt to push the hard-pressed beachhead forces.

took a deadly toll—nearly an entire platoon was blasted into the air while trying to cross the beach.

Among those killed by sand covered mines was Lt. Wojcik of El Guettar fame, commanding D Company in the Gela assault. A dozen more of his men were wounded on the barbed wire and all company officers fell as casualties. Command was then assumed by lst Sgt. Randall Harris, himself gut shot so badly that he had to hold in his intestines with his cartridge belt. He drove on, D Company right beside him, to carry the beach defenses. A man of high courage but low profile, Harris lived to receive a Distinguished Service Cross and a battlefield commission.

It took two hours for lst and 4th Battalions to break into Gela. By then the mix-up of units caused by mingling boats off the treacherous sandbar had been sorted out, and Darby's plan to take down the town began to work. This meant street fighting; and though such combat is not a standard ranger speciality, it had been practiced extensively during preparations in Africa. Gela is divided neatly into rectangular blocks: platoons knew from street signs (visible even at night in the glare of battle) which were their assigned blocks, and companies knew which row of blocks were in their sectors. Drills for moving house to house had been well rehearsed. When possible they were cleared from top to bottom after demolitions opened the tile roofs.

By daylight the Rangers on four axes had fought through Gela to its outskirts. First ID could now use the port, though sniper and indirect fire remained a problem throughout the day. Two hours before, A and B Companies had landed to take down the two gun batteries from the town side, the direction from which the Italians least expected an attack.

Under command of Lt. James Lyle, these Rangers got no support from planned 4.2 concentrations because the fire direction radio had been lost at sea. Practically lashing his men with orders to move faster, Lyle shouted that if they were caught in daylight it would be their last morning. They raced through the streets, overrunning road blocks, blowing up bunkers and giving no quarter. With barely a half hour of darkness left, they came upon their objective, the guns.

Not waiting for covering fire, they blew apart barbed wire like it was the last obstacle to home leave. Two platoons branched off to attack a mortar platoon that had been giving the landings trouble. It was wiped out to the last man. At the batteries, Rangers found the crews stunned by naval gun fire. Most had already fled; those remaining surrendered without a fight. Three 77mm field pieces were thus captured but their sights had been removed by retreating Italians. Even without sights, as Robert Black described it,

"The varied background and cross training of the Rangers was immediately put to good use. Three crews were formed and spent two hours in gun drill. Lyle's position was astride Highway ll5, and the innovative move proved worthwhile."

Extremely worthwhile, beginning that very morning . . .

MG Omar Bradley commanded II Corps for the Gela landings. These were his impressions the next morning:

An invasion beach the morning after landing is a dismal sight. This one was no exception. More than two hundred assault craft wallowed in the surf after having burned out their engines in crossing the runnels while coming ashore. Bulldozers churned through the soft sand, dragging pallets of supply from the water's edge to be piled in dumps behind grassy dunes. A fleet of more than seven hundred DUKWS moved from ship to shore, ferrying in the bulk tonnage that would be needed for support. Everywhere along the fifteen-mile length of beach to Gela belly-wrapper lifebelts littered the sands where they had been discarded by assault troops. Behind the beaches A.A. crews were digging deeper into the dunes, anticipating another noisy night of bombing...

My radio was malfunctioning so I decided to run down and see Terry Allen in person. There was too much smoke and dust coming up from his beachhead, and a lot of artillery going in both directions. It looked like there could be serious trouble...

Dog tired, he waited for me in a makeshift C.P. by the beach. His eyes were red from loss of sleep, and his hair dishevelled.

"Have you got it in hand, Terry?" I asked.

"Yes, sir, I think so," he sighed, "but we're going to have a helluva time stopping them till we get some anti-tank stuff ashore."

The "serious trouble" was a series of increasingly heavy counterattacks by Axis armor, including the vaunted Hermann Goering Panzer Division. Allied air strikes and ambushes by paratroopers had delayed commitment of German mobile reserves during D Day, so the first probes of the shallow Gela beachhead were piecemeal and undertaken by Italian light armor. More of a problem was the Luftwaffe which raided Patton's supply ships almost at will but luckily without much accuracy.

Allied fighter strips had not yet been established on the island, and anti-aircraft batteries were late coming ashore. Anti-tank guns were in equally short supply as most of them were to have come in over the Gela pier, now destroyed. The single tank battalion available to Bradley was still with the 45th ID in floating reserve. It was during this 'longest day', D+1, when combat power was so scant in the beachhead that the Rangers were gripped in the toughest fighting on Sicily (7th Army took more casualties than in any other 24-hour period in the campaign). They had to hold or the more successful landings at Licata, fifteen miles away, would be negated.

Here 3rd Battalion (Maj. Dammer commanding) had come ashore at 0300 with all six companies abreast. The greatest difficulty they faced were steep bluffs behind the beaches, strewn with wire entanglements. C Company Rangers used pickets in the wire to haul themselves to the top, while supported by a base of fire set up behind rocks that at high tide had been submerged.

The Italians weren't ready or alert. Some Rangers reached enemy positions before the defenders could occupy them. Mortar emplacements were overrun almost as easily, and by daylight 3rd Battalion had moved through all coastal obstacles to establish a perimeter l500 yards deep. Well camouflaged pill boxes and machine gun nests still harassed the Rangers but the Italians had made a mistake by encircling these positions with wire that gave away their locations.

Thus, with 3rd Battalion paving the way, Truscott's 3rd ID came ashore with negligible opposition and promptly secured all their D Day objectives. 3rd Battalion went into division reserve in Licata which they found to be a more dangerous place than the field, for in daylight the Luftwaffe struck the port often and hard, causing more Ranger casualties than they had sustained in the assault.

The next objective was to link up with the Gela beachhead. Dammer tried to monitor radio traffic from his sister battalions over there but there was little to hear except the building thunder of naval gunfire off Gela. Dammer assumed that they were receiving plenty of fire missions from shore, and he was absolutely right. First and 4th Battalions were fighting for their lives and to prevent lst ID from being cut off, exposing the Gela beachhead and opening a huge in the Allied lines. If this happened, German armor could turn upon Seventh and Eighth Army separately and the invasion be pushed back into the sea.

Here was Darby's situation at the beginning of this crisis:

His two battalions having cleared Gela except for snipers, he had set up his CP in the Cathedral Square, a decision Darby would later change because the cathedral spire was the tallest point in town and made an excellent reference point for enemy artillery. Lyle's company held Allen's left flank, using the guns he had captured and their emplacements to defend the roads leading to Licata and inland. 1st Battalion was strung out on the outskirts of town alongside 4th Battalion which tied in tenuously with lst ID at a shallow river just east of town.

During the morning of D Day, Italian probes had been turned back, the first by 14 light tanks that C Company engaged with their captured field pieces. Nine tanks were stopped 5,000 yards from Gela but the remaining five made it into town, only to be set upon by Rangers attacking from basements and rooftops with 'sticky grenades'—concussion grenades with adhesive surfaces—satchel charges and bazookas. Four tanks managed to back out of town; but the last one roared into Cathedral Square where the Ranger CP was protected by Darby himself manning their only 37mm anti-tank gun. As Patton described the action, "[The Rangers] hurried down to the beach under fire, got hold of a 37mm, split a box of ammo with an axe, hurried back to the square, and went into position with the gun less than a hundred yards in front of a tank coming down on them."

Darby and Capt. Shunstrom disabled the tank, then other Rangers swarmed over it and blew the hatch off with satchel charges. The 4.2s chased the retreating Italian armor, killing one tank and discouraging any more.

That concluded the first counterattack of the day; the next would fall on the 1st ID as the enemy probed for soft spots in the thin defenses of the beachhead. Throughout the night, both sides patrolled feverishly to determine how best to fight on D+1. Without sufficient LSTs to bring in US armor, Allen's hopes were in the flotilla of DUKWs that ferried back and forth all night, mostly carrying anti-tank guns and artillery that could be used for direct fire in an emergency. There would certainly be such an emergency because German panzers could be heard assembling far off in the darkness.

The Luftwaffe continued to hamper ship to shore movement, but to balance the odds naval gunfire was an equally tough interdictary problem for German General Conrath commanding the Hermann Goering Division. He knew he had to get his infantry together with his tanks or else they would be ambushed inside Gela as the Italians had been the day before. But he also knew the American beachhead was growing stronger by the hour. Should he attack while he had such a huge advantage in armor or wait till infantry arrived to join his tanks and form effective combined arms teams? Conrath opted for the first alternative—and at first it looked as though his gamble would win.

His first blow hit the 1st ID sector at dawn of D+1, with a supporting attack by the Italian Livorno Division against the Rangers. Their commander apparently forgot to tell them which war they were in: a battalion came marching down the road toward C Company like some formation out of the 19th Century. C Company's captured guns broke them up, then A and B Companies tore them up.

Meanwhile Conrath's panzers had swarmed into lst ID's positions, penetrating so close to the beach that naval gunfire could not be employed. Murray began to shift his small reserve toward the river where he might have to refuse his flank if the Germans reached the sea in force. Using artillery as anti-tank guns, The Big Red One managed to hold on. But Conrath's next assault included mechanized infantry (panzer grenadiers) that had finally reached the fields of Gela. This blow was aimed at the Muroglio River, the boundary between lst ID and the 4th Rangers.

Darby had no reinforcements to send Murray, as the whole Ranger sector was under a coordinated attack. Lyle scooped up quartermaster and whatever other troops on the beach that he could to back up his line that was now down to 120 Rangers. Trailing a plume of dust two-miles wide, the Germans bore down on Gela as waves of Luftwaffe fighter-bombers struck the Allied fleet.

With his captured pieces and 4.2s, Lyle was doing his best to slow the onrushing attack, but the Germans were softening up Gela with their own artillery that landed so close to Lyle that he loosened the chin strap of his helmet so a nearby concussion would not break his neck. At that moment he met the most unexpected reinforcement to come up from the beach, Gen. Patton himself, whose aid to Lyle for the moment was to point out that his chin strap was unbuckled—contrary to 7th Army policy.

Having tended to this important business, Patton moved over to check out Darby's CP and found that he was cut off from the lst Division on his right by seven Mark IV tanks which now closed in to one thousand yards of the town.

Darby had the roads patrolled by a few half-tracks. They were not designed for combat, but with medium caliber weapons they greatly annoyed the Italians whose attack seemed to get stuck at about 1150.

Over on the right, Germans 88s opened up. They hit the building we were in twice, and also made holes in the roof of the building across the street, but no one was hurt except a few civilians. I have never heard such screaming.

By noon the Italians had run out of steam, but the Germans were just getting rolling. Gen. Omar Bradley picks up the story of the battle from here:

Forty tanks had cut across Allens' front towards Gela. If those columns were to converge there and break through to the water's edge, lst Division would not only be cut off, but the whole beachhead would be endangered.

Now in desperate need of artillery pieces with which to fight those panzers, every gun on our front was rolled into position to meet them at point blank range. Trucks raced to the beach to drag up additional artillery as fast as it came off landing craft. At the same time, fire-support parties called for help from naval guns.

Though sometimes overrun, our infantry did not fall back, but instead burrowed into their foxholes to let the panzer wave wash over while they waited to repel the grenadiers advancing behind it. Fortunately that additional artillery enabled the line to hold, and most of those tanks were stopped on the plains outside Gela. Of the sixty panzers committed to that counter-attack, more than half were destroyed... But Allen had barely squeaked through, for those tanks had advanced to within two thousand yards of the beach before they turned.

Later that afternoon the enemy resumed his attack, though this time with fewer tanks. As salvo after salvo split their armored hulls, [Conrath] wisely concluded that 26-ton Mark IV's were no match for cruisers.

The Rangers' favorite cruiser was the USS *Savannah* whose six-inch guns staggered the counter-attacks and left bodies dangling in olive trees. After a battering of 500 rounds from the *Savannah*, 400 dazed Italians gave up to C Company without the Rangers having to fire a single shot. The naval support, the artillery's direct fire, and most of all the dogged backs-to-the-beach defense by the infantry had turned the battle, and Patton was quick to seize the initiative:

I happened to look out to sea. From a Liberty ship, which the Germans had bombed about a half-hour earlier, smoke was escaping. Then before my eyes a tremendous explosion threw white and black clouds several thousand feet into the air. The ship literally blew in two though the rear half stayed afloat...

About that time an officer from the 3rd Division came in with ten tanks, having made the trip by the shore road from Licata. I told him to close the gap between Gela and the lst Division and send come tanks to help Darby. This was done, and the Rangers moved out at once, bagging five hundred prisoners.

With the battle and the Sicilian campaign moving inland, Rangers could take stock of lessons learned from Gela and at a cost of 31 dead.

The first regarded rehearsals where the essential submarine guide had been missing as it indeed was in the actual operation. This absence was beyond Darby's control but it reinforced the lesson that everyone involved on D Day has to be present for D Day rehearsals. Perhaps, even though the submarine didn't show up at H Hour, if the skipper had been at the rehearsals some 'no sub' contingency could have been planned. As it turned out, Darby had to take matters into his own hands at the very last minute. Incidentally, it was the day after Gela that Patton offered Darby a promotion and command of an infantry regiment, "But," Patton noted, "he refused to take it because he wished to stay with the men he had trained."

Second, essential commo equipment was lost at sea, this time preventing support from the 4.2s in the early hours of the assault. Rangers were already overloaded, but some form of commo redundancy was vitally needed.

With good commo, and able coordination from naval observers, Ranger training in calling for fire support paid off richly when the Germans made their go for broke counterattack.

Finally, some form of anti-tank capability had to be available to Rangers before they met enemy armor in daylight. At night, tanks are not very effective, especially if they are not executing a pre-planned counterattack. But daylight brings visibility and swings the advantage from the infantry to the armor. To even the odds, a shoulder fired anti-tank weapon was essential for ranger survival, a development beyond Darby's control but a problem so common in the war against Germany that a higher caliber bazooka became a priority project.*[*FN: The Germans countered technologically by developing the more heavily armored Tiger tank and, ultimately, the Tiger Royal that by the war's end could withstand not only 3.5 inch rockets but also fire from the main gun of a Sherman tank.]

With the beachhead secure, Patton's first offensive objective was the Biscarri airfield so that Allied fighters could neutralize the troublesome Luftwaffe in the combat equation. He now had tanks from the 2nd Armored Division ashore and they spearheaded his attack with the lst and 4th Rangers attached.

From his African experience, Darby knew that pressure on the withdrawing Germans had to be constant and relentless or they would organize the high ground inland and check the American drive. Consequently, he pressed a night march on the Gela-Butera road, rolling up the first enemy defenses with an attack at 0430. An armored infantry battalion from the 2nd AD could not make as good time at night and were short of their objectives at daylight, leaving them exposed to heavy machine gun fire and salvos from five artillery batteries. Once again the trusty cruiser Savannah saved the day, drumming the Germans from twelve miles away with devastating counter-battery fire. By 1400 hours, Darby's force had broken through and seized the first range of high ground overlooking Gela.

Patton described the aftermath of this action:

On the field south of Biscarri, where we had quite a fight, I could smell dead enemy while driving for at least six miles along the road... The enemy has been booby-trapping his dead, firing on us from the rear after we have passed through him, and used dum-dum bullets. The

Germans, on several occasions, put mines behind the Italians, so that when they attempt to run they get blown up. This naturally does not cause the Italians to love the Germans.

As darkness fell, the Rangers were reeling after another exhausting day of combat. Few had more than four hours of sleep in the last forty-eight. But Darby knew that they could not rest because they could not give the Germans any rest. As Patton noted, the Germans were ruthless experts in delaying tactics and pastmasters in the use of mines. True to form, they sent out mining parties that night to hold up the Americans the next day on the Butera road. Rangers sent out ambush patrols to stop the miners and killed eleven. The next morning lst Battalion, unhindered by mines, continued their advance toward the mountaintop village of Butera, an enemy stronghold which from all appearances looked unassailable.

Meanwhile, 3rd Battalion had been covering a lot of ground. Released from the 3rd ID reserve, Dammer was ordered to reconnoiter north and develop the situation around Campobello. A seventeen-mile forced march closed the Rangers on the crossroads town by 2000 on 14 July, the day lst and 4th Battalions were getting their first look at the heights of Butera.

Dammer rustled some Italian stragglers out of Campobello, then asked Truscott for further orders. They were to turn left toward Porto Empodocile which would be the westernmost port in Allied hands, if the Rangers could take it, and open the coastal road to Palerno, Seventh Army's new objective. Dammer was to make his recon by way of Naro and clear the railroad enroute.

3rd Battalion made good time, slowed mostly by hordes of Italians surrendering. Such EPW's*[*FN: For clarity and consistency in this book, enemy prisoners are henceforth designated by the current term, EPW's, and friendly troops captured by the enemy designated POWs though in all wars prior to Gulf, prisoners of both sides were called POWs.] had, throughout Sicily, become a major logistical and mobility problem. Eventually it was eased when Sicilians in the Italian army were merely disarmed and paroled to return to their homes. By another night march, again about seventeen miles, 3rd Battalion headed for Porto Empodocile from the north, but stopping to check an important crossroad outside Agrigento.

Here they came upon a strongly fortified and well defended roadblock that was reported to Truscott. His orders were to eliminate it if possible; otherwise, bypass and hurry on toward Porto Empodocile. A blown bridge prevented Dammer from being reinforced by tanks, so he decided to take on the roadblock with another night attack. But at 0300 it was discovered so he ordered the entire battalion forward, A Company up the road, D Company around a flank to key terrain.

The two prongs hit the Italians simultaneously, and Dammer ordered C Company to pass through A Company to maintain the momentum of the attack. For an hour there was violent close combat, with grenades and bayonets used liberally. The Rangers were able to overlap the Italians and come at them from all sides. A hundred and fifty were captured, another seventy killed, as 3rd Battalion overran its objective without a pause from the time the first shot was fired.

The coordination and perfect execution of battle drills required would have been admirable for a well planned daylight attack. Considering that Dammer launched it hastily, unsupported and at night, almost from a march column, makes this sharp action outside Agrigento a graduate level example for infantry texts.

With dawn, 3rd Battalion pushed on, scouts well ahead of his column. They signalled back that a convoy was advancing northeast toward Agrigento. Dammer had to guess if it was a relief force for the roadblock he had just taken out, or unsuspecting Italians evacuating Porto Empodocile. Either way, he felt he could handle three trucks loaded with infantry and the accompanying ten motorcycles with sidecars.

The Rangers scurried off the road into a hasty ambush, concealing themselves on grassy hillocks. Fire discipline was perfect. Every driver was killed by the opening bursts, their vehicles crashing into gullies. The entire enemy column entered the kill zone and none escaped, even though Dammer had not had time to block either end of the ambush with obstacles. The only problem again was prisoners, forty of them this time with several wounded. At the conclusion of this ambush and from earlier actions, Dammer had about a half dozen wounded Rangers.

Proceeding toward Porto Empodocile, 3rd Battalion came under artillery fire. Luckily a US spotter plane happened overhead. Dammer was able to raise him on the radio and received an estimate of where the guns were firing from. Supported only by 60 mm mortars that distracted the Italians, two companies maneuvered through valleys that provided sufficient concealment, to take four batteries from the rear and capture them. The sudden attack left enemy artillery ammunition burning and exploding as if the batteries were still firing.

From the action on Sicily so far, Dammer could reflect that the 'old' lst Battalion could hardly have done better than his 'new' 3rd Battalion was doing. And Dammer's day was hardly over.

Sneaking down toward Porto Empodocile, the battalion sprang another hasty ambush, that led them to locate the Italian command group for the entire Agrigento valley. An EPW said it was in an OP-radio relay station on a peak that dominated the countryside for twenty miles. Soon the peak came into sight. Dammer diverted C Company to check it out—destroy what was on it if possible—then rejoin the battalion which would continue the reconnaissance toward Porto Empodocile.

An Italian lookout spotted the Rangers approaching the peak which leveled off on sheer-faced cliffs about a hundred feet high. They discovered an opening that turned out to be the mouth of a tunnel at the foot of a long flight of stairs leading up inside the peak. To climb the stone stairs would be to surely enter a trap, for a few grenades tossed down would kill everyone enclosed in the staircase. Yet that was the only way to the OP, so C Company's CO led the way, tiptoeing with a squad of tommy-gunners to the top.

Wheezing with relief that he had made it, the CO was amazed to find the OP, a complex of sheds and tents, deserted. Then, just as he signalled his bodyguards to join him, Italians appeared—all with hands in the air. As they did, a sniper's bullet chiseled stone and whined away. He was quickly found and killed, the only casualty on either side in this bizarre, and almost eerie, capture of a colonel and his staff.

Rejoining 3rd Battalion, C Company had trouble convincing their comrades that it all had happened. But the whole advance from Compobello had resembled an adventure movie. Men began to think they had been dreaming from their fatigue—but the 'dream' was not over.

For now in front of them, finally, was their objective, the key port of Porto Empodocile. The battalion arrived around 1400, with plenty of time for a leaders' reconnaissance to study the approaches to the town and lay their plans. As they did, their men threw themselves down for the first rest in 36 hours.

Situated on hills overlooking the sea, Porto Empodocile was divided by a deep ravine. Taking the town could not wait for a night attack that would probably be safer for the nocturnal Rangers. Using what little concealment there was available, Dammer sent three companies echeloned on each side of the ravine so that the battalion formed a 'V' or spearhead centered on the ravine and pointed toward the sea. He expected that as the lead companies approached Italian positions, they would probably be taken under fire, revealing the enemy strong points. Then the echeloned companies would have maneuver room to get around the positions. With the EPW's guarded by wounded Rangers, the 60mm mortars set up in the ravine where they could support both flanges of the spearhead.

The attack went pretty much the way Dammer had planned it though Italians on one side of the ravine put up less resistance than those on the other where a walled cemetery was like a small fort. Inside, the Italians had anti-aircraft as well as machine guns that kept D Company's heads well down. E and F Companies were able to fight house to house till the cemetery was encircled, then white flags went up.

By 1600 Porto Empodocile was secure, providing Dammer buildings for a compound that held the day's bag: 91 Germans and 675 Italians. 3rd Battalion's blitzkrieg had been faster than Truscott could imagine. As Rangers outposted the town, they were mistaken for the enemy garrison. The US Navy began to shell them, luckily without the effectiveness they had demonstrated off Gela. Frantically searching through radio frequencies, Dammer was able to contact the spotter plane who flew over to identify the Rangers' uniforms. The spotter soon returned with a seaplane that landed in the harbor, taxied to a jetty where Dammer hopped aboard for a short flight out to the cruiser *Philadelphia*. With ship to shore commo, he was able to report to Truscott that Porto Empodocile was in Ranger hands. The CG was delighted, and to ensure continued possession of the port, he dispatched an armored cavalry squadron to reinforce Dammer's perimeter.

The capture was also very good news for Patton who immediately began to shift forces west to exploit the seizure of this forward port. The German high command (now the German General Hube, a one-armed veteran of the Russian front) took notice as well, scraping up a crack company of light tanks from his dwindling mobile reserve and sending them ahead of a battalion of infantry to recapture Porto Empodocile, or at least deny its use to Patton.

This combined arms force was able to move out undetected from Agrigento. That night they ran into a Ranger signal party of three men accompanying a troop of the cav squadron at a blown bridge over a deep ravine just north of Porto Empodocile. If the enemy surprise attack gained the south bank of the ravine, it could continue to the port, as Dammer's perimeter was very thin.

The three Rangers with half of the cavalrymen were on the north bank when the Renault tanks rumbled out of the night. Both sides were surprised as close range fire broke out. The Rangers did what Rangers do in a confused situation, they took the initiative and attacked. Sometimes a major action hinges on a single small incident that steers the outcome one way or the other:

The tanks were deployed to sweep up to the north edge of the ravine and from there they could support their infantry (who had not yet arrived) to clear the south bank. Then up stepped Ranger commo sergeant, Bob Holiday, out of the darkness and coolly shoved a grenade into the commander's port of the lead tank moving slowly toward the rim of the ravine. The muffled explosion killed the crew, but the Renault moved on under the dead driver's foot to tumble into the ravine where its ammo blew up in a thunderous fire ball.

The fate of that lead tank unnerved those following it. They halted and buttoned up, allowing the Americans to scramble down the ravine and up the other side where they were able to hold off the Italians till dawn when Allied tac air arrived to scatter this last attempt to regain Porto Empodocile.

Within two days 3rd Battalion was relieved, with hearty thanks and congratulations all the way up the chain of command including Patton, who like all other senior commanders had been particularly pleased with the important intelligence the Rangers had provided in the fluid situation that arose after the breakout from the Licata beachhead.

Haliday's action was a fitting culmination of 3rd Battalion's epic exploits between Licata and Porto Empodocile, almost all conducted independently ahead of friendly forces and largely without outside support. Meeting engagements like Haliday's, raids like C Company's on the peak, ambushes like the one outside Agrigento, attacks as at the final objective and then its defense—all separated by grueling marches, dangerous recons, while operating by day and night—were triumphs that made Darby immensely proud when he received the reports. It was his training programs that had enabled these feats, plus the intangible ranger spirit that had clearly infected his new battalions.

The Sicilian actions definitely proved how quickly Darby had developed three battalions of the same high caliber, validating his method of cadreing 3rd and 4th Battalions. Yet there must have been a feeling in the 1st Battalion that they could still do their sisters one better—that they were first among equals. Maybe Darby allowed them to demonstrate that they were, for confronting 1st and 4th Battalions was the most awesome objective that had yet been assigned to an American force, the mountaintop fortress of Butera, blocking and overwatching Patton's advance into the interior. An avid historian of the lands through which he campaigned, Patton noted that

The most striking characteristic of Sicily is the many towns that cling like limpets to the tops of precipitous peaks. On closer examination you will find that the highest building on these peaks is the ruin of a Norman castle... They had not fallen to an assault for centuries.

Butera was one of these town, and none was more formidable. Darby decided to let 1st Battalion go for it.

There was only one approach, a single-lane series of switchbacks that climbed vertically up the towering mountainside to the thick-walled Norman citadel 4000 feet above the plain where advancing forces were completely exposed. Sharp ridges and steep cliffs prevented a supporting attack, though Allen offered tanks if Darby felt he could use them.

He did, but only to give the impression that the Americans would try to bypass Butera if possible. This deception plan required intense artillery preparation as well which would also serve the purpose of making life miserable for the Butera garrison, a force of German-led Italian fascist troops. Dependent upon Mussolini's survival, these troops would fight harder than the run of the mill Italians who only wanted the war to go away.

Naval guns, firing at nearly maximum range, pounded the fortress with their heaviest shells, then 1st ID artillery took up the slack with battalion concentrations. The howitzers could be adjusted right from the guns, so conspicuous was Butera on its peak. The 1st ID DIVARTY commander joked that these were the longest range direct fire missions in history—but that did not make Butera an easy target. The naval guns were very precise for deflection but just a slight error in elevation sent the huge shells roaring over Butera like express trains to detonate far inland. The problem for the howitzers was the extraordinary tightness of concentrations required so that most shells did not land harmlessly at the foot of the mountain.

All this softening up of Butera was, in Darby's plan, more for show than for real because he realized that the fortress might be taken by stealth where it could not by strength. Indeed when Patton asked what would be needed to storm Butera, Allen's answer was his whole division. Darby decided to go at it first with just a company— E Company of 1st Battalion led by surely one of the best and brightest Ranger captains, Chuck Shunstrom.

His most recent feat was loading the 37mm gun under fire while Darby knocked out the tank that had penetrated to Cathedral Square in Gela. As a 2nd Lt., Shunstrom had gone on the Dieppe raid and was in the Commando mold. Unlike most of his peers who preferred stand up shoot outs, he was a stealth fighter, particularly fond of knives and silent killing. A night on Butera mountain was just the time and place for Shunstrom. Most units reflect the personality and preferences of their commander, so E Company were the right men for the mission.

Which was to explore the single approach to the fortress from the foot of the mountain—a company-size reconnaissance patrol, really, to learn what chances were to reach the top. But if Shunstrom found a way, Darby had 1st and 4th Battalions ready to follow on a moments notice. From days of continuous fighting, E Company's strength was down to fifty men, but Shunstrom didn't want more. Speed and silence would get his two platoons up the mountain—if that were possible.

At nightfall they began the three-mile approach march. Previous reconnaissance by 4th Battalion had disclosed locations of protective minefields which E Company was able to move around. Shunstrom had warned them that wounded would have to left where they fell till Darby could send up a litter party. This eventuality, plus the early warning that exploding mines would provide the Italians, were extra strong incentives for E Company to watch their steps.

Quietly on the radio, Shunstrom kept Darby advised on his progress. Considering the minefields and lack of concealment (there was an eighth moon) E company moved pretty fast, partly thanks to marker rounds mixed in with H&I fire on the mountain. Preplanned concentrations by every battery in lst ID were available to churn up the mountain if Shunstrom got in trouble.

But to the base of the mountain, so far so good. Then the climb began. The troops were lightly loaded, but the terrain was nearly as tough as at El Guettar. The winding road took up all the gradual slopes, and the Rangers of course had to stay off it, climbing instead between the switchbacks.

In this way they got behind the first outposts undetected after three tense hours of sneak and creep. Italians farther up the mountain apparently relied on warning from these outposts. For whatever reasons, no one seemed to be expecting a visit that night.

Even superb stealth and good luck could not get the Rangers all the way up to the fortress. Inevitably there was a clash. Fortunately it was initiated by Rangers who fell upon off guard Italians who saw them and reached for their guns. The first burst of fire came from Tommy-gunners, turning the Italians' surprise into shock.

Darby could see the muzzle flashes and grimaced. With lst ID, artillerymen's hands were ready to pull lanyards or to plug ears for a thunderous first volley.

But Shunstrom's fire mission never came. Rather than alert the fortress that a major storm was brewing, he hoped to give the impression that the Italians were experiencing a minor probe. E Company continued a torturous infiltration around blocking positions till suddenly a machine gun opened up on them, wounding the lead platoon leader and his RTO.

Discovered, the Rangers still had an advantage: the Italians manned different levels on the steep mountainside while E Company was mixed between them. The Rangers knew what they were shooting at; the Italians weren't sure.

Taking advantage of the enemy's uncertainty, Shunstrom maneuvered the platoon, the one that had been hit, onto a knoll next to a switchback where they could provide a base of fire up and down the mountain. The other platoon took to the road because speed, more than stealth, was the now the most important ingredient for success. When they reached the next roadblock, which was covered by machine guns and anti-tank guns, they were able to flank it by climbing as fast as mountain goats. This is hardly an exaggeration, and an example of how *extraordinary* physical conditioning—admittedly aided by supercharges of adrenaline—enable feats beyond any expectations of the enemy. The Rangers destroyed the Italian positions and everyone in them.

This eye opening success gave Sundstrom a hunch that E Company might be able to continue up the mountain, even without supporting artillery fire. He was encouraged by a startling group among the EPW's. They were hard core fascists, armed with flame throwers they were to use in a kamikaze defense of the switchbacks. If the most fanatical of Bureta's defenders had been overwhelmed by Ranger aggressiveness, maybe there was hope to reach the top tonight before the defenders got their act together.

Yet Shunstrom must have been tempted to call Darby for backup from lst and 4th Battalion in their assembly area. They could prevent E Company from being cut off by the Italians they had bypassed, and half way up the mountain, surrounded by Italians, was no place to be at dawn. But Shunstrom decided to drive on unaided. There would be more chance for success if the fortress garrison continued to believe that the firefight they heard down the mountain was just the sound of a small probe being turned back.

Hooking their way up between the switchbacks, the platoons alternated fire and maneuver. Then as the slopes rounded off as they approached the top, Ranger scouts came upon a large group of Italians and Germans assembling, apparently to check out what was going on below. Boldly, Ranger Constantine shouted in Italian for them to surrender. This command resulted in an intra-Axis scuffle as the Italians tried to take up Constantine's offer while the German officers demanded that they turn and fight. The Italians won the argument, throwing down their arms; the Germans ran back to the shelter of the fortress.

E Company was right behind them, rushing through the gate in the massive walls, shooting at any hostile movement. Street fighting was wild until the Germans were cornered in the citadel. Shunstrom demanded their surrender but the reply was machine gun fire. With only these die hard defenders left, he signalled success to Darby who then mopped up on the slopes below. Somehow, perhaps through medieval secret passages, the last Germans managed to escape, for the next morning they were gone and Butera belonged to the Rangers. The lst ID historian described the feat as "one of the most brilliant actions of the war."

Butera was the crowning achievement for the Rangers in the Sicilian campaign which now became a race for Messina, the choke point for German withdrawal off the island. Unfortunately for the Allies, Hube was able to pull three of his best divisions out of the tightening noose, evacuating them to Italy where their presence would repeatedly punish Alexander for his cautious pursuit on Sicily.

Gen. Bradley reproached himself for not more fervently urging a regimental-size landing on the Italian side of the Straight of Messina to cut Hube off. Darby's Rangers were marked to lead this assault, for he engaged in some preliminary planning (with Patton's blessing), but the British were unenthusiastic, citing troop fatigue and the shortage of landing craft. Patton argued that indigenous boats would suffice for a voyage in calm seas of no more than ten miles, and Darby began to look around at the native fishing fleet.

Though such a spurt of the moment amphibious operation would not initially be able to put any armor ashore, naval gunfire was abundant as well as field artillery firing from Sicily. Bradley, Patton and Darby all felt the golden opportunity justified whatever the risk. But the two generals could not convince Eisehower to lean on Alexander to give Darby a go. Sweating with relief, Hube made good his escape. Never again in the Mediterranean would the Allies have the Germans in such a vice.

To keep up with Patton's whirlwind advance during the last weeks on Sicily, Rangers became mule skinners as they used these pack animals to haul equipment over hundreds of miles of rugged terrain. The three battalions split up for attachment to various divisions whom they served mostly as screening forces and connecting files. All drew praise from their respective commanding generals, and the Rangers future seemed as bright as the new luster they had added to their reputation from North Africa.

Yet Darby had not been assured that his command was permanent, despite enthusiastic endorsements from every American general in the Mediterranean theater. In August he felt compelled to write the War Department (through Patton and Eisenhower) requesting clarification of the Rangers status and firm guidance for future employment. His order of preference was: l. Formalization of the present organization with an organic regimental-type HQ. 2. Reorganization into a conventional regiment (though reserved for ranger-type missions) assigned at corps or field army level. 3. Reorganization into a separate reconnaissance regiment.

This last alternative is interesting for the fact that except for 3rd Battalion's recon from Compobello to Porto Empodocile, and the screening missions performed by Rangers in the last weeks of the Sicilian campaign, Darby's units were neither designed nor equipped for, or experienced in, reconnaissance as it was performed by cavalry regiments. Reconnaissance would become a Ranger staple in Vietnam, and perhaps the first serious inkling of this forthcoming function is indicated in Darby's letter to the War Department.

The letter made no mention of the original "training and demonstration" mission for Rangers. He probably did not wish to revive a ghost in the memory of anyone in the Pentagon.

Unbeknownst to Darby, a similar request for Ranger permanence was being forwarded to Washington by the G-3 of the ETO in London. When the greatest show on earth began in France, BG Cota wanted regular Ranger units in his tent. He proposed that Darby bring a cadre of veterans back to England and start training at least two new battalions.

Though Darby no doubt would have resisted conversion to the hated 'school troops' mode, many a Ranger would have lived to see the end of the war if Cota had his way.

Organizing and Training

The short month between the end of the Sicilian campaign and the commencement of the Italian was too brief for Darby to set up a training camp as he had in the Atlas Mountains. He received about 300 replacements who were not kept together but rather dispersed through his three battalions.

The twin focus of battalion-level training was on night and amphibious operations. A small vocabulary of flashlight signals was developed that proved most useful in both types of operations. Leaders down to section level carried these red-filter flashlights, and Ranger-wide recognition signals were adopted (e.g. the letter 'R'—presumably for 'Ranger"— in Morse Code).

During this month of September in Tennessee, 2nd Battalion completed its basic ranger training and decamped for Port Pierce, Florida, to the U.S. Navy's Scout and Raider School. The Rangers may have expected that they would eventually be fighting Japs as they repeatedly practiced scrambling down cargo nets into rubber boats and assault craft the way the Marines were doing in the Pacific. The jellyfish, mosquitoes, and steamy temperatures of Florida seemed better preparation for the jungle than for Europe. But reading Darby's after action reports from Sicily, 2nd Battalion noted one disturbing similarity—malaria, which was beginning to add non-battle casualties to losses in lst, 3rd and 4th Battalions.

As 2nd Battalion moved out on trains, 5th Battalion was activated at Camp Forrest with 34 officers and 563 EM. The new requirements added for Ranger eligibility were qualification

Gen. George Smith Patton, Jr., Commander, U.S. Third Army, 1944-1945.

for parachutist training, participation in a division-size maneuver, and proficiency in all ranger arms.

Back in Sicily a new arm was being added to Darby's arsenal: four 75mm howitzers self propelled on half-tracks, which became known as Cannon Company, commanded by Capt. Shunstrom of Butera fame. His mobile CP was his jeep named The Joker, while his little battery were the aces of the four suits. Besides providing indirect fire to complement the excellent work of the 4.2s, the 'aces' were expected to perform an anti-armor role . With thin skins and no turret, the half-tracks could not go one on one against a tank but were fine for ambushes and pretty good at hit and run.

In the upcoming Italian operations, the Ranger infantry preferred that Cannon Company support from the rear. That was because Shunstrom remained basically an infantryman whose place he felt was on the front lines. Indeed he delighted in slipping one of his aces in among the foxholes, putting out some H&I fire or sniping with a 75mm, then careening away before the Germans reacted, usually with artillery counterfire. So the ace slipped back into the deck, but the infantry had to stay and feel the enemy's revenge!

With mutual regret, the Rangers said goodbye to the lst Infantry Division and Seventh Army. The latter was reorganized into the Fifth Army, commanded by Gen. Mark Clark. Patton was in disgrace for having slapped a soldier he believed was malingering. Later restored to command by his West Point classmate, Eisenhower, he rehabilitated himself in England by preparing for Operation Overlord under the command of his previous subordinate, Gen. Bradley.

A victim of burnout, Gen. Terry Allen relinquished command of the Big Red One who too relocated in England for Overlord. There was no bigger booster for Rangers than Allen. They were all very pleased when, after D Day in Normandy, he was given command of another infantry division. What these changes meant for Darby's Rangers in September 1943 was that they would be fighting alongside and coordinating with new people. And it is the author's opinion that these people were not on par with those departing for the ETO. At Salerno, the Rangers were attached to a British corps in which none of the divisions had seen combat in Sicily. Gen. Clark was an unknown quantity as an army commander. His only resemblance to Patton was in the size of their egos.

In the broad military picture of the Mediterranean, there were major changes that would affect the Rangers. The fall of Sicily also brought Mussolini down. He was replaced by Field Marshal Bagdolio who entered into clandestine and intricate negotiations with the Allies to surrender all his armed forces. To do so, however, would cause Italy to be occupied by their erstwhile allies, the Germans. The political-military ramifications of this development were many, involving much perfidy and double crossing, but for the invading Allied troops there was one major outcome: they would be up against the hard bitten *Wehrmacht*, not the second string Italian army.

Good Missions:

VI. Maiori, Italy, September, 1943

The Italian campaign was the result of another Allied compromise, with again the British getting the best of the bargain. The seizure of Sicily fulfilled the essential strategic requirement to open the Mediterranean. Now shipping could proceed unmolested from Gibraltor to the Suez Canal, greatly shortening the supply line to the Far East where British forces precariously held back Japan on the Indian frontier with Burma.

Consequently, there was no strategic imperative for the Allies to invade Italy. But if they did not, what else would they do in or around Europe for the rest of 1943? Stalin was still screaming for a second front.

Dieppe, though more than a year in the past, demonstrated that the Allies were not ready to take on Hitler's *Festung Europa* before 1944. Gen. Marshall, however, had a bottom line for which he would make some significant concessions: if the British would sign on—writing in blood if necessary—to a cross-channel invasion in early 1944, the United States would entertain the notion of fruitful diversions till then. Churchill, in particular, had a fascination for the Balkans which he described as "the soft underbelly of Europe". Without trying to understand what that meant, suffice it to say that an advance into Italy was essential for its fulfillment. Post war objectives were always clearer, more focused, for him that for the Americans. In any case, Marshall was prepared to concede some merit for entering Italy provided he was assured of the quid pro quo early next year in France.

Thus the compromise was reached: seven of the most experienced divisions, plus the top commanders, would be devoted to Overlord. In return, twelve Allied divisions, a few of them veteran, would invade Italy for two persuasive reasons:

1. Tie down significant German forces that were much needed on the Russian front, as well as prevent their transfer to France where they could oppose Overlord.
2. Seize the great airbase of Foggia, from where strategic bombers could cover the gaps of those based in Britain, and thereby subject the entire Third Reich to aerial destruction.

Additionally, there would be immense morale benefits in capturing Rome, the capital of a previous Axis power, and of course the foremost city of Western civilization. Beyond those objectives, continued campaigning in Italy was dubious.

But the campaign was continued to the bitter end, and it is the saddest irony that the father figure of the Rangers, William O. Darby, lost his life during its last days.

But at its beginning, a professional tip of the hat must be paid to the staunch German defenders whose situation in September 1943 was far more dire than that of the Allies who, though short of shipping, could deploy twelve divisions almost anywhere along hundreds of miles of coastline. Only three panzer and three panzer grenadier divisions, plus two airborne divisions, were responsible for defending the entire Italian peninsula. They were under the command of Albert Kesselring, oddly enough a *Luftwaffe* rather than an army general, who nonetheless had a genius for making the most of what he had on the ground.

His skills were in exploiting the advantage of interior lines— how to create and move 'fire brigades' with a speed that multiplied their numbers. Kesselring knew what terrain was key and what was strategically worthless. He reacted swiftly and surely to a multitude of threats. He had a campaign plan, rather than a series of battle plans, and a keen appreciation of how to trade space for time. Additionally, and perhaps most importantly, Hitler gave Kesselring latitude to fight the Italian campaign as Kesselring saw fit, a unique concession for a commander in the German army now dominated by Nazi amateurs.

Alexander's invasion plan assigned the Ionian and Adriatic coasts to the British Eighth Army (after they put two divisions across the Straits of Messina); while the main attack, supported by a naval diversion, would be by the US Fifth Army across the Tyrrhenian Sea with the strategic objective of capturing Naples as a base for further operations. Clark's tactical options were limited by lack of shipping, some of which had been diverted to the Pacific, but more so by the combat radius (200 miles) of fighter cover from Sicily. This marked the Gulf of Salerno as the logical place to come ashore though every effort was made to cause Kesselring to believe the invasion would be on Sardinia, Corsica or the Gulf of Gaeta.

Spread very thin, Kesselring initially could commit only the 16th Panzer Division to cover Salerno. This they did energetically and ingeniously: mining likely beaches, constructing road blocks on beach exits, plotting artillery concentrations to cover the entire beachhead, and dispersing their armor to camouflaged assembly areas for preplanned counterattacks.

The blow for which the 16th Panzer was well prepared fell at 0330 on 9 September. Two British and one American division splashed onto the broad beaches between Salerno and Agropoli. The objective for the three Ranger battalions, for the first time fighting side by side, was to anchor the left flank of Clark's beachhead by seizing the high ground north of the coastal village of Maiori.

Their short voyage from Sicily had not gone well. Before sundown on D-1, German reconnaissance plans had spotted the huge invasion fleet in the Tyrrhenian Sea. Bombers attacked within an hour. They were driven off by concentrated anti-aircraft fire, but the *Luftwaffe* continued to track the convoys, dropping flares all the way to the Gulf of Salerno.

Under this unnerving illumination, 4th Battalion loaded into their assault boats at 0230, an hour ahead of Clark's main landings to the south. Then the gods of war once again smiled on the Rangers: the short, narrow beach at Maiori was not mined. The efficient Germans had planned to do so the next day but till then there were higher priorities in their preparations. Actually the Rangers' apparent good luck was more a result of Darby's doctrine—always hit where you are least expected, especially when coming ashore.

This was also the reason 4th Battalion was able to move inland rapidly while the Allied divisions to the south were immediately in a fight for their lives. German batteries thundered in reply to the naval bombardment, while in Darby's sector Rangers silently crossed a sea wall, then clambered up the hills overlooking Maiori. Quickly they linked up with a Commando battalion moving north out of Vietri, enroute capturing a very surprised German officer on a motorcycle.

Both elite forces were heading for Nocera, the Commandos to force open Highway 18, while the Rangers hit the town from the southern flank. The Commandos ground ahead slowly till an advance guard of the Hermann Goering Panzer Division reached the scene, and after daylight the Commando attack stalled.

Pushing cross country with only the rugged terrain to fight at night, 4th Battalion— with lst and 3rd close behind, fanning out to eliminate pillboxes and OPs they had studied on aerial photos— were ahead of schedule, and by daylight occupied a salient six miles deep overlooking the plains of Naples. Barely twenty miles away, the great city—and the theater objective—was clearly visible with only scattered and surprised Axis elements in between. Yet there would be six weeks and thousands of casualties before Clark secured that objective.

Perhaps the single most important reason for the costly delay was his inability to exploit the Rangers' startling success. It is obvious that Clark never imagined it to be possible that no more than a regiment (in numbers) could break out of a tiny penned-in beachhead, then race through 3000-foot hills covered with brush and along hairpin roads to jab the mighty German's supply line in the course of a single night. This is what Darby's Rangers did, but instead of piling more troops into their salient, Clark battered ahead against the growing strength of Kesselring's panzers which he dispatched from all over Italy now that he had determined the true location of the Allied main thrust.

It would fault Clark too much to suggest that he blew a clear opportunity to snip Kesselring's jugular in the first 48 hours of Salerno. A five-division amphibious invasion force can not change directions on the water like a speed boat. Nor was there anything like a corps MSR to be had between Maiori and Nocera—as the Rangers were soon to realize. Nevertheless the VI Corps CG, Dawley, was soon relieved for lack of aggressiveness in punching out of his beachhead; if there had been the desired aggressiveness (one must speculate if the Army CG and been Patton rather than Clark), more would have been made of the Rangers' remarkable breakthrough to positions that interdicted Highway 18.

Instead it was the Germans who immediately realized the importance of Darby's easily won positions. Kesselring threw the best of his forces at the Rangers, to remove their threat to his northern MSR.

This threat became stronger when Darby swung 4th Battalion up through the Lattari Mountains, cutting off the Sorrento Peninsula and controlling the approaches between Castellammare and Salerno. Here Murray dug in, separated by three miles of rugged brush from lst and 3rd Battalions who defended Chiunzi Pass from the same heights that commanded the Nocera defile.

On their right were the Commandos who could be counted on not to leave them in the lurch; in every other direction were crack German troops, highly motivated to kick the Rangers off Kesselring's MSR. It would be the best going up against the best in some of the fiercest fighting of the war.

Outnumbered at least three to one in manpower, and without armor to counter scores of the tanks from the Hermann Goering Panzer Division, Darby could count on no help from Fifth Army which was totally committed in holding its shallow beachhead south of Salerno. Exploiting the excellent defensive terrain that he had moved onto so unerringly would be his only hope and key for success.

But the 4000-foot heights upon which the Rangers stood had two important disadvantages: They were very rocky, hard enough to break entrenching tools and caused dangerous ricochets from German fire that multiplied the amount of metal in the air. The arid hills ranging for seven miles back to Maiori were crossed only by a single, narrow dirt road. All supplies, including water, had to make this torturous trip, as did the evacuation of wounded. No more than a dozen vehicles of all types, most of them commandeered from the enemy, were sporadically available for such treks, though a number of mules were extremely valuable.

Of the problems caused by terrain, ammo resupply was the most serious because for the next two weeks the Germans attacked lst and 3rd Battalions constantly—4th Battalion only a little less—with increasing strength until Clark could bring sizable reinforcements up Highway 18. Hauling 4.2 projectiles up mountain tops was especially difficult, but without them the Rangers could not have held on. From such elevations the mortars could shoot 2000 yards beyond their rated range and did so with great effect.

There was very little tac air to support the Rangers but naval observers joined them to call in fire from cruisers and even a battleship.

It would all be needed in the desperate days ahead.

Ironically it was the devastating fire from off shore that prompted the Germans to redouble their efforts to clear the Rangers' heights. Darby's OP's called in mission after accurate mission all along the road and railway across the plain from Nocera to Pompeii and even into Naples. The Germans had to withdraw forces from their thin line opposing Fifth Army to reduce the Ranger salient.

That was because the first onslaughts by a parachute regiment with the Hermann Goering Panzer Division decidedly failed. The twisting road into Chiunzi Pass was covered from all sides by squad-size Ranger outposts on spurs of hills. The paratroopers reeled back before they ever reached Darby's main defensive position at the top of the pass. They tried again at night, this time supported by panzers, but when the first few were knocked out by Canon Company, the disabled panzers became a road block.

Next the Germans, now reinforced by mountain troops that had fought in the Caucasusses, went at the Rangers by infiltrating up the steep slopes. This resulted in unending small unit combat as patrols and counterpatrols ran into each other amidst the crags and gorges that echoed with small arms fire at close quarters. Darby could not let the infiltrators isolate his positions even if this meant the Rangers would have to leave cover and search them out.

This went on day and night while indirect fire punished both sides. Casualties mounted, but the Rangers had far fewer replacements. Gen. Clark, appreciating the growing importance of the Ranger salient, reinforced them with a company of paratroopers from the 82nd Airborne Division, an RCT of which had been held in floating reserve. They were not enough to strengthen Chiunzi Pass sufficiently, so Darby put them with 4th Battalion which he ordered to advance on Castellammare—an unexpected action which hopefully would take pressure off his defense of the critical pass.

Half of 4th Battalion moved down on the large town on a moonless night. One company took up a fire support position while the other maneuvered to the outskirts. Rangers slipped through the German outpost line, taking the security force for a battery of 88s by surprise. But the Germans were well entrenched and their reaction swift. 88s and heavy caliber machine guns began to plaster the hills. In the dark their aim was poor, but it was soon clear to Murray that Castellammare was too big a nut for 4th Battalion to crack. Yet his mission was to take pressure off Chiunzi Pass; to best do so he had to deceive the Germans that they were being probed by a far larger force. He ordered his attacking company to load some tracers, something usually not done in night combat because it gave away positions.

Scattering like guerrillas, the Rangers moved around after every burst they fired, creating the impression of a regimental reconnaissance in force. The Germans believed it and fired back with everything they had. Then Murray called for salvos from naval guns as if they were final preparatory fires before a big attack. The Germans got their heads down while Murray exfiltrated his attacking company in pairs. By dawn the Rangers were gone, but the Germans reported that they had held Castellmarre against a division.

Whether reserves were diverted to Castellmmare is unknown; in any event there were still plenty of crack troops to assault lst and 3rd Battalions. But now the Germans resorted to a little different tactics that contained an element of siege warfare. Whenever there was any movement at all from the Ranger positions, 88s sniped at them relentlessly. In between, pre-registered artillery fire thundered down concentrations without warning. The tactic was to grind the defenders down, interdict their supply line, and sap their resistance while inching up casualty lists like Chinese water torture. The Americans were impressed by the effectiveness of the tactic and used it later during the siege of Mount Cassino.

Now only able to use the road to Maiori by night, the stranded Ranger positions became increasingly threatened. And night was hardly a safe time because of plentiful German H&I fire. By day, counterpatrols against infiltrators was extremely dangerous because of the 88s If a Ranger ventured onto the Chiunzi road itself, he did it on the dead run or become one of the dead. The top of the pass was so hot with fire that it was called '88 Junction'. Only one place seemed immune to complete destruction and that was because of its stone walls three feet thick. This was a shepherd's house, built from granite, its roof utterly demolished, but nevertheless a haven for 3rd Battalion's aid station called Fort Schuster for the medical officer, Emil Schuster.

Shivering with cold on the windswept heights, beleaguered, nearly starving, with dwindling strength and climbing casualties, lst and 3rd Battalions seemed to be teetering on the top of the pass where one more hard push by the Germans could dislodge them. Seven strong attacks had swarmed up at the Rangers, each turned back after sleepless fighting. Between attacks, German artillery had been incessant. Siege warfare was not a Ranger capability, but Darby had bought Clark many precious days and held back a major counterattack against the northern half of the Salerno beachhead.

But the Rangers could not hold out indefinitely. The Germans seemed to sense that and on 16 September launched a regimental, all out, do or die effort to break through the pass. The regiment was SS and fresh, its plan well devised to split lst and 3rd Battalions, then defeat them in detail. A company of panzers would apply pressure up the road, and every German piece on the plains of Naples fired in support. Rangers in shallow foxholes were shaken like moles in an earthquake.

Hidden by sharp bends in the road, the 'four aces' awaited the panzer thrust. Shunstrom darted out in The Joker. When he spotted a target, he signalled forward one of the aces to take a shot while his jeep roared back into defilade. A round, maybe two, from the 75mm and the ace pulled back just as fast. Such snap shooting was not always accurate but still effective because there were so many targets and nowhere for the tanks to pull off. Burning panzers began to litter the much cratered road, their smoke clouding visibility of the heights. German artillery observers on the plain had no such problem and brought down deafening barrages whenever an ace appeared, no matter how briefly.

Pummeled in Shunstrom's gauntlet (which was well complemented by the 4.2s), the panzers could only move up the twisting road like an earthworm in a maze. They could not even keep abreast of the German storm troopers crawling through the brush as they scaled outcroppings that rose to the Rangers' strongpoints. 1st and 3rd Battalions countered first with 60mm mortars, perfect weapons for getting behind knolls from where the Germans launched their probes. The Germans had their own mortars which they fired from gorges, so that both sides slugged it out even before they engaged with small arms.

When that stage of the battle arrived in the early afternoon the Rangers found that every strongpoint was virtually encircled by relentless Germans. Each position had to do or die because Darby could not get reinforcements to them—not that he had many! The Rangers fired selectively to hoard ammunition. The favorite weapon on both sides at this range was the grenade, which could deliver death without giving away the thrower's location.

Another weapon that was a Ranger favorite in the hills was the Springfield "03 with a mounted telescope. From long range, a single shot would ring out from a Ranger outpost, and among the boulders a German squad lost its leader. Sniper victims often toppled down into chasms so that the marksman could mark them up as KIA, even though initially only wounded.

Not that they had time to keep score of the dozens of enemy eliminated. They were taking too many casualties themselves, and night was falling (late at this time of year in this part of the world). In most situations, darkness is an advantage for attackers but not when the defenders are Rangers. Now they could slip out from their OP's, out from under the constant 88 barrages that had located them so well. It was out into the night to meet equally stealthy moves by German infiltrators, a deadly game of cat and mouse with Germans and Americans in the role of both animals. Short bursts from the ammo-short Rangers versus heavy fire from the climbing storm troopers. It was kill or be killed, and there was plenty on both sides.

Walleyed from lack of sleep, Darby and Dammer clung by their radios to learn which of the positions would survive till morning. All of them did. Battered, bullet ridden, and down to single magazines for Garands, the Ranger strongpoints held out and held on to their key terrain. Chiunzi Pass was still in their grip, and by dawn's early light they were once again able to adjust naval gunfire on German reinforcements caught in the open on the road from Naples.

Yet by now lst and 3rd Battalions were almost a phantom force, nearly delirious from lack of sleep and showing many symptoms of what in World War I was called shell shock, in WW II, combat fatigue. The exhausted Rangers were incoherent.*[* FN: This was the opinion of some war correspondents who rushed to Ft. Schuster when word of "the Rangers' Alamo", as one of them put it, reached Clark's HQ.] Dehydration, and something close to famine, had reduced the survivors to robotic fighting machines with scarcely a trace of humanity left. Representatives from Fifth Army's staff were understandably shocked. Apparently they had less than full appreciation of the ordeal that went into holding Darby's crucial salient in its flanking position beside Nocera.

Just in time, Clark became convinced that Darby needed major help, which was now available after the Salerno beachhead was secured by a force equivalent to five divisions. Indeed the Germans believed that one of them was holding Chiunzi Pass—when in fact it was two Ranger battalions. Darby was beefed up by more paratroopers and the British 23rd Armored Brigade, a total of 8,500 troops under his operational control. He immediately pulled lst and

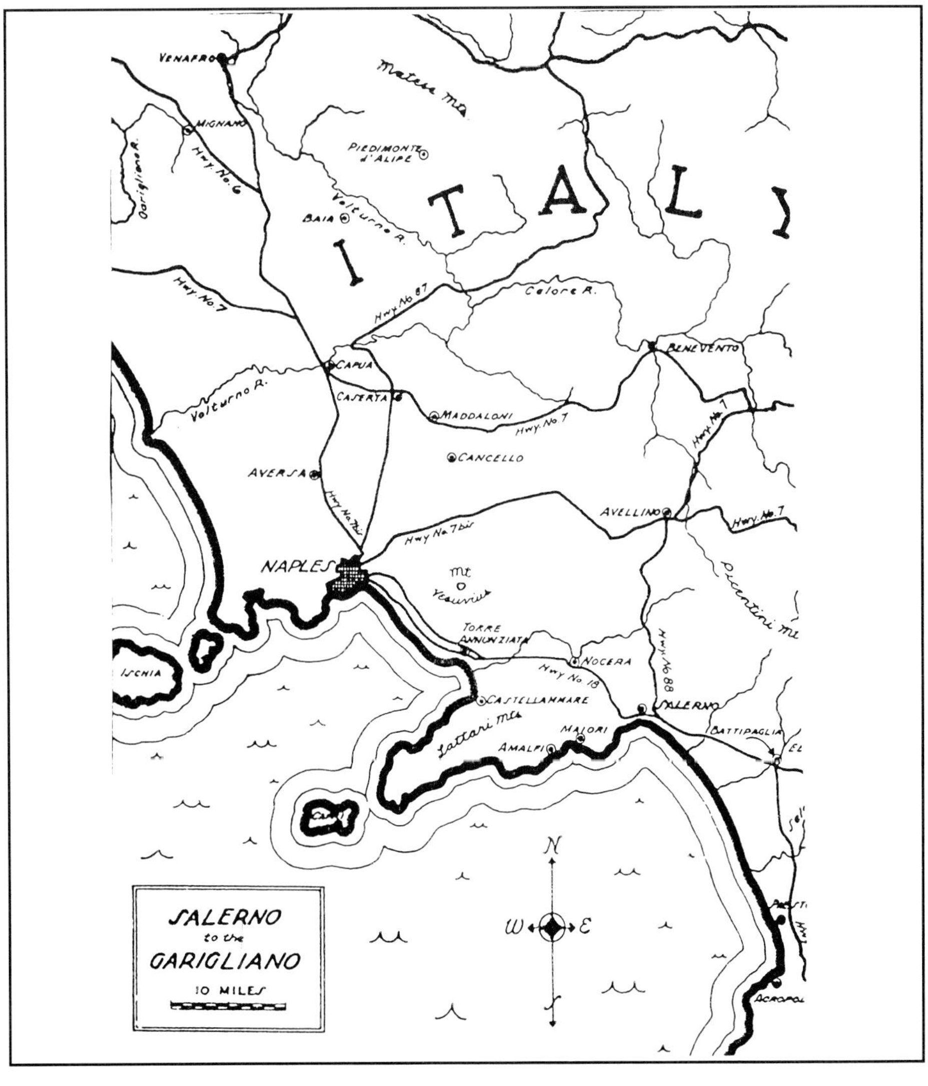

3rd Battalions off the pass before they toppled off the heights from sheer exhaustion. 4th Battalion continued to threaten Castellammare in brutal, bushwhacking combat between patrols.

The respite for Darby's Rangers was brief, hardly time to count their losses which were thirty-two dead, seventy-three so seriously wounded that they were evacuated out of the-ater, and nine MIA's, presumably captured. Without a permanent TO&E, Darby had no reliable pipeline for replacements. The Ranger battalions simply shrank until there was a long lull in the fighting when he could dunn Fifth Army for volunteers.

Such a lull was not in the foreseeable future. From Chiunzi Pass the Rangers could see far to the north and how the Appenines dissolved into ranges increasingly faint. They seemed to pile up upon each other, each range higher than the one to the south. Next time it would be the Germans who held the commanding heights. They would be defending, the Americans attacking. Nothing but bloodier war was ahead for Darby's Rangers.

As they considered this prospect and moved inland, Darby could take stock of the lessons learned from the action that started at the Gulf of Salerno. The first lesson was very positive, the location and method for getting ashore. Maiori was reminiscent of Arzew as an example of a surprise landing at an unexpected point. That such surprise was maintained even after German planes tracked Clark's invasion fleet with flares is all the more remarkable. The Rangers were lucky that the Maiori beach was not mined as the Germans had planned, but Darby made the most of his luck by rushing his battalions into the commanding hills and never looking back. Incidentally, the hills cramped the town right down to the waters edge, so there was no alternative to crossing the sea wall and beach in front of Maiori.

The second lesson was one that had repeated since Arzew, the utter necessity for Rangers to be good forward observers. That had been important in the interior of Tunisia and especially at Dernia Pass where the fire was army artillery. But it was naval gunfire, some of it British, that did so much to save the day at Gela and on Highway 18, thus requiring Rangers to be adept at adjusting flat trajectory missions and using unfamiliar fire control procedures. To date they had little experience in controlling tac air strikes, primarily because the Allies had yet to wrest the sky from the *Luftwaffe*. This unhealthy situation would continue till the Allied air forces became fully operational at Foggia.

The third lesson was one first learned in the closing weeks of the Sicilian campaign—the value of pack mules. The rougher the terrain, the farther the objectives, the worse the Ranger MSR, the more essential were the mules. After the hill fighting of September, Darby always maintained a stable of long-eared cavalry. The other innovation that endured was Canon Company, Darby's pocket tank-destroyers. Indeed from their origins as light infantry the Rangers had come to resemble a regimental combat team.

The last lesson learned was the dark side of beefing up toward becoming an RCT. Fifth Army took to thinking that they <u>were</u> an RCT. Holding key terrain for two weeks along a continuous front, as the Rangers did so admirably around Chiunzi Pass, was outside the missions and capabilities for which they were designed. Rangers came out of the hills grumbling about misuse. But OK, so Fifth Army couldn't relieve them as soon as possible—Clark had a lot of problems with a green army in its first amphibious operation. The Rangers' complaints were brief because Salerno seemed like a unique situation that was not likely to be repeated.

But they were wrong.

Finally pulled out of the line, they had only about two weeks to refit, rest up, and integrate replacements before moving to Naples (captured on 1 October) where their scroll shoulder patch was now famous. Replacements were all volunteers, but that was not enough to entitle them to wear the scroll. That was only permitted by the veterans after the cherries proved themselves in combat. That opportunity would first arrive for 4th Battalion a month later.

During that month Fifth Army had slugged ahead, and in early October reached the snaking, bridgeless, rain-swollen Volturno River. Though slow to gain much ground, their drive had certainly been a strategic success: German panzers and mechanized troops were pouring down out of the Alps to delay an Allied advance on Rome. These were precious formations that Hitler sorely needed on the Russian front.

Eisenhower and Alexander would not allow a stalemate to set in; otherwise Kesselring could grow strong enough to counterattack and negate the Allies' hard won gains. So Fifth and Eighth Armies pressed on, though the Germans could fall back on ideal defensive terrain.*[*FN: Over on the Adriatic, the Commando Brigade helped the British Eighth Army advance with a landing to flank German defenses on the Trigno River] The rivers on the way to Rome flowed east-west, dominated by high ground on their northern banks. As Kesselring improved such positions on what was known as the 'Gustav Line', he held a strong forward trace called the 'Winter Line'.

The Germans were wondrous in field fortifications. On the Winter Line, disguised by chilling sleety weather, craters and mine fields blocked every approach; machine gun and indirect fire positions were blasted out of rock and thoroughly camouflaged; artillery was registered on all likely trails and assembly areas. In short, the Germans knew where and when the Americans would be coming. Surprise, the Rangers' trump card, would not be in the hand Darby was dealt.

Good Missions:

VII. Mignano Valley , November 1943

Remembering 3rd Battalion's fine work in the drive for Porto Empodocile in Sicily, the 3rd Infantry Division asked for Rangers to spearhead their drive from the Volturno to the Winter Line. 4th Battalion was in the best shape of the three after the arduous fighting north of Salerno, and therefore drew the mission.

Fortunately the German defenses on the Volturno were relatively porous and, with the aid of Italian partisans, Murray's scouts were able to guide 4th Battalion past enemy outposts on the night of 3 November and ford the icy river. Strung out single file on a slim slippery trail, the Rangers climbed thousands of feet before reaching a ridgeline, the first leg of an expedition that would take them far behind enemy lines.

This was an unusual and hybrid mission that did not fit standard terminology:

It was not a raid because Rangers would stay in the objective area rather than return. With fixed objectives, it could hardly be called a reconnaissance in force. Though there were expected to be ambushes, there did not have to be. In a way 4th Battalion was 3rd ID's advance guard, but the division would not take the same route. Though in many ways the mission was to attack, we'll call it an expedition.

Murray could not expect linkup with the 3rd ID for several days, so the Rangers carried double rations and extra ammo. Because of the need for concealment in very barren terrain, pack mules could not be used and so 60mm mortars would be the only indirect fire support available. The enemy would be bypassed, if at all possible, in order to reach the deep objective, high ground dominating the neck of the Mignano Valley where the headwaters of the Liri River rushed north toward Rome.

In this valley was Highway 6, the best axis for Clark's advance. The German Tenth Army opposed him and Montgomery now, comprised of one mountain, one parachute, five infantry, three panzer, and three panzer grenadier divisions.

In the rear of such forces, Murray could hardly avoid contact. In the first 24 hours, his Rangers bumped into a number of patrols, killing or capturing twenty of the 34 Germans engaged. So far, so good; but the EPW's were from the 15th Panzer Division, a sign that the expedition would no longer be just a hike in the hills.

By 1430 on 4 November, D, E, and F Companies commanded by Capt. Walter Nye had gained, without opposition, the summit of Mt. Cavalle jutting into the plain almost half way to Mignano, VI Corps' objective. Mt. Cavalle was the place where 4th Battalion hoped to interdict German reinforcements moving up toward the Volturno. A, B, and C Companies, plus the command group, were concealed in a forest grove on the eastern slope of the mountain where they had jumped three small patrols and annihilated them without drawing attention from larger enemy forces who were maneuvering to take on 3rd ID.

Misty rain had been a godsend for the Rangers so far in their infiltration, but as the weather cleared somewhat that afternoon, Nye could see Germans and they could see him.

What he saw was a wire-stringing detail with twelve mules and about sixty men. Probably believing that the Rangers were only Italian guerrillas, the German force attacked; but Nye quickly got the better of the meeting engagement by swinging a company down the hill to take the enemy on the flank. About a third of them and all the mules fell. Nye continued the attack as the German survivors fell back onto prepared concealed positions on Mt. Rotunda across a long saddle from Mt. Cavalle.

Here Nye's pursuit brought grief to the aggressive Rangers: a platoon was caught on open ground in the saddle as two companies tried vainly to storm Mt. Rotunda. The Germans had called in final protective fires from well sited artillery. Though presently behind the lines, they had expected Mt. Rotunda to be attacked some time in the future, and so were ready to defend it.

Murray poured in all the 60mm's he had but the platoon from E Company, acting as rear guard for the two attacking companies, was pinned down in the saddle and had to exfiltrate individually, losing two killed, seven wounded, and nine missing. The Germans lost 24 KIAs and six EPWs in this action. Murray had to assume his missing were POWs, and that the enemy over on Mt. Rotunda now knew that a sizable number of Rangers were outposting Mt. Cavalle.

That night he had to decide what his actions and orders should be the next morning. The most pressing and personal factor was that he had WIAs who needed immediate evacuation or else they would die of wounds. He had radio contact with 3rd ID, but they were not yet within supporting distance, even with artillery. So as priority #l, Murray decided that the wounded would be taken back to friendly lines, even if it meant detaching a Ranger company to do so.

His next most significant decision is less understandable—though, considering ranger doctrine and spirit, it is comprehensible—to continue the attack on Mt. Rotunda which he estimated was held by 200 combat troops. Murray's plan of attack was to send C Company around the saddle by way of a draw, with D and F Companies overwatching with suppressive fire. A Company would reinforce C Company if the attack seemed to be succeeding. B Company carried the previous days wounded back over the rugged route to the Volturno, while E Company took a flanking position off Highway 6.

Rain clouds drifted between the two mountains, providing enough concealment so that a dawn attack was not necessary. Instead A Company used the morning hours to carefully recon their route, hoping to fall upon the Germans by stealth. However, this was not to be.

E Company caught a platoon of paratroopers moving up the slopes to reinforce Mt. Rotunda. It seemed more would be coming, so Murray ordered that the attack kick off before the objective was further strengthened. The Germans were ready, replying first with artillery, then machine guns as A Company came up the steep slopes. The enemy was familiar with the approaches to their positions, so they dropped plenty of mortars on Rangers in the draw.

Before the situation got any worse, Murray called off the attack. As A Company struggled back, the Germans counterattacked. Now the terrain favored the Rangers who had already lost five killed, eight wounded, and six missing. Regrouping on Mt. Cavalle, they inflicted triple that number of casualties on the Germans who did not seem aware that the remainder of Murray's battalion held the heights. It was payback time for the Rangers who could now shoot down on the Germans.

But ammunition and daylight were running out. The Germans withdrew to Mt. Rotunda while Rangers used the falling darkness to search for wounded and missing. Murray took advantage of the night to regain the initiative, probing around the saddle so that the Germans would expect a major attack in the morning. They reacted by swinging a company out of the force holding up 3rd ID, moving it up the backside of Mt. Rotunda. This lessened the resistance for the Third Infantrymen who next morning battled through to take up strong positions near Murray's right flank. Most welcome of all was the robust fire support now available from 3rd ID's artillery.

Murray used it to break contact and carry out his wounded. On 6 November, 4th Battalion was relieved of its mission for 3rd ID, a mission considered by VI Corps as having been accomplished with great success: the Germans had been amazed and confused by the sudden appearance of a ferocious battalion deep behind their Volturno line. Briefly they imagined that the Rangers had been dropped by parachute, and this possibility changed their defensive calculations and reactions. Consequently, VI Corps' advance into the Mignano Valley was significantly aided by the battles around Mt. Cavalle and Rotunda.

But, for the first time having not seized an objective, the Rangers were less than content with the outcome of their expedition. The author has designated this mission as 'good', though it was really somewhere between good and bad. This mixture results from the Rangers having only half of the vital combat support they needed: accurate intelligence and nonorganic indirect fire. The intelligence was there for Murray, thanks to the Italian partisans who showed him the goat trails that enabled his remarkable infiltration. However, when 4th Battalion became decisively engaged, there was no artillery on call.

So staying within range of big friendly guns was a lesson underscored rather than learned for the first time. During the infiltration artillery support was not so important because Murray wisely bypassed German strongpoints enroute. But once in the objective area he had to use the element of surprise as much as possible, and this meant moving on Mt. Rotunda before friendly artillery had displaced far enough forward to support the attack. With hindsight a critic could fault him for not waiting, but he was uncertain about the enemy strength on Mt. Rotunda, and he developed the situation in the best way—the Ranger way; i.e., by aggressively attacking.

Secondary lessons included more coordination with partisans. They had been invaluable as guides but also might have been some help in evacuating wounded, or at least concealing them until contact was made with advancing Allied forces. But for the average Ranger in 4th Battalion, the primary lesson learned was that 'sunny Italian' was a propaganda term to draw tourists! The Italian mountains in the rainy season of winter is shivering cold. Clothed with only lightweight tanker jackets, trying to sleep on a single ground sheet, Rangers could get no rest; and this seriously degraded both their efficiency and endurance. No man is so tough that he does not need rest. Protection against the cold was more than a matter of comfort, it had become essential for successful combat. Like everything else the Rangers needed, Darby had to scrounge them winter clothes any way he could. His S4, Capt. Joe Fineberg, was famous for obtaining wool socks from the British in exchange for captured weapons and other war souvenirs. These the Rangers had in abundance.

Bad Missions:

II. Venafro-San Pietro, November, 1943

By 8 November, VI Corps' attack ran out of steam at the Mignano defile. The 3rd and 15th Panzer Divisions had orders to hold it at all costs. The mission for the US 3rd and 45th ID's was to batter through to seize Mount Cassino and open up the Liri Valley. As LTG Lucas, the VI Corps CG, would soon learn, this was a mission impossible for the forces he had available though X Corps would be making a strong supporting attack. Indeed, Mount Cassino would not fall until the next year after some of the most bitter and costly fighting of the war.*[*FN: To provide stateside replacements with some idea of what fighting in the Italian campaign was like, the Army asked the famous movie director, John Huston, to film the battles leading to Mount Cassino. Huston's masterpiece, 'San Pietro', is surely the most realistic training film ever made, but it was so grim and frightening that the Army decided not to show it right away to the replacements.]

From craggy heights, German OP's called down pre-registered fire on anything in the valley, thereby restricting American resupply and reinforcements to slow, nighttime movements. First and 4th Battalions were tasked to remove those OP's while the two American divisions attacked toward San Pietro on a narrow front. The Rangers would be between them as connecting files, lst Battalion attacking east and 4th Battalion north. In addition to eliminating the German OP's, the Rangers were to block the mountain road between Venafro and San Pietro, preventing lateral movement of German forces. The Rangers would be supported by corps artillery, and a battalion from the 82nd Airborne would be under Darby's command.

To begin his missions, he was first to relieve a regiment of the 45th Division in the line. This was the first occasion for such a large-scale relief in place, and it was ominous for what it implied: that two Ranger battalions, with attachments, were the equivalent of a conventional infantry regiment. Corps G-3 was apparently looking at map symbols and not real numbers. As Bob Black observed of this occasion, "[Conventional] commanders and operations officers tend to associate the terms 'company' and 'battalion' with a certain strength. This often works to the disadvantage of Ranger units that are used in line-infantry circumstances. At full strength, Ranger units are considerably smaller than their line counterparts."

And Darby's battalions were far from full strength, owing to the grinding fighting so far in Italy and the scarcity of replacements. For example, A Company, 4th Rangers, had but three officers and 43 EM, while E Company was only a bit larger than a platoon with a morning report strength of one officer and 34 EM.

Considering that Darby still had his warrant from Eisenhower to recruit anywhere in the Mediterranean Theater, why weren't the Rangers brought up to strength? The first reason is that Darby didn't have an administrative pipeline to do so. Theater repo depots were still mostly in North Africa or Sicily, and he could ill afford to detail any officers or NCOs to recruit there or anywhere else. Second, he would not dilute the quality of the Rangers in order to put more of them in foxholes. Darby kept the ranger role in mind: that of shock troops whose specialty was amphibious operations. So far the Rangers had not been seriously misused, but that was to change as the battalions were sucked into the maw of line combat where they were expected to function like conventional counterparts—and do more besides.

No doubt Gen. Clark would have preferred to hold out the Rangers to train for upcoming amphibious landings—Darby certainly argued the case for such essential refresher training—but Fifth Army needed everything it possessed to crack the Winter Line. Rangers had shown that they could slice through enemy positions like no other spearhead, so they were thrown at the toughest places to penetrate. But Clark did not seem to appreciate that though sharp, a Ranger spearhead is light and brittle. By mid-November it was also nicked and worn.

So, with little pause to catch their breath or even to acquire winter uniforms, Darby's Rangers had to conduct on the job rather than amphibious training. As one replacement who joined the Rangers in Naples put it, "It was unique training. They took you into German territory. The only thing you could be sure of was that you were going to be engaged." Most casualties were men new to their jobs, but this bloody and wasteful Ranger school was all Darby had in Italy.

It can now be assumed without repeating that the 4.2s of the 83rd Chemical Battalion were backing up the Rangers once again. With the punch of a 155 howitzer, these mortars were ideal delivery systems for indirect fire in the mountains. They'd certainly be needed around Venafro because Rangers would not have the element of surprise going for them. With the 3rd and 45th ID's moving stuff forward for their attack, the Germans knew something big was coming and soon.

They countered preemptively with air strikes (500 lb. bombs) on 9 November followed by constant shelling all the next day. As they did, Ranger recon patrols searched out minefields and scouted the German positions above them. 4th Battalion started up the slopes at 0430 on 11 November (Veterans Day) as the valley rumbled and echoed with preparatory artillery fire.

Showers of shells had denuded the hills of trees, leaving little concealment for the Rangers. And the 3rd Panzer Grenadier Division was ready for them. Dodging through enemy mortar barrages, the hard-to-kill men of A Company had to crawl after being taken under fire by machine gun nests emplaced in solid rock. From here on up, the 4.2s lead the way, walking explosions and white phosphorous into enemy positions less than a hundred yards ahead of advancing Rangers.

In a low crouch, Tommy-gunners crossed the summit with such a rush that the Germans could not man their reverse slope defenses. The rest of A Company came on, widely dispersed because already German artillery was zeroing in on the key terrain that so lately had been theirs. The Rangers had to dive into the enemy dugouts in order to use the overhead cover. There was no time for niceties: whatever moved was shot. Nevertheless, five German officers and forty EM were captured in addition to a dozen killed.

But about thirty had gotten away and fired back from the reverse slope. Before A Company could signal Murray that they had taken their objective, they knew there would be a counterattack, and there was. It had been a sharp attack but the objective had been won more easily than expected. Rangers could imagine some German general furious when he heard that the high point on his ridge was lost, and bellowing at his colonels to get it back now matter how.

Unfortunately for the Germans, they had strongly organized the mountaintop for all around defense, so from the enemy's dugouts and bunkers, the Rangers were easily able to beat back the initial counterattacks. Then the Germans attempted to blow the Rangers off the mountain with the sheer weight of artillery.

As they hunkered down under head rattling barrages, A Company looked around at the equipment they had so recently captured. It included much wire, many radios and batteries, plus enough supplies and ammo to hold out a month. Clearly this position (Hill 630) was a vital link in the German's defenses. The Rangers were sure the 3rd Panzer Grenadiers would be back in force to retake it. A Company asked Murray to send up some friends, but the rest of 4th Battalion was busy gaining positions farther out on the ridge. Perhaps two reserve companies could provide some help next morning, but till then A Company would have to hold the fort alone (though they were granted priority for 4.2 missions).

Multiple rockets and 170mm artillery prepped the Rangers for an hour before the panzer grenadiers started a day long counterattack. The supporting concentrations shifted to A Company's rear to cut off reinforcements, but fortunately Murray had changed his mind and sent his two reserve companies to join A Company during the night.

They had spread out along the ridge and poured fire into the flanks of the climbing Germans who fell back and regrouped to consider how to take on the new defenders. By probing and patrolling they determined that the Rangers were very thin along 1500 yards of ridgeline. There had to be gaps the Germans could exploit. They did so with great persistence, skill and coordination.

Meanwhile Murray was exploiting his possession of the high ground by directing crushing barrages on the Venafro-San Pietro trail. Motor pools, supply dumps, assembly areas were all hit to the consternation of German rear area units who had never expected to be under observation. Indeed 4th Battalion had more targets than fire missions, as the two American infantry divisions were also into their all out drive to control the Mignano defile. The Rangers had already seized their objectives so priority of artillery support went elsewhere.

What 4th Battalion got was priority for German artillery! It came down on the ridge in larger caliber and quantity than Rangers had ever experienced. When it finally shifted, Murray's men were joined by 1st Battalion and Darby assumed command of the forces defending the critical ridgeline. A battalion from 45th ID moved up along side just in time to help turn back wave after wave of German infantry rising like a tide behind their mortar fire. The 4.2s replied in kind, and the decisive battle began to determine which side would control this key terrain on VI Corps' front.

That afternoon provided the keynote to the sort of fighting that would go on constantly for days and nights. Knock down, drag out—pitched battle—no quarter given or asked—these are some of the ways the fighting was described, but most of all the battle of Venafro ridge was a test of wills and guts to decide who would share the heights with the dead.

Every Ranger of both battalions was now on line: Darby had no reserve left. Two German battalions pushed in two Ranger companies till the line bowed, but cooks, clerks, litter bearers, ambulance drivers—everyone wearing a scroll fought like hardened infantry till the onslaught was turned back. The only men not in actual contact were ammo bearers who round tripped over thousands of feet of elevation, up and down the mountain all day to keep their buddies' guns firing.

Hundreds of Germans lay in heaps or writhed wounded on the ridgeline. If a panzer grenadier made it to the top he soon realized that he might as well fight to the death because Ranger machine guns and 4.2s cut off all routes of withdrawal.

When the day ended it belonged to the Rangers—and so did the ridge—but with darkness the Germans resorted to infiltration. Experts at that game themselves, Rangers set little ambushes from behind the many large jagged rocks that strewed the ridgeline. Ranger strongpoints engaged groups of well-led infiltrators who deployed to rush the strongpoints. Rangers anticipated enemy lines of departure and when they were crossed, the ambushers raked the Germans with flanking fire.

On both sides of the mountainside, the groans of wounded filled the night. Medics were not spared though when they went out on missions of mercy. Some German mine laying teams were able to get behind the lines, and the next morning a 4th Battalion ambulance was destroyed with its patients when it ran over a Teller mine.

That same morning the Rangers had to conduct a police call to eliminate infiltrators and snipers who had survived the night. Somehow a few had reached the highest point on the ridge. Moving out under German artillery and mortar fire, Rangers set about to encircle and annihilate these lingering threats. This police call was the only lull in the heavier fighting that had allowed no rest.

But the Germans weren't through. Again following a creeping barrage, by midday a fresh battalion managed to seize two knolls on the left of the front from which they could not be ejected.

But they were neutralized by constant Ranger probes that were like small Indian raids, keeping the defenders exhausted and harassed. Moving like reptiles, Rangers crept up behind rocks, lobbed grenades over on the Germans, then crawled away before they received fire. By such tactics and accurate sniping the Germans on the knolls could not be reinforced and their strength attritted. Nearby the Rangers could hear them dying of wounds.

But Ranger strength had also waned. In holding the ridgeline for forty-eight hours against the best the Germans could muster, 4th Battalion lost 24 killed, sixty evacuated with wounds, and three missing. Murray's men were at the end of their string, his battalion virtually combat ineffective when finally relieved on the ridge by infantry from 45th ID.

So it was up to lst Battalion to take on the last mission assigned to the Rangers around Venafro: clear two bald hills onto which Germans had penetrated and from where they were calling fire on 45th ID's rear. During their occupation, these panzer grenadiers had blanketed the hills with minefields, covering them with fire from adjoining knolls.

First Battalion did not have to climb to reach their objectives, instead attacking down 4th Battalion's ridgeline though it was so narrow only two companies could be deployed side by side. This attack was generally north while Murray's had been west. Shrewdly, Darby had mounted his 4.2s in a central position where they could support both battalions without displacing. Now with plenty of shells stockpiled, the base plates were pounded deep into the ground as salvo after salvo preceded lst Battalion's assault. White phosphorous shells were especially effective in reaching the dug-in defenders and blinding fortified OP's.

There was pretty good concealment on the rocky slopes of the hills, but the tops were bare and exposed so Darby did not try to occupy them; instead Rangers worked around them till they were encircled, then sent search and destroy teams up at night to root out German die hards. More than 80 of them died or were picked off next morning. There were no prisoners.

First Battalion lost 27 dead during the two weeks around Venafro. An abnormally large number were hit above the waist, and this was true of German casualties as well. This was because the fighting was done mostly from prone positions. When men looked up to pick out targets, that's when counterfire or shell fragments got them.

As usual, the Germans counterattacked lst Battalion after a two-day bombardment. On 19 November a composite battalion of panzer grenadiers and paratroopers struck at 1800 hours, hoping to orient on their objectives before dark. C and D Companies not only turned them back but pursued with patrols that worked in among the withdrawing Germans and started fire fights between them. Unprepared to defend their line of departure, they ended the night farther back than where they had started.

The Rangers' aggressiveness proved to be an effective spoiling attack for the Germans' plans the next day. A number of EPW's were taken in C and D Companies successful action. The prisoners revealed that the Rangers' tenacious defense had sapped the Germans' determination. It looked as though they were finally willing to concede the heights of Venafro to Darby.

Though his men were depleted in numbers and numb with exhaustion, he would not let a stalemate arise from the smoke that lingered over the hills so bitterly contested. The best way to keep pressure off the hard won hills was to apply it to the enemy. It took strong leadership but Ranger patrols now took the fight to the panzer grenadiers while they were back on their heels.

Typical of these patrols was one from B Company that discovered a platoon bivouac in a cave. The patrol leader sent back for some gasoline and wine bottles the Germans had left behind on the ridge. With some cotton wicks made from undershirts, he prepared Molotov Cocktails. The patrol snuck toward the mouth of the cave where German OP's detected them, but with heavy fire support his fiery grenadiers got close enough to serve the Germans cocktails for breakfast.

Satchel charges were heaved in too; the resulting flame and smoke drove the Germans out into an ambush. With all hell rising around the cave, other German positions took notice and drove off the patrol, but one platoon of panzer grenadiers stayed in the cave that had become their mass grave.

By such probes, Darby turned the ridge into a reconnaissance and surveillance line that gave the Germans no rest. Some patrols were so deep that they had to be resupplied by parachute drops from single engine observation planes normally used as artillery spotters.

More than any other sector on their front, the Ranger positions were a painful thorn in the Germans' side. They resumed their attacks in company strength; when one was turned back, E Company followed the rout and penetrated the German lines and occupied a ridge on their outpost line. Darby didn't have the strength to reinforce that success so E Company could not hold it, and withdrew by night exfiltration.

The next day panzer grenadiers countered with a limited objective attack that succeeded on lst Battalion's left flank. D Company moved down the line to retake the ground that had changed hands several times during the last two weeks. Like well conditioned heavyweights in the last of a fifteen round championship fight, with mortar fire and counterfire, raid and revenge, the Rangers and panzer grenadiers slugged each other wearily but neither side would go down for the count. Bob Black best described the result:

The battles of the Venafro were different than amphibious assaults and night reconnaissance the Rangers had trained for. At Venafro they experienced the strain of constant combat. Men who endure prolonged combat often suffer its many effects. Dr. Sheldon Sommers of the lst Battalion found brave men suffering from battle burnout. NCO's who had distinguished themselves in combat would come up to him and say, "Doc, I can't take it any more."

Sommers would keep these men in the medical area about two hundred to five hundred yards behind the rifle pits. The battle-weary men would make coffee and perform chores. After a week or so of rest, the recovered their stability and went back to their units.

The Rangers were experienced soldiers and knew the tricks of reducing casualties from small-arms fire. Artillery, however, was another matter. Dr. Sommers found that the overwhelming number of casualties he treated were from shell fragment wounds, especially from the German 88 gun . . .

Strike and counterstrike, the deadly action continued in this manner for the remainder of November. Each day the casualty list mounted. Darby's superiors had no understanding of the capabilities and limitations of a Ranger unit. The knowledge and experience of the Rangers was being shot to pieces in line combat.

But like a bloodied pit bull, the Rangers would not stop fighting as long as they were in contact with the enemy. Finally, after Thanksgiving, lst and 4th Battalions were pulled out of the line. They had been in savage combat for twenty days.

Now it was 3rd Battalion's turn in the cauldron. On 2 December they conducted a reconnaissance in force to determine the enemy's strength and positions around San Pietro. Heavy fighting had raged along Highway 6 for three weeks, and the Germans now laid low until they were sure a main attack was coming. This tactic made 3rd Battalion's already difficult task much harder, as they had to be right on top of enemy positions before the Germans opened fire. Nonetheless, by 7 December Dammer had developed the situation sufficiently to be assigned an attack mission: seize Hill 950, key terrain overlooking San Pietro and a lynch pin for German defenses against 36th ID.

Hill 950 was a sawtooth of sharp peaks that could only be approached from the flanks. By night 3rd Battalion started up a long slope, evading listening patrols, snipping barbed wire, and feeling for mines. It was slow going and squads in file for much of the way. They moved up like Indians, quiet as shadows. 4.2 spotter rounds marked landmarks until the Rangers were within a thousand yards of the peaks. It had been a remarkably secret climb for which the Germans were not prepared, but then alert sentries must have taken notice of some movement to their front.

Their probing fire was not returned, but the Rangers spread out into skirmish lines and kept climbing, E and F Companies' scouts leading. At this point veterans of the Sened raid must have felt a sense of deja vu. White phosphorous shells from the trusty 4.2s were dropped on two suspected machine gun nests that so far had remained silent. Howls from burned Germans confirmed these positions, and stealthy Ranger grenadiers drew close.

They were spotted moments before they threw. German machine guns and Ranger grenades sounded almost together, and a platoon assaulted each nest within seconds. The action was fast and furious but quickly over. Rangers were among the Germans on the peaks before they could call for final defensive fires. Once again Rangers had succeeded by using terrain the enemy had considered too difficult for an attack.

But after their initial surprise, the Germans reacted with artillery and mortars; Dammer did the same and a counterbattery duel developed as the Rangers pressed on over their objectives. Satchel charges blasted Germans out of their bunkers or killed them by concussions within. Ranger combat patrols dashed down the reverse slope to break up any counterattacks before they formed.

The 36th ID was as amazed as the Germans when Dammer, at 0600, signalled back that Hill 950 was in Ranger hands. But panzer grenadiers still controlled the peaks from three sides. Their blanketing artillery prevented 3rd Battalion from being reinforced as planned. Then the Germans turned their full fury on the Rangers, raking the peaks with 88s and heavy caliber machine gun fire from higher ground to the north and east.

Dammer could let his battalion stay and die or pull back. Under a smoke screen fired by the 4.2s he decided to save his strength by moving out from under direct fire to Hill 773. With his machine guns covering the withdrawal, he got two-thirds of the Rangers onto the new position, but the rear guard platoon took fifty percent casualties as the Germans counterattacked with two companies. Nor was the movement to Hill 773 without cost: enroute E Company had to destroy a German strong point that was covered by numerous secondary positions.

Fog shrouded the mountains till mid-morning, and Dammer used the concealment of the air to organize his new position in depth. Ambush OP's were set all the way down the reverse slope of Hill 950. The Germans swept over their objective with too much confidence, running into concealed Rangers at close range. This had become their favored tactic, as it had with lst and 4th Battalions—to wait till counterattacking forces were within grenade range, then blast them from positions so close that enemy artillery could not help.

After a brief flurry of fighting, an ambush OP would fall back behind another, each time forcing the Germans to deploy and readjust artillery. By such delaying actions, Dammer was able to bloody the attackers again and again while keeping them away from his main defensive positions. The time gained also allowed him to evacuate his many wounded down the mountain.

The Germans turned to dive bombers to get at Hill 773 as the artillery duel between 36th ID and the 29th Panzer Grenadiers began to turn in the Americans' favor. But with a battalion of Germans on Hill 950 overwatching Hill 773, Dammer was still at a disadvantage. He decided to do the unexpected and retake 950.

Behind non-stop barrages from the 4.2s and 36th ID, 3rd Battalion jumped off at 0530. The terrain was familiar to the Rangers, and their vigorous night patrols had kept the Germans from mining the forward slopes. Out of the fog the Rangers came once again with marching fire but without quarter—only three EPW's were taken while two dozen panzer grenadiers killed. The weary German line buckled (they had many shell shock casualties of their own), then broke into fragments that 3rd Battalion encircled. Short of ammo, Rangers dispatched the remaining forces with bayonets and Sykes knives.

Though they launched several strong combat patrols at 3rd Battalion, the Germans principal revenge was artillery as thick as the rain that pelted the frigid mountains. Casualties from cold and sudden artillery barrages were the most frustrating for Dammer as the Rangers were powerless to protect themselves except by trying to digging deeper into nearly solid rock. So it was with grateful relief that they were pulled off Hill 950 on 13 December, replaced by regiment of the 82nd Airborne. Once again Rangers had done all that was asked of them and much more.

Their lessons learned? That Rangers can be exceptional light infantry in a conventional role—but can only do so for a limited period. The Venafro-San Pietro missions were not 'bad' by the two principal criteria used in this book: that either combat or intelligence support (or both) were inadequate. Instead the missions were bad because they were extended too long, grinding down the Rangers' finite capabilities to sustain themselves. At about 65% strength when they came out of the line, the three battalions were certainly experienced in mountain warfare, but at a higher cost to their NCO leadership and company grade officers which contained the greatest expertise in ranger craft, especially amphibious operations. To regain those unique capabilities was Darby's first priority as Alexander sought to break open the Italian front with a landing at Anzio.

Organization and Training

Fortunately Darby's command would have nearly three months to reacquire classic ranger skills. Clark had become as much a believer in the Rangers as had Patton.* [*FN: On 30 December, after a sluggish performance with Eighth Army in Italy, Montgomery departed for England to prepare British forces for the invasion of France.] The Fifth Army CG himself pinned on Col. Darby's new eagles, and with the promotion came the creation of what amounted to a regimental headquarters: the 6615th Ranger Force (Provisional).

This was certainly not the desired final status, but it was sure better than what had existed before, merely an ad hoc CP for "the senior ranger battalion commander' as Darby had been designated since North Africa. Murray—trying his luck where Darby's had failed—had written directly to the War Department, requesting "a clear-cut directive" that would recognize the Rangers as a permanent force, entitled to a hundred graduates per month from the Camp Forrest Ranger Training Center. With this allocation, he wrote, "our replacement problem would no longer exist, and we would continue to be an effective fighting force without interruption."

None of these, or any other of Murray's recommendations, were approved. It is curious that a battalion commander would go so far out of channels to petition the commander-in-chief of all US ground forces—perhaps Murray had known him previously—but this illustrates the frustration of the Rangers' bastard status.

Bastards or not, on 16 January 1944, Darby had himself an HQ to fill out. He brought up Dammer to be his XO/S3, and brought in a West Point classmate, Maj. Jack Dobson, to command lst Battalion. Murray continued with 4th Battalion, and Maj. Alvah Miller, an original Ranger from the Scotland days, moved from XO to CO of 3rd Battalion.

Other replacements were not so easy to come by because the Camp Forrest pipeline was filling 2nd and 5th Battalions in the Overlord buildup. 2nd Battalion arrived in England just before New Years, and 5th Battalion reached Liverpool on the ocean liner HMS *Mauritania* on 9 January.

Their training emphasized scaling cliffs

PFC Hall, Fourth Ranger Battalion.

The Anzio Beachhead: Encircled in a small area by a ring of steel, the beachhead forces battle desperately to repel the enemy, the Ranger's greatest test.

while 6615th's focused on night landings conducted on the relatively flat shore around Naples. Besides the forthcoming Anzio operation, more 'left hooks' up the Mediterranean coast were planned. Darby was glad to see these developments because they pointed to more of the kind of combat for which Rangers were designed and trained. His men had enough of slugging matches in the line against much heavier opponents, and looked forward to once again becoming the Army CG's lightning left hand.

Though the interval between Venafro-San Pietro and Anzio was as long the Rangers ever had for filling out and training up, several factors caused this period not to be as effective as would be desired.

First, in-theater replacements were few and far between; and, as has been noted, there were no trained Rangers coming to the Mediterranean from the stateside pipeline. In-theater, every other infantry commander suffered from manpower shortages and took any measures to keep the men he had.

Second, the lure of the Rangers was not as strong as it once was when Rangers were the first units in combat. Now in Italy there was war aplenty for any infantryman who wished to experience it—he didn't have to volunteer for the Rangers to find it. Moreover, it was common knowledge in Fifth Army that Rangers drew the toughest—often nearly impossible— missions with the highest anticipated casualty rates. By now the spring time of war had been chilled by the grim Italian winter. Patriotic bravado had made way for realism; for example, a slogan in Fifth Army before Salerno had been "End the war in '44!" In front of Cassino, it was replaced by "Stay alive till '45."

This understandable phenomenon did not mean that Darby could not get good men—far from it. Indeed, the replacements he obtained between Salerno and Anzio, though fewer in number than he needed, were mostly combat veterans, not just enthusiastic cherries.

The third factor in their inadequacy was the training they received as Rangers. For one thing, there were too few hard core veterans to train them—too many had been lost during the previous three months in the mountains. Those who remained were war-weary, their powers of recuperation drained by ordeals of relentless combat. Of course they were motivated to pass along their hard won experience to the new men who would be fighting beside them in the future, but something was missing, something intangible, something between zest and evangelism like that demonstrated in Africa when 1st Battalion veterans greeted the initial truck load of replacements with small arms fire at their feet. Gunplay was not something anyone took lightly any longer.

Furthermore, training facilities were not as instructive as they had been in Africa. Landing craft were much more scarce, and difficult to obtain for practice landings. Artillery adjustment, the ace in the hole for Rangers in trouble, could not be practiced as much because the guns were on the front firing live missions. Consequently, adjustment of indirect fire was a skill that did not reach as far down the Ranger ranks as it had before.

Finally, though three months seemed an adequate length of time for the Rangers to get back in shape, more of it than expected was required for just plain rest and recuperation. Having been shaved by death so frequently in Italy, the veteran Rangers were intent on enjoying the social life of Naples as much as possible. And has been true in every war, with every unit, the longer men fight the more they drink. Training runs at 0500 seemed like harassment for combat centurions who had bested the best of the *Wehrmacht* in some of the hardest fighting of the war. From EPW's they felt the Germans were not training like that, so why should the Rangers.

Darby and his chain of command must have sensed that the 6615th had not developed the ideal edge for what lay ahead, but the original officers had grown extremely close to their men and had every confidence that the now famous units would acquit themselves with further brilliance when the chips were down. This was certainly true but was also a rationalization for easing off in training. Nothing seemed broken with the Rangers, so nothing appeared to need fixing—nothing, that is, that a little more rest couldn't fix.

Among incoming Rangers, two attitudes prevailed:

The EM felt that if they just imitated the veterans, things would go all right. "Just watch me, pal, and do what I say," must have been the counsel they heard frequently from battle scarred warriors. For new Ranger officers there was a similar attitude. What did a green lieutenant have to teach a grizzled platoon sergeant with two Silver Stars and three Purple Hearts? Not much. Instead, the platoon leader's overriding ambition was to show he could cut it in such an elite organization. It was still elite but also increasingly brittle, like fine steel that is too often subjected to extreme heat and cold. Italy now was extremely cold and the fighting had been extremely hot.

Good Missions:

VIII. Anzio, Italy, January, 1944

An envelopment of the German flank from the sea had become an obsession for Churchill, who, as he put, wished to "loose a wildcat" in Kesselring's rear. The alternative, all agreed, would be a bloody stalemate around Mount Cassino where the Germans gave no indication of withdrawing.

For the upcoming Anzio landings (Operation Shingle), there were major changes in the strategic picture. First, Italy—that third of the peninsula not still under fascist occupation—had declared war on Germany and provided a division of mountain troops to Alexander who was also strengthened by a small corps of French colonial forces, among which the Moroccan 2nd Infantry Division was really first rate. With these reinforcements, Alexander was able to weight Clark's main attack on Cassino which would be launched concurrently and in coordination with Operation Shingle.

But Kesselring was able to match the Allied buildup. Von Manstein had been able to stabilize the Russian front, so eight more divisions could be released for the defense of Italy. The balance of forces, in other words, had not changed significantly though there were more formations on both sides.

Landing craft, it cannot be too often repeated, were the crucial factor in providing the mobility differential that Alexander had to have to make use of the seaward flank where Allied warships ruled over negligible opposition. US industry was turning out landing craft at an stupendous rate but priority for them was first for Overlord, then the Pacific, and lastly to the Mediterranean. The timetable for Overlord was such that much of Alexander's naval assets had to shift to British waters, leaving him with a window of opportunity for Shingle that would close on 1 February.

Eisenhower was in the curious position of being the losing theater commander for landing craft, then becoming the gaining commander when he switched jobs a few weeks later. To his credit, he did not slight Shingle and relaxed Alexander's deadline by two weeks. Given the limitations for Shingle, Ike was skeptical that it would have the 'wildcat' effect that Churchill hoped for:

I agreed to the general desirability of [Shingle] but pointed out that the landing of two skeletonized divisions at Anzio, a hundred miles beyond the front lines as then situated, would not only be a risky affair but that the attack would not by itself compel withdrawal of the German front... The Nazis had not instantly withdrawn from Africa or Sicily merely because of threats to their rear. On the contrary, they had reinforced and fought the battle out to the end.

Indeed, those were exactly the orders Kesselring received from Hitler—fight for every inch of Italy and concede nothing.

The last element in the Shingle equation was the air. By now the Allies enjoyed air superiority but not supremacy. Anticipating another landing like Salerno somewhere along the Tyrrhenian coast, Kesselring husbanded his fighters and dive bombers frugally, holding them back from the Cassino front where foul weather and mountains limited their effectiveness anyway. Instead he based them within striking distance of likely landing sites, and planned to employ them in mass at the decisive moment.

Commanding the Anzio landings was the VI Corps CG, John Lucas who was not a proponent of the operation, and for good reason: shipping was inadequate to land more than the equivalent of two infantry divisions (the British 1st and the US 3rd) initially. The necessary follow-up forces, including almost all the armor, had to wait till the convoys round tripped from Naples. Given this slow buildup of his forces, Lucas could only hope for a cautious and gradual breakout from the beachhead.

The strongest component of the VI Corps plan was surprise. In contrast to Salerno, Kesselring did not know where to expect the landings. The British, masters of ruse, developed a naval deception plan that kept considerable German forces north of Rome.

The fire support plan also emphasized surprise: no off shore bombardments were scheduled until the first wave of troops was on the beach. The sole exception was rocket ships firing for fifteen minutes before H Hour.

The Ranger Force landings were unsupported. For the purpose of seizing Anzio's port facilities, a battalion from the 509th Parachute Infantry plus a company of combat engineers were attached; and as always the 83rd Chemical Battalion came along with their reliable 4.2" mortars. After the port facilities, Ranger Force's priority missions were to destroy coastal defense batteries, clear the beach between Anzio and Nettuno for following waves, and link up with the two infantry divisions on their flanks. Once this was done, Ranger Force would come under command of 3rd ID.

Though no one knew it at the time, this would be the last Ranger amphibious operation in the Mediterranean. Fittingly, and by coincidence, one of the vessels on which they embarked was the HMS *Royal Ulsterman* , the troop ship for lst Battalion at Arzew, a time from the past that must have seemed an eternity for the Rangers who made both landings. No more than twenty percent of them could claim that distinction. Nor could either climate or locale have been more different as the Rangers set sail on the raw, cold afternoon of 21 January 1944.

In contrast to the voyage to Salerno, the ten-hour passage to Anzio was uneventful with no indication that the Germans had any inkling that Italy was about to be invaded once again. Rangers took this to be a good omen that another Gela was not in store for them.

Like Lucas, Darby could not bring all his forces ashore at once: the landing craft that delivered the first wave of Rangers would have to go back to pick up the second. The problems and danger inherent in this situation did not materialize; instead good omens continued as the Rangers at sea could see headlights moving along the coast roads, indicating that the Germans would be taken unawares.

They were. Despite the warning provided by the rocket ships' fusillades, no opposition was encountered by the first wave (two companies from lst and 4th Battalions abreast) who were placed on shore right on time by guide boats. This was another good omen because the rocket support that had been planned was postponed because the ship was out of position to fire. The Rangers moved swiftly and accurately onto their objectives, knocking off rear echelon Germans like ducks in a shooting gallery.

Ranger Force HQ came in with the second wave. Darby had planned to use the waterfront casino as his CP, and his boat landed only a few feet from the entrance. As radios and field phones went up on roulette tables, Rangers looked around at their surroundings and remarked that urban warfare wasn't so bad—and sure beat rooting panzer grenadiers out of the mountains.

Indeed the Rangers—at last back in their element—completed their landings and missions like clockwork, causing Lucas to report that even though the Germans had wired the port facilities for demolition, fast action by the Rangers prevented much destruction.

As lst Battalion had at Arzew, 4th Battalion swept through the unlighted town as if they had trained in Anzio. Their prime objective was a bridge on the railroad that led to Rome—the 'Eternal City' only thirty miles north. Hastily the Germans had set up machine guns to defend the bridge. In the darkness Rangers worked around the nests, moved in, and took them out—but at the loss of a platoon leader's life, the first KIA at Anzio.

A few stray German armored cars roamed the streets till Rangers riddled or ran them out of town. Nine vehicles were captured and put to use. First and 4th Battalions reported crossing phase lines ahead of schedule and establishment of contact with friendly forces on the flanks. At 0300, 3rd Battalion came ashore and headed for their objective, the coastal batteries. Dobson's men were ready to fight for the guns, but the crews had vanished. Surprise had been complete, mishaps few, and Churchill's 'wildcat' metaphor seemed apt as dawn broke over VI Corps' beachhead. The Germans had been caught with their pants down but no one, least of all Darby, was writing them off.

Even the first wave of Rangers had carried 4.2 shells which they deposited on the beach for use by follow-on forces. This lesson learned from previous landings enabled Darby to quickly reply to indirect fire that began to dribble down on the port. The reason for the growing accuracy of German artillery was evident when daylight revealed the hills that cupped Anzio-Nettuno: so long as fire could be adjusted from those hills it would spoil every day in the beachhead. A vital function in the buildup was to produce counterfire: many Allied artillery batteries fired 80,000 rounds on such missions.

Kesslering decided that now was the time and Anzio the place to throw in his air force before the beachhead grew stronger. He sent wave after wave of dive bombers at the landing fleet. Stukas were decimated by the navy's antiaircraft fire and swarms of Allied fighters. The sky was crisscrossed by smoke from burning planes and constantly speckled by bursts of ack ack. Nevertheless, the *Luftwaffe* was able to sink several vessels— one of them an ammo ship that blew apart with a volcanic explosion— before they could make the turn around for Naples to pick up more of VI Corps.

This loss of precious shipping was daunting for Gen. Lucas who was less encouraged by the successful first two days of the landings than he was apprehensive about the delayed reaction of his counterpart who had a curious surname for a German, Mackensen. He was scrambling to throw up a perimeter around Anzio while Lucas, according to plan, built up but did not expand his beachhead. Thus, both sides consolidated their positions during 24-25 January, a lull before the storm.

Ranger Force was now controlled by an old acquaintance, Gen. Truscott, CG of the 3rd ID. He wanted more real estate to push German artillery back where they could not hit the unloading areas. This reasonable desire put him at odds with Lucas who felt the beachhead perimeter would be too thin if it bulged out to include the first range of hills which Truscott wanted to seize or at least deny to the Germans. Lucas certainly faced a dilemma that few generals would savor. He decided to temporize and let Truscott conduct a limited objective attack, something more like a reconnaissance in force up the Albano-Anzio road.

Truscott led off with 4th Battalion and a platoon of tank destroyers (TD). On their flank the British 1st ID would move on Carroceto while 3rd ID developed the situation toward Cisterna. All these attacks gained the desired ground on 25 January, though unfortunately the advance fell short of Cisterna. On the 26th, Truscott held back as he could not persuade Lucas to probe further to the Alban Hills. The northernmost extent of the Allied advance was that of the Rangers on 27 January when they reached the crossroads near Campoleone. This was as far as Fifth Army would go until May.

3rd and 4th Battalions led the way against the roadblock protected by thick-walled stone farmhouses. F Company of 4th Battalion was taken under fire from long range by 88s emplaced in an anti-tank mode. But one of F Company's difficulties was that they had no tanks—the principal reason Lucas felt his beachhead could not be expanded. So it was for the infantry to break through stiffening German resistance, but F Company had few men to do it. Guts and guile succeeded when numbers and armor support were lacking. But at cost more valuable than the casualties lost. It took even the company command group, including the CO and 1st Sgt., to carry off the charge that overcame a platoon of well entrenched panzer grenadiers. Killed was F Company's XO, Lt. Howard Andre, who had taken part in all five amphibious operations including Dieppe. Such leaders were irreplaceable.

Rumors of impending German attacks preceded their actuality by a week. One premature scare came when Nazi paratroopers were reported to have dropped into the beachhead bent on sabotage and destruction. Darby mounted a motorcycle to see for himself. Indeed there had been a jump—but only that of a *Luftwaffe* bomber crew shot from the sky by ack ack.

Far more serious than rumors were German mines, and not just those on land. An LST delivering two batteries of 4.2s to support Ranger Force struck a mine in the bay. The survivors transferred to another ship, but it too detonated a sea mine. Total losses for the Rangers' favorite chemical battalion were nearly 300. The well knit teamwork that had previously developed with such soldiers was also irreplaceable.

On 28 January Ranger Force turned back numerous patrols from two infantry divisions that had begun to hem in beachhead from the north. Heavy German artillery, as well as the ubiquitous 88s, were massing near the railroad that ran east-west beside the Apian Way. One such bombardment killed the CO of A Company, 4th Rangers, and badly wounded his first sergeant. More irreplaceable losses.

That night the Rangers were relieved by a British reconnaissance regiment, and placed in 3rd ID reserve. Darby went over to Truscott's CP to hear about his next mission. It would be to slide east about ten miles in back of 3rd ID's front lines to assemble for a coordinated attack that was to push the Germans off the Apian Way. The objective for Darby's Rangers would be the small town of Cisterna, a name that meant little to them at the time but would become the final chapter of their history, written in blood and tears.

Bad Missions:

III. Cisterna, January, 1944

With Truscott's attack scheduled to kick off in less than 24 hours, the first order of business for Darby was to study the intelligence available. The enemy was identified as the Hermann Goering Panzer Division—old acquaintances from Africa and Sicily—recently arrived at Cori. Their outpost line was believed to extend a few miles into the flatlands between the railroad and Mussolini Canal, but the Germans main defenses were assumed to be in the hills just north of Cisterna. Truscott's G-2 opined that the enemy's most likely course of action was to delay in front of Cisterna, then defend it with perhaps as much as a reinforced battalion of panzer grenadiers.

Time did not permit Darby to verify this intelligence with his own recon patrols. He had no choice but to act upon the intelligence provided him by higher headquarters.

The true facts were that Mackensen had coiled three infantry, two panzer, and one panzer grenadier division for an all out counterattack to crush the beachhead. Only Allied air interdiction and artillery counterfire had prevented him from launching his attack on 28 January. But now his divisions were ready to move out, combat support well forward, with strong counter-reconnaissance patrols to prevent snooping behind their outpost line.

3rd ID's operations order simply assigned Ranger Force the town of Cisterna as their objective and a sector about four miles deep and two miles wide in which to maneuver. Leaving the briefing, Darby remarked that "This isn't my plan."

Indeed his only input was the tactics he would use to seize Cisterna, an attack that had to be coordinated with infantry regiments on both flanks who were also going for objectives across the railroad. Their tactics would be similar to his: lead battalions should cross the flatlands by night infiltration, with heavier attack forces attacking down the roads at dawn. This general plan was adopted across the 3rd ID front.

With the toughest objective, priority of artillery fires went to Darby; but not many panzers were expected to be encountered, and with US tanks so scarce, none were attached to Ranger Force. Only five tank destroyers plus Canon Company and 4.2s would support their attack. No tac air was planned because of the identification and coordination problems for such support during a night infiltration. This and the lack of much anti-tank capability did not overly concern Darby as German armor would be restricted to the roads because of the marshy ground on all sides, crisscrossed by irrigation ditches. For such conditions, a cool bazooka man in a roadside ditch at night could be as effective as a Sherman tank or a P-47.

3rd ID's attack was to have kicked off on the night of 29 January but the British division on their left requested a 24-hour delay which Lucas granted. Mackensen used that period to bring down the 26th Panzer Division into the Cisterna area to exploit the penetration expected when the Goering panzers jumped off on 31 January. Despite US aerial reconnaissance, Mackensen's reinforcement was accomplished with complete secrecy. Consequently, and unbeknownst to the Rangers, their objective now stood on the line of departure of two full panzer divisions, both locked and loaded to attack.

At 1800 on 29 January, Darby briefed his battalion commanders at the Ranger Force assembly area. First Battalion was to cross the line of departure at 0100 to advance on Cisterna by infiltration through the flatlands. Fifteen minutes after they cleared the LD, 3rd Battalion would follow and support through the same terrain. At 0200, 4th Battalion with 4.2s would advance on their left, up one of the two roads that converged on Cisterna, clearing land mines as they went.

So the general scheme of maneuver was for 1st Battalion to bypass everything and get into Cisterna an hour before daybreak with time enough to set up defenses against counterattack. The sister battalions' mission was to prevent enemy interference with 1st Battalion's strike.

Speed and stealth were essential, Darby emphasized, especially for 1st and 3rd Battalions. Avoid contact at all costs and kill silently if at all possible: the most the Germans should fear was that a strong combat patrol had penetrated their outpost line. Seize Cisterna and organize it for defense before they realize scope of the attack.

The town was built of stone, an ideal stronghold once it was in Ranger hands. Five tank destroyers and the four aces of Canon Company would be coming right up the road behind 4th Battalion; when they got into Cisterna the Rangers could handle anything the Nazis threw at them.

To end the briefing, Darby announced that the first mail received since the landing would be distributed in the assembly area. This probably boosted Ranger morale more than the prospects for the attack. The plan sounded OK—not a walk over but with the thin German forces expected, the mission seemed within Ranger capabilities. The infiltration would also be favored by the absence of a moon.

Rangers probably grumbled less about the mission than the chill of the night, the half-frozen marshy land they would cross, and the icy water they could anticipate in those irrigation ditches. 4.2 crews knew there would be problems setting up in the mud. And infiltrating flatlands was something knew, even for the Ranger veterans. Craggy mountains had been tough but at least they offered concealment and some cover, plus the big advantage of coming at the Germans from unexpected directions. Cisterna would be straight ahead, right up the middle, and with open fields of fire that would aid the defenders. There had better not be too many of them . . .

Ranger Force HQ moved up to a farmhouse just behind the LD where common checks were conducted with the battalions. At this short range, all tactical radios seemed to be in working order. Radio silence would be in effect until 1st and 3rd Battalions crossed the phase line about two-thirds of the way to Cisterna.

In a wooded assembly area, the Rangers went through pre-combat routines without much conversation. Bed rolls were stacked for movement by vehicle to Cisterna the next morning. Packs were stripped down to combat essentials: K rations and extra ammo. Detailed operations orders were briefed down through squad leaders—every man had to know the overall mission as well as his own.

A cold breeze came up in the faces of infiltrating Rangers as they silently deployed to cross the LD. A good omen, some veterans thought, because a head wind carried any nose from the Germans toward them. However the loudest sound they could hear was the slight crackle of the ground under their boots: the night had frozen the fields hard with a thin layer of frost. For veterans of many previous infiltrations, the enemy territory in front of Cisterna seemed unnaturally quiet.

All the more important for the Rangers to move even more quietly. Some recent replacements seemed to think they were doing so, but the veterans moving with them could hear sounds that they couldn't. This concerned the veterans because silent movement was easier here than in the mountains where a slip or dislodging a stone gave loud warning of approach. The cherries should have seen Guettar Pass if they thought this infiltration was tough.

With about four miles to cover over this flatland, five hours should be enough time to reach Cisterna before daybreak—though the veterans wished the attack could have jumped off earlier. It could not because of the need to synchronize with adjacent units. That was a factor which made the veterans uneasy: a lot of coordination was required with conventional units, and this limited the sort of independent action at which Rangers were so successful.

Tentative fire by some Germans popped from the left, then later from the right, coming from the sectors where battalions of 3rd ID were infiltrating. Ranger veterans cursed under their breath: no matter how stealthy they were now, the Germans had been alerted.

Closer fire broke out within an hour and a half after 1st and 3rd Battalions crossed their LD. It was in back of them where 4th Battalion ran into a roadblock. Murray had his companies 300 yards off the road but even out there the Germans were thick. This was the first ominous sign that there was more than an outpost line confronting the Rangers.

Within fifteen minutes, 4th Battalion was in a full scale fire fight while trying to fan out and flank German strongpoints. But the enemy positions controlling the road were solid, tied in with interlocking fire from machine guns dug in every hundred yards, with rifle pits at ten yard intervals in between. Murray pounded them first with 4.2s, but base plates shattered on the frozen ground. He called for artillery but didn't get the expected priority of fire because the infiltrating battalions from 3rd ID had run into a buzz saw.

Now all Murray could call for was the skill and experience of his veterans to press the attack. This they did—like Lt. George Nunnely, commanding C Company, who led the maneuver element while the rest of his company returned fire at full volume. It was not enough to cover his dash to an irrigation ditch. Nunnely and his RTO were cut down, and the company radio could not be recovered from under ceaseless grazing fire from the well concealed Germans.

He was among several key veterans to go down in the first hour of the attack. Two squads of cherries were without experienced leaders and became ineffective. Murray reported to Darby that 4th Battalion had knocked out two entrenched German positions but this did not produce a gap through which he could proceed. The German strong points were mutually supporting by crossfire: at least three in a row would have to be eliminated before their defenses were dented. Murray needed armor to do that. Until armor could help out (and battle through thick minefields), 4th Battalion was pinned down by fire and unable to move under the numerous flares the Germans had launched into the night.

Darby rang up 3rd ID G-3 to get some tanks up but none were forthcoming. The situation for 4th Battalion was growing grave, but at least no news was good news from 1st and 3rd Battalions. Apparently, despite the roaring firefights elsewhere, the Ranger infiltrators were making headway undetected.

This was true for the first hour or so. 3rd Battalion was echeloned to Dobson's right while he guided toward Cisterna by orienting on a deep irrigation ditch, and were more than two miles ahead of 4th Battalion on their left when Murray became locked in battle for the Feminamorta-Cisterna road.

The Germans' attention was focused on that firefight which they apparently believed marked the farthest American advance (the infiltrating battalions from 3rd ID had been stopped cold and badly shot up). 1st Battalion did all they could to keep the enemy in the dark. Coming upon stray Germans or small groups, the Rangers knifed or blackjacked them before they could sound an alarm. Such stealth and stalking moved the infiltrators so near artillery batteries that Rangers could hear the cannoneers' firing commands. It would have been easy to destroy these batteries but the mission was to seize Cisterna before daybreak—nothing had higher priority or could compromise that mission.

The violent clash of arms nearby had disadvantages, however, as well as distracting the Germans and creating a lot of welcome noise. Their flares were being fired to the infiltrators' rear where the swinging lights silhouetted them, causing many 'freeze' stops. Further delays arose as German reinforcements moved toward the fighting. Dobson had to detour around a column of panzers, infiltrating his battalion across the road section by section. It is a high tribute to the Rangers' stealth that they penetrated two miles into a very thick German positions before taking them on in decisive combat.

As a new commander, Dobson had conflicting factors to consider while under the immense stress of life and death responsibilities as well as personal danger as he slinked along as an infiltrator himself. What he knew was this:

1. The pre-operational intelligence was all wrong. There was a helluva lot more than an outpost line between the LD and his objective.

2. 4th Battalion's supporting attack was not progressing. Though radio silence was still in effect for the infiltrators, Dobson could monitor 4th Battalion's traffic with Ranger Force HQ, and he had heard nothing good.

3. From the sounds on his left and right, the infiltrating battalions from 3rd ID had been discovered and were also in serious trouble.

4. Darby must be aware of the general situation; if it warranted aborting the Ranger infiltration he would radio instructions to do so. Unless and until he did, the mission was still to seize Cisterna.

What other factors must have mixed in Dobson's mind can only be conjectured. He might have noticed that the Germans dispositions seemed to indicate an intention to attack: for example, the forward positioning of artillery and deep bombardment of Anzio itself; the panzer columns where G-2 had said panzers were scarce; some of the German dead were paratroopers—there weren't supposed to be any in front of 3rd ID's sector. All of these factors warned Dobson that he was in too deep. Perhaps some Ranger veterans thought so, but none evidently approached him to express that opinion when it could have been acted upon. What the survivors all said they were most worried about was running out of darkness.

That's exactly what happened. But initially the seeping light was encouraging. First, the low stone structures of Cisterna became faintly visible. Second, a friendly armored column was scheduled to blast up the road to sup port the Rangers at first light. Air cover also was to become available then. Third, and most gratifying, dawn found 1st Battalion on top of a reserve company of German paratroopers as they began stand to. The Rangers fell upon them like banshees, bayoneting them in their sleeping bags, grenading their bunkers, blasting their vehicles, destroying their radios in a fifteen minute slaughter that wiped out the German company entirely.

Sweeping through this blood soaked bivouac, 1st Battalion headed toward the outlying buildings of Cisterna. Speed was of the essence, but they were met with fire that halted the advance after a few hundred yards. Germans now were everywhere: tanks, flak wagons, squads and platoons of paratroopers rallying from initial surprise, rising like an awakening army of ghosts in the mists of early dawn. Fighting broke out on all sides—the action quickly became a brawl with guns.

At first the advantage in close combat was with the Rangers: the Germans were unprepared for 400 ferocious fighters in their midst as if dropped by parachute. Rangers cut down the milling Germans by the score, but within an hour, the first hour of daylight, Dobson was trying to organize a hasty perimeter.

Echeloned to his right rear, 3rd Battalion had also caught companies of paratroopers unawares and closed in with devastating fire that eliminated those elite Germans from the contest. Miller surged ahead but ran into fire from all sides as he advanced to Dobson's aid. The two commanders could not see each other, but both must have been thinking how their situation was beginning to resemble Gen. Custer's—aroused Indians were everywhere and on all sides.

Dobson got onto his radio to contact the Ranger Force CP. The best course of action he could recommend was to hook left toward the road and have 1st and 3rd Battalions fight their way back to 4th Battalion's stalled attack. But because of concussion, cold, water, or whatever, Dobson's radio malfunctioned. Almost as tragic was his consequent inability to get artillery support. With only shoulder fired weapons, he might have started toward 4th Battalion anyway but the result would have been far from certain—the high rate of fire was using up ammo at an alarming speed. He didn't have much time to do anything as the vengeful Germans closed in.

From this point, around 0700, the vicious fighting broke down into scrambled actions by small groups and individuals on both sides. Artillery did not come into play immediately for two reasons. The Rangers' commo was sporadic due to inexperienced RTO's and battle damage to radios. Furthermore, the fighting was at such close quarters that Rangers could rarely raise their heads to adjust artillery fire, and there was high danger of casualties from friendly fire with the Germans and Rangers so mixed together. Neither Dobson nor Miller could get a good sense of how their battalions were deployed. Initially the Germans had this same problem locating forces on both sides, but eventually their commander could get around the periphery of the fighting and designate targets.

It was when the German artillery, unopposed except by Darby's 4.2s, entered the battle that it began to tip against the Rangers. Maj. Miller was killed around 0800 by one of the early barrages, and Dobson wounded later. Safety from artillery could only be obtained in the scattered stone buildings near the outskirts of Cisterna. These became Ranger objectives, some won, some not, but with the effect of dispersing the Rangers still more as they assaulted such cottages outside their ragged perimeters, causing more gaps to open.

Before the loss of their CO, 3rd Battalion continued with their mission to support 1st Battalion's drive toward Cisterna. With Dobson stopped and recoiling from counterattacks, Miller tried to take pressure off him by advancing with three companies that leapfrogged their platoons forward. But a huge German combined arms force rose up on their right, causing the other three 3rd Battalion companies to take up a blocking position so that the attacking companies could press on. A deep stream bed became the blocking position, the only piece of terrain to favor the Rangers. The attacking companies battled across Highway 7 and the railroad, the only units on the Allied front to make it that far; but panzer grenadiers had by now occupied Cisterna in such strength that Rangers could not break into the town. This was their high water mark. No additional amount of courage, skill and resourcefulness could take them any farther.

Help, if there was to be any, had to come from the rear, from Darby; but his headquarters itself had come under heavy artillery fire that killed several key members of the CP. His 4.2s were also hit by counterfire. Darby tried to revive hope in his beleaguered battalions by radio, urging everyone to rally on the veterans and fight in an organized defense. That was good counsel but unnecessary: the Rangers were already doing their utmost to plug holes between the increasingly scattered elements but now, in full daylight, movement was virtually impossible under crisscrossing German fire of every caliber. Their ammo supply was near at hand and inexhaustible—another indication that a multi-divisional attack against the beachhead had been scheduled for that very morning.

So the Rangers, for once, had inserted themselves into the wrong place at the wrong time—just the opposite situation from their many successful operations. The penalty for this misfortune would be annihilation.

Such an outcome did not seem inevitable in the early morning. The promised infusion of US tanks had been delayed and reduced by action elsewhere, but Darby was pleading with Truscott for armor support and lashing Murray to open a road for them. 4th Battalion knew the increasingly desperate situation of their sister battalions, and at dawn struggled on to relieve them. The lead company, shot up badly during the night, infiltrated back from where they had advanced to within fifty yards of the German main line of resistance. Now three more companies moved out with two tanks, two TD's, plus all four aces of Canon Company.

The morning thaw prevented much cross country mobility by the tracks. They were confined to the road and its low embankments, both thoroughly covered by German self-propelled guns and 88s Murray's attack carried two hundred yards farther than the night before, but the lead tank then hit a Teller mine, blocking the road. Undaunted, Rangers fanned out to press on alone except for the Ace of Diamonds and one TD. These vehicles too were soon knocked out, and the armor team commander, Ranger Otis Davey, killed.

In contrast to the two trapped battalions, Murray had commo and therefore adequate artillery support. This he used to the maximum but the Germans had more, as they had concentrated their pieces for their forthcoming attack. Moreover, the Ranger penetration was now the only threat—3rd ID's advance had already been turned back— to Highway 7, the German's lateral line of communication. They poured in everything they had to eliminate that threat. An additional incentive was to take revenge on the Rangers who had so tormented them on two continents. An horrific fight to the finish was under way.

It was far from finished because 4th Battalion had battled forward to narrow the distance to 3rd Battalion down to a mile. All three battalions could hear the others firefight—sort of a mutual signal that one would fight on so long as the others could.

Though pinned down and under relentless fire from all sides, 1st and 3rd Battalions now at least had the advantage of defending rather than attacking. 4th Battalion, on the other hand, was trying to advance into a devilish complex of integrated German strong points. Actually, the deeper Murray's men pressed their attack the more entangled they became in enemy defenses, an enemy comprised of long time veterans of the Russian front as well as extended combat against the Allies. Consequently, their positions showed sophisticated planning; e.g., farmhouses were usually not occupied but strong points dug in and cleverly camouflaged nearby or at the intersection of ditches. They had to be precisely located before US artillery could be brought to bear. Pinpoint dive bombing might have been effective, but all Allied fighter-bombers were committed to defending the supply ships off Anzio harbor.

In World War II, the Japanese were famous for their snipers, but they had nothing on the Germans who used probably the most accurate sniping rifles in the world. Mine clearing teams on the Feminamorta-Cisterna road were favorite targets as 4th Battalion tried to punch through an opening for the skimpy armor now available to Darby.

His German counterpart, whoever he was, evidently perceived the situation as well as Darby did. The key to the encirclement was the density of German forces between 4th and 3rd Battalions. So though it subjected them to heavy 4.2 and artillery fire, the Germans slid more troops onto that ground. Temporarily the Rangers had the pleasure of seeing rather than being targets, and German infantry reinforcements were shot to tatters. But despite heavy casualties, the encirclement of 1st and 3rd Battalions thickened.

Then panzers moved in for the kill, attacking the two battalions from the direction of American lines. Out of commo for most of the morning, the surrounded Rangers first believed the tanks were US armor breaking through for the long awaited linkup. But then the Rangers who tried to signal them were cut down by coaxial machine guns. Rather than the direction of salvation, the road from Feminamorta became the path of destruction.

With the fury of terrible disappointment, Rangers rose from roadside ditches to take on the armored column with bazookas and sticky grenades. If the Germans thought their thrust would finish off the Rangers, the next disappointment was theirs. The seventeen panzers were road bound and attacked from many directions by Rangers dodging through machine gun bursts to leap on tanks and spray tommygun fire through apertures. One after another fifteen Mark VI's succumbed to heroic assaults onto their hulls. It was a pitched battle, and one in which the unprotected Davids slew the armored Goliaths.

But shortly triumph turned into tragedy. Four Rangers managed to capture two of the Mark VI's. Suddenly, 3rd Battalion had some armor of its own, courtesy of the previous owners whose next vehicles would only be hearses. The four Rangers were enlisted men, but they had never lost sight of the infiltration mission which was to capture Cisterna.*[*FN: Throughout the twelve hours of the raging battle, Darby never changed the mission. This was no indication of rigidity, but rather a change in the significance of Cisterna. Originally the town was a key in expanding the beachhead; after the ill-fated attack began, Cisterna became the best defensible feature for 1st and 3rd Battalion to hold out while waiting to be saved by friendly armor. If they could have seized the massive stone buildings of the town, some Rangers probably could have survived though their shortage of ammo was critical.] They drove their prizes up the road, probably preoccupied in trying to remember how German tanks worked, to break through to the town which was the objective. But all 1st Battalion saw was Mark VI's with black crosses on them. They, and the Rangers in them, were destroyed by bazookas.

Nevertheless, smashing the panzer column was a morale spike for the surrounded Rangers who had seen nothing encouraging since they crossed their LD eight hours earlier. They still had ample life and fight—though very little ammo—as the Germans soon realized. The paratroopers and panzer grenadiers shrewdly called off attacks and clamped down a siege they now knew could not be broken by 4th Battalion alone, then let self-propelled guns, 88s, and multiple rockets work over and wear down the dwindling Ranger strength. First and 3rd Battalions were about to make their last stand. Darby could only pray in frustrated agony as the sounds of German weapons swelled while Ranger counterfire dwindled.

For the hours around noon in the Ranger pocket—a rough oval now only a few hundred yards long— history was individual rather than by unit. The situation was not every man for himself but more every man for every other man. The nature of the last ditch fighting can only be described by the men themselves.

T/5 James P. O'Reilly, B Co. 3rd Battalion:

Back toward the port our ack-ack was raising hell with [German] night raiders. Pieces of shrapnel were landing all over our area.

An ack-ack shell which hadn't exploded in the air landed smack in the middle of our company and wounded several fellows [including the company CO]. That left Lt. Newman in command.

We pushed off about 2000 that night. Though we were all pretty tired from lack of sleep during the last seven days under fire, most of the fellows were in good spirits. The hike to the [Mussolini] canal wasn't too rough, as the bitter cold made us feel like walking to keep warm. It's funny how you can forget how tired you are when you start out on a job. All the fellows seemed to be quite confident we could do the job—right up our alley...

A couple of hours after we pushed off, things began to happen. Over on the left and a little behind us, we heard several short [German] machine gun bursts. Then complete silence. We began to wonder if the 4th Battalion had run into trouble on the road.. Suddenly flares started to burst all around us. But before they could fully brighten the fields, we were all flat on the ground. When the flares died out we would run as fast as the terrain would permit, then flop on the ground... We could hear mortar and artillery barrages landing on our left flank. Someone whispered, "the 4th must be having a bad time."

At first light we heard a lot of shooting up ahead. The 1st Battalion was having quite a fight. Then suddenly everything happened at once. Hidden in a haystack on our left, a Jerry flak wagon opened up with a point blank barrage of 20 mm. Then sniper and machine gun fire began cracking at us from both flanks. They were so close we could throw grenades at them. One of our sections went after a bunch of Jerries in a farmhouse about fifty yards to the left of the ditch. They fought their way into the building and killed ten Krauts who had several machine guns set up inside and mortars beside. Out company was right alongside the road running north into Cisterna. We were fairly well dispersed along a deep irrigation ditch. Now that the fighting had actually started, the tenseness that we had during the night seemed to leave. I guess sweating out the fight is always the worst part of a battle.

Before long the flak wagon which had been firing at us tried to pull out but was hit by our mortars. The first shell landed smack in the middle of it; two more shells burst less than three yards from the front end. We could see the crews' bodies blown skyward with parts of wreckage. We all cheered loudly and felt much better.

Moving up the irrigation ditch, we advanced about a hundred yards, wading through knee-deep water. Then three more [self propelled] guns came down the road firing their canons into our ditch. Our rocket gunners waited until they got close then let 'em have it, stopping them cold. Some of the boys crawled up with sticky grenades and finished them off. The crew of the first one tried to get away. We waited until they all cleared the flak wagon then raked them with BAR's and rifle fire. It was just like shooting ducks but with much more satisfaction. [The other crew surrendered] but while being marched down the ditch they tried to make a break—so we had to shoot them down.

I remember crawling up the road as another flak wagon came tearing down to help the others. I threw a sticky grenade but it rolled off. I threw another and it landed right inside the gun platform. The flak wagon kept going for about ten yards before the explosion [detonating some ammo]. It was a awful mess. The road was splattered with gnarled wreckage of twisted steel and broken bodies.

All around us the fighting was getting real hot. Some of the companies broke up several attempts by Jerry infantry to close with us. But most of the attacking was done by Rangers—trying to get into farm buildings the Germans held. From what I could see of the battle, each company seemed to have a separate fight on its hands. We were too busy with our own fight to notice much of what was happening up and down the line—if we could call it a line, we were so scattered. We knew by then that we were completely surrounded but we still had hopes of our support troops reaching us. Sure I was scared and I guess a lot of others were too, but the way those guys were in there slugging, you would never know it...

As we hit a ditch, Lt. Newman's carbine was shot out of his hand, leaving him armed with only a pistol.. While working our way up a small draw we surprised a Jerry machine gun nest. They never had a chance. We let them have it, first with grenades. After that Lt. Newman and a few of the boys went after another machine gun nest in the middle of a vegetable patch farther to the left... He broke his glasses when he jumped that nest with only his pistol. Then he couldn't see worth a damn. That's why he was captured. He was a little guy with big thick glasses, but boy he loved a fight. He never sent his men where he wouldn't go. And there was never a place, however hot, that he would hesitate to go.

After Lt. Newman left I continued toward a small vineyard where I'd seen small puffs of greyish smoke. It was too noisy from other sounds of the battle to hear it, but I was sure that a machine gun was firing from that vineyard. I made it up a small ditch until the fire got so hot it just about dug me out of the ground. I got up and ran the last yards and almost fell on top of two Krauts with a neatly camouflaged light machine gun. They must have been changing clips because I was able to shoot them both before they could get their gun into action. On the way back to the road I saw another one crawling down a ditch. The place was lousy with them. I jumped him with my bayonet and pinned him to the ground—right through the gut.

As I neared the road I ran into some remnants of B Company going for a house about 200 yards nearer the town. Sgt. McCallum, our topkick, yelled for everyone to form on him. We had about a platoon spaced on both sides of the house. He went into the house to see who was in it, and arrived just in time to stop a whole platoon of Germans coming from the other side. Single-handedly he killed about fifteen of them, then organized the defense of the house. His platoon was all there was to hold the flank in that area. The 1st Battalion men who had been there earlier in the battle had all been either killed or wounded...

*By this time our ammo was running low and our men were being thinned out rapidly. Mac sent me back down the road where several of our companies were holding out to see if I could get some more men and ammo. I returned without any luck, as they too had been hit hard. Captain Shunstrom * [*FN: Shunstrom had moved from Canon Company to 3rd Battalion.] told me that reinforcements were expected at any time.*

Sgt. MaCallum again sent me back but it was no use; what few Rangers there were left were needed there... There was little organization anywhere. It was a section here and there, badly shot up and holding off the Jerries as best they could...

We were all certain by then that the trap had closed, but we were still hoping for reinforcements. We had the Jerries where they could not move in front of us, but far to our left we could see a helluva lot of them going around to our rear. Shortly after that we heard many roaring engines which sounded like our tanks. We were happy as hell, thinking our armor had broken through.

Our elation gave way to bitter disappointment because they were panzers coming up on our rear. We were lucky to hold out much longer against another full scale attack. I had one clip left in my belt. I had fired my own two bandoliers and a beltload, and also two bandoliers and a beltload I'd taken off a dead Ranger. The rest of us were as bad off as I was. The BAR gunners were out completely.

Shortly the Germans, advancing behind some Mark VI tanks, attacked what remained of E Company. They captured some Rangers and marched them up to a strongpoint where one of our sections was causing the Germans a lot of trouble. Using the prisoners as hostages, they called out for that section to surrender or they would shoot the prisoners.

[The Ranger section] refused and told the Jerries to come and get them. Then we opened fire, killing quite a few Germans so some of the prisoners could escape. It didn't work. The Germans shot seven or eight of them to show they meant business. Our guys couldn't keep on shooting and have the bastards take it out on Ranger prisoners. The section had no alternative but to come out...

We could hear more panzers coming down from the Appian Way. We only had a few shots left apiece, and a lot of our boys were out there in front of the Germans with their hands up, but they didn't ask us to surrender and save them. I guess they wanted the decision to be ours and ours alone. Again a German officer, who spoke perfect English, yelled to us to come out or our comrades would die. We came out.

Maj. Jack Dobson, CO, 1st Battalion:

[About 0600] we set up our radio and tried to contact headquarters... but the radio failed and I had no alternative but to follow the original orders.

By this time, it was quite apparent that there was no one on either our left or right, and firing about two miles to our rear indicated that the [3rd ID] regiments were way behind.

We crept on through some unoccupied German trenches for about a quarter of a mile, then came out on the very level expanse leading to Cisterna. From the trenches, we started to run, hoping to reach the town before the sun rose. Approximately 600 yards from the outskirts, we passed what appeared to be a German bivouac area. At this point, things began to pop... In the ensuing melee, using only knives and bayonets, forward elements of 1st Battalion killed approximately 100 Germans as they tried to get out of their foxholes.

We continued on for about 400 yards more, trying to reach a small ridge along the edge of town. Sgt. Heiser of A Company crawled forward and knocked out three successive machine gun posts by creeping up, tossing a grenade, then going in with a bayonet.

But by then we were running into real trouble. The enemy fire was too hot for us to keep moving forward. We consolidated a position astride a lateral road outside the town parallel to an [irrigation ditch]. It was without cover, but it was the best available...

[Panzers] overran our position, but we knocked out fifteen of them with bazookas, grenades and about everything else we could lay our hands on. All these tanks and guns were burning and exploding in the middle of our position—a beautiful sight.[*FN: It was a flying piece of wreckage from a tank that caused the second of Dobson's two wounds.]*

Then Germans tried a series of infantry attacks from all sides. This continued until noon, and all the time the entire position was swept by point-blank artillery fire, two batteries of multiple rockets, and fire from 88s nearby. Meanwhile, two companies of 3rd Battalion had come up to our left. The remainder of that battalion was cut off about 1000 yards to our rear.

At no time did the Germans succeed in punching into our position by fighting. However, at 1230 hours, we saw a line of American soldiers forming up on the road off to our left.

They were the remainder of 3rd Battalion. The battalion commander had been killed and they were caught in daylight in a ditch. Casualties were extreme, and finally those still alive were overpowered.

The Germans were pulling those left onto the road that ran through my position, with two tanks and an armored car behind them and about 100 German infantrymen closely pressed alongside. They started about 800 yards away and moved toward us. The column built up as it moved along, pulling more Rangers out of the ditch.

As they approached our position, we arranged ourselves to dispose of the armored vehicles with the last rounds of bazooka ammunition and the 1st two or three rounds from our [60mm] mortars. Riflemen were stationed to pick off the German infantry.

We opened fire when they were 300 yards away. What was left of 3rd Battalion came toward us with their hands over their heads. Nevertheless we opened fire. We killed several Germans, and we killed or wounded some of our own men in an attempt to stop the column.

The Germans then machine-gunned and bayoneted some of the helpless prisoners. They had already shot the wounded left in the ditch. Then they got behind those who remained and forced them to march right into our position. Our ammunition was finished and we had no way of stopping them. Portions of my battalion tried to escape at this time, but the wide-open nature of the terrain left them exposed.

Approximately 50% of both 1st and 3rd Battalions were casualties, either killed or wounded. About 450 were captured. Later I was told by a German interrogation officer that they found 400 dead Germans just in the immediate area where the last fighting took place. They were all paratroopers.

Postwar information set the total number of German killed and captured by the Rangers at more than 1700, with triple that many wounded; but no matter how dearly 1st and 3rd Battalions sold their lives, or how valiantly 4th Battalion struggled to rescue them, it was a staggering and utter defeat for the Rangers— so much so that Gen. Clark ordered an investigation of its causes.

These turned out to be the two cardinal elements of every bad Ranger mission. That is, inadequate intelligence and/or support; in the case of Cisterna, both, resulting in double disaster.

Intelligence: 3rd ID's attack plan was hastily put together because Truscott was only able to convince Lucas of its viability shortly before it was to kick off. Darby's Rangers slid over laterally from another sector barely in time to receive attack orders in their assembly area. Therefore, the scheme of maneuver relied upon intelligence acquired by the previous unit on the Ranger line of departure.

This meant that Darby, because of lack of time, could not conduct Ranger patrols to verify the extent of German preparations in the Cisterna sector. All investigators agree that the Germans cleverly and clandestinely prepared the battlefield; but if two Ranger battalions could, and indeed did, infiltrate it so deeply before significant contact, there is no doubt that Ranger recon patrolling before the attack would have revealed crucial factors which Darby surely would have considered and developed counters.

But time to patrol was not available. Should Darby therefore have demurred from participation in Truscott's attack? The answer is in their respective ranks. Darby was heard to lament that the plan was not 'his'. His or not, he had to carry it out or fall on his sword in protest. That was not his way nor the Ranger way. Indeed occasions where subordinates have refused to conduct perilous attacks ordered by generals are virtually unheard of in military history— whether the charge of the light brigade or Pickett's charge at Gettysburg, though both were opposed by the commanders who had to execute them.*[*FN: In Pickett's case, his corps commander, Longstreet, also objected to the frontal attack.]

The author does not suggest that Cisterna was comparable in recklessness with the previous two examples. Nothing in the record indicates that Darby was wringing his hands about the risk his men would be taking. Instead he was probably pleased and proud at how the Rangers had been on a roll at Anzio: the landings had been nearly flawless, the expansion of the beachhead successful, and many difficult objectives attained without undue cost. As O'Reilly noted in the recollection of the mood in the assembly area, the troops were tired but in good spirits; moreover, the Cisterna mission seemed "right up our alley". It was for Darby to perceive—as it had been for Lee at Gettysburg—that the objective might be beyond the capabilities of his command. If true, Darby should have verified that his Rangers would have adequate combined arms support to compensate for inherent 'lightness' in Ranger capabilities.

Support: Could Darby have guaranteed that adequate combat support be available if the Rangers ran into more than was expected? Nothing is guaranteed in war, but there is reason to ask if Darby was insistent enough or rather that he was assured of support that never materialized. Clark's investigation did not resolve those questions. Lucas claimed that tanks were attached to the Rangers. Truscott's division order does not mention such attachment.

Priority of artillery became a somewhat moot point because of the infiltrating battalions' faulty commo and spotty forward observation. Which is to say, 1st and 3rd Battalions didn't get the artillery fire they needed not only because there were too few guns in direct support but also because their fire could not be accurately adjusted. On the ground in faint light and hot fire fights, friend and foe were mixed: always a dilemma for forward observers. It is also true that indirect fire adjustment was not on par with previous Ranger operations. For a variety of reason discussed earlier—mainly the attrition of experienced leaders from extensive previous combat— some Ranger skill levels had degraded.

Moving to other aspects of the Cisterna operation, first mention must be made of the infiltration commander. Maj. Dobson, a man of high competence and courage, had been acquainted with Darby for decades; but he was not so well known by the Rangers, as he had only become one recently. A commander in such a position feels impelled to 'win his spurs'—show he deserves to wear the Ranger scroll, in this case—and is therefore inclined to persist with an evidently impossible mission where a veteran commander might more likely use his own judgement. If, out of commo as Dobson was, Murray or Miller had been leading the infiltration force, either might have aborted the mission, confident that Darby would back him up later and no face lost would be lost with the troops. With 20/20 hindsight, we see that what Dobson should have done was to convert the attack on Cisterna into a limited objective raid, taking out those German batteries then turning around to link up with 4th Battalion. New to his command, Dobson felt—understandably—that he did not have that discretion.

Time was scant in the assembly area, but it was not too scant for Darby to have at least discussed abort criteria with his battalion commanders. Maybe he knew them so well that he felt such a discussion was unnecessary.

It must be noted that the disastrous consequences of Dobson's decision not to abort the infiltration would have been obviated if Darby had led it. Normally the commander's location is with the bulk of his forces which that tragic night was the infiltrators. If Darby, leading the way as he surely wanted to, had come upon the situation that Dobson faced, no doubt the outcome would have been radically different.

It is equally true that Darby had to control the entire Ranger operation, including 4th Battalion's crucial role, and to do so would have been well nigh impossible with his command group slinking through the fields and ditches. Yet the experienced and accomplished Dammer was available back at the Ranger CP, an XO par excellence to whom Darby would have gladly delegated operational control.

Naturally this was Darby's fervent preference, but at the time of Cisterna he was no longer merely 'senior Ranger commander' but a 'full bull', CO of 6615 Ranger Force, the equivalent of a regimental commander. In the conventional army, regimental commanders were not expected to rub burnt cork on their faces and sneak around like Apaches behind enemy lines. Nowhere does the record indicate that Darby requested to lead his two battalions; however it can be assuredly assumed that he asked to do so, but was refused by Truscott who was not about to let the most singularly important and celebrated colonel in the theater risk death or capture as an infiltrator.

This is a major problem, as a generalization, when Rangers work in a conventional role for conventional commanders: the probability of serious contact with the enemy is assumed to occur considerably lower in rank with conventional units than with Rangers. For example, O'Reilly cites 1st Sgt. McCallum's personal killing of fifteen Germans as a matter of course (as well as a demonstration of typical Ranger aggressiveness), not something expected of conventional 'top kicks'.

Lesser lessons learned from Cisterna include, once again, the need for redundancy in commo. Absence of reliable radio contact was a serious problem as early as the Arzew landings, recurring in later operations. Dispersed at 1st Battalion needed to be as the infiltration penetrated deep into German positions, Dobson could not be accompanied by an moving array of antennas like some infantry company commander in Vietnam. When his tactical radio malfunctioned there was no back up from perhaps his fire adjustment radio. Eventually commo was established with Darby but it went in and out.

Commo can always be a problem for which there is no guaranteed solution. Foot mobile Rangers especially can not carry all the back up that could be desired. Yet some predetermined signals might have provided certain essential communications: like a red flare to request the mission be aborted, and an answering green flare to grant permission.

Another SOP signal that would have been invaluable is a means of marking captured vehicles so that they would not be destroyed by other Rangers as so tragically happened on the Feminamorta-Cisterna road. Perhaps long, light, red-white-and-blue streamers could have been carried in pockets to attach to the antennas of captured armor.

Like radios, not all the ammunition that could possibly be needed can be carried by infiltrators. For Cisterna, they were stripped down to combat essentials (no packs, parkas, K-rations or even extra canteens). O'Reilly relates that he crossed the LD with two bandoliers and a cartridge belt—about 35 clips, or 280 rounds—for his M-1. He also carried a number of grenades, so he certainly was not lightly loaded. Midway through the morning, he reports that the Rangers' BAR's (their substitute for machine guns) were out of ammo. Indeed in the end, lack of ammo contributed more to the disaster than any other battle factor, even more than the monstrous German practice of using POWs as human shields.

The only hint of a possible (and partial) solution to the ammo problem is in early and maximum use of captured weapons, thereby saving US ammo for emergencies. Obviously such use is inherently dangerous because of misidentification of friendly units as enemy—especially at night— so extensive training and caution are required. There is a compensating factor in confusing the enemy.

Lastly, the tragic outcome of Cisterna includes a bitter irony. Had 1st and 3rd Battalions not been such skilled infiltrators, they would not have been trapped so deeply. The conventional battalions on their flanks were stopped much earlier and were therefore extricated.

The remaining wartime history of Darby's Rangers is quickly told. Mackensen's assault on the Anzio beachhead was preempted by Truscott's attack. Thereafter a state of siege set in around Anzio. There was no part to play in siege warfare for the 6615th Ranger Force, now a shattered shadow of what it had been until the final 24 hours of January 1944.

Struggling toward Cisterna, 4th Battalion had fought to the point of combat ineffectiveness. Ten days later it was attached to the 504th Parachute Infantry Regiment and briefly went into reserve, though for two more months it would see much hot defensive combat. The praised and prized 4.2s of the 83rd Chemical Battalion ended its long association with Rangers and was attached to the 45th Infantry Division.

Thus, with no combat elements to control, the 6615th headquarters was dissolved. On 17 February, the Germans were hitting the 45th ID hard, so Lucas named Darby to be one of its regimental commanders. Two days earlier Murray, stricken with malaria and exhaustion, was hospitalized and relieved from command of 4th Battalion. Miller was already dead and Dobson captured. Like the majority of Darby's Rangers, he was subjected to a final formation—a march of humiliation through the streets of Italy's capital, showered with garbage by fascist Romans, and under the guard of Nazis who proclaimed, "The Rangers have at last entered Rome, not as conquerors but as our prisoners."

Their subsequent story is one of individual captivity and often escape, some in Italy where they fought as guerrillas with the partisans, some in eastern Europe where they made their way to Russian lines, demonstrating that Rangers remain Rangers under any circumstances. They lived or died with a creed expressed by Kipling: "Deal with triumph and disaster, and treat these two impostors just the same"

They had fought to many narrow triumphs, and suffered one narrow defeat that resulted in disaster. Neither outcome changed the core of their spirit.

They were either in POW camps or being hunted as fugitives in the Appenine Mountains when, on 26 March 1944, their brothers in 4th Battalion ended its history as a combat unit. 19 Officers and 134 EM were sent back to the United States, most to be assigned to infantry training centers as instructors. The remaining Rangers, those who had not accumulated sufficient overseas or combat time, were added to the 1st Special Service Force in Italy, a combined unit of Americans and Canadian commandos. Then, on 26 October 1944, the original Rangers battalions were officially disbanded—the first units to be completely inactivated in World War II. Shattered at Cisterna, then scattered throughout Europe and the America, Darby's Rangers were no more. Never accorded a permanent status, they were erased from the books—after writing some of the most shining pages in the history of the U.S. Army.

It would be for their young sister battalions—the 2nd, 5th and 6th—to take up the name Ranger: to continue the reputation, add to its honors, and carry on inimitable standards. It was as if the gods of war realized what a cruel fate they had meted out to Darby's Rangers at Cisterna; for the next time, six months later, Rangers attacked it would be their greatest triumph and one that opened the way to victory in the decisive campaign for Western Europe.

Good Missions:

IX. Omaha Beach, Normandy, June 1944

"Almighty God: our sons, the pride of our nation, have this day set upon a mighty endeavor." - Franklin Roosevelt, 6 June 1944

"These are the men of Pointe du Hoc. These are the men who took the cliffs. These are the champions who helped free a continent. These are the heroes who helped end a war." - Ronald Reagan, 6 June 1984

"These men gave us our world. The simple sounds of freedom we hear today are their voices, speaking to us across the years." - Bill Clinton, 6 June 1994

Even before it received its name, Operation Overlord, the cross-channel invasion of Hitler's 'Fortress Europe', was acknowledged by all to be the main event for the second half of World War II. If Overlord succeeded, the Germans would be in a two-front vice with the inevitable outcome of being crushed between the Russian and Allied armies. But if Overlord failed, Hitler could then shift fifty divisions east to beat back Stalin's counteroffensive that had been rolling west from Stalingrad since 1943. In sum, Hitler could still win the war if the Allied invasion was turned back on the coast of France.

To do so, he personally designated Rommel to be the tactical commander. As Montgomery noted of his old adversary,

It is now clear that *[Rommel's]* intention is to defeat us on the beaches. He is an energetic and determined commander who has made a world of difference since he took over *[in February 1944]*. He will do his best to 'Dunkirk' us. We must blast our way ashore and get a good lodgement before he can bring up sufficient reserves to turn us out.

Rommel himself acknowledged the utter importance of the fighting for the beaches by predicting that the first 24 hours of the invasion would be "the longest day" of the entire war. Who would prevail at the end of D Day was a question of who controlled the beaches over which more than five infantry divisions would be coming ashore. At the most difficult beach, code-named Omaha, and a few miles away at Pointe du Hoc (translation: Hook Point), Rangers would lead the way against the strongest resistance to be encountered.

Two months earlier 2nd Ranger Battalion, under command of Lt. Col. James Rudder, had celebrated its first birthday. The Second's history has been noted in this book, consisting of thorough training at Camp Forrest, Tennessee, and in amphibious operations at the Navy Raider School, Florida. Training in this specialty continued in England.

Its sister battalion in England, the 5th, was only nine months old, having also been born at Camp Forrest, and was commanded by Lt. Col. Max Schneider, a veteran of both 1st and 4th Battalions in North Africa, Sicily and Salerno. So Schneider owed the Germans a payback for Cisterna, and by relentlessly training his battalion made sure they would collect on 6 June 1944. Up until Cisterna, Rangers and Germans had pretty much afforded each other professional respect. After that, the gloves were off and no holds barred for the Rangers.

On 6 May, 5th Battalion moved to the coastal town of Swanage to hone final skills in cliff climbing. Cliffs were critical because, as the Commandos had taught and Rangers had demonstrated repeatedly, the most difficult terrain is the most lightly defended. This was true as well for certain cliffs in Operation Overlord, except that as key positions for German heavy guns they represented a strong means of strategic defense. That is, they were dominant for control over a major invasion beach, but vulnerable for defense of the guns themselves.

The most important of such coastal batteries were the guns at Pointe du Hoc, which had not only dominance over Omaha but clear observation and range to cover Utah Beach, the site of the other multi-divisional American landings. It is no exaggeration to say that these five 155 mm guns could stop the landings of two US corps. It was solely the job of 2nd and 5th Battalions to take out those batteries and control their heights, for the Germans would surely reinforce them as soon as they realized that the linchpin in their coastal defenses was being pulled out.

Readers will recognize many similarities to the Ranger missions at Dieppe, Arzew and Gela. What made Pointe du Hoc's unique (besides its importance) was its niche within the largest amphibious invasion in history. The Ranger force would come ashore in a few dozen landing craft among thousands that plied the choppy waters on D Day. Three airborne divisions would have already been fighting for hours behind the Allied beachheads. A titanic shore bombardment by six battleships, 20 cruisers, 68 destroyers, and 10,000 warplanes would have already plowed the soil of Normandy, while a dense flotilla of invasion vessels stood off shore, filling the sea from horizon to horizon and the sky with streaking rainbows of huge shells. Maritime approaches to the beaches would be jammed with troops from two infantry divisions assaulting concurrently with the Rangers. All of which makes the point that except for their daring objective, they would not have their most valuable ally, the element of surprise; nor could they exploit their highly developed skill in maneuver because there would be no room for maneuver while confined to a few hundred yards of sand.

So to succeed the Ranger assault plan had to be perfect. Perfection is the case of Pointe du Hoc meant enough flexibility to allow for changes and contingencies that would surely arise from the stupendous amount of coordination and number of moving parts involved in the Omaha operations—even before considering the inevitable effects of enemy intervention and vagaries of weather.

The plan was perfected within this command structure:

The US landings were under Gen. Bradley's First Army controlling VII Corps (LTG Collins) at Utah Beach and V Corps (LTG Gerow) at Omaha. Along side the Rangers on the beach would be a regiment, the 116th of the 29th ID, plus most of the Big Red One, shortly followed by the rest of 29th ID. The 2nd ID was scheduled to land on 7-8 June; but not until 10-13 June would the 10th Armored Division enter Normandy. That meant no armor to support the Rangers except for a few experimental 'amphibious tanks' on which Rudder, CO of 2nd Battalion, could hardly rely.

He would also be commander of the entire Ranger force for Overlord, beefed up by artillery/naval gunfire teams and an company of amphibious tanks. His deputy was Schneider who retained command of 5th Battalion. Overall command on Omaha Beach was with 29th ID's Assistant Division Commander, BG Norman Cota.

The plan divided the Ranger force into three elements, A, B and C. Force A, leading off with D, E. and F Companies in ten boatloads, would be in the first of all waves to hit the beach about 1.5 miles west of Omaha. They were to get up the 100-foot cliffs and take down the German battery: that was Priority Number One and the essence of the entire Ranger mission, so Rudder himself would lead Force A.

Force B was C Company of 2nd Battalion, commanded by Capt. Ralph Goranson. They were to land on the extreme western edge of Omaha Beach to secure the flank of the amphibious tank company landing with them. Then C Company would set out to scale a secondary promontory, Pointe de la Percee, destroy any enemy emplacements on it, before heading west to link up with Force A.

Force C—5th Battalion plus two 2nd Battalion companies under Schneider— were backup if the first wave of Force A failed. In floating reserve until H+1, these follow-on forces were available by the ninth wave. The eight remaining companies would land whenever Pointe du Hoc was secured, then break out to the south to help block German counterattacks from the only hardtop road that ran laterally behind the beachhead.

Such was the plan whose execution was awesomely difficult even on paper, still more so considering the lack of surprise and that the assault was to take place without the Ranger's second most important ally, darkness.

But the two other vital ingredients for success—and a 'good' Ranger mission— were present: ample combat support from other services inaccurate intelligence.

<u>Combat support</u>: The crucial battery on Pointe du Hoc had already been worked over considerably during the aerial preparation for Overlord. Medium bombers had cratered the point and its access roads, forcing relocation (no one knew if temporary) of the guns which had not yet been fortified by casemates that would make them armor-plated pill boxes. But during April and May bombing could not concentrate further on Pointe du Hoc in order to deceive the Germans about the true locations of the landing beaches, or even that the Colvados coast (the name of the region where the American operations took place) was the main invasion site. In accordance with a masterful deception plan that was hoped to draw Rommel's focus on the Pas de Calais, Allied air forces would not return in numbers to the Colvados coast until D-1, at which time heavy RAF bombers would add a final massage on Pointe du Hoc; then medium bombers would rake it once more at H-20 minutes.

For indirect fire on a big scale, the 14 inch guns of the battleship, *Texas*, stood by to support the Rangers after a preliminary bombardment of their objectives. For pinpoint targets, destroyers were prepared to practically run aground if Rudder needed close range fire. Naval support would not be timid as it had been at Dieppe when capital ships declined to maneuver in shallow waters. Everything was on the line so everything would be thrown in by the Allies, especially at key terrain like Pointe du Hoc. There was not a piece of earth more important to the invasion than the high bluffs atop vertical cliffs.

<u>Intelligence</u>: Before departing from England, Rangers down to squad leaders could draw accurate sketches of the main bluff, its cliffs, and all the ways it could be reached—which were very few. Such detailed knowledge resulted from the strategic decision to land on the Colvados coast made by Eisenhower's combined staff in 1943; from then on all intelligence efforts had been exerted to learn the lay of the land and the German dispositions on it.

To study the topography, maps were plentiful and accurate. The Normandy beaches, though for most of the year rather cold and rainy, were nevertheless a tourist site; so postcards and commercial hiking maps showed virtually every significant land feature and trail. Similarly, the beach hydrography was well known from yachting charts.

However, these of course did not show the underwater obstacles Germans had so ingeniously and industriously planted to protect the beaches nor the minefields. Normally, for a typical Ranger amphibious raid, determining beach conditions and hydrography would be a Ranger responsibility with help from specialized naval assets. But for the rush hour water traffic of Overlord, higher headquarters were responsible for selecting amphibious routes. Rudder had to take those assigned to him and hope higher headquarters had done their job.

To supplement good maps, he had almost all the aerial photos he could use to determine German dispositions on the terrain. But one crucial photographic angle was lacking—how the cliffs actually looked from below which was how the Rangers would approach them. Photo reconnaissance planes had made some water level passes at the cliffs; and though their pictures were helpful, they did not show vital details like hand holds and outcroppings that could hold grapnels. Nor was the composition of the rock understood to a satisfactory degree. Finally, a critical question could not be answered—how would the cliffs stand up under the stupendous pounding they would receive from the naval and air forces?

The best Rudder could do to prepare for Norman terrain was to train on the most similar cliffs in the U.K. By oversight or in understandable deference to the British landscape, these cliffs were not subjected to bombardment. This lack of realism proved nearly fatal to the Rangers when they had to scale the deformed and blasted heights of Pointe du Hoc. Long crevices they expected to use for major ascents had been blown away; massive rocks they planned for holding ropes were loosened or had caved into the sea. Because of the devastation from their supporting arms, the Rangers' objectives turned out to be quite altered from what they had anticipated.

Intelligence about the enemy was largely produced by the French resistance, a delicate matter because they could not be told when or where the invasion was coming. Who and how many Germans were around Pointe du Hoc was simply an item French agents were asked to investigate. They did, reporting that the Rangers' stretch of coast was manned by the 726th Infantry Regiment, part of a static infantry division that had occupied the area since 1941, and contained a high percentage overage troops as well as many non-Germans. For example, there were hundreds of Russians who had been captured on the eastern front, and preferred to serve Hitler's dictatorship rather than Stalin's. Others were less willing, and the Germans used them as slave laborers.

Pointe du Hoc itself was garrisoned by the six-155 mm howitzer battery and a 125-man security company. Most of their transportation was horse drawn, so they had participated in very few mobile maneuvers, but instead spent most of their time strenghtening the already 'impregnable' natural position.

To begin with, at the very precipice of Pointe du Hoc, they had constructed a reinforced concrete observation post that would have the best view anywhere of the invasion. Buried land line connected the OP with two guns in casemates and four in open emplacements.

Barbed wire was piled all along the summit, and command detonated artillery shells hung on ropes of varying length so as to explode anywhere on the face of the cliffs. Also a stockpile of point detonating shells was available to be dropped by hand into minefields on the rocky beach below. Probably to protect against an airborne assault, which the Germans believed to be the only likely threat, four antiaircraft machine guns ringed the battery. And in case paratroopers closed from inland, one heavy and two light machine gun nests provided all around defense, connected by communication trenches with overhead cover. Rifle pits and dugouts completed a perimeter that provided interlocking, overlapping fire in all directions. The sector most lightly covered by infantry was the 100 degree arc facing north over the top of the cliffs. A machine gunners' view from there was of nothing except the distant horizon of open sea.

The Germans could not imagine ever seeing the face of an enemy soldier come over those cliff tops. The company commander considered that sector as dangerous only because one of his men might slip and topple to his death a hundred feet below. One of the reasons the CO placed barbed wire on the cliff tops was to prevent such an accident.

So except for sleep-walkers, the Pointe du Hoc garrison probably felt as secure and safe as any Germans on the entire thousand miles of French coastline. Even at high tide they could peek over the cliffs and see mine-tipped underwater obstacles like steel poles and tetrahedrons that would stop and skewer landing craft. These thin-skinned vessels could then be destroyed at leisure with mortars fired from Vierville a mile away, or by direct fire from an 88 rolled out of a nearby cave in what was known as the 'Vierville draw'.

The howitzer battery on Pointe du Hoc was for more important targets like LST's and concentrations of troops farther down the beach in both directions. Fire discipline had been emphasized to the gunners: they were to wait patiently for lucrative targets to come well within their ten-mile range. Moreover, until a huge ammunition magazine could be excavated on the point—a project to which Rommel gave high priority— howitzer shells would not be as plentiful as the Germans might wish.

Below them, the water line at high tide was only ten yards wide, composed mostly of shale shingle, rising steeply to dunes or sea walls built by the French long ago to stop beach erosion. Neither provided cover or concealment from Pointe du Hoc. So commanding were the cliffs that any tanks—if they were so lucky as to get ashore— could not elevate their guns high enough to fire at the cliff tops.

Apart from bombardment, the garrison's chief concern was being encircled by forces landed far down the beach. Relief by German reinforcements was expected from the alert 352nd Infantry Division, one of the best units in France, stationed near St. Lo about twenty miles away. Unbeknownst to LTG Gerow, in June the 352nd was conducting a counter-invasion exercise right behind Omaha Beach.

Because of Hitler's vacillating micro-management of his 'West Wall', panzer reserves were more distant from Omaha. Before counterattacking, Rommel had to hold them back until he was sure where the Allied main landings were. He, and no doubt the Pointe du Hoc garrison as well, reflected that, for the first twelve hours at least, the battle for the beaches would be decided by infantry.

This too had been Rudder's assumption during the months he prepared the Rangers for their mission. No immediate need for an antitank capability like 37mm guns or TD's—a good thing because there was neither space nor terrain for them. As at Gela, fighter-bombers and naval gunfire would have to be relied upon to take out panzer formations approaching the beachhead, though the Rangers would carry a few bazookas in case some tanks broke through.

So Omaha Beach would be one of the Rangers' rare operations where enemy armor was not a major initial worry. Instead, it would be the terrain which, as at El Guettar, was most problematic. The Germans were certainly positioned to make best use of the terrain, but they were a secondary concern compared to the cliffs themselves. Rudder was confident that if the Rangers— and it would not take all of them— could get at those Germans, Pointe du Hoc would be taken.

How to get at them was the crux of preparatory training. In the parlance of professional mountain climbers, the cliffs of Pointe du Hoc are judged to be 'Class 4' grades. Class 6 is the ultimate in difficulty, requiring a full array of specialized equipment like ice axes and crampons. Class 5 is a vertical climb for experts only, usually at altitudes where bottled oxygen is used. Class 4 is for experienced climbers with ropes, pitons, and snap links. There is no class for climbing while under fire!

Rudder's force did not contain any professional mountaineers, and no instructors were available from the Commandos who had their own cliffs to climb on the extreme eastern flank of Sword Beach. Consequently the Rangers had to teach themselves and come up with expedients that would get them to the top. There were two means to do so: by unassisted ('free') climbing and grapnels rocket-fired up the cliffs. Both means were employed, but grapnels were primary with free climbing literally the fall-back method.

From photographic study, there seemed to be no more than four possible grapnel ascents up the cliffs, and maybe one more for free climbing. On D Day Rudder would try them all because he could not count on success on any.

The free climbers, even though roped together for safety, would do very well to get up at all carrying anything more than carbines and extra clips plus a few grenades. The firepower Rudder needed would have to reach the top on rope and/or light metal ladders. Experimentation in England came up with several special purpose types all, except for the metal ladders, attached to rocket-fired grapnels. All ten landing craft in Rudders' first wave were equipped with six grapnel launchers (a pair at the bow, stern, and amidships) for a total of sixty, or fifteen launchers targeted on each of the four routes. Rudder could not have too much redundancy when it came to ladder launching—an indication of the failure rate he expected.

One type of ladder was just a single-strand length of three-quarter inch rope on which Rangers could pull themselves up hand over hand. A second type required a more powerful launcher because the rope was studded with foot-wide wooden rungs that would aid men climbing with cumbersome loads. A third type was an all-rope ladder. At high tide, landing craft were expected to get close enough to the cliffs to reach the tops by firing the grapnels.

But there was backup in the form of shore-fired grapnels attached to lighter half-inch rope to be carried across the beach to the foot of the cliffs. This way too was how the metal ladders were to be used. Broken down into four-foot sections of tubular steel with universal fittings, they could be assembled and reassembled over and over again as the climb advanced toward the top.

Finally, and if all else failed, there were four amphibious DUKW's mounted with extension ladders like those on fire trucks. But instead of water, at the top of each ladder was a pair of machine guns to hose what would surely be the most surprised Germans on the continent: suddenly, quite out of the blue, two fearsome Rangers rear up from below the horizon like the unexpected appearance of window washers on some skyscraper, startling office workers a few feet away. Before the Germans can say, "*Gott in Himmel!*", they perceive that the window washers are not equipped with squeegees but rather hold machine guns and are in a hurry to fire them!

The 'deadly DUKW's' were comic book stuff, and positions on them were not sought after by Rangers who realized that if success of the mission depended on such bizarre contraptions, Pointe du Hoc would be a lot tougher even than it looked. It was tougher because the cliffs could only be reached by window washers or grapnels if the Rangers got through two types of mines.

At sea, many of the Germans' underwater obstacles were mined. Gen. Bradley had no means or plans to do anything about that—landing craft would simply have to take losses with the hope that enough of them would get through to secure Omaha Beach. In a morbid sense, the first waves would be mine sweepers for later waves. There just wasn't any other way to get enough force ashore in the short time required.

As for land mines, Rudder had a general idea of their location from information provided by French agents. One minefield was believed to be squarely in front of the four ascent routes. Like Bradley, there was nothing he could do about that except advise the Rangers that their brothers faced the same situation at Gela and yet accomplished their mission. If it was any consolation, the beach—and hence the minefield—was narrower at Pointe du Hoc.

Narrower even than anyone expected, because due to a climatic quirk, the tide on D Day would be considerably higher than predicted. This unsuspected fact, plus the misknown whereabouts of the 352nd Infantry Division and the alteration of the cliffs from bombardment, would be the wild cards in Rudder's game plan for Omaha Beach. Otherwise, even in hindsight, the plan was as sound as it could be and rehearsed to the greatest extent possible. Victory, if it could be attained, would be within a slim margin of time and terrain. By rational measures, defeat was likely; and if defeat was the Rangers fate it would not be marginal but on a scale of disaster that would surpass Cisterna.

Such thoughts must have been in Rudder's mind when, on the eve of D Day, he heard Field Marshal Montgomery's recital of a verse by the Duke of Montrose:

He either fears his fate too much,
Else his desserts are small,
Who dares not put it to the touch,
To win or lose it all.

It was surely win or lose it all for Rudder's Rangers. Darby's had lost. It was for 2nd and 5th Battalions to change the Rangers' luck.

As they rolled toward their marshalling area on D-5, the Rangers could also feel how dice were about to be rolled. Once the dice struck and tumbled, guts and grit would take over the outcome of generals' plans. What was vital was to keep the plan in mind while shells and men were screaming. The objective of the plan, as every Ranger in Force A had to and indeed did know, was to put about fifty men atop Pointe du Hoc before the Germans could effectively react.

Because of anticipated casualties:

Fifty men meant two hundred starting up the four ascents.

Two hundred at the base of the cliffs meant three hundred Rangers coming ashore in the first wave.

Three hundred ashore meant four hundred starting out in landing craft.

If this, Rudders' force, was insufficient it would be for Schneider with all remaining Rangers to assume the mission.

Almost a perfect analogy for the Rangers' plan is a mountain climbing expedition. A base camp is set up requiring the labor of a large party. Then advance camps, successively smaller, are established higher on the slopes until, finally, a few climbers make a dash for the summit. The top climbers are reserved and rested for the summit assault, but for the Rangers everyone had to be capable of doing it because no one knew who would be lucky or who would be lost in the earlier stages.

Such expected attrition was the basis, as noted, for the redundancy Rudder built into his plan. Redundancy included, indeed emphasized, commo especially from shore to ship. Schneider had to know the outcome of Rudder's initial efforts to take Pointe du Hoc: whether 5th Battalion should take over the mission or go for secondary objectives after landing on Omaha Beach.

To assure that the two commanders could communicate, LT James Eikner, commo officer for 2nd Battalion, was prepared to use everything but semaphore—a means he wished to include, but the navy couldn't spare any flagmen. However, in addition to several radios with alternate frequencies (very scarce in so huge an operation as Overlord), there were pre-designated 'blind' transmissions, a variety of visual signals, bull horns aboard landing craft, couriers, blinking lights in Morse Code, and even carrier pigeons.

From nearly six months of unremitting preparation, all that could be done or anticipated, had been accomplished by planning and training. With very few exceptions, every Ranger felt ready for the expected situation. It was grim, even grisly, but it could be imagined and thought through in private moments:

"If I make it to the beach I go left and look for that crevice shaped like a lightning bolt. I look around for the grapnel rocket Sgt. Jones is carrying. If he's dead, I find it. Go through the firing procedure just like we did a hundred times in training. Shoot it 20 degrees above the where it should catch. If I'm the only one around to go up, I go up. Others will see me and follow."

What this typical Ranger was not ready for was the totally unexpected. No one is, ever. Veterans have experienced the stomach wrenching sensations before, but it emphasizes the powerlessness of any soldier caught in a catastrophe, and the God-bestowed luck if he is extricated from it. This reality was probably the basis for the famous statement by Bill Mauldin, the foremost artist/cartoonist of World War II:

"Look at an infantryman's eyes and you can tell how much war he as seen."

Veterancy is an advantage before any combat, much more so before a mega-operation like Overlord. But of the Rangers to go ashore on Omaha Beach and Pointe du Hoc, however, few were veterans—indeed almost all were cherries. Well trained cherries, to be sure, but none who had seen a white-hot muzzle flash from in front of an enemy weapon. Professional, to say nothing of personal, psychological pressure was more than extreme for the D Day Rangers— it was unprecedented in American history. They were rookies whose first day in the big leagues was the seventh game of the World Series. They had to step up to the plate.

Among them, only a handful of veterans like Schneider felt they were as ready as they could be. That is as much as any Rangers, before or after them, could hope for. After that, each was face to face with high-speed death on the most personal terms, with only a bedrock of faith in the men around them—and a hope that the Almighty would let him live beyond D Day.

Things would go wrong, maybe badly wrong, but the plan was flexible enough to allow for failure in its components. Only the objective was immutable, and it could be reached by either Force A, B, or C, singly or in any combination if Rangers could keep moving over the sand and think on their feet.

Overlord was probably the most secret operation ever undertaken, but paradoxically the whole world knew it was coming, approximately when but not where. In the United States, news of the airborne landings were flashed in the late evening. The next morning, businesses and offices closed so that Americans could stay by their radios. In cities, minute-by-minute reports were read on moving lights watched by standing crowds.

Sporting events were cancelled, schools sent kids home, while churches and synagogues opened for constant services and communal prayer. In Philadelphia the Liberty Bell was rung for the first time since it cracked in 1835, followed by the tolling of more bells all across the land. At dusk in blacked out New York harbor, the torch of the Statue of Liberty was lit for fifteen minutes. In Columbus, Ohio, all traffic stopped for five minutes for people in the streets to pray to God.

God seemed to be withholding judgement. A wild storm in the English Channel forced postponement of D Day which was originally scheduled for 5 June. That same storm, however, convinced the Germans that the threat of imminent invasion was off. Rommel even believed he had enough time to visit his wife in Bavaria. His absence during the first hours of the longest day would be a factor in its outcome.

With control of the North Atlantic, and weather stations in Greenland, Eisenhower's meteorologists had better information than did Hitler's. The Allied forecasters predicted a 36-hour break in Chanel storms beginning around midnight on the 5th, a narrow weather window that Eisenhower had to get through or put off the invasion for perhaps a period of months. Hitler's forecasters could not see that window. After the most agonizing decision of his life, Ike ordered the invasion to proceed on 6 June. On four British and two Belgian ships, the Rangers set sail at 1630 on 5 June.

British correspondent, Alexander Baron, described what the passage from England to France was like:

The men felt the ship begin to vibrate—it was underway. The coast moved slowly past them, the men looked at each other meaningfully, then bent again to their card games.

Before evening the entire convoy was on the move, the endless lines of ships easing slowly to sea while Spitfires from airstrips which were plainly visible back on shore took up their swoop and snarl overhead.

A couple of hours later they were out in the open Channel. Some men stood listlessly at the rails, watching *[English headlands]* recede till they were indiscernible in the dusk. There was a strange, flat feeling, as if life had been suspended between two planets, while the ship crawled with maddening slowness across the grey, bleak sea that seemed merged with the grey, bleak sky . *[Then]* walls of purple dusk crept in from the far horizon. Nothing now could be heard but the thudding of the engines below decks, and the sound of weaving Spitfires in the clouds that thickened as the frightening night closed in.

Through this overcast night, soldiers on

board could hear and at times see the airborne divisions winging to their drop zones. The last of the waiting would soon be over. The Rangers turn would be next. Robert Capa, a war photographer, described their thoughts:

The men were assembled on the open deck. The invasion barges were swinging on cranes ready to be lowered. Waiting for the first ray of light, we stood in perfect silence. Whatever men were thinking, it was some kind of prayer.

One of the features of Omaha Beach that made it the toughest in the invasion was the long, gradual slope here of the continent out into the English Channel. This required landing craft to be launched ten miles from shore, and had allowed the Germans to emplace their widest strip of underwater obstacles. The first wave or Rangers lowered away at 0430, about fifty minutes before first light. But there was plenty of flashing light in the sky from the blistering naval bombardment, which hardly paused except for fleeting minutes when attack aircraft went in to add their devastation.

For the soldiers of D Day, the thunderous naval gunfire brought cheers and reassurance. But shortly the navy advised Gen. Gerow of a serious problem: the unusual tide (at its lowest ebb at H Hour) was running laterally onto Omaha Beach, from west to east with such strength that gunfire ships had to steam into the current if they were to stay on station. Gerow was advised to expect dispersion and deflection by landing craft that lacked sufficient power to head straight for their assigned beaches. This caution proved to be almost disastrously accurate.

At first, for the troops hunched in their landing craft, disaster seemed the fate of the Germans as the shore winked and erupted with shell bursts, but no return fire. The Americans' problems appeared minor by comparison—just seasickness from the nauseous up and down of their vessels breasting the choppy tide. Not enough daylight had seeped onto the invasion scene for the boat commanders to determine if they were on course for their assigned beaches.

With Pointe du Hoc as the most conspicuous landfall, Rangers were first to notice that their boats were off line. Tense confrontations with coxswains began to develop even before the Germans opened up. When they did—following standing orders not to fire until targets were well within range—boats immediately began to veer away from where water spouts were thickest. Rudder was in the lead LCA with ten more guiding on him, but his coxswain was drifting east of the Ranger beach. Rudder ordered the course corrected, presenting a dreaded dispute that had not been entirely obviated in planning. He was the commander, empowered to designate the landing site; but the coxswain commanded the vessel and was responsible for its employment. Rudder's coxswain acquiesced though the Rangers' heading pointed straight into harms way while portions of Omaha Beach were under less fire. Other coxswains had to be convinced by pointed pistols of Ranger boat commanders.

Lest the sailors be faulted for lack of determination, the record shows that most pressed on over mines and underwater obstacles, through shot and shell in the best navy tradition of 'Damn the torpedoes—full speed ahead." Across ten miles of open sea, they could watch other brave boats swamp, sink and destroyed—as reporter Irwin Shaw described:

Perhaps the worst beach was Dog Green, directly in front of [Pointe de la Percee] where strong points guarded the Vierville draw with heavy flanking fire from emplacments to the west... Company C of the 2nd Rangers came in on that target after passing an LCA from the First Infantry Division that foundered, the men jumping overboard and being dragged down by their loads...

[Next to C Company] at H+6 minutes, the remaining First Infantry craft grounded in water four to six feet deep, about thirty yards short of the outward band of obstacles. Starting off the craft in three files, peeling left and right, the men were enveloped in accurate and intense fire from automatic weapons. Order was quickly lost as the troops attempted to dive under water or dropped over the sides into surf over their heads. Mortar fire scored four direct hits on one LCA which disintegrated. Casualties were suffered all the way to the sand, but when survivors got there, some found they could not hold and came back into the water for cover, while others took refuge behind the nearest obstacles... Leaderless men gave up any attempt to move forward and confined their efforts to saving the wounded, many of whom drowned in the rising tide. Some troops were later able to make the sea wall by staying in the edge of the water and going up the beach with the tide.

The small 64-man Ranger company, carried in two LCA's, came in to the right of the Vierville draw. Shells from an anti-tank gun bracketed Capt. Ralph Goranson's craft, killing a dozen men. An enemy machine gun ranged in on the ramp of the second LCA and hit fifteen Rangers as they debarked.

Without waiting to organize survivors, sections set out immediately across 250 yards of sand towards the base of the cliff. Too tired to run, they took three of four minutes to get there, wounded men crawling behind them, and few made it. When the Rangers got to the shelter of the cliff they had already lost 35 men.

[Elsewhere] an LCI carrying the alternate CP of the 116th RCT was struck by artillery fire as it made a first attempt to get through the obstacles. Backing out, the craft came in for a second try but it could not get past. The ramps were dropped in six feet of tossing water, then an artillery shell hit the crowded forward deck, sending up a sheet of flame. Clothes burning, men fell into the sea and tried to swim in under continued artillery fire.

Shaw's comment about how the Rangers were "too tired to run" is surprising in light of the fact that they were among the best conditioned soldiers in any army. With every incentive from German bullets to get to the cliffs as fast as possible, no doubt their legs were working at top speed but this exertion only produced what appeared to be walking rather than the sprint that was intended. Seasickness and cold water immersion had chilled and weakened the Rangers while the soggy sand and heavy loads of ladders and grapnels further slowed their top speed.

Not even their hard physical training was enough to prevail over these unique factors and the debilitating exhaustion of vivid combat. Marathon runners have a term for the similar sensation when all energy is expended while the will persists to keep going: it's called 'hitting the wall'— the victory of matter over mind, a singular illustration of how 'the spirit is willing but the flesh is weak.'

Like hitting the physical wall from glycogen famine, there can also be muscular paralysis from horrific psychological impacts like those experienced by Force B during the dawn of D Day. In the rip tide they had come upon boats of the 116th Regiment impaled on obstacles, their passengers being picked off or disappearing into the grey sea. Through a burning maze of sunken craft, only half of the sixteen amphibious tanks made it to land, there stranded and immobile as an 88 came forward to finish them off. Under fire themselves all the way in, the Rangers became prime targets of mortars, the 88 and a torrent of machine gun fire. The ramps were dropped, exposing the men tightly bunched together to direct automatic fire. Flesh exploded from heavy impacts. For every man it seemed he had reached the last minutes of his life, so minds as well as muscles were paralyzed.

So in the watery hell between ship and shore, Force B had reached their physical and psychological limits even before they engaged in the combat required by their mission. It took the utmost in leadership to refocus them on what they'd trained for and come to do. Like that of platoon leader Sid Salaman, blown off his feet by a mortar concussion and sprawling in the frigid surf. Reeling like a drunk, he staggered forward toward a group huddling under the cliff of Pointe de la Percee. They needed someone to tell them what to do. Salamon, his ears still ringing, did not know for the moment but realized that when his head cleared something would come to him.

Before Force B began anything that deserved the description of 'maneuver', they instinctively and intrepidly started in the right direction—up. Enemy fire was from the high ground. Get around it if possible, get at it regardless, and take it out no matter what. Because someone had to cause the Germans to start defending their lives rather than just spray, with endless ammunition, the landings where cowering troops on the beaches had yet to return effective fire.

Ranger lieutenant Bill Moody, two sergeants and a PFC began a new form of free climbing, using bayonets to jab their way up a crevice where so much rock had been blasted away that a marble pattern of clay had appeared in the shale. They had to negotiate ninety feet of height, but a considerably longer climb because to follow the clay required several traverses. The 'bayonet method' (never contemplated in training) amounted to scores of one-armed chin-ups by each Ranger. Pants became gasps, and none of them believed they could get to the top, but wouldn't stop trying so long as their buddies were.

They were watched intermittently by Rangers on the beach, scrambling through deadly chaos to position grapnels and assemble ladders, though enough ladder sections had not made it ashore. If the five climbers could reach the top they could grab ropes fired up to them. Only three hand-fired grapnels were present, such had been the attrition between ship and shore. Neither of C Company's LCA's reached the beach, so grapnels fired from them would have been of no use.

Goranson was beginning to get a picture of what his survivors faced. Platoons were mixed and riddled with casualties, but everyone else was taking up the slack by working twice as hard at two activities: either attempting to get up the cliffs or blocking German patrols from coming onto the beach from the Vierville draw. Ranger guns were angry and accurate, the first though tentative sign that American infantry could begin to trade punches in what so far had been an utterly unequal fight from the surf line to the fringe of Omaha Beach.

Somehow adrenaline, fright and desperation supercharged quivering muscles till 'the Moody five' crawled exhausted over the lip of the crevice. They were the first Americans on high ground. From here on, the Germans would have to turn some attention from the slaughter they were inflicting on the beach below.

Two grapnels were fired up toward Moody who then anchored them by hand. The remainder of C Company started up. Goranson, who till then thought he would have nothing but Ranger casualties to report, gratefully set up his radio in a shell hole near the lip of the 'bayonet' crevice, and called Rudder with the news that half of Force B (that was all that was left of it) were on Pointe de la Percee and about to move out to their objective.

That would have been encouraging news for Rudder if he had received Goranson's message, but he did not because Force A's landing craft were still at sea dodging German shells: part of a traffic jam like that caused by a multi-vehicle collision in fog on a highway. LCA's were milling off shore in fearful confusion.

This was the longest hour of the longest day, a situation of imminent disaster such that Bradley seriously considered ordering Gerow to move over to Utah Beach.[*FN: The 4th ID's landings at Utah Beach had gone off virtually without a hitch, primarily because they were not dominated by cliffs as at Omaha, and also because the US airborne divisions had already secured the beach exits.]*

Rudder (an appropriate name for the occasion) had forced his coxswain to make for Pointe du Hoc despite German fire and Allied confusion, violent tidal current, haze and gunsmoke of naval bombardment—all factors not confronted in training. In general, these many negatives had caused the Omaha landings to slip east. Fighting the tide, the boats of Force A wallowed toward their landfall. As they did, one DUKW was hit by artillery, killing five Rangers and two crewmen.

Fire from Pointe du Hoc was so heavy that two destroyers moved in to soften it up, further delaying Rudder's landings by a half hour, a period during which the Germans recognized that their 'impregnable' position was targeted for a frontal attack, thereby forfeiting Rudder's hope for tactical surprise.

What's more, though headed in generally the right direction, a British guide boat began leading Force A toward Point de la Percee rather than Pointe du Hoc—a difference of over a mile. The second required change of course lost more time during which naval gun fire lifted as scheduled, allowing the Germans thirty minutes of respite to man their defenses.

Rudder's final mishap at sea was in Force A's arrival at the beach. Havoc amid underwater obstacles caused the touchdown of his nine boats (one never made it in) to be on the eastern flank of Pointe du Hoc rather than on both sides of the tip as planned. Some beached, some broached in the surf—all disgorged Rangers into sand and water peppered by fire of all calibers. Death was everywhere, but at least the Rangers were no longer in the maritime shooting gallery, and were relatively grateful to have finally reached their element where they could take some counteraction against the enemy who had things all his way so far on D Day.

Force A was strewn along rocky beach 500 yards long and thirty wide. They scrambled into the many large shell holes for protection, finding the pits so deep that instead of sand the sides were clay toward the bottom, making them difficult to climb out when Rangers tried to dash forward. A few disbelieving Germans at first poked their heads over the cliff tops; after one was picked off and fell like a Hollywood stunt man down to the beach, Germans were rarely seen and their indirect fire, though professionally pre-registered, was not effectively adjusted. The Rangers found that the closer they could get to the cliffs, the safer they were from the only effective fire which came from both flanks about a quarter-mile away. Rudder soon found a little room for his CP in a small, shallow cave, then began to appraise the situation.

By occasionally sweeping the cliff tops with fire (particularly from Rangers who remained with crew served weapons on their landing draft) that drove off enemy patrols, the Rangers had little to fear from shells dropped on them by hand. The artillery shells the Germans had suspended by ropes had long since been blasted away by naval bombardment. Indeed, the cliffs no longer resembled those studied so meticulously in training. The four potential ascent routes were a memory. Rudder immediately began directing his jumbled companies toward new climb sites.

All landing craft not still afloat had beached too far from the cliffs to make use of their grapnel launchers, though several tried. Two LCA's that had not come ashore bravely maneuvered closer through the swells, then fired eight grapnels in succession. The heavy hooks lofted listlessly, fell scraping the cliff faces, but none reached the top or caught on the sides.

Captured Nazi Flag from Maizy Battery near Point Du Hoc, D+4, 1944 (Normandy). Pfc. Edward Rutkowski, Sgt. Paul Lutick and 1st Lt. Jay Mehaffey, 5th U.S. Ranger Bn.

Looking west from 5th Ranger Position, Omaha Beach, H-Hour, D-Day, June 6, 1944.

The Rangers immediately—that is concurrently—began the fall-back method of hand-fired grapnels. From the sloshing ride to shore, ropes were water logged and electric wires shorted. Sgt. John Cripps had to stand amid sand-kicking machine gun fire and hot wire a launcher. It ignited, and the grapnel rose—but the soaked rope was too heavy. The grapnel fell short of the cliff. Cripps tried with another: up it went and was the first to stick.

Encouraged by Cripps' success, Lt. Bob Arman, commanding an LCA that had turned back from the beach, steered to another location where his men, under increasing enfilade fire from the east, were able to detach four shore-mounted rockets and wrestle them to the foot of a cliff. Hand-fired, all four caught and held.

Lt. Jacob Hill's boat also made second run at the cliffs (no Ranger boats ever gave up and returned to the fleet), and four of his ropes held. But they were exposed to deadly fire that wounded three of his men before they could start to climb. Circling off shore and providing fire support, was CPT Otto Masney's boat, dodging through water spouts and the traffic jam of 1st ID landing craft. When he saw Arman's and Hill's ropes hung on the cliffs, Masney ran into the beach, ordering his reluctant coxswain not to fire his grapnel-rockets until the boat was firmly grounded. This was done, and five more ropes draped down from Pointe du Hoc to the cheers of Rangers all along the beach.

Ten ascents were now possible—the other ropes were too intensely covered by fire—and now a battle broke out for the future of three perilous ladders. Small groups of Germans tried to get at the cliffs to cut ropes or dislodge grapnels; about half the ropes and half the Germans were cut down. Rudder had to get men to the top or his lifelines would be gone soon after they had been put into place at the cost of so many lives and wounds.

The Rangers did not need his urging, and surged toward the ropes as if there was a contest to see who would be first on top. Hand over hand, feet kicking the rock like a reverse rappell, men, then sections, then squads worked their way up like ants on a sticky straw. At places they were aided by huge mounds of clay and rock—some twenty-feet high— where the cliffs had crumbled under naval bombardment. The mounds became platforms where steel ladders could be used.

The Rangers daring progress was the only faint sign of potential success on Omaha Beach which was described this way by writer, Paul Fussell:

Men were so scared they lost control of their bladders and bowels. Leaders discovered that their water-soaked radios didn't work and that the noise of naval bombardment kept shouted orders from being heard at all. Men carrying explosives off the ramps were hit and blown up; lucky ones struggled through a red foam to crawl onto the beach...

A glance out to sea and to all sides indicated utter failure: in the waves as far as sight could reach, half-sunken ships, disabled landing craft scattered among floating bodies. On the edge of the beach, a litter of smashed radios and weapons, abandoned packs and helmets, gas masks, life preservers, bits of uniform and bandages, together with pocket-size New Testaments, mushy cartons of cigarettes—and a guitar. Washing back and forth in the shallow water were hundreds of bodies and body parts: heads, torsos, arms and legs.

Amid such carnage, Force C was awaiting news that would direct them either to follow Force A and assume the mission of seizing Pointe du Hoc, or head instead to set up a blocking position behind it that would cut off German reinforcements. It was Schneider's decision, but he had no information on which to make it.

Despite all the multiple means of communication, word had not been returned from shore that Force A had or was about to handle their mission. Actually the case was the latter, but Rudder's Rangers were nearly an hour behind schedule. Salt water had caused his radios to sputter, signal lights on his LCA's weren't working because of dead sailors or battle damage, and only God knew what had become of the carrier pigeons. On various channels Rudder tried blind transmissions but none of his messages were acknowledged. In fact, only one was received and so garbled that it told Schneider nothing. All Rudder had which he could count on was flares, and they could only convey a signal message—that Force A was handling Pointe du Hoc. At the time Schneider had to know that, Rudder's situation was still too uncertain.

With powerful naval binoculars, Schneider studied the cliffs for signs that Force A was mounting them. His tossing boat, shell bursts, gunsmoke and early morning mist obscured Pointe du Hoc too much for him to be sure. He could not keep Force C together at sea much longer. The entire Omaha landing flotilla had become stretched and bunched like an accordian as boats were diverted, delayed and generally mixed up under the giant patter of German guns firing all along 6000 yards of coast line. Schneider had to rely on his instincts, and they told him that Force A was doing the job. He was ultimately correct, and his was one of the most important tactical decisions made on D Day.

It was made as time and tide were running against him. The tide was rising, beginning to submerge mine-tipped underwater obstacles that soon could not be detected by coxswains. Every minute under fire at sea meant more landing craft hit and men killed. With arm and hand signals, Schneider ordered his boats to turn away from Ponte du Hoc; his Rangers came ashore on Dog White, a few hundred yards east of where the 116th Regiment was pinned down and being slaughtered at the water line.

Force C's location turned out to be a godsend because the unexpected Ranger landing soon took pressure off the 116th by engaging the Germans all along Dog Beach which became their most threatened sector, the place where they directed reinforcements needed to hold back the 1st ID forces pushing onto the Easy Beach.

The first wave of Force C came in on six boats; one was shattered by a mine, but the others grounded in surf that was not above chest depth. As the ramps went down, Rangers could hear bullets striking their last metal shield of protection. Force C stormed out, hitting the water and being hit by direct fire. The boat that struck a mine emptied in deeper water, the men jettisoning all equipment and even their weapons in order to flounder ashore, which most of them did. The LCA continued to float, broaching in the surf with several Rangers still aboard who had been knocked unconscious by the mine's explosion.

They were among only a few Rangers who had temporary safety. Leaders were at the front of their boats and so met the first focused bursts of fire. A and B Co., 2nd Bn. had a commander, four platoon leaders, and a sergeant major killed in the first minutes of landing. Schneider was struck on the helmet by a mortar fragment. His ringing ears caused him to wonder if landing here was the right choice. He signalled for his second wave of seven LCA's to land slightly east. One swamped when it struck an obstacle, but by 0740 Schneider had 450 very determined 5th Battalion Rangers ashore and leaning forward to do battle.

In small groups they rushed for the intermittent seawall, which here was more than a thousand yards from Pointe du Hoc so was only subject to enfilading fire. Besides, 2nd Battalion was giving the Germans on the cliffs all they could handle. Thus, the initial encouraging developments on Omaha Beach were beginning to take shape while Schneider took a first look inland:

A mile down the beach to the east, a few tanks had finally made it to shore, and were providing some fire support for 1st ID. What the situation was on Pointe du Hoc and Pointe de la Percee, Schneider still didn't know; but it was clear to him that Force C could provide them maximum assistance by moving out for Vierville, blocking reinforcements, and possibly help take Rudder's objectives from the rear.

Schneider was beginning to execute this maneuver—actually a contingency, only starting from a different point than had been planned—when a party of Americans came dodging up the beach from the east where the battle was still greatly in doubt. To the surprise and admiration of the Rangers, that party turned out to include BG Cota who was ground commander for the initial Omaha landings. Huddling in a shell hole CP with his company commanders, Schneider was pointing out objectives for their first attack when Cota appeared, standing erect despite spurting sand and mortar rounds landing uncomfortably close by.

Schneider first gestured for Cota to join him in the protection of defilade, but the general seemed to be making some point by exposing himself to enemy fire. Not wishing to appear any less courageous or confident, Schneider arose and saluted—while hoping for a <u>very</u> brief conversation...

"What outfit is this? "asked Cota belligerently, as if he didn't already know.

"Fifth Rangers, sir!" Schneider shot back.

"Well goddamit, if you're Rangers, lead the way!"

Schneider had only to nod to his company commanders, then the Rangers rose in waves—to the delighted surprise of Cota who, in trying to be dramatic, was treated to a drama himself by watching Force C move out smartly as if they had just been waiting for the word. Their mortars began to thump, pumping out shells to cover the advance, while machine guns on the flanks provided fire for what seemed a well rehearsed maneuver. Cota was agog—some reports were that he had tears on his face—and Rangers forever after had a motto born on the day of their most famous triumph: RANGERS LEAD THE WAY!

With Rangers in the lead, other good soldiers followed, specifically those in the 116th on the beach beside Schneider. Informally, they became a task force of hard fighting infantry and fearless demo men who covered then blasted through rows of barb wire entanglements protected by well sited German machine gun nests. The attackers' first phase line was the road running laterally to Vierville. To get there, 150 yards of flat, open ground had to be crossed and taken in a rush. Schneider felt momentum building, but so too did the grim Germans who realized that now was the time to stop the invasion or it could carry on to Berlin.

The open ground was grassy. In spite of recent heavy rains, the Rangers began to load tracers into their BAR's in hopes that the grass might ignite. If it did, the prevailing wind would blow smoke toward the German positions, causing their fire to be inaccurate. Schneider could not wait to see if this effect was achieved, but pounding German mortar fire from the bluffs ahead caused him to pause anyway. As he regrouped his force for a double envelopment of the bluffs, Schneider heard enemy fire building. But it did not seem to be directed at his men. Apparently, Force A and B were in a helluva fight around Pointe du Hoc and the Vierville draw.

In fluctuating combat, Goranson had the survivors of Force B (about thirty men with various light weapons and little ammo) on the lip of Pointe de la Percee. They were too few in number to take that objective, but fortunately Goranson was able to signal to a company of the 116th who then came up the Rangers' ropes.*[*FN: For easier understanding of the battle, composite groups of Rangers and men of the 116th will still be called Force A, B, or C because they were commanded by Ranger officers. The author would like to make clear that though this narrative is focused on Ranger accomplishments, laurels are surely to be shared with the 116th who fought side by side and with a level of courage that can not be divided.] Even then, the composite force was not enough to kick the Germans off; nevertheless, Goranson could keep them occupied and make their lives very difficult—while ending many of them.

However, Force B continued to take many casualties even after Lt. Moody made the 'bayonet' ascent and opened the lid on Pointe de la Percee. That dauntless officer himself was killed by a sniper within a half hour after leading the way. This happened on one of the many patrols Garonson sent across the bluffs, searching out German strongpoints, harassing their guns that were sweeping the beaches to the east.

The Germans counter-patrolled, touching off fierce meeting engagements around stone cottages that had been reduced to rubble by bombardment. Short of ammo, Rangers picked up enemy weapons and used them with as good effect as their previous owners, though such gunfire created problems for both sides in determining who and when Americans or Germans held certain important positions. These changed hands back and forth, neither side strong enough to hold a fixed perimeter.

Typical of these actions on a battlefield in flux was the one in which Moody was killed. He, another platoon leader, and a half-squad of Rangers probed toward the center of Pointe de la Percee only to find themselves in too deep. Under fire from all sides, Moody was killed and three wounded, causing the patrol to pull back to the crevice. There the patrol leader persuaded Coranson to send out a beefed up combat patrol. Though there were few Rangers available, they had newly acquired knowledge of the terrain. This patrol went out and got the Germans who had gotten Moody—burned them out of a ruined cottage with thermite grenades. Screaming Germans, stumbled out aflame, to be cut down by vengeful Rangers.

Though he could only probe at his objective, Goranson nonetheless shared with the Germans a vital piece of high ground at the mouth of the Vierville draw. They could stop him from seizing Pointe de la Percee but he could stop them from attacking down the beach. The German commander realized this and, as mid-morning approached, he ordered up a reaction force from Chateau de Vaumicel; infiltrated them—in some places using underground passages—to encircle Force B and drive them off the bluffs. Some of the increased tempo of fire that Schneider had heard came from this action that was placing Goranson in his second crisis of the morning.

The rest of the stepped up fire was nearer by from where Force A had gained the cliff tops and were locked in a pitched battle for their bluffs. As Bob Black points out, "Though the climbing of the cliffs has captured the imagination of the public, it was the manner in which the Rangers conducted themselves on reaching the top that best shows their uniqueness."

Thanks to suppressive fire from off shore, and the audacity of the ascents, Force A had taken only sixteen casualties since they'd gone up the ropes. Upon reaching the lip of the cliffs, they scurried like roaches toward any place where they heard fire. They moved out individually, not waiting for platoons, squads or even sections to form. Shock and speed had been the watchwords repeated all through training—shock and speed was what they achieved.

What was inapplicable from training was their expectation of the terrain on Pointe du Hoc. Big bombs and huge naval shells has gouged, twisted and tortured every feature, including the howitzer emplacements. Peeking out of shell holes, the Rangers could see no sign of the battery. Had it been blown into unrecognizable pieces or had the howitzers been moved permanently? If the latter, they could be in alternate positions and Force A still had to get them at all cost. The search for the guns began the moment the first Ranger reached a grapnel and laid down his rope.

For the dazed German company defending Pointe du Hoc, Rangers seemed to be coming at them from everywhere except the rear. Shuttling between shell holes and along collapsed trenches, searching for demo, looking for more linked ammo, pointing out pockets of Germans, and flinging grenades between short burst, Rangers swarmed and stung like hornets. The Germans gave as well as they took, so that bullets crisscrossed the bluffs in many-sided firefights.

The first objective to fall to the Rangers was the concrete OP at the very tip of the point. It was taken by an assault so hasty that Rangers came at it from two sides simultaneously without knowing of the others presence. This was no mix-up, but rather aggressive initiative for the purpose of speed and shock.

When they saw boats come ashore below, the Germans had added a machine gun and rifle security to the artillery observers manning the OP which consequently had weapons firing from every aperture. After shooting off the OP's radio antenna, Rangers chipped away at the apertures with suppressive fire. No satchel charges were initially available, so no one tried to get on top the thick concrete roof of the OP. But grenade after grenade was heaved up by Rangers clinging to the steep rocks with one hand while throwing with the other. Eventually three grenades found their marks and the German machine gun went silent.

But other apertures still blazed with fire, and grenades could not neutralize them. Rangers fell back with head wounds as they tried to crawl closer. Luckily a bazooka man was found among men who were arriving on the ropes, and he was rushed to the best firing position, temporarily preempting a BAR there. From about fifty yards away, a hit on the OP would be good shooting; to put a rocket in an aperture would be miraculous. The first shot sailed over the OP and the cliff. The bazooka man steadied his tube on a notch in the rocks and fired again. This time an aperture was hit and smoke bulged from the others. Rangers closed in like filings to a magnet.

From the other side, another group of Rangers was also ready to storm the OP. Minutes earlier, they had engaged Germans even before reaching the top of the ropes, driving the enemy to cover. But from that cover, the Germans continued to lob grenades down the cliffs at the train of climbers. Though most of the barb wire entanglements had been blown away by bombardment, clumps of it held up Rangers trying to get at the German grenadiers. Finally, a BAR man and a buddy managed to wriggle around to where they could see where the grenades were coming from. The buddy rose for a second as a decoy. As the Germans opened up on this new threat, the BAR man fired two 20-round magazines into the position where later three dead Germans were found with grenades in their hands.

As if it were a coordinating signal, which it was not, the bazooka hit on the OP launched assaults from both sides. Of course the Rangers could not force open the massive door, but they pushed grenades through the apertures till they were sure no Germans were left alive inside. Fumes from seven decaying bodies drifted from the OP all day until enough demolitions reached the top to blow the door open.*[*FN: Inside, Rangers found a highly valuable code book for the German Navy.]

With hardly time for a handshake, the Rangers rushed on to help with other objectives.

This was typical action on D Day: overcome resistance and move on till more comes along; organize and reorganize on the move.

As Bob Black described it, "As small groups encountered one another, they flowed together as raindrops become rivulets then streams."

The river master was Rudder who had now displaced his CP from a cave at the foot of the cliffs to a huge shell hole at the top. He liked the sound of the back and forth battle as his Rangers began to fight on more even terms. Those who had lost their weapons when their landing craft were sunk or swamped became sherpas constantly hauling ammo up the cliffs and helping the wounded down.

Most WIA's had to bandage and treat themselves. No place on the top was safe enough for an aid station but there was now some shelter under overhanging cliffs. The beaches were strewn with dead and moaning wounded, some bodies bloated from drowning, others mangled on shore by fire from the cliffs. Ironically, medical supplies were plentiful on the bloody sands because so many medics had been killed on the beach. Any landing craft heading back out to the fleet was used to evacuate the seriously wounded. There were scores of them and now a dozen dead on the bluffs, but Rudder could not yet evaluate Ranger losses. Not until he found those howitzers, the guns of Pointe du Hoc.

Though that objective was far from being secured, Rudder launched the flare signal for Force C not to assume Force A's mission. Schneider never saw it. He was already inland, regrouping to attack across the burning grasses and seize Vierville. Goranson saw the flare signal: a great relief for him. He now realized that he would not have to fight his way over to Pointe du Hoc to help out Rudder. Instead force B would contribute most by holding what they had and contesting control of the mouth of the Vierville draw. To do so, he created a sort of roving defense, probing here, setting up a strong point there but moving it if the Germans threatened elsewhere. In so doing, he simulated the strength of a larger force, much the way Murray had done at Castellammare.

Goranson began to feel that the Germans no longer held the whip hand. Their confidence now seemed subdued as if they were growing uncertain that the invasion on Omaha Beach would be flung back into the sea —though at noon that report was relayed back to Berlin by the German Seventh Army.

So far as Rommel knew, as he raced back to France, the points and Vierville were still in German hands; moreover, their strongest defenses were designed against an airborne attack from inland, so the Americans had not yet seen the worst he could do to them. Unbeknownst to him, the Rangers had taken on the worst of the terrain and were moving closer to victory.

For Force A, an anti-aircraft emplacement, secured by remnants of a German platoon, was the last tough nut to crack on Pointe du Hoc. Naval observers had been so sharp that Rangers had taken no casualties from friendly fire at sea, so the guns of a destroyer were called upon to help with the anti-aircraft position. Rangers pulled back a bit to see if the pin point target could be eliminated. Deflection was slightly too low for the destroyer, however; though its fire was helpful, Rudder had to move his infantry back in and root the Germans out. Some were deaf when captured, their ears bleeding from concussions.

Schneider noticed where the destroyer's shells were roaring in. From monitoring the naval fire direction net, his own gunfire shore party confirmed that Rudder's target was an AA gun which Schneider knew was the most inland on Pointe du Hoc's defensive perimeter. Greatly relieved that Force A had progressed so far, Schneider was now confident that Force C was doing the right thing by going for Vierville rather than diverting toward Pointe du Hoc.*[*FN: It remains a mystery why radio contact between Force A, B and C took so long to establish, although a mix-up in the SOI of the 116th, who controlled the Ranger net, surely contributed. It was not until the afternoon of D Day that the three Ranger forces became fully aware of the others situation, and that was done more by contact patrols rather than radio.]

He waited only long enough to shift his axis of attack farther south and call in artillery on the bluffs where the Germans could enfilade his right flank. Naval gunfire was still vital but now, though much of it had been lost at sea, some of 1st ID's artillery was ashore and accepting fire missions. Their higher angle fire could be brought in closer to the advancing Rangers. They set out across the burning grass, echeloned right to keep suppressive fire on the bluffs. The lead companies moved straight ahead as skirmishers, double timing to get out of range of the bluffs. They were spurred on by occasional rows of machine gun bullets stitching the grass, fired from the bluffs—for the Germans realized what Force C was attempting to do. Rangers dropped, shot in the side, but their squads kept going.

They were coming upon the first wooded areas encountered behind the Omaha beachhead: clumps of trees, many splintered by bombardment, marking the approaches to the little farming village of Vierville. From there, German mortars ranged in on Force C and interlocking, heavy caliber machine gun fire reached out to them from scattered copses. This was not going to be easy but Schneider saw he had room for fire and maneuver. He called in artillery concentrations on the machine guns and suspected mortar locations while urging his company commanders to close quickly because supporting artillery was not plentiful.

A major worry for Schneider was land mines. If they ringed Vierville, capture of the town would be costly.*[*FN: There were many fewer mines than minefield signs, which were white with a death's head stencilled on them in black. Rommel had ordered that literally millions of mines be planted but his troops were too overworked to reach their quotas. To avoid the wrath of Rommel's inspectors, the Germans often placed signs even where there was no actual minefield. Just the signs themselves, however, slowed the Americans with caution.] Living cows were an indication. The French would have kept them out of minefields, so if there were any cows grazing, they would indicate the way to go. But Schneider could see only bloated bovine carcasses. Had they been killed by shell fire or mines? There was no way of knowing except by close and slow inspection, and there was no time for that. He drove his companies on, spreading them out to the maximum, to develop the situation.

His hunch was, and again it proved correct, that though the German defenses were comprehensive and provided all around security, they were oriented against a main (airborne) attack from the south; whereas, Force C was approaching from the northeast. And even more significant, if Force A was making the headway that had been indicated, they were attacking from due north, right into the rear of German defenses west of Vierville.

This was in fact the situation, and the history of its subsequent development is written by individual Ranger actions. The tide had now run high at sea, corresponding to the onrush of Rangers submerging and overcoming all obstacles to the accomplishment of their mission.

Rudder too had good instincts. They told him that someone—he assumed Force C—was bedeviling the Germans on his left flank as he began to punch south to establish blocking positions. Having overrun the Pointe du Hoc garrison, Force A was about to bump into the stronger reserves who so far on D Day had been trying to reinforce both Pointe du Hoc and Pointe de la Percee, but with marginal success because of horrific bombardment and tac air. Now the fighting was about to be brought to these reserves—an understrength battalion but well trained and totally familiar with the terrain and defensive plan—whose morale remained high in their ability to hold out until a combined arms counterattack was launched by the 352nd Division that would relieve Vierville and punch through to the sea.

That hope was vain because of the Allies' horrendous interdiction, but the Germans, especially those forming to counterattack Point du Hoc, still believed Hitler's promise that the invasion would be stamped out on the beaches.

Rudder could tell that Germans were in no mood to give up as their resistance stiffened against Force A's advance south. Artillery zeroed in on two Ranger companies forming up to attack, while a cluster of farmhouses on west of Vierville was like a porcupine stopping the probes of scouts from another company. Momentum could be lost— seven Rangers were killed and eleven wounded even before Force A could begin to maneuver effectively.

Naval gunfire and Ranger flanking thrusts caused the Germans to evacuate that farmhouse complex, but Rudder could sense they were but an outpost. Beyond was a field 50-yards wide at its narrowest, where Germans took the attack under fire from a trench line with overhead cover. The navy blasted away and succeeded in keeping the Germans heads down (and ringing), while sections rushed the trenches like they would be the last obstacle to ending the war. Once in the trenches, Rangers fired laterally, ducking around turns preceded by grenades, and gunning down defenders if they did not surrender in a microsecond. Unlike Ranger raids of the past, EPW's were not a priority, and few were taken on D Day.

Now a concrete roadblock on the main trail

to Pointe de la Percee confronted Force A. The Germans had placed it to block egress from the beach, and Rudder's Rangers were attacking from another angle, but the defenders had orders to hold at all costs, which they did— at the cost of all their lives. From a hundred yards away, Rangers swept around the roadblock and blasted it from the flanks, with so much ammo that the Ranger company commander cautioned that there was much more fighting ahead and a few rounds into a German was enough.

Opposing artillery was now dueling, not with each other but pounding the other sides infantry. More exposed than the German defenders, the Rangers had to change this aspect of the battle from one of attrition. For attrition would not eliminate the howitzers that remained the ultimate Ranger objective. Rudder and Schneider would drive on, at all costs and to St. Lo if necessary, till those guns were found and eliminated.

The cost grew as Rangers entered rings of mines protecting dug-in machine gun nests. Because they were attacking from the rear of the Germans main defenses, Force A was not confronted by a continuous defensive line but rather a chain of strongpoints. Between them Rangers tried to infiltrate or bypass with fire and maneuver.

Their advance then became uneven: penetrations at some points, halts at others. Aggressiveness could lead into destruction, as once when a machine gun-killer team of eight men probed so far ahead that they found themselves in a pocket, cut off by German artillery and mortars. They were in a small field, pocked by shell holes that had collapsed a German trench line. Indirect fire was pre-registered on them, so they had to spread out to avoid shrapnel, and in this dispersal some lost sight of the others. Then German infantry closed in through a spider web of trenches. Without interlocking fire, the Rangers were further separated and isolated. One by one they were hunted down, killed or captured. Only one escaped.

He came upon another group of Rangers leapfrogging forward and warned them of the trap. They divided into two sections, each skirting the field but then running into German strongpoints. Two close range firefights broke out under blasts by artillery from both sides. Trying to escape bombardment, a German stumbled into three concealed Rangers who captured him at knife point. He then 'volunteered' information the Rangers needed: an overlooked length of abandoned trench that was not covered by fire from the strongpoints.

The three Rangers gathered five more, found the trench entrance, and slipped along until they could no longer hear German automatic fire crisscrossing over their heads. When they peeked up, the German machine guns were looking in the opposite direction. From a few yards away, the Rangers surfaced and blew a machine gun away. The other one turned in surprise, but the last thing they saw was a frag grenade spiraling down. Unfortunately the EPW was also killed in this action.

By such encounters and episodes, Force A pried their way toward the hardtop road from Vierville. Between Force A and C, Goranson noticed that pressure on Force B was beginning to ease. He was then able to turn his attention to an 88 well to the west that nevertheless was pinning his Rangers down. Through the 116th, he was able to bring naval gunfire to bear and the 88 displaced or was destroyed.

Meanwhile, Force C was closing in on the eastern outskirts of Vierville. Here extensive barb wire entanglements required bangalore torpedoes which fortunately were among the munitions which reached shore. Two-way fire was the most intense of the day as both sides fought for the lives of the brave demo men trying to slide their torpedoes under the wire. Just when it was most needed, a US tank rumbled onto the scene to add the necessary support. Allied air power had been so effective that panzers had yet to make an appearance near the beaches.

Lead elements of Force C broke through to the hardtop road that ran through Vierville. Here they were stopped by crossfire from dogged Germans holed up in stout buildings, but friendly ground fire was directed at them from the northwest. Schneider looped a contact patrol in that direction; within an hour they exchanged recognition signals with a combat patrol from Force A. The two main Ranger forces had not yet linked up but they could begin coordinating their actions.

Under Schneider's control, A and B Companies of the 2nd Rangers had been on the spearhead since coming ashore at Dog Green. Of the 136 men in the two companies, 31 had been killed, 66 wounded, and five MIA. Schneider pulled them back to reorganize into a composite company which became Force C reserve— the first time that day the Rangers had the luxury of a reserve.

They needed it because even harder fighting was coming up as both forces strove to link up west of Vierville. At the eastern edge of the village, clusters of mines now thickened into continuous belts. On two prongs, Force C had to advance with companies in column until gaps were opened, and the German rifle pits covering the minefields showered with grenades. Force C did not lack for grenades or ammo because BG Cota had teams from the 29th bringing up combat supplies up like a gang of coolies. However, only two tanks could be spared from the other beaches. Schneider had one overwatching each of his prongs; they were too valuable to risk being knocked out in leading the attacks through the minefields.

Force A had no tanks and less ammo, but thinner minefields to penetrate. Rudder advanced on the broadest front possible. His Rangers had the most trouble from plunging fire from buildings the Germans had converted into blockhouses. Force A went after them one by one, often having to just exchange fire until naval guns scored direct hits. Shortly all habitations around Vierville were in flames, the drifting smoke providing some concealment for the Ranger advance. Too thin to exploit the elimination of scattered blockhouses, Rudder's attacks were generally repulsed. He had to settle for destroying strongpoints one at a time while holding down casualties.

Meanwhile, Force B's fight was diminishing; indeed the Germans began to pull back, leaving 69 bodies. Goranson, however, did not have the strength to exploit the withdrawal. Force B leapfrogged forward, keeping an eye and accurate fire on the Germans who were trying to break contact by liberally strewing their route of retreat with 'shoe' mines. The Germans were able to re-establish strong defenses down the Vierville draw, but at least the long see-saw battle for Poine de la Percee was over.

The situation for Force C began to improve as well when Schneider was able to shove a company west down the hard top, isolating German fortifications north of the road. Fighting hard on both flanks, the company was not able to storm to the center of the village, but by forcing the German reserves to defend Vierville, pressure was somewhat relieved on Rudder Force A so that Rudder could enlarge his beachhead. At the same time, however, Force C was counterattacked by infantry from the southeast, and Schnedier had to turn his attention in that direction. Rudder was handicapped on his flank of the battle after being shot in the thigh. Thus the morning wore on, exhausting and deadly, as fortune swung this way and that.

So far on D Day, fortune had not swung favorably for 1st Sgt. Leonard Lomell of Force A. He had been hit in the side as he debarked his landing craft. His medic had been wounded, and only a field dressing, now sopped in blood, covered a wound that on any other day would have required Lomell's immediate evacuation. His company had been decimated fighting south from the bluffs and, with all the officers down, he was leading the largest group, about twenty still hard-charging Rangers.

Then fortune beamed down on Lomell like the sun suddenly opening clouds. The Germans in strong positions in front of him just pulled back or pulled out—Lomell didn't know where or why, they just disappeared with a departing burst of fire.*[*FN: The Germans' situation was nearly as confused as the Americans'. On several occasions defenders were observed pulling back while reinforcements moved by them in the other direction.] He waved his group forward; the Rangers rushed by two burning blockhouses, to find themselves on a torn up black top road where they paused to look around. Sniper fire pecked at them from the south, but by comparison to the rest of the morning this was a lull in the fighting. Lomell knew where he was: there was only one black top in the area, and it ran from Vierville to Grandcamp. Which way now should he turn his force?

East could result in linkup with Force C and speed the capture of Vierville. However, Lomell did not want to walk into Force C's supporting artillery fire which was drawing closer; nor was he about to become targets for fellow Rangers who were not aware that friendlies were now in front of them (Lomell had no recognition panels with his platoon).

To continue south might mess up the Germans communications, and possibly get at their rear area installations. Most of the Rangers with Lomell liked that idea: Chateau de Vaumicel, just a couple of miles southeast, was believed to be the German CP and probably the location for their fire direction center. Wouldn't it be fun to bust in on them?

But Lomell remembered that Force A's mission emphasized blocking possible counterattacks from the west. That's the direction he turned his men, paralleling the black top, when he noticed Germans moving west from the Vierville draw where could hear fire from Goranson's standoff. A 'kraut' machine gun crew was spotted withdrawing from this action, crouched over, carrying their damaged weapon and escorting three walking wounded. Lomell dispatched five men to ambush them. This was done with total success, the five Rangers hurrying back to rejoin Lomell who had continued to patrol west.

He found a country lane in a north-south gully intersecting the road that his men could cover from clumps of hedgerow. Now he worried that no one knew he was here, and the danger that meant from naval gunfire and allied bombers. All he had was a flashlight with which to make the 'R' signal in Morse Code. He dispersed his men into a goose egg oriented west. Their heaviest weapons were three BAR's, so Lomell could only hope for light contact.

While shallow foxholes were being dug, Lomell decided to do a little personal recon. He needed to see what was going on farther west where the Germans were being forced back from Utah Beach, but first he wanted to know what was in the nearby trees to the south. Lomell took along another NCO. Less than a quarter mile down the lane, just before they were to turn around, the two-man patrol found the prize: the guns of Pointe du Hoc.

Five of them, they were laid to fire on Utah Beach but could traverse to cover Omaha.* [*FN: An unfortunate and long-standing misperception arose from Cornelius Ryan's famous book, *The Longest Day,* where he gave the impression that Force A's mission was ironically unnecessary because the guns were gone from Pointe du Hoc. But if Lomel's search had not found the battery, it could have been firing within minutes and greatly affected the outcome of D Day.] Neat pyramids of shells were already fused, and firing charges cut and stacked. The sights were mounted, the howitzers well maintained and cleanly lubricated. All that was missing for them to go into action were firing stakes—and the crews.

It was not for Lomell to speculate as he rushed back to his patrol asking if anyone had more thermite grenades, but history has always wondered why the guns were where they were, the way they were, and fifty years later there still are only theories.

In a spreading, shallow gully the howitzers had defilade from naval gunfire—no accident at all but rather a demonstration of shrewd planning. For a number of reasons, the author believes the gully was the battery's supplemental position where they made a pre-planned displacement no later than 3 June when German meteorologists predicted good invasion weather for the next couple of days, a period in which the German Seventh Army went to their highest state of readiness that meant placing main guns where they wanted them to fire if the invasion came. Allied bombers had already worked over Pointe du Hoc, convincing the Germans they needed a more protected place from which to fight.

Trees in the gully offered some concealment from aerial attack, and camouflage nets were nearby. Lomell noticed vehicle tracks indicating recent use, but pre-invasion intelligence reports from French agents were that the guns were subject to being shifted around by hand as their emplacements on Pointe du Hoc were being hardened. So the tracks were probably not those of prime movers; maybe the vehicles had hauled in the howitzers' ammo though the tracks did not look deep enough for heavily loaded trucks.

Germans who could explain the mystery of the guns abandoned location did not survive the war, but there are clues from three Ranger engagements that morning which at the time did not seem significant. Off a trail to Vierville, eight Germans, who did not appear to be infantrymen, were flushed out and killed by Force A. These Germans apparently had been trapped by a wall of naval gunfire that had prevented their earlier escape south. They could have been the crews for two guns. Later, farther south, lead elements of Force C caught six Germans fleeing west on the hard top, riding bicycles. They too fulfilled their promise to Hitler to die to the last man.

Then, about the time Lomell was making his discovery, more of Force A had fanned out to the southwest. One patrol noticed a horse drawn wagon emerge from a farmhouse about halfway to the chateau. The wagon turned back under fire, spilling some artillery shells. Later reconnaissance revealed that the it was on the lane that lead to the guns 200 yards away. Land line was also found, leading from the chateau to the battery.

The Germans were careless, too confident of the security of the guns; apparently never worried that a patrol like Lomell's could slip through their layers of defense and steal the bacon. They apparently never believed they would ever be attacked by the likes of Rangers.

Lomell wasted no time in disabling the guns. Thermites fused recoil mechanisms, rifle butts smashed the sights, and frag grenades destroyed the tubes. The Rangers then left the guns because a company or so of enemy with half-tracks was approaching the roadblock from orchards to the south and west.

German scouts were driven off with three dead, but then the formations behind them deployed in a professional manner that convinced Lomell to send runners to Col. Rudder informing him that the guns were destroyed. They returned with orders to "hold out," which they did until relieved on D+2.

As he delayed back toward Force A, a final salute sounded for the guns of Pointe du Hoc. Screened by Lomel's roadblock, a five-man patrol led by Sgt. Rupenski had scouted south toward the country lane between the howitzers and the chateau. In a deep natural depression they came upon the main howitzer ammo cache: about a hundred shells with their firing charges.

This location—only 200 yards from the gully but more than a half-mile from the empty emplacements on Pointe du Hoc—further supports the theory that the gully was where the Germans planned to employ their howitzers, and on Pointe du Hoc were only dummy positions. Why the crews had not assembled at their battle stations in the gully can be explained by the combination of Ranger aggressiveness, naval gunfire, broken German commo, and possibly sheer fright: to put those howitzers into action would have attracted every fighter and bomber at Bradley's disposal. The Germans had a good dose of Allied air power in the previous three months, so they knew what to expect if they became the priority target.

We may never know the full story, even fifty years later, but that morning Sgt. Rupenski knew what he had to do.

"Fire in the hole!"

It was a good thing he used a long fuse because even fifty yards away the mighty explosion blew off his helmet.

So important were the guns to Eisenhower that news of their fate was flashed to him within the hour. He expressed great relief, in his memoirs mentioning how, "everything from then on seemed to ease up." Nothing, however, eased up for the Rangers as D Day was a long ways from being over or decided, with some of their hardest, bloodiest fighting was still to come.

Elements of the 352nd Division recaptured the now useless guns. These Germans also prevented linkup between Force A and C; but adjusted from Pointe du Hoc, six-inch gunfire was brought to bear by a British destroyer that ran dangerously close to shore. Such naval support, supplemented by Allied fighter-bombers, prevented the 352nd from significantly exploiting a numerical advantage, though their arrival heated up the battle to a boil. It would rage on into the night.

In the early afternoon was the only instance of friendly casualties from naval fire. This occurred as the survivors of Force B moved over Pointe de la Percee and were mistaken for withdrawing Germans. Two Rangers were wounded as a result, and one so badly wounded that he couldn't find his rifle when it was slung on his shoulder.

Darkness did not arrive until 2000 on 6 June, plenty of time for fortune to swing till the longest day became the longest night. Like brush fires igniting the sixty miles of countryside, small and large scale combat sprinkled the Covados coast. For the Rangers, the action shifted generally from the attack to defense as they struggled to consolidate their beachhead against menacing probes by the 352nd Division.

There is an audible difference in fire when darkness intervenes between forces in battle. Targets fade with visibility; movement supersedes fire as the prime element of combat. Around Vierville, individual Germans and Americans gratefully seized the opportunity to reconnect with friendly forces. Often they passed close in the night, setting off short but furious firefights. Rudder, weak from his serious wound, and Schneider worked almost frantically to create an organized defense. Their German counterpart tried just as hard to form up a coordinated attack. But the real story of the night is of individuals moving in opposite directions:

Late in the day, LT Jacob Hill and PVT Jay Bacho had come upon a squad of Germans on the other side of a hedgerow. Hoping to hit and run, the Rangers threw grenades but none detonated, and the Germans overwhelmed the Americans with firepower that mortally wounded Hill and hit Bacho in the hand. Falling over Hill's body, he feigned death. When night closed in he made it back to Force A's positions. Three misdropped paratroopers from the 101st also made a dangerous passage into Ranger lines at twilight. In the confused mixture on the ground, death from friendly fire was as great as from enemy.

Bypassed Germans began to rise from hiding and rejoin their forces to the south. They funneled toward the inland side of the Vierville draw where Sgt. Elrod Petty manned the most lethal BAR on the beachhead. In the attack, he had already accounted for dozens of Germans, and now others tried to slink past his defensive position about a hundred yards away. Short quick bursts brought more down one after another. A fifteen-minute pause, then more tried to slip along the sides of the draw. By the light of a three-quarter moon, Petty did his best shooting, and morning revealed thirty of his targets to be buried.

Due south of Vierville, a returning Ranger patrol met exfiltrating Germans at the corner of a hedgerow, five men in each group. It must have been a meeting engagement like those that would occur twenty years later in the jungles of Vietnam. All on both sides were killed, the last two evidently dying of wounds they inflicted on each other.

Both sides patrolled vigorously, the Rangers to locate enemy concentrations, the Germans trying to get the Rangers to reveal their positions by fire. The Germans seemed to be looking for machine gun nests, and no doubt were encouraged to find so few. Rudder and Schneider tried to move machine guns constantly, but by 0100 the 352nd realized that only a light force covered Pointe du Hoc. German artillery, now that darkness protected them from accurate counter-battery fire from the navy, began to prep Force A's positions for a night attack.

It struck just as a composite platoon from Force C achieved linkup and moved in to reinforce Rudder's thin line. Before they could coordinate, two companies of fresh Germans had used the shadows of hedgerows to infiltrate within a hundred yards. An exposed and understrength Ranger company quickly came under a three-sided attack supported by heavy German machine guns. Confusion with the reinforcing platoon resulted in both Ranger units believing the other was falling back.

It was a disorganized withdrawal— really a shabby retreat— conducted by exhausted, famished men, many wounded earlier, low on ammo and even water, jarred by incessant hard fighting; but most of all they were sleep deprived and not functioning at all normally.*[*FN: Such abnormal behavior resulted in the atrocious slaughter of some Russian slave laborers in German uniforms.] Most had been awake and in action continuously for more than twenty-four hours of shock and shell. As one described their state, "All civilization had been peeled away; we had become fighting animals."

An impending disaster was prevented by luck and individual heroism. Sgt. Petty, for one, held off a charge of whistle-blowing Germans, and small unit leaders ran around through the night calling for their men to rally. The Germans were slowed by mines laid by the original defenders of Pointe du Hoc, their own inaccurate mortars, poor use of marching fire, and the Rangers' use of captured German weapons that confused the attackers. Nevertheless, they killed or captured twenty Rangers and seemed about to move onto Pointe du Hoc.

But daring, small-scale counterattacks by Rudder on the shoulder of the penetration stalled the thrust, and by morning both sides were battling with standoff fire. That morning the Rangers were not nearly the force that had scaled the cliffs. Shaken and battered, bloodied and uncertain, they had to rely on personal guts and inspiring leadership to survive D+1 which in many ways would be tougher than D Day.

About fifty men, including the invincible Sgt. Petty, had been cut off by the night attack. At dawn they began planning how to make it back to Pointe du Hoc without being killed by fire from either side. Rudder was down to three companies that in total did not number more than a hundred men, many of them wounded.

During the night, Force B linked up with Force A, but Goranson's command had taken the heaviest casualties of any comparably sized unit on any beach.

Force C was still battling Germans for control of the center of Vierville, now just a low silhouette of shattered trees, destroyed and smoking buildings. Weapons were jamming from overuse in the dirt; ammo and morale were low, casualties high (more than half of the total Ranger force). Replacements and reinforcements were desperately needed or the victories of D Day would be wiped out on D+1.

But there were no promises of relief. Rudder's request to Cota for reinforcements was answered by a message that "all Rangers had been landed." Rudder did not have to be told that—what he had to have very soon was fresh troops from any source. The 116th was a shattered force itself, having suffered 340 casualties.

But with its own CP on the beach threatened by counterattacks in the Ranger sector, the CG of 29th ID scraped up ten tanks and 250 men to join two Ranger companies from 5th Battalion under CPT Edgar Arnold. Though this task force was not able to break through to Rudder (largely because of increased German artillery fire), they took some pressure off Force A and bought time enough for some ammo to arrive on the cliffs, carried up the ropes by thirty men gathered by MAJ Jack Street, a 1st Battalion veteran of Sicily, who knew how to get help to Rangers in trouble. He brought in two LCA's full of ammo, grenades, food and water, then carried seriously wounded men back to the landing craft before returning to become Rudder's first reinforcements.*[*FN: Though Rudder medically required evacuation, he stayed on the bluffs.]

For the rest of D+1, the Rangers held on by guts and by God. Never have cherries become hardened veterans so fast. Slowly the invasion swept inland from the other beaches, diverting the 352nd Division from Omaha, where the 1st and 29th ID were able to claw forward, incorporating the tattered Ranger positions. Ghastly, ferocious fighting filled the following days, but the Allies lodgement in France was secure—in no small measure to the performance of the Rangers at a beach the world would remember forever.

Their victory was a very close thing, the margin extremely thin. How close is described in a message Eisenhower had already written when he contemplated the difficulty of the Omaha mission:

"Our landings have failed to gain a satisfactory foothold, and I have withdrawn the troops."

It was because of men like the Rangers that Ike never had to deliver that message to the waiting world.

The operational lessons learned by Rangers in Overlord were largely unique because they applied to amphibious operations on a uniquely enormous scale. But paradoxically, Overlord was an amphibious finale: the longest day would be the last time a Ranger battalion went ashore under fire.

The general lessons from Overlord were that training can never be too realistic, and that options and alternatives must be maximized, even if not promising.

Realism. The cliffs were so altered by bombardment that they hardly resembled those on which Rangers had trained in the U.K. Short of blasting away some of the white cliffs of Dover, more realistic training could not have been attained; but the *expectation* of greatly altered terrain could have been reinforced in the Rangers minds, so that they would have been anticipating it more.

Many men who made it in to Omaha later remarked that in training there had been underwater demolitions to simulate German artillery, but they were nothing like the shuddering geysers that actually erupted around their landing craft on that final ride into the beach. Also, though they had trained with wet ropes, none had been so heavily water soaked as those they tried to launch up the cliffs. Perhaps more powerful charges would have been used if the ropes in training had been similarly soaked.

Probably the ultimate in realistic training would have been to have balky coxswains, leaders removed, scrambled units, unavailable support, severed commo—then see how the Rangers reacted to totally confusing and unpredicted situations. This, of course, was the reality on Omaha Beach where general training in aggressiveness and initiative saved the longest day.

Multiple options. The 'deadly DUKW's' sounded like a far fetched idea when it was planned, and indeed turned out to be. The DUKW's never made it to the base of the cliffs, no doubt to the relief of the 'firemen' who were to hang suspended with their machine guns high in the air. But during hours when little was going according to plan, who could predict what might work in certain circumstances? The DUKW's were a fall-back, and Rudder could not have too many fall-back options. His plan is a model for estimating losses at every stage of an operation and providing options in case of failure at any location.

Flexibility (the current term is 'branches and sequels') can be factored into a plan; however, combat instincts, like those demonstrated by Schneider, Goranson, Arnold and Lomell, can only be assumed but must be relied upon when a plan merges with the din and death of battle.

There is no certainty that such instincts can be developed. They seem to derive—when they do appear—from personality rather than position or rank. Though elusive, such instincts must be stimulated through training that spares nothing in order to achieve the closest approximation to combat reality. Training without casualties is probably not tough enough: a dozen Rangers were killed on training climbs in the U.K..

Organization and Training

D+2 saw 5th Battalion pressing on to the west toward Grandcamp which they seized along with forces of 29th ID. This ended the hardest fighting for the Rangers in Normandy, for 2nd Battalion staggered off Pointe du Hoc to join the 5th in a bivouac by the Grandcamp sluice gate. Hot chow, field showers, mail and issue of new uniforms began the miraculous recuperative process by which young men can snap back from ordeals. Nevertheless, after evacuation of the wounded, the combined morning report strength of both battalions was between 30 and 40 percent. Chronic fatigue and post-combat nerves reduced their effective strength much lower.

Several days were well spent in rest and reorganization with replacements arriving in driblets, most of whom were Rangers who had never reached shore because of damaged landing craft or who had become separated and fought with the infantry divisions. There was no established pipeline for qualified Ranger replacements, and many candidates who were sent over from Britain were rejected after interviews.

Before the state of the Omaha Rangers bottomed out, it received another set back when Rudder reverted to CO of 2nd Battalion in order to make way for a full colonel to take temporary command of both battalions. He was Eugene Slappey, recently relieved from command of a regiment in the 29th ID. He had been transferred to First Army who in turn sent him down to replace Rudder. Slappey's command tenure with the Rangers was brief, ending in a couple of months when they were attached to VIII corps in Brittany. Until then, the Rangers' activities in Normandy were principally low intensity patrolling and guarding EPW's.

With experience in four campaigns in two theaters, Max Schneider was sent home to a training command. Beginning with Darby in Northern Ireland, he had risen from 1st lieutenant to lieutenant colonel in two years, one of the first (and fastest) Ranger career patterns on record. Schneider was replaced by Maj. Dick Sullivan who had led the attack in response the BG Cota's demand that, "Rangers, lead the way!"

Training in Normandy was ad hoc and improvised, the most valuable form being OJT while running patrols. New men had to be checked out, with veterans going along like lane graders. Forced marches were also the order of the day to restore stamina. By July, the battalions were able to reach their pre-D Day march rate of thirty miles in six hours with combat loads. Aerobic sports like football and a wild version of 'capture the flag' further conditioned the Rangers.

During this period, the Allies were breaking out of the Normandy beachheads, and when Patton's new Third Army spilled into Brittany, army group planners began looking up and down the English Channel for ports to support subsequent operations. The Ranger S-2 received intelligence folders for several possible landings to seize ports. They were told the most likely one had a name from their history—Dieppe.* [*FN: Fittingly, the capture of Dieppe was a mission given to the Canadian First Army who took it virtually unopposed in August.]

Though Rudder was now a renowned officer in the ETO, he was still only a lieutenant colonel without the clout and prestige of Darby, probably the only man who could have swayed Bradley to keep Rangers at their amphibious speciality or use their new airborne capability. But Darby remained in Italy when the 82nd and 101st Airborne Divisions were alerted for drops (which were aborted because of the speed of the Allied advance) between Orleans and Paris. The Rangers were never contemplated to jump, an indication that their importance had receded in the minds of high level planners. This is hardly surprising, considering that a million soldiers were pouring into France. Ranger operations in the ETO became a sideshow. Though they did not know it at the time, and much intense fighting lay ahead of them, Pointe du Hoc was the Rangers' high water mark.

When the battalions were alerted for the Brest mission, training changed from individual to company-size, featuring attacks in built-up areas that progressed from dry to live fire exercises; however, there was a theater-wide ammunition shortage, and ammo for training was a low priority. Making do with what they had, night operations were emphasized by Ranger S-3s, and procedures for coordinating artillery and air support reviewed.

The battalion staffs certainly had opportunity for practicing coordination. During their last two months on Normandy, the battalions moved around like handymen, at various times being assigned odd jobs for an army, two corps, and four divisions.

Bad Missions:

III. Brest, France, August, 1944

The author departs from his previous standards for a 'bad' mission to label Brest as such, though the Rangers had adequate support and intelligence. 2nd and 5th Battalions' operations in Brittany were simply inappropriate—siege warfare, the last sort of mission for which light infantry is equipped, minimizing their singular maneuver capability, reducing their potential to achieve surprise. But needless to add, the Rangers accepted the mission with utmost resolve to carry it out and lead the way.

The mission's success proved to be largely useless, but that outcome was not obvious when the Ranger battalions began a wild night time truck ride west across France that made them think the war depended upon them getting to Brest by morning. But the situation was hurry up and wait as they went into reserve for VIII Corps.

Brest was the largest French port on the Atlantic. Allied logisticians believed it to be utterly necessary for supplies. Hitler was equally convinced, therefore designating Brest (along with LeHavre, Boulogne, Calais, and Dunkirk) a 'fortress' to be held to the last man. To carry out this mission in Brest, he had MG Hermann Ramcke who had held off the Allies for months at Monte Cassino. Ramcke commanded about 50,000 men— some of them garrison troops, marines and shipless sailors—but with a hard core of well supplied infantry including the 2nd Parachute Division.

MG Troy Middleton's VIII Corps was comprised of the 2nd, 8th, and 29th ID, plus the 2nd and 5th Ranger Battalions—a force smaller than the Germans in well fortified entrenchments that ringed Brest in a perimeter about thirty miles around. Ramcke was trapped with no hope of relief, but with excellent underground facilities and ample supplies, his troops' morale was high: the worst they could expect was eventual capture, followed by a comfortable life as a POW in the United States. Ironically, they would be back in America before the Americans who fought them.*[*FN: A post-war joke about two *Wehrmacht* veterans reminiscing in the rubble of Germany went like this:

Hans: Life is so hard now. Do you think we will ever see good times again?

Fritz: *Ja.* Perhaps if Germany starts another war we could be recaptured and sent back to Kentucky.]

Morale among the besieging Americans hinged upon rumors of imminent surrender by the besieged Germans. This developed an attitude that no one wanted to be the last KIA before Brest caved in. It took stern discipline and gritty leadership to counteract this cautious frame of mind.

Usually 'siege guns', the heaviest of artillery, belong to the attacking force but this was not true at Brest where Ramcke had dismantled navy guns and revetted them ashore. He also had the heaviest caliber—280 mm railroad guns—weapons on the battlefield. So on the ground, Middleton was outgunned, but the air belonged to the Allies, and close support was his trump card.

Three companies from 2nd Battalion were attached to 29th ID who employed them in counter-patrolling. In order to keep up an aggressive spirit and train his garrison troops, Ramcke had his paratroopers counter-patrol with the REMF's. These were big patrols, fifty men or more, and not difficult for the Rangers to track with the help of the French resistance (called the FFI) who were natives of the land.

A typical action was when a platoon of Rangers with a few FFI were put on the trail of about thirty Germans conducting a daylight security patrol from a headland. A Frenchman informed the platoon leader that the Germans' probable route would take them by a good ambush site, and he could lead the Rangers there ahead of the 'Boche'. The problem was that the site was subject to direct fire from 88s on the other side of a narrow bay. The Rangers decided to shoot and scoot rather than police up the kill zone of the ambush. This was done. Three Germans were killed, a number wounded, and the Rangers got out unscathed before 88s tore into their ambush site.

Intelligence from the FFI was so important

that 2nd Battalion was given a unique mission: a composite company, including most of the S-2 section, were to infiltrate behind German lines to set up an intelligence collection point where FFI agents could report prior to a two-regiment attack by 29th ID. Valuable tactical intelligence was radioed back to the Division G-2.

For the attack itself, four companies from 2nd Battalion, augmented with scout cars and light tanks, screened a flank which was open country. German artillery improved the speed of the Rangers' double time, so they were able to keep up with their armor. The task force ran over outposts, killing eighteen Germans who could not pull back quick enough. Holding up in front of a series of small forts, the Rangers dug in fast because indirect fire superiority belonged with the Germans.

"Stay in your holes," was the command that night as Germans came out to patrol. "Anything moving gets shot."

The Rangers' probe built up the next day into a regimental reconnaissance in force as 29th ID hoped to jam a crowbar into Ramcke's perimeter. The ad hoc regiment was called Task Force Sugar consisting of a platoon of FFI, elements of both Ranger battalions, a cav troop, plus two infantry battalions from 29th ID, supported by an artillery battalion and a 4.2 mortar battery. The Rangers had found the most promising location to break through, and Middleton was prepared to give them all the backup they could use.

They would need a lot because they came into range of the 280 mm's Ramcke had named *Batterie Graf Spree* for a famous German cruiser that had raided Allied shipping very successfully before being sunk off Argentina. Ramcke expected the same of the battery: to inflict maximum damage before going down. It lived up to his expectations, holding out for another month. The battery introduced itself to the Rangers with shells that sounded like freight trains overhead.

The Ranger component of TF Sugar was designated TF Slater, named for 'Duke' Slater, XO of 2nd Battalion, who had never landed on D Day because his boat swamped. Slater was the kind of man who wanted to make up for any war he'd missed.

Under him was CPT Arnold of Omaha Beach fame, with a two-company command. He pushed them forward relentlessly to close fast with the German infantry in order to cause the *Graf Spree* guns to hold fire. His aggresiveness resulted in fifteen EPW's. The entire Brest campaign had this feature: that Germans would surrender with the knowledge that their deaths would not mean much in the result of the siege which, like all sieges, is a matter of time and attrition.

With TF Slater's success, TF Sugar's prospects became so important that 29th ID placed the division chief of staff in command. Tank destroyers were added; with them Arnold attacked at dawn the next morning, killed eleven Germans and captured 94. The *Graf Spree* battery, still within range and protected from land attack, cost the Rangers two dead and five wounded. These big guns became a visceral Ranger mission like those at Pointe du Hoc, but satisfaction would not come soon or easily.

Surging on deeper into Ramcke's defenses, TF Slater came up with a surprise prize: 162 Russian slave laborers, POWs from the Soviet Army, who quickly converted into a valuable fighting force against their previous captors. It was Arnold's good judgement that permitted them to help considerably in subsequent operations.

In the afternoon the Rangers teamed up with tank destroyers to take out another strong point, but supporting artillery was not heavy enough and the attack stalled. That night, to test their effectiveness, Arnold sent Russians out on a ambush patrol. Just before dawn, he heard a flurry of gunfire. Shortly a messenger from the Russians reported to Arnold with a note from the patrol leader. It was an apology for having wiped out eight Germans and not saving one for interrogation.

The Rangers understood.

Through pelting rain, TF Sugar pressed on, bypassing strong points where possible, then with loudspeakers and leaflets urging them to surrender. Fearing that they would, Ramcke sent out a relief force that collided with TF Slater and resulted in a standoff by nightfall. Using TD's to hold off the relief force, the next morning Arnold turned back on a strong point, sending an EPW toward it to encourage surrender. Seventy-two Germans did, then Arnold reversed direction again to take on another strong point.

But this time the Rangers came under fire again from the 280 mm's. The next day was spent digging into mud. By dark TF Sugar's attack had petered out.

Fifth Battalion went into action on 29 August with two companies attached to 2nd ID. In order to mass for breakthroughs, Middleton had to accept a wide gap between the 2nd and 8th ID, so Sullivan had a new role for Rangers: economy of force for higher headquarters. He accomplished it by screening and flank security, not normally appropriate missions for a foot-mobile battalion. 5th Battalion spread out into an archipelago of outposts; none were mutually supporting so Sullivan's reserve company linked them with visiting patrols.

Ramcke knew he could buy the most time with shrewd pre-emptive raids and limited objective attacks that would delay VIII Corps from massing at decisive points. Middleton was also aware of that threat, and so he promised Sullivan very little support except from the air. 5th Battalion was pretty much on its own, stretched out in a thin defensive stance.

Probing for gaps in the front, German paratroopers with some panzers kept Sullivan's Rangers very busy beating off one small attack after another. This would not have happened had not the stubby US fighter, P-47, proved to be an excellent tank buster and troop strafer.

Around Brest, air-ground coordination developed to a high level of proficiency, really the first time Rangers and fighters operated together significantly. But air support at night or in rain never worked well enough, and the Germans learned to use those periods for their hardest probes. Night time is Ranger time, however, and though outposts became encircled they stung the paratroopers like hornets nests so that no Ranger positions around Brest were ever overrun.

Sure to be outnumbered by the size of German patrols, the men in 5th Battalion OP's shifted to alternate or supplemental positions at nightfall. Fire discipline was especially important so as not to reveal true locations. The Germans were allowed to approach to where Rangers could meet them with grenades. Every rifle pit was loaded with them.

Visiting patrols often looped out toward German lines before arriving at Ranger OP's. This way patrols might come across a German probe and at least provide early warning or even break up the German force before it reached its objective. These were tense actions that featured tommy-guns to immediately gain fire superiority. Tension was hardly less when such patrols arrived for a 'visit' to an OP where every man held a grenade with the pin pulled. Cupped, red-filter flashlights were the usual recognition signal, but an additional safeguard against fratricide was Sullivan's effort to take visiting patrols from the same company which was manning the OP's that night. Often a whisper by a familiar voice was reassuring verification of the recognition signal.

There were occasions when far flung OP's had to fight an all-around defense and ran low on ammo; but none screamed 'Mayday' on the radio, triggering a reaction force. A relief patrol was sent out, not to break through to the OP but to set up an ambush on the Germans' expected route of withdrawal. Two hot and heavy contacts came about from this tactic, both an hour before BMNT when the Germans pulled back under covering fire from their artillery. Eleven dead paratroopers and nineteen EPW's were the result of these ambushes, at the cost of one Ranger WIA. Thereafter, the paratroopers had to set out security for the withdrawal of their combat patrols.

Without significant success elsewhere on his front, Middleton ordered 29th ID to rev up TF Sugar again and blast through to a piece of key terrain, Hill 63. Confined to a narrow sector, TF Slater moved out, a perfect example of a bad Ranger mission and unsuitable employment. Their's not to reason why—and Arnold made up for tight maneuver space by accelerating maneuver speed. In fast forward, Rangers darted across the landscape with only one killed. The Germans reaction was not as quick, causing them to lose twenty-nine men.

Pouncing into the former enemy positions, the Rangers were paid back in full by German artillery, mortars, and a 20 mm rapid-fire cannon blazing away from another fort. Digging the German rifle pits deeper, TF Slater's casualties still began to climb. The Germans fire was pre-registered so it continued effectively through the night.

However, darkness allowed a platoon of TD's to beef up Arnold's position. He decided it would be beat on all the next day unless he could knock out that 20 mm cannon. So before dawn the Rangers teamed up with the TD's to drive onto a hill that overlooked the cannon. This they were able to do undetected in the dark and rain; at BMNT the Germans were shocked to taken under fire from higher ground. Especially unhappy was the cannon crew which had no cover in an open emplacement. The crew was killed or scattered and

Arnold believed he'd put the cannon out of action for good.

Artillery and mortars prevented him from overrunning the cannon, and forced the Rangers to dig in again for a night. The rest of TF Slater patrolled out from Hill 63.

The next day, 31 August, paratroopers counterattacked supported very effectively by the pesky 20 mm. TF Sugar consolidated beside the Rangers, and both sides slugged away with increasingly heavier calibers. Despite direct fire by 75 mm howitzers bravely brought into position by Arnold's men, a stalemate ensued watched mostly by infantrymen in helmet defilade.

On 1 September, Arnold got priority for the really big stuff: 210 mm howitzers from corps artillery, 105s with concrete-piercing rounds, and P-47 fighter-bombers. The Rangers wondered what good was their infantry MOS? They were now essentially forward observers and air-ground controllers!

Rudder tried to find them another role. Pulling out of the line, TF Slater divided to loop around to into two gullies that led down to the coast. If both Ranger probes could reach the sea, the German garrison and their 20 mm cannon in the fort called Kergolleau would be isolated. By hard fighting all the next day and night, the Rangers forced their way down the gullies in short rushes. German infantry did not present impossible difficulty—a dozen were killed and a hundred captured— but indirect fire from Kergolleau took its toll, wounding eleven Rangers. TF Sugar did not follow on with enough mechanized troops, so Rangers in the gullies were nearly as isolated as the Germans in Kergolleau.

Though the 20 mm's strongpoint was now cut off, it was well stocked with ammo and provisions. TF Slater would have to dig them out, attacking from two sides while also receiving fire from two sides itself. To better the odds, 5th Battalion came on from a third side, killing three and capturing three Germans in the first contact. The Ranger battalions were now together for the first time around Brest. If their coordinated, supported attack couldn't carry that Kergolleau, this would indeed be a long siege.

Rudder directed Sullivan to keep the German company in the strongpoint occupied while Slater tried to take two field forts whose capture would complete Kergollau's encirclement. Battling through a village, supported by a platoon of FFI and tanks, forty of Arnold's Rangers made good progress until counter-attacked by a relief force of 150 Germans sent out from Brest. The FFI were excellent scouts and pretty good part time guerillas, but they had no experience in pitched battle; so when the German company struck, many FFI disappeared. Arnold had to fall back and set up a night defense around the village.

Sullivan, meantime, had coordinated his attack with a larger one of the 116th Infantry (old friends from Omaha Beach), with the mission of reducing a string of five small forts that would aid TF Slater while allowing 29th ID to split Ramcke's perimeter and penetrate to the Bay of Brest. This objective was a tall order. 5th Battalion tripled its combat load because they would be firing an awful lot of ammo.

The bay was seven miles away with the five forts in between. So that the first two would have trouble mutually supporting each other, Sullivan decided on a night infiltration preceded by a daylight feint by two companies in another direction. Their initial reconnaissance immediately ran into a kill zone of long range fire from the forts, adjusted by hidden OP's. It took all day to knock them out. That night the Germans counterattacked to reinstall them. They ran into 5th Battalion's scouts guiding its own night infiltration. Heavily outnumbered everywhere, Sullivan had to commit Headquarters Company to help pull out some of the feint force.

By morning, it was clear to him that the original plan was just wishful thinking. None of his companies had made it to where they were going; and though all were engaged, none made much headway until the 116th came abreast with armor and priority of fire from div arty. Ranger maneuver hadn't worked in their narrow sector, so it seemed Sullivan faced the dreaded prospect and misuse of undertaking frontal assaults.

If this had to be, 5th Battalion would do their damndest; but Ranger resourcefulness seeks solutions from all directions—including up. Teething problems in air-ground coordination had been frustrating so far, but aerial support would be decisive now or Sullivan would be looking at another Cisterna.

As at Cisterna, daylight could betray the Rangers if they got in too deep, so he began at dawn when tac air could do its best work. The scheme of maneuver was straight ahead, six companies in a double column. Five P-47s went at the second fort while friendly artillery massaged the first. The initial bombs were a quarter mile off mark. The lead Ranger companies stayed dug in at their line of departure to adjust the next strike. It was closer but ineffective.

Patiently the air-ground control team critiqued the last strike, then marked the target with a purple smoke round. It didn't detonate, so a white phosphorous was used next. It was a little wide. The P-47s swung around again, waiting for the next WP which hit squarely on where Sullivan wanted bombs.

This time the fighter-bombers were right on, hitting with sixteen 500-pounders that bulged the fort, and was also the signal Sullivan used for his men to move out. They swept forward, just twenty yards behind strafing runs by the P-47s 50 calibers who made their runs perpendicular to the Rangers axis of advance so that artillery could keep suppressive fire on the fort.

The three-way coordination worked to perfection. Five officers and 242 EM in the fort surrendered while the Rangers' casualties were only two WIA's.

The close teamwork had been slow, taking most of the day (2 September), so the air-ground attack could not turn to the next fort till the next day. It succumbed, though with a harder fight, by noon and Sullivan, rotating his lead companies, went after the third fort. Its garrison, having watched the fate of their comrades in the first two forts, used the night to evacuate.

The fourth and fifth forts were manned by paratroopers and stood astride a road to the bay which the Rangers could now see. The forts would have to be taken simultaneously. Shrewdly, the Germans realized that a forward defense would help their cause. If they could battle the Rangers in the field, the American tac air would be less effective. Using underground passages, paratroopers slipped out to ambush approaches to the forts.

Sullivan sent a reserve company up to recon those approaches. In a sudden firefight during late morning, the company met thirty paratroopers moving into an ambush position. The Rangers were quicker on the draw. In a sudden and brief firefight, five Germans were killed, six captured and the remainder scurried back to their fort. The Rangers lost three lightly wounded.

As the company approached the fort, fire pinned it down so P-38s came in for close support. The Rangers took off their white undershirts to mark their position. Under protection from bombs and 105s, the company was able to secure a defiladed assault position for a night attack by two other companies. Sullivan moved them into position but decided to postpone their attack for 24 hours till, with the rest of 5th Battalion, he could support from previously captured forts and some high ground he had not yet secured.

At midnight the two-company attack jumped off, guided by tracers from the Ranger forts. A supporting attack by a company of the 116th kept one of the objectives occupied while the two Ranger companies concentrated on the other called Fort de Mengant. It proved to be too tough a nut to crack without help from 500-lb bombs that became available to Sullivan the next morning.

Stone, mortar and steel showered from the fort, some pretty big pieces striking the Rangers as they cheered the P-47s work. This was really close support: some bomb blasts lifted the Rangers out of their foxholes but they came down smiling. Not so for the Germans. Three hundred of them staggered out badly dazed. To accept their surrender, the Rangers had to gesture for them to get out of the line of fire from the other fort. It seemed their comrades there were willing to shoot them down to discourage more surrenders.

The last objective for 5th Battalion was Fort de Dellec. The fighter-bombers were ready to help take it down but Sullivan was delayed by a number of problems. First, his foxhole strength had been reduced by more than a third—luckily most casualties were WIA's who later returned to duty—while effectiveness had degraded from five days and nights of big time, full time fighting. 5th Battalion had become so run down that a platoon of Arnold's Russians were put into American uniforms to deceive the Germans about Sullivan's waning strength. Fourteen of these 'Soviex' soldiers fought well and bravely with the Rangers for the rest of the Brest campaign.

Another problem, an unexpected one, was with weapons. VIII Corps had plenty of ammo, and Rangers had been shooting it up so much that mortar, machine gun and BAR bores needed replacement. Middleton was very pleased with Rudder and Sullivan's work so this problem was rectified in short order.

While 5th Battalion had been taking down their five forts, Arnold's two companies from TF Slater had tightened the noose on Kergolleau. On 5 September Arnold tried the sort of close air support that had worked so well for Sullivan, but the fighters were P-38s new in this role. One of them dropped a bomb on the Rangers, luckily with no casualties. Because of manpower shortage, Arnold had to severely economize his holding force around Kergolleau in order to strenghten his maneuver element. With only one Ranger officer and five EM beside them, the Russians filled in admirably with US uniforms.

September 6 was a day of heavy fighting, with Arnold's companies launching a diversionary attack around Kergolleau while TF Slater (-) recaptured Hill 63. Both forces were hit hard by 88s and SP artillery as Ramcke realized that his defensive crust was in danger of breaking. The battle turned that day through the work of 4.2s firing WP shells, the only thing that could penetrate through apertures in Kergolleau. Inside, 26 Germans died of burns. Slater's attack took a thousand yards of terrain and Hill 63, killing sixty and capturing forty Germans in the process. One Ranger was KIA, 19 WIA.

Pausing that night only to bring up more ammo, Slater and Arnold continued their attacks the next morning with the objective of cutting the road from Brest to put Kergolleau out of reach from of support adjusted from other German strongpoints. After inflicting 31 casualties, the two Ranger forces closed a pincer by midafternoon. This was a discouraging sign for many Germans. During the rest of the day, 144 surrendered to TF Slater's patrols; thirty more were killed, with no Ranger losses.

Middleton's plan really began to jell when 5th Battalion fought their way down to tie in with 2nd Battalion. The 116th closed in on Rudder's other flank, so that a strong spearhead was in place to go for the *Graf Spree* battery at Lochrist, neutralizing Kergolleau.

The Ranger plan was for Rudder to make the main attack by dispersing as widely as possible but converging on Lochrist before the Germans there realized they were his objective. It was hoped Ramcke would perceive the principal threat was to the town of Plougenvelin, the 116th's objective, which was more vulnerable, or to Le Conquet, 5th Battalion's objective. The plan was a gamble with high stakes: if Lochrist fell, German morale elsewhere would shatter at the loss of their strongest position, but if the 280's could be defended and continued to fire defiantly, it would be the Americans whose spirits suffered—with costly casualties as well.

Initial signs on 9 September were that the Germans were beginning to believe their time was running out. 2nd Battalion's first flanking maneuver of a village produced a hundred EPW's. They and many more would become an administrative problem as the attack, which relied significantly on speed, was delayed by wide-scale German surrenders. Rudder was also slowed down by the grassy terrain which was mined more heavily than had been expected.

Like scattered brush fires merging into a flame front, 2nd Battalion closed in on Lochrist against increasingly demoralized resistance. What fight Ramcke had left in his command was elsewhere with his paratroopers who themselves had been attritted down to half strength. He did not think that the Americans would go right for his hardest target, the *Graf Spree* battery in Lochrist, but that's just what Rudder did in broad daylight. Squad-size combat teams, built around flame throwers, satchel charges or bazookas, found holes in the German perimeter that was constantly bombarded, bombed and strafed. Upon achieving penetration, the Rangers main problem then was identifying themselves to friendly fire support. With a month of practice in such coordination, no Rangers were killed by friendly fire on Lochrist.

The *Graf Spree* battery, and supporting 88 and 20 mm positions, did not have local security—all German infantry was out on the crumbling perimeter of Lochrist. Lt. Edlin with four other Rangers from A Company, snuck through the fort till they found the thick-walled revetments for the 280's. The crews were prepared to shoot, awaiting the signal from the battery commander to pull lanyards, but he seemed reluctant to give the command for fear of attracting attention and Rangers. He was too late. Edlin ducked out of a passageway, a grenade with the pin already drawn in his hand. The startled German commander did not need to speak English to understand his choice. When he ordered his crews to go out of action, Edlin replaced the pin and took them prisoner.

Edlin quickly made his capture known to Americans and Germans alike by running up the stars and stripes above *Graf Spree* battery. Seeing that the heart of their defenses had been cut out, the defenders gave up in droves. More than a thousand Germans surrendered to 2nd Battalion on 9 September 1944, the largest number of EPW's ever captured by Rangers on a single day.

At other forts, Ramcke had more strong-willed garrisons than the one at Lochrist. 5th Battalion had tough going against paratroopers defending the town of Le Conquet. Vicious fighting for a series of pillboxes required help from tac air which was right on the money. For their final objective, 5th Battalion had to cross an inlet on rubber boats, using smoke and air support to neutralize German fire. Quickly the Rangers prevailed, killing five and capturing 130 EPW's while suffering only five casualties themselves.

The last significant fight for Rangers in the Brest campaign was against their nemesis, the 20 mm rapid-fire cannon, and the die hard garrison of Kergolleau. Fittingly, Arnold's force drew the mission, kicking off their attack on the morning of 10 September. With little fighting going on elsewhere, tac air and artillery support was luxurious. The Germans could answer only with mortars and machine guns because Arnold's preparatory fires finally knocked out the 20 mm that had bedeviled the Rangers so long.

They hugged the creeping barrages that rolled unrelentingly over Kergolleau. Just the smoke alone from the bombardment seemed enough to suffocate the defenders. Ranger uniforms were pale with stone dust when the first white undershirt appeared in front of them. Arnold ordered a cease fire at 1500. Fifteen minutes later Kergolleau surrendered; 74 EPW's marched out. Mopping up operations continued for another week until Ramcke surrendered Brest, resulting in the freedom for 400 Allied POWs, most from the Normandy campaign, including one Ranger from 2nd Battalion. He had gained weight in captivity, so well had the Germans stocked up for the siege they knew was coming.

Hitler had expected Ramcke to hold out for at least another month, but the mission was as futile as Middleton's. Brest was no longer significant as a logistical base for the Allies whose advance across France had been so rapid that it cleared Channel ports much faster than any of the Overlord planners had anticipated. So the siege of Brest turned out to be a sideshow of no strategic importance—small reward for the high casualties and hard fighting by men who blasted each other with the heaviest caliber weapons used in the war.

Untrained and unfamiliar with siege warfare, the Rangers learned many lessons during their unique operations around Brest. First and foremost was the high proficiency in fire support coordination required in frontal assaults, the undesirable but most typical offensive operations during a siege. The experience gained in air-ground coordination would serve the Rangers well when they came up against similar static defenses on the Siegfried Line.

Second, once again as 4th Battalion had used Italian partisans, Rangers found valuable adjunct forces in the FFI and Russians. Not often are they an important factor in combat, and rarely can they be employed as if they are Ranger units of equivalent size; but for intelligence, security and economy of force, local irregulars can be of much more worth than might be assumed from their numbers and weapons. Going back to their earliest history, rangers were originally irregulars, and the affinity continues so that Rangers should always look around for indigenous allies.

Training and Organization

There were no significant organizational changes resulting from the Brest campaign. Often parcelled into task forces with conventional units during that period, a coordination element was put together at VIII Corps headquarters so that the Rangers as a whole would be represented. The assets for this element had to come out of the hide of the maneuver battalions, so it was temporary, very small-scale, commanded by a captain, and manned mostly by Rangers who were not fully fit for duty, like those recovering from wounds in Normandy.

So, with regard to establishing a permanent TO&E Ranger 'regimental' headquarters, 2nd and 5th Battalions were behind what Darby had achieved five months earlier in Italy when HQ 6615 Rangers operated briefly. At Brest, Rudder had no more authority—and no more of his own regimental headquarters— than Darby had before Anzio when he was merely 'senior Ranger commander' in previous operations. Organizationally, Rudder had to keep trying to reinvent the wheel when it came to a regimental HQ. For the rest of World War II there would never be a hub in that wheel.

This absence was more than a slight or a matter of prestige; indeed, the lack of a regimental HQ affected the appropriateness of Ranger missions. Every unit—Rangers are no different in this respect—needs a permanent parent organization to speak in its behalf to higher headquarters, to exert influence and provide input for planning operations, as well as conducting administration and handling logistics. Even with the best will and cooperation by both parties—an exception rather than the rule—attachment to a bigger organization is no substitute for a parent organization. Without a regimental headquarters, Rangers remained bastards. Since D Day they had been handed around to numerous foster parents— attached for varying periods to three corps, five divisions, and too many regiments to remember. Sometimes the Rangers received good treatment, often not.

That's the way bastards live. That may be one of the reasons behind the expression, "to fight like bastards." The other organizational gap was in the replacement pipeline. Good intentions for a steady flow of cherries from Camp Forrest were interrupted by manpower requirements throughout the Army which could see the bottom of the barrel in the draftee pool. At the time the Rangers were trucked away from Brest, eight out of ten American males between the ages of 17 and 25 were in uniform or had received draft notices. The other two enjoyed exemptions or had been classified 4F (unfit for military service).

There was no Ranger with stars on his shoulder pounding tables in Washington to keep 2nd and 5th Battalions up to strength with qualified men. Rudder and Sullivan made do with what they got, and they no longer could be very selective. Yet there would be plenty for new Rangers to do, even if they weren't cliff climbers or fort busters, because the functions of a regimental headquarters had to be done though no headquarters was there.

There was a form of internal replacement that Rangers practiced that amounted to robbing Peter to pay Paul. NCO's were discharged, then immediately commissioned within the battalion. Why they were not just awarded battlefield commissions is a question that baffles the author, but that's way the way things were done. Sergeant Len Lomel, the man who found the guns of Pointe du Hoc, became a 2nd Lt. this way in the month after Brest.

Both battalions were earmarked for unspecified assignments in Belgium when they went into a comfortable bivouac near Kirbilben in eastern France. Because cherries were scarce and further movement would be on short notice, a comprehensive training schedule was infeasible. Physical conditioning through aerobic sports was the general guidance, and one major contest was fought, the result of which appeared in 2nd Battalion's journal but not in 5th Battalion's; namely, softball games between the 'home' team, Rudder's Runners, and 'visiting' Sullivan's Sluggers, in which 2nd Battalion swept a double header, its NCO's prevailing over 5th Battalion's, with the same outcome when the respective officers played. Both games, though umpired by chaplains, were decided by close scores and disputed plays.

Sullivan demanded an immediate rematch on his home field. The games had to be postponed till after the war, the next occasion when the battalions were together again. By that time, only eleven of the 36 players were present—the rest either dead or sent home with serious wounds.

Bad Missions:

IV. Hill 400, Germany, December 1944

Breaking out of Normandy, Eisenhower's armies attacked on a broad front after trapping the German Seventh Army around Falais, then cutting off the remaining forces in France by a secondary invasion through the Riviera (Operation Anvil). This campaign which lasted throughout the fall was known as the Pursuit to the West Wall.

Because the Allied front moved rapidly east, Ranger participation in the advance was scant owing to lack of organic tactical vehicles. 2nd Battalion was attached to several units in First Army while 5th Battalion was with Patton's cavalry in Third Army. Both battalions had some short, sharp actions but pockets of German holdouts either dissolved or were bypassed. The hardest skirmishes occurred when the Rangers were introduced to the still dangerous panzer formations counterattacking to cover withdrawals. Till then, panzers had not been the Ranger nemesis they were in the Mediterranean.

By November, the Allies had outrun their logistical capabilities and run into the fixed fortifications of Hitler's West Wall, also called the Siegfried Line. Pursuit turned into slow pounding ahead as the air chilled and the Germans caught their breath to prepare for winter warfare in which they had immense experience from the Russian front. V and VII Corps attacked on a front as narrow as that they assaulted on D Day into terrain just as formidable as Omaha Beach, the dense and dark Hurtgen Forest. Attached to 8th ID in V Corps, the 2nd Battalion became a reaction force against counterattacks. Casualty rates in 8th ID climbed faster than the temperature dropped. Soon Rudder was called upon to assume the mission of a line regiment that had fallen to half strength.

2nd Battalion's foxhole strength was not much better because of the shriveled pipeline for replacements. Nevertheless they went into an assembly area for an attack to seize heights three times taller than the cliffs of Pointe du Hoc, against Germans fully alert and prepared to defend what they called Castle Hill, and what the Americans designated HIll 400 (meters).

A combat command from the 5th Armored Division had penetrated into the German frontier town of Bergstein, a key position in the West Wall's defense of the Roer River valley. German counterattacks had retaken part of the town, leaving the tankers exhausted, exposed and depleted. Neither were they suited for urban combat nor to take the precipitous high ground that dominated Bergstein. They were told to hold in place and support a Ranger night attack by fire which was scheduled for the third anniversary of Pearl Harbor.

When 2nd Battalion received that mission, their most difficult since D Day, Rudder had been called away to First Army HQ. Execution of the attack then devolved to his XO, Maj. George S. Williams, recently promoted from Battalion S4.

Both he and 8th ID would have preferred a night attack, but the Rangers—perhaps remembering Cisterna—had not had time for their own reconnaissance, so the attack was scheduled for 0700, about first light for that time of year.

Williams' plan was for half the battalion to seize adjoining ridges, then provide a reserve against inevitable counterattacks against the rest of the battalion which would move through Bergstein, cleaning it out, enroute to the face of Hill 400. No maneuver room, no 'impassable' terrain over which to take the Germans by surprise—just take the high ground any way you can. Moving out at 0130 on 6 December, there was only time for a leaders' recon which was largely useless because of no moon and much cold rain.

Huddled in the mud, the Rangers gloom deepened when in the darkness Lt. Col. Rudder moved through them. His return was unexpected, and the news he brought even more so: against his wishes he had been given command of a regiment in the hard hit 28th Division. He went around exchanging handshakes and quiet farewells; he would be gone hours before the problematic attack on Hill 400. A more mistimed change of command was impossible to imagine.

Guides from the combat command did not arrive in the assembly area, so Ranger officers set out on their own to lead the way. H&I fire from both sides thundered above the rain and set numerous buildings in Bergstein aflame. Slinking by flickering shadows and the hulks of knocked out armor, the Rangers faced a black void, the looming silhouette of their objective, Hill 400.

A, B and C Companies infiltrated along the edges of the town to start up to the adjoining ridges. In the rain storm, German indirect fire searched for them but with little success. 'Kraut' infantry seemed to have pulled back: B Company gratefully occupied their trenches and rifle pits. Williams was as surprised as he was pleased to learn. C Company also reported success in setting up a trail block that would take under fire any counterattack from the other flank of Hill 400. Now all that was left to do was take that hill.

From their line of departure, a sunken road on the edge of Bergstein, D, E and F Companies (under Slater) sent out recon patrols to find the best routes up the hill. They returned around 0500 with nothing but bad news: after crossing a mined, open field, the slopes beyond were so steep and slippery that they would be an almost impossible climb even if undefended. But as at Cisterna, there was an ominous silence on the approaches to the objective. Veterans like Lomel, who had led one of the patrols, could 'feel' a trap above them.

Nothing could be done now except tighten chin straps and ease forward with utmost caution. Speed and surprise are Ranger trump cards but they weren't in the hand the Rangers held as dawn seeped through the rain storm of December 7th. This might well be another 'day of infamy'.

To have maximum fire forward, Slater had his companies advance as skirmishers. To maximize the possibility of surprise, Hill 400 was not prepped by artillery, but B and C Companies on the ridges called in fire on the sound of German vehicles, believed to be SP artillery. This distraction to the Germans (elements of a

Volksgrenadier—'home guard' division) was welcome as Slater's men charged across 75 yards of open terrain before beginning the ascent of Hill 400. A thin layer of snow showed footprints to where mines had been laid, so they were easily avoided.

The foot of the hill was terraced, the upper levels thickly forested, with bunkers constructed in the ruins of the castle at the peak. At 0730, a 'rolling' barrage of artillery kicked off the attack, the Rangers moving behind it as closely as possible. I would have been better to begin the barrage at first contact, but commo was poor to 8th ID's artillery (land line was constantly torn up by German shelling; radio line of sight masked by hilly terrain), so Slater's ground attack had to follow a predetermined schedule in order to stay right behind the rolling barrage.

After the first concentrations, a German popped up from a listening post and raced up the hill. He was not taken under fire because the Rangers felt it would be best not to fire until fired upon. This German, however, apparently got word to his supporting artillery because they began firing pre-registered barrages soon after. US batteries replied with counter fire.

But their scheduled fires were about to move up the hill, and the Rangers were stuck by the slopes and direct fire from the flank (which turned out to be a heavily fortified machine gun complex). Moreover, German mortar rounds were creeping closer till there were only about 200 yards between the US rolling barrage and the German protective barrages. Ranger veterans could see what was coming and it looked awfully bad.

One of Slater's lieutenants, a new replacement, wanted to send scouts ahead to probe the next terrace line. Sgt. Elrod Petty of D Day fame said that would be crazy with tons of artillery in the air, and so advised the lieutenant in the favorite GI vocabulary. The order was repeated to Sgt. McHugh who similarly refused. The lieutenant then ordered a private over the top where he was immediately shot down. At great risk, Petty and other vets crawled out and dragged him back to safety.

On Omaha Beach an officer got the initial attack going by stating the obvious to the men pinned down around him: "There are only two kinds of men who will stay here—the dead and those who will die." This was identical to the situation at the base of HIll 400 as heavy mortar barrages crept down hill onto Slater's force. It was better to climb up into the Germans' small arms fire than be torn apart by mortar shells without fighting back. Sgt. McHugh led the ragged charge by waving his tommy gun, leaping to his feet, and yelling "Let's go get the bastards!" Slater's Rangers swarmed up behind him. From this point, the battle for Hill 400 was largely led by veteran NCO's.

Cpt. McBride, commanding D Company, launched his men to keep abreast of McHugh, but was soon wounded and out of the fight. The Ranger assault hit the first line of German positions just as the rolling barrage shifted. In close combat, the *Volks grenadiers* were outmatched despite their strong positions on commanding terrain. Most died in their holes; those who tried to flee up hill were cut down by the Rangers or the rolling barrage. The Rangers were in a state of mind something like sharks in a feeding frenzy. Bayonets plunged for blood, and at least one EPW was shot after surrendering. To accomplish superhuman feats, some men become inhuman, marking the thin line between heroism and horror, audacity and atrocity.

McHugh's attack of rage carried about a third of the way up Hill 400 till it was raked by machine guns from the wood line near the top. McHugh and other sergeants were hit; it was time for junior leaders to step up as the attack lost control. One of them was Pvt. Bill Anderson who had been busted down from sergeant after a fist fighting incident. Fighting was what he was about, and Hill 400 on 7 December the place for him. He was known for sharpshooting mistakes by his leaders. Now he was about to lead, which he did magnificently.

As the attack paused, he sent down hill for 81 mm mortar FO's who so far had not been able to adjust fire. The feeling of the Rangers on the hill was that those still back on the sunken road (the attack's LD) had better join the party because everyone was needed. Slater was eager for this as well, but the attack had disappeared into the mist, interdictory indirect fire was tearing the lower slopes to shreds, effectively cutting off the attackers from their support.

Rain, cold, and battle damage had disrupted the radio net to virtual ineffectiveness. Slater did not learn the situation on the hill till some medics made their way down with badly wounded Rangers and amazing news: the attack had carried the peak!

No one knew quite how, but individuals and small groups had just dodged ahead, firing the whole way, picking up and using German weapons when their own ran out of ammo, grenading everything that moved until nothing did. Slater sent a wire team back up with the medics; before German artillery cut the wire, Slater learned snatches of what was going on up there:

It was a confused and nearly desperate situation because casualties had been heavy and unit integrity shot to hell. With Cpt. Masney (who had hung some of the first ropes on Ponte du Hoc), Lt. Lomell (who had found the guns), Petty and Anderson trying to set up a defense on the forward slope, Hill 400 was in their hands but subject to murderous indirect fire that had denuded the peak of it previous tree cover. The ground was too frozen and rocky to dig in. Fortunately, a German troop bunker an OP bunker provided some overhead cover for the wounded who were multiplying by the minute. One of them was a shell shock case. There would be many more unless Slater could get reinforcements onto the hill fast.

Williams got a better picture of the situation as small groups of walking wounded stumbled down to the sunken road. He called for A, B, and C Companies to slide platoons over from the adjoining ridges, but the land line was severed once again so this order had to go by messenger and was much delayed. He also reported the objective secured to 8th ID but a plea to reinforce the Rangers success brought no reaction.

It was the Germans who reacted first. Gen. Walter Model, CG of Army Group B (the equivalent of Gen. Bradley who commanded 12th Army Group), was stunned that the strongest point on his front had fallen so suddenly. Because of Hill 400s natural strength, he had manned it with overage and understrength *Volks grenadiers*. That turned out to be a mistake, but Model—known as 'the Fuehrers fireman' for patching up huge holes on the Russian front—would rectify that by committing his best reserves (paratroopers) supported by corps and army level artillery. As added incentive, Model promised Iron Crosses and a week's leave to each man who took part in the recapture of Hill 400.

Against this effort, Cpt. Masney, senior man on the hilltop, was down to a hundred effective Rangers. By mid-morning the rain of shrapnel was reducing that number like Chinese water torture, and fresh German paratroopers, behind creeping barrages, were infiltrating between scattered Ranger positions. Slater did not seem to understand (though he certainly did) that one strong counterattack could knock 2nd Battalion off this hard won objective. Masney decided to go down and tell him and Williams himself. He was the first to discover that German infantry had gotten behind the hill top. On the way down he was captured.

Lt. Lomell was now the senior officer, and his hand was mangled, one finger attached by a single sinew. In great pain, with great courage, he slipped around the ragged perimeter to tie in fields of fire. It held as the first wave of German counterattacks confidently swept up hill behind constant bombardment that tore and shook the earth so much that aiming rifles was nearly impossible. As the Germans recoiled, retreating under US artillery, Rangers grabbed their abandoned weapons. An unforgettable picture in Lomell's mind is that of Sgt. Secor, a machine pistol in each hand, chasing Germans down hill with their own weapons.

Both Bill Anderson and his brother Jack died of wounds. Many more Rangers were bleeding to death, and those who watched them were raging with frustration. If they had fought their way to the top, what was keeping battalion HQ personnel from joining them? Didn't Williams realize that every man who could pull a trigger was vital up here? Even the dauntless Lomell seriously considered pulling everyone off before they were all killed. He was dissuaded by NCO's who would not leave the wounded. Lomell argued that the Germans were noted for giving needed medical attention to enemy wounded.

Too weak to hold everywhere, Lomell had to learn where the Germans were building up to attack. He could slide some BARs around to block them. He boldly sent out two-man recon patrols to check out likely enemy assembly areas down hill. These crafty patrols were eminently successful, so Lomell was ready to meet each thrust with what little strength he had.

During a brief lull, Sgt. Petty went down to plead for reinforcements. Upon entering the Battalion CP (a very battered church in Bergstein), he saw a senior NCO, kicked and cursed him for not being on the hill. Officers intervened, and explained what was being done to help.

It wasn't much, through 2nd Battalion now had priority of fires. Williams had decided to move half of A, B, and C Companies off the ridges to the foot of Hill 400 as a counterattack force, though this repositioning seriously reduced the strength on the hill's flanks. Hell, Petty told him, infiltrators already had the hill top almost encircled. And why prepare to counterattack when you can reinforce and maybe hold the hill?

The only encouragement Petty took back up the hill was that 8th ID was trying to put together a task force to relieve 2nd Battalion. Petty said they might as well be a burial detail. Slater had something for him though: a litter jeep that was willing to try driving to the top. Halfway up it was destroyed by German artillery. Two more volunteer jeps followed, however, and made it to the top.

Evacuation of wounded was now Lomell's obsession. If jeeps made it up here, why not TD's, or, praise God, some of 5th Armor's idle tanks? If Hill 400 wasn't important enough to commit them, why was it worth all the Rangers' lives? December 7th was indeed turning into a day of infamy.

Lomell sent down every wounded man he could. Maybe if HQ counted how many he had left, they might finally appreciate the situation. Rage grows in a man at times like this, a rage equally vented on higher headquarters as much as on the enemy.

Lomell's men were now killing machines, bent upon selling their lives dearly because none expected to survive. At 1300 a strong German attack from the east was turned back; at 1450 another from the north. Germans and Americans screamed at each other as they fired from grenade range and closer. Despite their huge numbers and ruthless determination to expend as many as necessary, the Germans could not overcome Rangers who were now nearly maniacs.

Dusk came early, about 1600. Lomell had to draw in the perimeter for a night defense. An icy rain resumed, as if to mark the Rangers funeral. Exhausted snores began to sound on the darkening hill, a double danger because snores attracted German patrols and showed that the Rangers could not stay awake.

At 1900, six TD's reached Williams who attached them to the depleted companies on the flank ridges. It seemed too risky to send them up the hill in the darkness to integrate into Lomell's patchwork perimeter. The first really good news for him was when Lt. Wintz, new CO of F Company, slipped a section of E Company onto the hill top—just a dozen men but with them was an FO from the 5th Armored who had good commo with his field artillery battalion. From then on, the weight of indirect fire tipped the scale in the American's favor.

But there was still plenty of incoming. Around 2100, while helping the wounded, Lomell was hit again and was evacuated himself. He had done much to save the day and part of the night. Senior commander on the hilltop was now Lt. Wintz. D Company and only sergeant left.

At 2345, rations, water, ammo and replacement weapons came up the hill despite continuous German artillery fire. At 0333 a company from 8th ID reinforced 2nd Battalion, but at the same time the combat command from the 5th Armored was withdrawn from Bergstein. What Williams now had was all that he would get, and the Germans resumed the attack at dawn.

Their supporting attack was at the hill, while a combined arms team broke over a ridge and entered Bergstein. Williams' TD's on commanding ground knocked out the enemy armor and 2nd Battalion headquarters personnel hunted down the infantry. A heavy machine gun at the CP itself delivered defensive fire until the Germans were driven back.

At the conclusion of the dawn attack, the SITREP from the hill reported that D Company was down to ten men. E and F Companies combined had seventeen.

Weather cleared around noon, and P-47s worked over suspected German assembly areas. At 1330, officers of an infantry regiment came into Williams' CP to arrange a relief in place. It would not occur till after the Germans made one last, and their strongest, effort to storm the hill. The Rangers had used the morning to tie in positions, and the attack came just where they expected, a place that was weakest the night before. Cloudbursts of artillery from eighteen FA battalions and dialogues of cross fire from Ranger strongpoints (now with plenty of ammo) sent the Germans reeling back down the hill that became dotted with their corpses in the slushy snow.

When the relief was conducted, 8th ID soldiers found each Rangers crumpled motionless in his pit, gazing blankly ahead, with an arsenal of weapons he had used: M-1s, BAR's, German machine pistols, grenades—and huge heaps of shell casings. Only twenty-two Rangers came down Hill 400 on their feet. 133 had been killed, seriously wounded, or captured. 2nd Battalion had barely survived their toughest fight of the war.

2nd Battalion's mission had been bad by both criteria: inadequate intelligence about the German strength around Bergstein and inadequate combat support from 8th Infantry and 5th Armored Divisions during the first day.

The time allowed for Ranger recon was inadequate. Only after crossing the LD were their scouts able to probe ahead toward their final objective. Even with time for a thorough reconnaissance, there wasn't much intelligence to be learned that would have changed Slater's tactics, but at least his force would have had a good concept of what they were up against. Possibly, recon could have discovered infiltration routes that would have been less costly than the frontal assault that Sgt. McHugh had to lead out of sheer necessity. Captain Morton McBride, CO, D Company, led his company assault with F Company across the field and up Hill 400.

Rangers keenly appreciated the value of meticulous planning. The lighter the force, the more thorough must be their planning; and planning can never really begin until the maximum reconnaissance is accomplished.

Lomell's brilliant defense of the hill top illustrates how recon and planning can be essential even while a force is in hot action. Most commanders in his situation would have conserved his men by letting them stay in the scant cover available on the blasted peak, then hope for the best while fighting statically. Instead, Lomell pushed out small recon patrols to learn where the Germans were building up. Had he not done so, he could not have rushed around a reaction force to threatened points in time to save them.

Initial combat support was inadequate for two reasons: faulty commo, and higher headquarters slow reaction to 2nd Battalion's success. It seemed that 8th ID, the higher HQ most directly responsible, were as surprised as the Germans when the Rangers took their objective in a couple of hours. Hill 400s principal importance was as an OP to bring observed fire on a large sector of the German front. Division FO's should have been within shouting distance of the Rangers as they struggled up the hill, but an FO team didn't come up till that night.

It was not until higher HQ became convinced that Hill 400 was about to be <u>lost</u>, that they provided the support Williams had needed from the <u>start</u>. The adage, "A stitch in time saves nine." is often illustrated in Ranger history, cropping up most often when the Ranger mission appears so difficult that higher planners at higher headquarters are less than confident of its success. This causes them to be more prone to contemplate the consequences of a repulse than they are prepared to exploit—or at least support—unexpected success.

No doubt Rudder, if he had remained in command for another twenty-four hours, would have seen to it that FO's and reinforcements were on the objective on time. Newly promoted and new to his job, Williams didn't have Rudder's stature to deal with division staffs. Thus, a weirdly inappropriate personnel action at army level nearly resulted in a second Cisterna. Even so it was a very close thing: Hill 400 was saved by brains and bravery at the junior level, and at a cost that put 2nd Battalion effectively out of action for a month.

They had been used as a spearhead, but with no spearshaft behind them. They had been like a high-speed drill bit, cutting through hard metal then discarded when burnt out.

Organization and Training

2nd Battalion was down to nearly cadre strength when they went into V Corps reserve near Eupen. Fillers, replacements, and wounded Rangers returning to duty built up strength in dribbles but never higher than 70 percent. Someone in Corps G-3 must have perceived that narrow-front assaults like Hill 400 might not be the best way to employ Rangers—if any Rangers were to survive the war. In any case, 2nd Battalion's training thereafter focused on recon skills and screening, an indication that they would be performing some security missions in the future. Until Arnold returned from detached duty, Williams commanded Force A (A, B, and C Companies) while Slater controlled the remnants of D,E, and F Companies.

Meanwhile, down in Third Army, 5th Battalion had been doing odd jobs for XX Corps as it approached the southern third of the Siegfried Line. Sullivan's Rangers guarded EPW's, secured CP's, tracked down bypassed Germans and acted as connecting files between US divisions. With such diverse duties, there was little time for training in Ranger craft.

Strangely enough, neither Ranger battalion took part in the largest pitched battle in the US Army's history—the Battle of the Bulge. With V Corps, 2nd Battalion was well north of the penetration when Von Runstedt's panzers roared across the icy Our River. V Corps moved out to contain the northern shoulder, but 2nd Battalion stayed behind to screen a twenty-mile sector that had been occupied by an entire division!

Similarly, 5th Battalion was left in the dust as Patton swerved his armor 90 degrees to attack the southern shoulder of the Bulge. Christmas Eve found Sullivan's Rangers happy to be in Metz. By New Years they, like 2nd Battalion, were outposting a twenty-mile sector. 5th Battalion's was along the Saar River just inside Germany opposite the Siegfried Line. Fortunately, that would be the site for their next operation.

Good Missions:

XII: Zerf, Germany, February, 1945

Though Hitler had shot his bolt in the Bulge, defensively the *Wehrmacht* remained strong. His West Wall was largely intact, while behind it between twenty and eighty miles, was the broad Rhine River, a formidable final obstacle before the Allies could cut out the heart of the Third Reich. The key to a breakthrough in Third Army's sector was the fortified city of Trier, a bastion in the Siegfried Line and gateway to the Moselle Valley, the high speed corridor to the Rhine at Coblenz.

To get at Trier, Patton first had to clear out the triangle of territory between the Saar and Moselle Rivers, a rugged, heavily wooded wedge about ten miles broad and twenty deep. At the base, was a mile-thick belt of minefields, pill boxes and obstacles called the Siegfried Switch Line. Third Army had been battering at it since November but, because of ten-thousand anti-tank 'dragons teeth' and the dense forest, it had proven impregnable to armor.

Gen. Walton Walker, XX Corps Commander, planned to take on the base of the triangle with an infantry and armored division, the 94th ID to pry open the Switch Line and the 10th AD to exploit the breach. On 7 February, 5th Battalion was attached to 94th ID as a counter-reconnaissance screen for the division's attack preparations. Sullivan used the mission to hone compass and navigational skills. They would need them because the Rangers were to perform the deepest penetration of the corps attack, through wilderness as thick as that which confronted the original 18th Century rangers.

The XX Corps attack kicked off with initial success. 94th ID infiltrated the Switch Line, and 10th AD revved up to shatter it. However, just east of the Saar River, the road hub of Zerf allowed the Germans to move reinforcements rapidly and hit the flank of Walker's penetration. Cutting this road net was therefore critically important. 5th Battalion was given the mission of getting through the Saar-Moselle triangle to block the roads out of Zerf.

To do so Sullivan had to pass through 94th ID's advance, cross the Saar, penetrate German defenses, then ambush and mine the Zerf roads. When this was accomplished, linkup was scheduled with task forces from both 94th ID and 10th AD. If all went right, 5th Battalion would be on its own (though with plentiful artillery support) for no more than forty-eight hours, beginning on 22 February.

Time was adequate, and Sullivan (like Darby) believed strongly in personal reconnaissance. He selected an approach march that required breaching minefields but afforded concealment and promised more surprise than if he moved out from where 94th had previously accomplished a breach.

Control was his primary concern. On the previous day he had received about fifty brand new replacements, bringing his total strength to 400 Rangers. The dense forests, dim light, the uncertainty of the new men, required a tight battalion formation and precise navigation. Speed was difficult in the tight woods anyway, so control was the watchword. Caution as well: 5th Battalion was not opposed by *Volks grenadiers* but rather the 136th Regiment of the 2nd Alpine Division, crack mountain troops who were very familiar with the rugged terrain.

Though the vegetation favored clandestine movement, the combat load for the Rangers, necessary for the extended mission, made them heavily clumsy. Six extra ammo boxes were lugged for each machine gun, and every rifleman carried an anti-tank mine. No doubt Sullivan wished for some of Darby's mules!

Before leaving 94th ID's lines, Sullivan consulted with their lead elements. They advised that the route he had selected (guiding on two dry stream beds) was well covered by the Germans, so he decided to move out on a compass azimuth after crossing the Saar on a footbridge the 94th had thrown up. He asked them for a diversionary probe as 5th Battalion set out at 2345 on 23 February.

The formation was two columns of three companies each, with the companies in file, and the columns separated by about fifty yards. That was about the best compromise between control and the ability to get sufficient firepower forward upon contact. The point squads were heavy in tommy guns and BAR's with FO teams from 94th ID artillery right behind them.

Contact of course, was the last thing Sullivan sought during the penetration. "Movement," he would say later of the entire operation, "was our best defense."

With each man holding on to the next ones pack, movement was glacially slow during the night. German H&I artillery fire seemed psychically accurate—six Rangers and an FO radioman were killed, along with twenty WIA's, by random fire before the infiltration got well underway. Evacuation of these casualties further slowed progress and resulted in the force being split.

To regain control, Sullivan called for a pause on a phase line. After a head count, it was discovered that a platoon led by Lt. Gambosi had become separated from A Company. Radio contact with him was achieved but in the dark forest he could not find the force. At the phase line, security patrols came upon an unwary German position and took twenty prisoners. EPW's would become an increasingly serious problem as the operation went on.

At dawn, Gambosi had run into heavy mortar and machine gun fire. 5th Battalion could not wait for him to break through so Sullivan ordered him to return to 94th ID's lines. With daylight, Sullivan spread the battalion into three columns of two rifle companies each, with headquarters moving with the center column. The route was changed 90 degrees to avoid German positions discovered Ranger scouts. However, this offset complicated navigation so much that Sullivan called for two howitzer rounds to be fired on the objective (the planned ambush site on the road from Zerf). The impacts seemed to be a couple of miles ahead and to the left. Slowly the Ranger formation swung around to the north and came on course.

Now avoidance of contact was all the more vital if surprise was to be attained. Four recon patrols, led by the craftiest Rangers were sent ahead to observe and report back, not even firing if fired upon unless absolutely necessary to escape. Movement in these circumstances was reduced to about a quarter-mile per hour. Around 1000, these scouts reported that three pillboxes stood across the route of march. It would take a couple of hours to bypass them, but the Germans seemed unwary and perhaps they could be overwhelmed without much shooting.

With the objective believed to be less than a mile away. Sullivan decided to take the pillboxes in a rush, withholding a base of fire unless the Germans reacted. They didn't. What little fire there was muffled by the dense forest. Six mountain troops were killed, fifty captured, but nevertheless Sullivan had to assume that his penetration had been compromised.

Indeed, while 5th Battalion closed on the objective, flank guards for the left rear company encountered a fifty-man German patrol sent out to investigate the action at the pillboxes. The flank guards fell back toward the main body. The Germans took them to be a small recon patrol and moved in to wipe them out. D and F Companies turned and set a hasty ambush with the flank guards as bait. Hit from the trees, both sides and center, twenty Germans were killed and the rest captured. This action resulted in a total bag of 110 EPW's (nearly a third of Sullivan's strength), and he was about two miles deep behind enemy lines.

Luckily, the 136th Regiment still remained unaware of the Ranger penetration. Germans had other threats as the 94th ID and the 10th AD were also probing ahead. Lady Luck smiled again when the 'lost platoon' of Lt. Gambosi met up with armor heavy Task Force Riley from 10th AD and was able to tell them about 5th Battalion's progress the previous night. Lt. Col. Riley told Gambosi to mount his Rangers in the task force's half-tracks and they'd go looking for Sullivan after completing some unfinished business. Linkup with 5th Battalion had been assigned as Riley's final mission.

Meanwhile Sullivan had arrived at the objective. His Rangers gratefully unloaded the backpacked AT mines and sewed the road and its embankments. By 1600, just before dark, 5th Battalion was dug into two separated ambush sites, awaiting either German visitors or linkup with Americans.

The former came first, because in the darkness several EPW's got away. An hour later scattered mortar fire fell around the ambush sites. Sullivan's recon patrols had some sharp encounters. Yet the first Germans to enter ambush kill zones were in a white ambulance; it was politely stopped as if the Rangers were MP's, and the doctor inside advised that he was a prisoner of war.

"No!" he exclaimed. "You can not be here! This is four-thousand meters meters within our lines..."

The doctor was soon assured of his situation, and thereafter he vigorously provided equal service to wounded of both nationalities.

Sullivan knew that if he passively sat on his ambushes, eventually the 136th Regiment would mount a coordinated attack, and that the Rangers could be in real trouble if linkup did not come off as scheduled. So to mislead the Germans of his strength and center of gravity, Sullivan had strong combat patrols radiate out from his ambushes, generally to raise hell, hit targets of opportunity, and create confusion—the sort of mission at which Rangers excel and love to perform.

One such patrol got plenty of such action at a four-building complex called Kalfterhaus. For its night attack, the patrol wanted an artillery prep, but locations were still unclear and it took an hour of registration fire all around the countryside before the howitzers could fire for effect. They were effective. The buildings fell and thirty prisoners taken at the loss of one Ranger KIA and one WIA.

Similar sharp skirmishes went on through the night. When patrol leaders reported back to the ambush sites, Sullivan began to feel that they might not be at the right places. Apparently the 2nd Alpine Division had decided to bypass the main road and was using a cut off trail to move reinforcements north to meet the thrusts of XX Corps.

The intent behind 5th Battalion's mission was to stop those reinforcements, so in the morning Sullivan decided to displace. His Rangers groaned at having to dig up the mines and shoulder them again. Guiding on the Zerf-Irsch road, they reoccupied the ruins of Kalfterhaus, then moved north looking for trouble or TF Riley.

Trouble arrived first but not before AT mines were re-emplaced and high ground organized to create a long, strong flanking ambush by the road. A single German SP gun was the first victim, then the Germans came on in company-size efforts to clear the road. At the north end of the ambush, A Company was hard hit where it didn't have supporting fire from the rest of the battalion, nor an FO with commo to 94th ID. For forty-five minutes, fighting was up close and personal before the Germans pulled back. 150 EPW's were taken but A Company had pretty much shot up all their ammo. To the south, Kalfterhaus had to be abandoned to consolidate 5th Battalion's position. Men without ammo guarded EPW's who now numbered over 200.

Though German attacks on the new roadblock had been driven off smartly, 5th Battalion was in real trouble unless they could get ammo resupply. Indeed the Germans seemed to sense Sullivan's main strain, calling from the woods for the Rangers to surrender or be annihilated while holding empty weapons.

The situation called for an improvised aerial resupply. Such a contingency had not been planned for, so it began with spotty results. The god send was good and constant commo with 94th ID Artillery. Sullivan's FO's persuaded them to divert two of their single-engine observation planes (Piper Cubs) from adjusting barrages to delivering the critical item, machine gun ammo, by free fall.

Creating a drop zone in dense forest was less of a problem than ground fire, as the Germans quickly realized what was being attempted. Unable to get under 1500 feet, the brave pilots first loads went to the Germans. After suppressive artillery fire the pilots adjusted themselves, they were able to drop crates on the Rangers from about 800 feet, crates which were lethal missiles that injured several men on the ground, and a frustrating disappointment when opened: the hard fall had bent or ruptured so many of the casings that only a few could be used and only in rifle clips. These rounds also had to be thoroughly cleaned of dirt from impact before they could be used.

Back went the pilots to refuel and reload with belts of ammo packed in metal 50 caliber boxes within the wooden crates. This ammo fared better from the drops, but little could be delivered in each pass. After a half-dozen drops, one of the planes was damaged by ground fire. Subsequent drops were cancelled because Sullivan felt he might have just enough ammo if strict fire discipline was enforced.

In case ammo ran out, Sullivan sent out two strong patrols to find exfiltration routes. Then the Rangers rearranged their positions for night defense, dug them deeper, set out LP's and awaited German reaction they were sure was coming.

It did at 0300— an attack by two battalions of the 136th, oriented on the road from both end of the ambush. Curled into a long oval, 5th battalion's position was thin but hard. Thick woods and utter darkness prevented Sullivan from moving a reserve to threatened points. Each company, platoon, and squad had to hold out with what it had, expecting nothing more except for priority of artillery fire—a tough call for the Sullivan and the FDC because the Germans coordinated attack hit just about everywhere simultaneously.

The Alpine troops, though admirably stealthy, were disadvantaged because of the unavoidable noise caused as they moved blind in thick woods. The Rangers learned to fire and listen, fire and listen, as only the sound of German and American weapons provided any indication of locations.

E Company was hardest hit, its machine gun ammo expended. In the most difficult maneuver— withdrawal under pressure in darkness— the Rangers pulled back to a firebreak in the woods where, at dawn, they would have at least a narrow field of fire. Still the Germans came on, using the last minutes of darkness to penetrate.

They were stopped at the firebreak, and with a little visibility Sullivan pulled a platoon from a less threatened position and had them sweep out to E Company's previous defenses. The Germans broke under the counterattack and accurate pounding from 94th ID's artillery. Fifty enemy dead were found in the sweep, a hundred more elsewhere around 5th Battalion's perimeter.

The Germans reported more than 50 percent casualties from their night attack. 136th Regiment was not much of a threat thereafter, but they would be very tough to dislodge.

At 1155, tanks from 10th AD made contact with 5th Battalion and took out Ranger wounded. Sullivan set off on a reconnaissance in force because XX Crops needed to clarify the situation which was still quite confused: American and German units of various sizes were colliding in the woods, positions were lost and won back and forth. In this fluid situation, the Rangers both attacked and were attacked. The most significant action of the day (26 February) was when a German battalion attacked where they believed the Rangers to be but Sullivan caught them in a cross fire from 90 degrees to their approach march, killing 150. The Germans didn't seem to have an ammo problem at all: every third Kraut in this group carried a machine gun.

Around noon on the 27th, TF Riley linked up, bringing a very happy Lt. Gambosi and his platoon, plus much needed belted ammo. Everywhere in the Saar bridgehead the Americans were breaking out but much dogged fighting remained to clear the triangle. 5th Battalion was attached to a 94th ID regiment, and for the third time cleared out Kalfterhaus. The EPW bag that day was over 200. Now with plentiful ammo and support from TD's, Sullivan was loaded for bear. However, he had trouble meshing with the TDs which attracted much German artillery, forcing the Rangers to disperse, thereby losing the impact of a combined arms team.

The last big fight of the operation was for an objective called Hill 4 (not its height), the CP of the 136th Regiment. The Germans defended fiercely with abundant ammo and rockets, and threw back three Ranger penetrations that could not be reinforced soon enough. These to and fro fights resembled the Battles of the Wilderness in the Civil War—short visibility and close range combat with no quarter. Combatants on both sides were often cut off, sometimes captured and escaped, only to be recaptured or sometimes taking their former captors captive. At one location the Rangers and Germans got so familiar with each other that they shouted each others names.

Finally, around 1500, the last German counterattack was repulsed and Sullivan moved into 136th Regiment's CP bunkers. The hundred EPW's taken reported that their regimental commander had been killed. Vengefully, the Germans pounded the location with unremitting artillery for 36 hours.

Both sides regrouped and caught their breath for the next two days; then on 2 March the Rangers led off again behind a barrage that started on the objective (another low hill), rolled back to the two attacking companies (A and F), then rolled out once again to the objective 300 yards away in thick woods. In that way, friendly artillery was both guide and guardian for the Rangers.

Considerably softened up, the defenders did not put up much of a fight. Two pillboxes were taken in a rush, then used as a base of fire as other positions on the hill were assaulted closely behind well adjusted artillery fire. The pounding had been so intense and extensive that the Germans were reeling from lack of sleep. A hundred were captured, fifty killed, without a single Ranger KIA.

Yet German artillery rained down on the now secure objective. Captain Charles Parker, CO of A Company, noticed that some German land lines were still in place. EPW's confirmed

27 July 1944. 75mm Mt. Pack Howitzer firing. Brass in foreground represents nine hours infantry support firing.

that the lines led back to a fire direction center miles away. Rather than cut these lines—German artillery would have continued pre-registered fire missions anyway—Parker moved the EPW's out into the open, and had their senior member call the FDC to lift their fire. The German officer consented to make the call, and artillery stopped bursting on the hill, a lull Parker used to dig in the Rangers.

The German officer was outraged, accusing Parker of giving him the impression that the lull would be used only for evacuating EPW's. That's right, Parker grinned: they would be taken to the rear soon enough, but in the meantime the Rangers would provide protection for themselves. It was a fine point in the international law of warfare; in any case, the EPW's moved out, the German bombardment resumed, but when it did the Rangers had made themselves some cover.*[*FN: Under the Geneva Accords, POWs are not allowed to be used as hostages, nor exposed to any more danger than their captors. Legitimately, Cpt. Parker proposed that the German artillery hold their fire so as not to endanger the EPW's. What was legally questionable was moving them out into the open where they were more exposed than before—but no more so than the Rangers. So the senior EPW's argument was that his men were, in a sense, being used as hostages. He had a point, but so too did Parker: overhead cover was limited, and the EPW's no more entitled to it than the Rangers.

On the question of using the lull to dig in, the Geneva Accords require that neither side take tactical advantage of a 'truce'. But a truce involves certain formalities like a flag of truce and negotiations. If Parker did not become a lawyer after the war, maybe he should have because he had a shrewd mind for legalities! There was no flag of truce nor negotiations. He was not negotiating with the Germans, he was allowing them to talk to each other. That resulted in <u>their</u> decision to lift their artillery, an action which produced a <u>partial</u> cease fire (presumably small arms fire continued to be exchanged) during such a lull Parker could take any steps he chose to protect his men and still be within the Geneva Accords.

It is wryly amusing to read how the Germans became such sticklers for punctilious observance of the Geneva Accords, in light of their recent massacre of POWs at Malmedy during the Battle of the Bulge. If the author had been in Parker's place, my one word answer to the German officer would have been "Cisterna"—reminding him of where Ranger POWs had been subjected to humiliation in Rome after inhumanity on the battlefield, such as the monstrous use of captured Rangers as hostages ('human shields') to force other Rangers to surrender.]

On 3 March, the Germans went into a delaying action as they evacuated the triangle. As Bob Black described the aftermath:

The 48-hour mission had lasted nine days, and 5th Battalion had been bled dry. A Company had one officer and twenty-four EM remaining. B Company had begun the operaton with a strength of three officers and fifty-nine EM; thirty-seven had become casualties. F Company had two officers and eighteen EM remaining, and so it went with the rest of the battalion. During February, Sullivan had lost 34 Rangers killed, 140 wounded and 12 missing.

Despite such losses, the Zerf operation was the best sort of Ranger mission, exploiting their unique mobility in difficult terrain that allowed plenty of concealment and room for maneuver. Intelligence—mostly gathered by the Rangers themselves—was adequate, while combat support was the best ever provided the Rangers in Europe. The results were gratifying: at least 300 Germans killed, 600 captured, the enemy lateral lines of communication effectively severed and movement of their reinforcements blocked.

So the lessons learned from Zerf were overwhelmingly positive: good recon, good planning that provided flexbility, and excellent integration of supporting arms. A small instance sometimes illustrates a large quality. Such an example was the use of howitzer concentrations to mark locations in dense vegetation. In wilderness like that of the triangle, sometimes artillerymen on the guns are the only soldiers who know where they are. They can sure help guide other soldiers! Another example is how Sullivan sent out recon patrols from his ambush site to find at least two escape routes if exfiltration became necessary. Planning like that was constant and ongoing, not something that ended when the operation began.

What no amount of planning could solve is what a long range penetration force can do with EPW's, especially in the numbers 5th Battalion captured them. The common expression, "take no prisoners" is not a legitimate order under the Geneva Conventions, but it is practical advice to preclude a dilemma when EPW's delay and endanger a force deep behind enemy lines.

The only serious planning oversight for the

Zerf operation was for resupply. The air drops were an expedient ginned up on the spurt of the moment and by necessity. It is no surprise that they didn't go well, though they worked after a fashion. Perhaps the reluctance of higher headquarters to 'misuse' observation aircraft for bundle drops prevented at least some paper rehearsals. It is curious though, that in their after action report Third Army almost bragged about the drops success, an enthusiasm not shared by the Rangers involved.

Zerf was a high note with which to mark the end of important Ranger operations in Europe. Within a week, Third Army had punched through the Siegfried Line and seized Trier. Farther north, Ninth Army broke out of the Roer Valley on a broad front, advancing with First Army to the Rhine between Rheinberg an Cologne, a span of seventy miles. Now all the Allied generals vied to be first across. That honor went to First Army at Remagen where they found an intact bridge on 10 March.

For the rest of March the Allies aggressively expanded bridgeheads against fitful and demoralized resistance. Germany east of the Rhine crumbled into pockets between lunging Allied armies and the onrushing Russian mass of force. In this scenario, the Ranger battalions followed mechanized units like kids trailing a parade.

These last campaigns against Nazi Germany were not easy or without danger, and often exhausting as attack became pursuit with no rest for the pursuers. Bypassed or cutoff SS units could, like hornets, sting before they died.

The last month of the war in Europe was more like an adventure than a campaign. One night would be spent in a castle drinking fine wine, the next in the Black Forest tracking down Nazi guerrillas. There were a number of exciting episodes, like ambushing some staff cars that tried to run a roadblock. Dead in one of them was an SS major general, CG of a panzer corps. No one bothered to bury him with any honors.

All along the way were German formations marching west with hands on their heads; thousands upon thousands of them, often guarded by no more than a squad. The dreaded Nazi war machine was a junkyard. In all directions came and went the former slaves of Nazidom, now known as displaced persons, searching for homes that had vanished.

Fifth Battalion was ordered to guard a signal dump near Lehesten, Germany. Patton had noticed that many Germans were ignoring or even denying the existence of concentration camps. One of the chores most rewarding for 5th Battalion was to escort thousands of citizens from Lehesten to the hideous camp of Buchenwald which still reeked with the stench of sub-human conditions and exterminated prisoners. What the Rangers had been fighting against could not be better illustrated than in this hellish visit.

The first week of May 1945 was the last of the European war, and found 2nd Battalion closing in on Pilsen, the beer capital of Czechoslovakia while 5th Battalion was on the Austrian border and the banks of the Danube River at Passau. Between them they had accounted for thousands of German troops killed, many more captured even before the rout during the last month. Ranger successes had not been cheap: their Purple Hearts also numbered in the thousands.

Second and 5th Battalions had been, so to speak, the second generation of Rangers, following the three battalions who had slashed their way through the Mediterranean. Like their predecessors under Darby (KIA on 7 May 1945), 2nd and 5th Battalions had lifted high the torch of courage and accomplishment that had been passed on to them. They, in turn, were about to pass the torch to men they didn't know in the Pacific, the third generation of World War II Rangers, 6th Battalion.

1945. Night tracer fire at Japs attempting to use Burma Road during night. Fire and smoke show successful effect.

Rangers in the Pacific

14 July 1944. PFCs Hugo J. Wagner, Albert M. Weaver, in pillbox taken from Japs with equipment left.

Introduction

The tropical islands of the war against Japan were better terrain for Ranger operations than Europe. From their birth in the wilderness of North America, boondocks had been the best setting for rangers, and the Pacific theater returned them to such surroundings. Indeed, for the rest of the 20th Century, Rangers would be fighting and stationed in the 'third world'— underdeveloped countries.

In contrast to all previous Ranger battalions, the 6th was a converted unit rather than a brand new one made up of volunteers gathered from elsewhere. Surprisingly, 6th Battalion's previous identity was as the 98th Field Artillery Battalion armed with 75 mm pack howitzers. They were all ranger volunteers in the sense that if they wanted to stay with artillery they could leave at the time of conversion. Their CO was Lt. Col. Hank Mucci who's previous job had been provost marshal of Honolulu. Surprisingly for an MP, he was not a strict disciplinarian; 6th Battalion uniforms often did not include shirts, and off duty fun was pretty much what a man could get away with.

25 September 1944—around the time their brother Rangers had cleared out of Brest—6th Battalion was activated (strictly speaking, 're-designated') in New Guinea where the 98th FA had fought during most of 1943. There was plenty of jungle and enough Japs left on that island for the new Rangers to get the necessary infantry OJT.

They were working for the Sixth Army Commanding General, LTG Walter Krueger, a rough, gruff infantryman who saw potential for Rangers in raids against the extended Japanese defenses that had to cover thousands of miles of coastline among thousands of islands. Consequently, 6th Battalion did a lot of training in amphibious operations. However, no amphibious raids were mounted before October 1944, D Day for the invasion of the Philippine Islands. Like Rudder and Schneider, Mucci would get into the war at its most important phase.

Good Missions:

XIII: Leyte Gulf, PI, October 1944

MacArthur's famous vow, "I shall return!," had waited two years for fulfillment, two years in which Japanese and Allied armies had not (except in Burma) seen much of each other. But in the air and at sea during that period the Pacific war largely had been decided. In this far-flung theater—covering more than a third of the planet—a game like 'scissors, paper, stone' was played. It went like this:

Who rules the sea can cut off the land (islands).

Who rules the air, controls the sea.

Who holds certain islands determines where air bases can be established.

Air power could also be launched from carriers, and this was the major weapon of the war in the Central Pacific, a vast sub-theater of its own under five-star Admiral Nimitz. In the Battle of Midway in June 1942, US air ruled and controlled the sea thereafter. By 'island hopping', Nimitz was able to move air bases ever closer to the Japanese home isles.

MacArthur's sub-theater was called the Southwest Pacific and included the vast Indonesian archipelago (then known as the Dutch East Indies), New Guinea, and the Philippine Islands. He relied less on aircraft carriers because with short-range amphibious landings, he could leap frog air bases forward. However, for the big 800-mile jump from New Guinea to the Philippines, he needed large carrier task forces which Nimitz loaned him.

Gen. Yamashita, the Japanese CG in the Philippines, had a similar but more difficult problem than did the Germans in planning his coastal defenses. The same basic question was whether to spread thinly along the beaches with the hope of stopping an invasion at the high water line, or holding strong reserves for counterattacks. Like Rommel, Yamashita faced overwhelming enemy air superiority. He also lost considerable inter-island mobility for moving reinforcements because of US naval power.

Yamashita's decision was to keep the strongest elements of his army on Luzon where they could deny MacArthur the vital port of Manila, and possibly defeat him in a major land battle. The rest of the islands Yamshita garrisoned lightly with instructions to meet the landing Americans early and head on. In conjunction with the Japanese Navy, he had a big surprise for them once they got ashore.

Thanks to Filipino guerrillas and small groups of US stay behind forces, MacArthur's knowledge of Yamashita's dispositions was very good. He chose the island of Leyte as Krueger's objective. Leyte was centrally located, offered good harborage, and once captured could provide air bases in range of Luzon.

Leyte was defended by the 16th Division, part of Gen. Suzuki's Thirty-fifth Army. With small intercoastal boats, he hoped to be able to reinforce Leyte with two more infantry divisions before Krueger could get all his force (four infantry divisions) lodged ashore. Of course Suzuki did not know if Leyte would be MacArthur's objective. Unfortunately, MacArthur did not have enough ships to pin down Japanese with feints at other islands. So where the Americans first struck would be the one and only battlefield.

First to strike would be the 6th Rangers. Indeed, they would come ashore three days before the rest of Sixth Army. Their mission was to seize one large island (Dinagat) and two smaller ones (Homonhon and Suluan) at the mouth of Leyte Gulf in order that mine sweepers could clear the way for Krueger's landings.

Dinagat, some fifty miles long and ten wide, was so jagged and inhospitable that it was practically uninhabited by Filipinos. Consequently they could not tell Krueger's G-2 much about Japanese strength on the island. The best guess was maybe 500. The Rangers would have to find out for themselves. Homonhon and Suluan were believed to be unoccupied by the Japs.

US warships had roamed freely around the Philippines since June when Japanese air power had been annihilated in the Battle of the Philippine Sea. So a small naval task force of ten ships was nothing new for the Japs to notice, or do much about. Six of those ships (the others were destroyers to provide shore bombardment if necessary) carried the 6th Rangers and hove to between Dinagat and Homonhon on the night of 16 October. At first light on 17 October, D Company launched for Suluan; A,C,E, and F Companies for Dinagat. B Company whose objective would be Homonhon, remained in floating reserve with Battalion HQ on the British ship, *Ariadne.** [*FN: For Rangers familiar with Greek mythology, the name of their vessel was not propitious: the goddess, Ariadne, had been a castaway on an island.]

Cpt. Leslie Gray commanded D Company. Their job was to secure a dilapidated light house at the south tip of Suluan, bypass any Japs, then re-embark and rejoin HQ at sea, relieving B Company as battalion reserve so they could go on to Homonhon.

D Company hit their beach at 0800 in a driving rainstorm. They were grateful for the concealment it provided, and the landing was unopposed though there were signs the Japs had taken notice. Local guides helped Gray scoot up the coastal trail. Enroute to the lighthouse, they flushed some Japs out of four storehouses which the Rangers destroyed. In this action D Company suffered the first KIA and WIA. Losing four men themselves, the Japs escaped into dense jungle.

Gray pressed on toward the lighthouse (the beacon was not working) as the storm grew worse. On a peak of sheer rock, the lighthouse rose to a total height of 300 feet, and was scaled by only a single flight of stairs. A few Japs and a machine gun could have made life very difficult for the Rangers but for reasons unknown, the inscrutable enemy had fled. Inside the lighthouse, they had left radar equipment which Gray destroyed. Mission accomplished, D Company countermarched toward the beach and radioed for their landing craft to meet them there.

Meantime the storm had grown into a near typhoon. The landing craft had broached in the raging waves and could no longer be reached by radio.

Gray set up a tight night perimeter, worried that the Japs could use the noisy storm to infiltrate the beachhead. Weather cleared somewhat the next morning, but the boats couldn't make it in. While he waited, Gray sent out two combat patrols in directions the Rangers had not yet traveled. They were to maintain radio contact with Gray so that they could be called in immediately if the boats reached the beach.

One patrol went half way around the island without contact. The other penetrated the interior and clashed with a small force in a large field of high grass. Firing on both sides was virtually blind. One Ranger died of wounds while five Japs were killed. The Ranger was buried in the center of a tiny native village. Without further incident or casualties, D Company returned to sea, leaving a radio with the Filipinos to report any Japanese activity.

B company, commanded by Cpt. Arthur "Bull" Simon had been delayed for a day by the storm when Gray relieved him.* [*FN: As a colonel in special forces twenty years later, Simon would lead the daring raid to free POWs from a North Vietnamese prison camp at Sonh Tay.] At 0900 on 18 October, they launched for

Homonhon in plywood landing craft. Their beach was narrow and surfy, allowing only two craft at a time. The landing was unopposed, Filipino guerrillas reporting the island to be unoccupied.

With this good news, Simon cut across the isthmus of the island to Colasi point, the southernmost tip of Homonhon where a 300-pound navigational aid was set up for Krueger's invasion force that would pass through the strait between Dinagat.

With the mission accomplished and the island secure, B Company had the honor of raising the first American flag on liberated Philippine soil. Simon ran up the Stars and Stripes on a stripped palm tree at Colasi Point. The after action report stated that "It was as if the pulse of each man's heart pushed the flag a little higher."

Meanwhile on Dinagat, most of the rest of 6th Battalion doing more exploring than fighting. The landings went as planned, and two companies secured a base for operations while two others pushed their way around the ragged east coast. The only casualties were seven Jap sailors believed to be coast watchers. They were killed before they could report that Dinagat was now Ranger country.

Sixth Army sailed into Leyte Gulf on 20 October, landing four divisions against light resistance. They had pushed west across Leyte for three days before Yamashita popped his surprise—an all out, do or die naval raid to destroy the invasion fleet. He was confident his gamble would succeed, so cock sure that he boasted he would soon be in a position to demand MacArthur's surrender.

In a clever maneuver, the Imperial Navy lured Admiral Halsey's big carriers north to engage the last of the Japs' carriers. This bait worked, leaving Leyte Gulf uncovered for three large surface fleets to break through a light screen, penetrating almost close enough to bombard Krueger's transports. The outgunned, outnumbered US Seventh Fleet fought magnificently, losing two escort carriers and three destroyers but driving off vastly superior forces, one of which almost made it through the Surigao Strait where Rangers had a ring side seat from Dinagat. From then on, they patrolled the shore extensively to round up shipwrecked Japanese sailors.

The Japs were lucky if Rangers found them before Filipinos did. The natives' hatred for the evil occupation was intense—the Japs had treated their subjects with the same cruelty as the Nazis. Wholesale murder, vast pillaging, enslavement, torture and rape were SOP's for the arrogant soldiers of the emperor who felt themselves to be the master race of the Orient. Even in defeat they did not stop their revolting practices as Simon found out when Filipinos on Suluan radioed that the Japs had come out of the jungle, where D Company had chased them, and were taking barbarous reprisals.

On 27 October, just after the US victory in Leyte Gulf, B Company scrounged up three sailboats and eleven canoes for a revenge mission to Suluan. Another storm turned them back after swamping seven of the canoes, but on the 29th Simon tried again and by 1100 , after ten hours on the choppy sea, Rangers again came ashore on Suluan.

Natives reported that most of the Japs were holed up around the lighthouse. They also had squad-size security forces on the approach trails. The only other Japanese activity on the island were foraging parties that robbed the natives of food, killing most of their livestock. Now it was payback time for the Filipinos who eagerly provided Simon with all the intelligence he needed.

The most efficient way to help the Japs join their ancestors was for a destroyer to stand off Suluan and blast that lighthouse into orbit. But the Seventh Fleet was scattered and fully engaged in the aftermath of the Battle of Leyte Gulf, so Simon had only his own resources.

His plan was take a squad of Rangers at night up the cliff on which the lighthouse stood while the rest of the company slipped around the enemy security elements on the approach trail. If Simon fired an orange flare, B Company (-) would storm the base of the lighthouse. A yellow flare meant Simon controlled the stairway so the assault should come straight up. A red flare meant his squad had captured the lighthouse itself, cancelling the supporting attack.

The moon was bright, and the native guides knew just where they were going. The planned route was so close to Japanese positions that the Rangers pulled socks over their boots to deaden their footsteps. With his company in place, Simon's team began the 240-foot climb. Cut by coral but aided by vines that grew on the cliff face, the first few men reached the base of the lighthouse around 0200.

On a narrow ledge were four small buildings, damaged by D Company's previous visit on 17 October. Simon's first sign of Japs was the stench of their latrine at the edge of the ledge. In the buildings, he could hear them talking. One came out to use the latrine. Unsuspecting, he ambled straight for a crouching Ranger who felt in the next seconds he would be discovered. Raising his tommy gun, he prayed, lead into the Jap, and the fight was on.

When it started, half of Simon's team was still climbing while those on the top raked the buildings with automatic fire, and hurled in WP grenades that belched out huge puffs of white, caustic smoke—quickly followed by Japs raving with pain. They were quickly cut down, but soon Jap grenades began falling from the lighthouse. Simon had only one light machine gun. In order for it to fire vertically up the lighthouse, he and the gunner dragged it to the ledge, Simon holing one leg of the tripod while the gunner simultaneously held another and pulled the trigger with the other. His fire was wild but added fright to the Japs' confused surprise.

Apparently only a few Japs were at the top of the lighthouse, the rest tried to get out at the bottom where they were met by fire from Simon's team. One by one, Japs toppled off the cliff. The fight was at very close range, seldom exceeding twenty yards. Sgt. Potts was shot in the shoulder while trying to throw a grenade. With the pin already pulled, he handed it to a Ranger beside him who threw it as intended.

Simon was too busy to launch a signal flare, but B Company could tell he needed help and assaulted. Japs had been sleeping in the jungle on the flanks of his attack. Their appearance

Rangers equipped with tommygun.

developed into a broad firefight as B Company spread out to prevent escape. By first light, they had the entire garrison penned in at the point where the lighthouse stood. Japs were literally driven into the sea, some of them choosing to jump to their deaths on jagged coral rather than accept 'dishonorable' capture.

The last holdouts were those at the top of the lighthouse. Simon's team was protected from their sporadic fire by occupying the ground floor. He decided to blow them away by destroying the lighthouse with demolitions which B Company got up to him by covering the demo team with suppressive fire.

Simon had consulted with the natives the night before. The lighthouse was important to them but could be rebuilt. They were enthusiastic that the Japs in it die, even if it meant loss of the old lighthouse. The bottom floor was rigged with demo and long fuzes; Simon's force evacuated, and soon the base bulged from detonations. The lighthouse wobbled, then toppled down the cliff, breaking in half before crashing into the rocky surf. The natives cheered at the symbolic disappearance of the Japs on Suluan.

B Company's losses had been two dead and seven wounded, while the Filipinos reported that the Japanese garrison had been a company. For the most part, they represented the last organized resistance to the Rangers in the Leyte Gulf operation. On Dinagat and small islands nearby it, 6th Battalion remained busy for another two weeks, not to seize objectives but rather as a posse for the Filipinos, tracking down bands of Japanese who turned to banditry for survival. They had no commo with their headquarters and no internal control. This would be a situation repeated throughout the Pacific war: isolated and independent groups of Japs with no military mission, who became essentially hunted outlaws. Some survived in this condition for decades, ignorant that the war had ended twenty years earlier.

Without radios, they never heard the Emperor surrender in 1945. Without his order to surrender, they never considered doing so on their own. Their loyalty and patriotism was admirable; their conduct as soldiers despicable, and an eternal blot on the reputation of the Japanese nation. Rangers never had the slightest qualms about killing them wherever found, and were more than accommodating to grant their wish never to surrender.

Lessons learned from Leyte Gulf were the importance of native guides and intelligence, and improvisation of water transportation, also a native asset. These aids could not be precisely planned for in operations orders, but had to be noticed and exploited when the opportunity arose on the ground. A curious feature of operations in such a vast archipelago as the Philippines was the necessity to plan for both maximum and minimum resistance. An example was the lighthouse, abandoned by the Japs during D Company's initial incursion onto Suluan, then reoccupied and defended fiercely when B Company revisited the lighthouse a week later. The reason goes back to the isolated nature of Japanese insular defensive operations. There was little coordination or even control, just inflexible orders (more a policy than an operational order) never to surrender.

Sixth Battalion learned of these Japanese command weaknesses in the Leyte Gulf campaign and would exploit them in the next operation, one of the most successful and gratifying of World War II.

Organization and training

Twenty-four days after coming ashore, 6th Battalion joined Sixth Army on Leyte where Krueger had pushed Suzuki into the northwest peninsula. The Rangers were tasked with security of Krueger's Headquarters. For a Japanese paratroop drop on an airfield some twenty miles away at Buraunen, Mucci was alerted to lead the reaction force, but US combat units closer by were able to wipe out the landing.

There were no organizational changes for the Rangers because casualties had been light. Joint exercises with the Navy were the major feature for training, as raiding opportunities abounded throughout the islands. None, however, were executed and Mucci's kept his men in condition with patrols of double digit mileage.

Good Missions:

Cabanatuan, PI, January, 1945

MacArthur brought in Eighth Army under Gen. Eichelberger to augment Sixth Army's main attack to reconquer Luzon, the prize island of the Philippines. After a rollicking New Years celebration on Leyte, 6th Battalion embarked on the invasion convoy for Lingayen Gulf. Initially, most of the battalion would stand off shore as part of Krueger's floating reserve, but one company went to secure an island at the head of the gulf for a radar site, and did so after struggling through a ten-foot surf that drowned one Ranger. The rest of the battalion came ashore on 10 January behind the 37th ID. Krueger's front moved south rapidly toward Manila; the Rangers were in a follow and support mission that involved mostly mopping up bypassed Japanese strongholds and patrolling into the hills in search of supply dumps. Many were found, the contents turned over to Filipino guerrillas.

On 27 January, Mucci received an exciting mission from Sixth Army G-3. The situation was this:

I Corps, consisting of three infantry divisions and an RCT, had pushed their beachhead south about seventy miles down the plains above Manila. Japanese resistance had melted into the hills to the northeast where Yamashita planned to fight the main battle. About seven miles ahead of I Corps' leading divisions was an infamous POW camp near Cabanatuan. Given the horrifying history of Japanese atrocities throughout the war it was clear that, as Yamashita continued to pull back, the emaciated prisoners would either be slaughtered in the camp or subjected to another death march into the hills, a march that no one could expect them to survive.

The apt term, 'death march', had originated in the Philippines and with the Japanese. After the fall of Bataan in the spring of 1942, the American survivors—tens of thousands of them, many sick and wounded, all malnourished and enfeebled— were ruthlessly herded fifty miles north without food or water, an ordeal in which hundreds who could not keep up with the forced pace the Japs dictated were bayoneted to death. Dispersed to foul prison camps, thousands of more POWs died of starvation, disease, mistreatment and often torture during the next three years. Cabanatuan was one such camp. By January 1945, gruesome death had reduced the number of its suffering inmates from 6,000 to about 500. These pathetic survivors were in desperate need of food, medicine, and every essential of life. They could receive them only by liberation.

Sixth Army had learned of the camp and its conditions from guerrillas, most notably bands led by captains in the Filipino resistance forces, Joson and Pajota. Also adding valuable intelligence were Krueger's Alamo Scouts, the long range reconnaissance patrols of their day. Acting on both sorces of information, G-2 had aerial photos taken of the camp, which at the time Mucci received the mission to liberate it was about seven miles in back of enemy lines. He was flooded with volunteers—every Ranger wanted to be part of the rescue. 'Rank hath its privileges', so Mucci selected himself to lead the raid.

Time was short and critical so a plan was devised in hours:

A squad of Alamo Scouts would infiltrate to the camp with a platoon of guerrillas, survey the area for 24 hours, and guide C Company, commanded by Cpt. Robert Prince, and a platoon from F Company under Lt. Frank Murphy. Total Ranger strength would be about a hundred, augmented by 160 of Joson's and Pajota's guerrillas ('G's'), lightly armed.

The Alamo Scouts reported a highly dangerous enemy situation but one that also offered encouraging possibilities. Near the camp ran Highway 5, the Japs' main supply line to their redoubts in the mountains. An infantry battalion and four light tanks secured the crossroads at the town of Cabanatuan. Traveling Highway 5, mostly headed north, were thousands of Japs in combat units but apparently with no responsibility for camp security which was that of about 200 rear echelon personnel. From his Leyte Gulf experience, Mucci determined that it would take unlikely coordination and unified command for the camp guards to be reinforced in time if the rescue was executed with speed and surprise.

A small Ranger TOC was quickly set up in Guimba (approximately 20 miles from Cabanatuan) with the 6th ID who loaned the raiders several rocket launchers and AT grenades. There would be a two-station Ranger radio net between the TOC and Mucci. At the TOC would be an Air Corps liaison officer because the camp was beyond the range of friendly artillery.

Mucci moved out his force before dawn on 28 January, wearing soft caps, no packs, and carrying only extra ammo and K rations. A couple of hours before dark, the Rangers arrived at Joson's G camp. Around 1830, the combined force crossed into Japanese territory between a security screen of guerrillas.

Joson's preparations had been admirable. Rural Filipinos along the approach march had been told to pen their chickens and muzzle their dogs so as not to alert the Japs. When crossing open areas or roads used by the Japs, lookouts had been prepositioned and signals coordinated. With such support the rescue force moved with unusual speed, sometimes double timing.

The route took them in a wide hook north of the camp where patchy jungle and enemy movements put Mucci behind schedule. He was short of the rendezvous point to meet Pajota's 90 guerrillas who had been continuously surveilling the camp. Eventually linkup was accomplished, and two Alamo Scouts guided Mucci's force into a clandestine assembly area closer to the camp where they waited out the last daylight hours of 28 January.

Pajota moved 160 unarmed G's into the assembly area; they were to carry the rescued POWs as many were not expected to be able to walk. For that purpose the G's had cut much bamboo to fashion litters. Also in the assembly area, Mucci received the surveillance reports. They disturbed him so much that he postponed the planned attack 24 hours, time in which the guerrillas and Alamo Scouts were to obtain specific information:

First, the information he had received about the layout of the camp had not included location of the POW stockade, the guard house, and what fixed positions the Japs used to secure the gates. Second, and perhaps even more important, what Jap units were transiting the area?

It was no easy decision to postpone the attack, risking discovery seven miles behind enemy lines, but without the EEI Mucci sought, it could be the Americans rather than the Japs who were surprised. As professional intelligence men, the Alamo Scouts understood what was now required. The G's were equally willing to find out; they only lacked the knowledge of what EEI entailed. They went about the intelligence gathering feverishly—it would be a tragic irony if the Japs evacuated the camp before the rescue was attempted.

By 0230 on 29 January the patrols had returned with the information Mucci needed:

The 90-man camp garrison unfortunately had been augmented by 150 infantrymen who were passing through and stopped over beside the camp. Exact locations of gates and guard towers were sketched for the assault leaders. Additionally, the G's reported that the Pampanga River was up and would have to be crossed to get the POWs out by the fastest, most direct route. Mucci asked his guerrilla commanders if they could help in fording the 300-yard-wide river. Yes, some ox carts would be assembled for that purpose, concealed on the near shore. Mucci emphasized that runners must be added to the camp surveillance teams so that they could return to the assembly area and advise him of any changes in Japanese dispositions.

The rest of 30 January was spent in briefings and brief-backs until Mucci felt sure that everyone knew what his mission was, and what to do if anything went wrong with the plan. At dark the rescue force moved out in three columns, G's securing the flanks, the Rangers in the middle.

They split about midnight, Joson's force to ambush the Cabanatuan-Cabu road 800 yards southwest of the camp. Pajota's G's moved off to keep an eye on Highway 5. Telephone lines to the camp were soon ready to be cut. Mucci wanted that done moments before the attack. With excellent guides, the Rangers infiltrated from the release point to within 700 yards of the camp.

Some of this distance was over open ground and there was a partial moon. To distract Japs in the watch towers, Mucci radioed back to the TOC for a P-61 (Black Widow) night fighter to make two low-level passes over the camp before striking elsewhere. Undetected, the Rangers reached their assigned positions.

Lt. Murphy's was a culvert where half of his platoon was to eliminate the command bunker at a corner of the camp. The other half was to assault the rear entrance by fire, blowing open or blowing down the gate. Initially, Cpt. Prince's job was for C Company to smother all the watch towers with fire, kill any sentries at the front gate, and seal off by fire the POWs stockade from the adjoining buildings where the garrison lived. Fortunately, the transient reinforcements evidently decided to stay within the camp too. If these Japs had bivouacked outside, Mucci would have had a very different situation and a much more dicey plan.

When the supporting fire proved effective, 1st Platoon of C Company would assault the front gate, closely followed by a weapons section to destroy all enemy vehicles. 2nd Platoon brought up the rear of the assault, and were to split, half heading straight for the POW stockade, the rest for the Jap officer quarters. The Alamo Scouts were Mucci's reserve with the additional mission of assembling the POWs, and getting those who could not walk onto the bamboo litters that unarmed G's would immediately start down the escape route.

Murphy's platoon had the most difficult route to their attack positions so when he opened fire that would be the signal for everyone else to do so. His M-1 cracked at 1945, a microsecond before two hundred weapons scoured the camp like hail from hell. Mucci had to have surprise and he got it fully. The watch towers were riddled, the guards within dropping dead on the floors or cartwheeling onto the barbed wire fences. Sentries reeled, spun around by multiple hits. The command bunker was the focus of Murphy's automatic and rocket grenade fire; nothing emerged except smoke. Both gates wavered on disintegrating hinges then clattered down in a cloud of dust.

In less than two minutes, the base of fire phase of the attack was 99 percent successful and the assault into the camp launched. 1st and 2nd Platoons of C Company raced into the compound; the Jap motor pool (it contained two light tanks in for repair) burst into flame as gas tanks on vehicle after vehicle ignited. By the light of the burning trucks, two Rangers shot off the lock of the stockade; others ran in, putting their weapons temporarily on safe, and calling for the POWs to come out for evacuation.

For minutes there was no reaction, so Rangers went in to see what was wrong—fearing that they had been too late to prevent a mass execution. They recoiled to see paralyzed ghosts strewn around the reeking, fetid prisoner area. Hardly any of the half-live creatures could get to their feet. The rest feebly waved and slowly began to weep. The only sound many could make was meowing like lost cats.

Rangers burst into tears at the sight of filth, horror and pathological brutality the world would not recognize again till Hitler's concentration camps were broken open. In assembly area briefings the G's had mentioned that the wretched condition of the POWs would be shocking, and Ranger commanders warned that such shock must not delay carrying out the rest of the mission. But when those commanders came up and glimpsed what Japanese captivity and depravity had done to fellow Americans (as well as some Australians and Dutchmen), they themselves recoiled into a long, stunned silence.

Hair raising rage reawakend them. They hardly had to give the order to hunt down every Jap in the compound. No revenge, however, could be as sweet as the poison had been bitter by which the POWs had been tortured for so many years. Mucci had not expected many EPW's from the raid and he did not get any.

What he got was a surprise from the outbreak of heavy fire coming from the direction of Pajota's roadblock. His G's were in the toughest fighting of the raid. By chance, some 800 Jap infantry bivouaced that night by a stream a quarter mile from the guerrillas' ambush site. A bridge crossed the stream. Clandestinely, Pajota was able to prepare the bridge for demolition. When fire broke out from Mucci's attack, the bivouaced Japs set out across the bridge to investigate. It blew up in their faces. More tried to ford the stream but instead turned it red under the G's machine gun fire. Tanks came up to support but Pajota was able to disable them.

At least five hundred Japs died that night. The only action the defenders were able to accomplish was drop some mortar rounds around the camp, wounding a half dozen raiders. Mucci pulled off his objective thirty minutes after hitting it. He launched a red flare to signal Pajota to break contact. He did so, and the G's melted into the jungle. The only delay at the camp was for Rangers who did not want to stop looking for prison guards to kill.

At Joson's roadblock, P-61s came to his aid as a column of Jap tanks and trucks veered off Highway 5 to investigate the flames climbing above the prison camp. The Black Widows cannons and 50 caliber machine guns turned the convoy into another bonfire in the night.

Only three hundred liberated POWs could stumble along on their own power. Others were so light from starvation that they could be carried in the Rangers arms. The rest were on litters. At the Pampango River, ox carts took them closer to safety. At daylight, an aerial evacuation of two seriously wounded Rangers and four dying ex-POWs was attempted but failed. All these men died.

Even with these sad losses, the spectacular success of the raid was hailed as the most heartening event in the reconquest of the Philippines. MacArthur described it as "magnificent", and encouraged Krueger to try an equally bold rescue of the civilian prisoners at a camp near Los Banos. Elements of the 11th Airborne Division pulled it off with equally gratifying results.

For such a feat of arms, it is difficult to find fault in any aspect of the Cabanatuan mission, except perhaps in the lack of specifics

Mucci initially asked for in his EEI. When the G's and Alamo Scouts did not first brief him with certain vital information, Mucci retasked them and they produced very well; but their second reconnaissance that was required delayed the rescue by 24 hours, fortunately without serious consequences.

Among the many positive lessons of the raid, especially noteworthy was the excellent work of the guerrillas. Not all Ranger commanders appreciate the value of irregular forces like Joson's and Pajota's. There is a tendency, frequently seen in Vietnam, to push irregulars aside with the attitude, "The Americans are here: just lock your weapons, bring up the rear, and we'll handle the job." Mucci, however, immediately recognized that the G's were the key to success. They knew the terrain, they knew the enemy, so Mucci deferred to them in these matters, using the Rangers as his strike force, while adopting the G commanders' recommendations as how their men should be employed for maximum contribution to mutual success.

US commanders sometimes half-joke that "The army is not a democracy". Of course this is very true and has to be, so subordinates are not often asked for their opinion regarding what their mission should be. However with guerrilla allies this is a very appropriate question, not indicating lack of professionalism by the US commander but instead his excellent judgement which acknowledges that guerrillas know better than anyone else what their capabilities are. It is a foolish Ranger commander who disregards their opinion or makes light of those capabilities.

The Cabanatuan raid was a fitting capstone to 6th Battalion's brief but brilliant history and also of all Ranger operations in World War II (though 5th Battalion's operations in the Saar-Moselle triangle came a little later in the war). 6th Battalion remained in the Philippines till after V-J Day, tracking down and mopping up fragmented Japanese units in some interesting operations but not of much significance after Manila fell and Yamashita's main forces were pushed back into the northeast corner of Luzon.

On 15 August 1945, after receiving two nuclear strikes, the Empire of Japan surrendered unconditionally to the Allies, ending the most devastating war in human history. On 20 August, 139 veterans of 6th Battalion embarked on the long and happy voyage back to America the beautiful.*[*FN: In WW II, the date for return home was individually determined by a system of 'points' based upon a veteran's records. A soldier earned a point for each month of Army service, another point for each month overseas. Combat decorations were worth a mere five points each (though the Medal of Honor was an immediate ticket home, as was POW status) with the Purple Heart carrying a three point bonus. This system was admirably objective until personal circumstances were factored in. Single guys got screwed, while married men got a 12-point credit for every legitimate child under the age of 18, with a limit of three such children. The 'point system' applied to all theaters, though the release of any soldier could be vetoed by regimental-level commanders if they deemed him 'essential for imminent operations'. On V-J Day, there were no operations imminent, so about seventy points were enough to assure a soldier a berth on a transport home.

The point system appears stringent—GI's scrambled for any way to short circuit it for compassionate reasons or through congressional intervention—but it was necessary to prevent a titanic demobilization, so drastic that the occupation of the Axis homelands would have been impossible except entirely with cherries. Consequently, most veterans did not return home till 1946.]

Thus ends the story of the Rangers in World War II, a saga of ordinary men—who by extraordinary esprit, teamwork, faith in each other, resourcefulness, and refusal to be intimidated by the odds against them—wrote some of the most inspiring history, not just in the annals of the US Army, but of all fighting men in all time. The names of their most celebrated battles—like Salerno, Pointe du Hoc, Cabanatuan—have rung out into the public domain.*[*FN: Books and/ or movies have been made of all these Ranger exploits.] But no less illustrious, if not as glittering, were battles virtually forgotten or unknown to the public. These are the private preserve of those who fought in them, emerging publicly only when grizzled veterans, old with years and slowed by ordeals of war from their youth, assemble in cities around America, in hotels where other guests wonder who these men are and what they are there for.

The answer is that the hotel, the city, the nation—America—might not be going about their business had not these veterans answered the call to arms as they did so long ago. Many of those not present at these reunions are said to have 'given their lives.' The living veterans know that this is not true. No one 'gave' his life; it was taken from him unwillingly. What he gave was his willingness to be where he was, doing what he was doing, when death arrived for its mysterious and personal appointment.

What he was more than willing to do was be among men who thought as he did. They didn't talk about it much, if at all. While alive, those killed would not break the faith with those still living. These reunions mean that the living have not broken faith with the dead.

Beyond these truths, applicable to all combat veterans, is one peculiar to volunteers. That's what Rangers of all six battalions were, volunteers. That is the single characteristic every man of them had in common: they not just accepted the possibility of life-ending combat, they volunteered for it—they asked for it.

So though the World War II saga of Rangers with a capital R is here concluded, the story of their universal characteristic can not be without including the tale of US Army soldiers with equal esprit and accomplishments, applying ranger tactics with ranger resourcefulness, and driving on against stupefying odds—'leading the way' where few others, if any, could follow.

And indeed Merrills Marauders, as they became known, were volunteers. It is time to draw back from inspiring sentiments and take realistic notice of two truisms. The first is a GI injunction to 'Never volunteer.' The second is a reflection upon the first: if volunteers are sought, it must be for such lousy a job that even higher headquarters doesn't feel right ordering someone to do it!

Yet there are always those who volunteer. If forced to explain why, one hears something like this:

If you're going to wear a uniform, you might as well be a soldier. If you're going to be a soldier, you want to serve with the best. Generally, the best soldiers are in volunteer units. Why? Well, maybe they figured that if they were going to wear a uniform . . . .

Marauders in a River Crossing.

Merrill's Marauders

Merrill's Marauders take time out to read old magazines less than 75 yards from enemy.

The US Army divided World War II into five major theaters: Europe (meaning northern Europe), the Mediterranean, Central Pacific, Southwest Pacific, and China-Burma-India (CBI). All except the last contained many American combat divisions and hundreds of thousands of troops—including Rangers who fought in every theater except the Central Pacific and CBI.

In the CBI, there was but one US ground combat formation, regimental in size, officially and drably named the 1688 Casual Detachment, later redesignated the 5307th Composite Unit (Provisional).

But they were to become known by dashing *nomes de guerre*. Operationally they were called Galahad (King Arthur's most sterling knight), while the press dubbed them Merrill's Marauders.*[*FN: Grammatically the name should have an apostrophe before the 's', and more authors include it than do not. The 5307th's veterans association uses the apostrophe but leaves it out on the unit crest appearing on the cover and within the book of their history. To be consistent with Darby's Rangers, this book should use an apostrophe but instead uses the title on the Maruders' crest.] For a unit so small, in campaigns so obscure, in a theater so remote from the main events of the war, the Marauders attracted inordinate attention and controversy.

For more than any other reason, their high public profile arose from the nature of the CBI theater where politics were as suffocating as the jungle.

Since early 1942, the British and Japanese had been in continuous ground contact, the longest period in which Allied and Axis armies faced each other. With the cream of their army fighting the Germans in Africa, Britain's road-bound troops were no match for the jungle-fighting Japs who quickly overran Burma. A static war then ensued on the Indian frontier. The first initiative taken by the British was in February 1943—a quixotically amazing expedition of 2,000 'Chindits' that even crossed and recrossed one of the mightiest rivers in the world, the Irawaddy, where it was miles wide and a hundred miles behind enemy lines.

The Chindits were commanded by an erratic and eccentric brigadier named Orde Wingate. It was his tactics that were the forerunners of the Marauders'.

Wingate was the father of long range penetration in modern warfare. He had a theory, pursued with fanatical conviction that, especially in regions of wilderness, the enemy rear could be attacked by fairly large regular forces supplied by air and directed solely by radio communications.

That had never been done before, but Wingate had already demonstrated major propositions of his long range penetration theory in a guerrilla campaign that, to everyone's astonishment, drove the Italian army—one of the most modern in the world— out of Ethiopia in 1940. Before that he was 'the Lafayette of Israel', a foreigner who raised and led the original Zionist forces which made possible the eventual birth of that nation.

In all his campaigns, Wingate's men were divided in their attitudes, either loving or loathing him; but his superiors were nearly unanimous in despising him. He felt that orders he received from them were at most general guidance, nothing to deter him from what he set as personal goals. His command style featured ruthless policies as well as bizarre behavior. He was not above building fires under pack animals, torturing prisoners for information, and abandoning his own wounded. He revelled in a boyhood nickname, 'stinky'; quite naked at briefings, he would scrub himself with a wire brush, a practice he also urged upon others to save them time from bathing. To greet inspectors from higher headquarters, Wingate kept a grossly foul and grease-stained field uniform balled up to don for their benefit. In this garb, he found that his appearance and odor were usually enough to send unwelcome visitors away fast. But it that didn't work, he had an alarm clock strapped to his wrist!

The senior American in the CBI was LTG 'Vinegar' Joe Stilwell, an officer not quite so weird or wild as Wingate but equally crusty and headstrong. What the two also had in common, and the reason their influences converged on the Marauders, is that they both were driven to the point of frenzy to take the war home to the Japs in Burma. They wanted to fight, and would fight anyone else in order to fight.

Not surprisingly, they fought over the Marauders, but that is getting a bit ahead of the story. As US representative for the combined theater called Southeast Asia Command—the acronym, SEAC, often derided to mean the supreme example of allied confusion—Stilwell wore many hats, none of which fit, because they required diplomacy. He had offices in three major headquarters, thousands of miles apart in India, Ceylon (today's Sri Lanka), and China. Stilwell scorned his Allied counterparts at all these locations but most of all Generalisimo Chiang Kai Shek, first among Chinese warlords opposing the Japanese armies gnawing on the vast country.

Chiang would do anything to obtain US money and support in order to line his pockets and secure his power base—anything, that is, except use his assets to fight the Japanese. To do so would risk defeat and consequent loss of face and image, as well as divert power he needed to continue subjugation of the areas in China Chiang still controlled (the Chinese communists under Mao Tse Dung were as much his enemies as the Japanese).

Gen. Frank Merrill

After two years of furious frustration with Chiang, Stilwell believed he had a way to get Chinese troops into the war. He convinced Chiang to lift two divisions (Stilwell provided the US transports) into northern India where they could train up for an advance down the Burma Road. Chiang agreed because if the road were opened he would get more supplies to convert for his own use.

To spearhead the Chinese drive scheduled for early 1944, Stilwell envisioned a regiment of US troops, penetrating somewhat like Wingate's Chindits. The call went out around the Pacific and army posts in the US: volunteers were sought for a hazardous mission.

As a composite regiment of these volunteers, the original battalions of the 1688th Casual Detachment each had distinct characteristics. It was the first job of the 1688th's CO to fit and forge them into a force. He was Col. Charles N. Hunter, a West Point classmate of Merrill, whose last assignment was head of combat training courses at Ft. Benning GA.

First Battalion was indeed a conglomeration of casuals, in various MOS's from all over the country, who assembled one by one at Ft. Stoneman, CA, outside San Francisco. On the first night they came together, the battalion was created on paper by spreading out all the individual records on a floor and slotting them into a table of organization. 1st Battalion would be led by Maj. William L. Osborne, a reservist who had commanded a battalion in the Philippine Army when the Japs invaded. Escaping into the jungle from Bataan, Osborne (with the aid of Filipino guerrillas) evaded capture for three months; then, with another fugitive, obtained a 22-foot launch they were able to navigate to Australia, a voyage of 2,000 miles and 56 days. Not bad sailing for a soldier! Osborne's ordeals on land and sea left him weighing 115 lbs as he lectured in Jap-fighting at Ft. Benning where Hunter soon recognized him as an officer who liked his subject so much that he wanted more practice at it—and revenge.

Second Battalion, commanded by a young West Pointer, Maj. George A. McGee Jr., came out of the Caribbean Defense Command, a majority from the 33rd Infantry Regiment stationed in Trinidad where companies had volunteered in mass. They arrived at Ft. Stoneman on a troop train as a unit. From previous acquaintance with each other they had the most initial cohesion of the three battalions.

The single luxury the Marauders ever enjoyed was the transport that carried them to war. It was the ocean liner, *Lurline,* impressed into service because the 1688th's voyage was off the track of the standard troop ship runs. The first stop from San Francisco was the French island of Noumea where the they picked up most of 3rd Battalion, to be commanded by Maj. Charles E. Beach, a former railroad policeman with a CIB from New Guinea. He was an enlisted man's officer, with an appreciation for the fact that to move fast in the jungle you had to go slow. His most conspicuous attributes were the combination of caution and courage.

From the decks of the *Lurline,* the men of 1st and 2nd Battalions watched their new comrades embark. The cherries were shocked to see the color of the veterans' skins—it was tinged yellow from an enforced diet of anti-malarial Atabrine. However, less than a hundred out of the thousand men in 3rd Battalion were *combat* veterans. The call for volunteers for hazardous duty apparently did not appeal much to men who had already experienced it. So the other 900 new men were veterans only of the Pacific War, yet they had a natural air superiority with the stateside cherries. To bring the vets down a notch, as well as spread around the thin level of combat experience, Hunter later (at Wingate's insistence) was obliged to exchange some of Beach's veterans around the regiment, a policy that initially did not go down well with either the gaining or losing battalions.

With a single Dutch destroyer as escort, the *Lurline* steamed around the south coast of Australia. These and the waters beyond were not seriously threatened by Axis submarines but, if the destroyer detected any, the *Lurline* would have to rely on sheer speed for protection. At Freemantle on the Indian Ocean, the 1688th had their only opportunity to go ashore, to march in the first and last parade in the history of the Marauders.

Then, under strict orders, the troops were to get back aboard ship without any post-parade festivities. Somehow the western Aussies understood this upcoming deprivation for their fleeting guests, so as the Americans marched by they were handed frosty mugs of Australian beer (some of the worlds best, in the authors opinion) that broke their alcoholic fast without breaking their ranks.

Because the *Lurline* drew too much draft to come into the port of Calcutta, she berthed a thousand miles away in Bombay on the other side of the India, requiring Hunter to accomplish a rail movement all the way to the middle of the sub-continent, and producing a lesson to be learned in this exotically unique theater of war: CBI (i.e., Stilwell) wanted the 1688th training and operational ASAP but had to accept a delay of weeks to accommodate SEAC's unhurried tempo of war, a delay no doubt preventable if lighters had been used to unload the *Lurline* in Calcutta.

Such small but constantly accumulating frustrations were driving Stilwell up the wall (up against or off the wall was somewhat his habitat), so that when he finally controlled a unit that would indeed respond— in the military manner he had come to expect from his previous experience and in accord with the immutable values he perceived to be those of the US Army— without demurrer to his orders—namely, the Marauders—he would exploit and abuse his jealously acquired authority and drive them unheeding to extinction. That tragic consequence is to be described, but the author wishes to flag its precursors.

In mid-October, 1943, all the 1688th knew as they came down the gangplanks at Bombay was that they probably wouldn't be fighting Germans. When they had embarked in San Francisco they became convinced that their destination was not Europe (though the army's illogic was never to be underestimated). What lay ahead could only be guessed, but with a world atlas they could determine that they were the only US ground force between New Guinea and Italy.

Aboard ship they had been taught something about long range penetration from newly published training manuals. Near Deolali in central India, the 1688th pitched their tents and began preparations for Galahad, but the surrounding terrain was not similar to Burmese jungle at all: the Indian hills were arid and largely denuded of trees. The platoon was the nucleus of training; conditioning marches and land navigation the subjects which occupied the most scheduled time.

The days were so full that critique of afternoon exercises had to wait till after dark. Then, within two weeks, daylight and nighttime training had about equal time. For weapons qualification, the 1688th spurned the British known distance ranges, and instead took the advice of its own jungle veterans to use pop-up targets concealed in the brush. Charlton Ogburn described the variety of other training activities in his book, *The Marauders:*

There was scouting and patrolling, squad and platoon attacks upon entrenchments, pillboxes and roadblocks; bivouacs, practice booby-trapping, taking airdrops, evacuating wounded, trailing and trail concealment, demolitions; approaching, withdrawing from and crossing rivers. There were attacks under varying conditions by half a battalion against the other half. And there were marches.., marches,.. marches...

Training had to be hard and it was, resulting in fatalities, most from drowning. Like the Rangers, the Marauders never made strong swimming an enforced prerequisite though, next to disease, drowning was the highest cause of non-battle deaths in the nine ranger battalions .*[*FN: It is an interesting question how the US Navy, in its UDT and SEAL programs, insisted upon, and certainly got <u>strong</u> swimming capabilities (not just minimal proficiency) from all its volunteers. The apparent reason is that swimming for SEALS is like hiking for rangers—an obvious and absolutely necessary skill to reach objectives. For rangers, excellent swimming is a nice to have individual ability, but not so necessary as to disqualify a soldier from being a ranger. That is the orthodox, but not the most enlightened view in ranger circles, the rationale for which is that talents on the ground often outweigh very much the raw physical ability to cross a body of water without a vessel.

The author concedes that reaching the far shore of a stream/inlet/river or the beach through surf is far less an athletic feat than navigating for miles on or under water, which is routine for SEAL's. But a concession that super-soldiers can not become strong swimmers is a different matter, as different as the two elements, earth and water.

The Army's reply (probably unofficial) would be that strong swimming is a secondary skill, the absence of which is not sufficient to disqualify a soldier outstanding in other more important respects. Very true. Just as true as the Navy's answer might be if asked why SEAls can't do very much a mile or more beyond the high water line.

What the Army's answer does not comprehend is that swimming can be taught to the extent of proficiency required. What is required is an attitudinal change by the pupil, a change as radical as the two elements involved: the land is rigid, to be confronted and climbed, but water is to be felt and accommodated on its own kinetic terms. It is easier to teach that than to teach an amphibian how to be a super-soldier.]

Selected Marauders were sent to British Intelligence School near the Taj Mahal, receiving a course from the few Chindit veterans—Wingate had lost a third of his men on the first expedition—a course that ended with a three-day survival problem in which the students were turned loose in the wilderness to live off the land.

Many Marauders were hunters in civilian life, and crack shots. They had little trouble stalking the small Indian deer who didn't have much concealment in the light forests. After the war, these veterans remembered how they ate better during this survival training than when their food supply in Burma was air dropped rations.*[*FN: Though there was some game hunting in Burma, it was curtailed when the Japs were close by—which was most of the time.]

By the middle of December, Hunter considered his force sufficiently prepared to maneuver against the Chindits. Though the two singular long range penetration forces had no previous acquaintance with each other, this exercise was to be something of a grudge match, as the Americans felt the British had little to brag about when it came backwoods (ranger-style) warfare. Marauders felt they had taught it to their former colonial masters in battles during the American Revolution—in which buckskin had prevailed over red coats.

For the first five days of the exercise, control was the obvious failing as both sides patrolled, penetrated, ambushed and bushwhacked so much that umpires were panting and nearly ineffective in efforts to prevent hand to hand combat. For the last two days the Marauders' mission was to hold an airstrip that the Chindits were to infiltrate and seize. Wingate's summary of the exercise concluded, "The maneuver was stopped when the opposing units came in contact before reaching the field, and it appeared that physical conflict might determine which side won."

Marauders and Chindits left the field with mutual respect and regard, as GI machetes were swapped for Gurkha knives—the Americans getting much the better of that exchange. Another enviable British asset was their 'rum ration': a fifth of whiskey per month for officers, and a shot of rum per day for everybody—courtesy of His Majesty, the King. Because the Marauders were receiving British rations, Hunter put in for the rum ration too. Wingate provided it.*[*FN: Long after the Marauders had entered Burma, word of their previous rum ration reached Cassandras in the states, and consequently Congress, where outcries were made against "army sanctioned alcoholism". Of course the rum ration for a Marauder squad did not total the alcohol content in a typical congressman's lunch with a lobbyist.]

He was relishing the opportunity to employ the Marauders, as he had already managed—in Wingate's inimitable way—to add them to the troop list of a Chindit division which would undertake the first air mobile operation in history, a series of landings 200-miles behind enemy lines, employing gliders to seize sites that would be improved to land C-47s. The Marauders would complement the air-landed Chindits with a large-scale ground infiltration into Northern Burma.

Thanks to spectacular lobbying by Churchill (using Wingate as his 'closer')this audicious plan had been approved by the Combined Chiefs of Staff at the Quadrant Conference in Quebec in October 1943. As that date approached, Churchill knew he had a hard sell to convince President Roosevelt, and especially Gen. Marshall, that the British were serious about aggressive war against the Japanese. So Churchill called Wingate home to advise how make that case. Wingate did not know the reason for the summons—but he knew how to make a first impression, as Churchill described:

I was about to dine alone on the night of August 4th at Downing Street when the news that he had arrived by air and was actually in the house was brought to me. I immediately asked him to join me at dinner...

He had just arrived, straight from the front, and with no clothes but what he stood in and still with the scent of the jungle. He plunged at once into his theme of how the Japanese could be mastered in jungle warfare by long range penetration groups landed by air. This interested me greatly.

I decided at once to take him with me on the voyage [to Quebec].

With only that grungy uniform, Wingate had to brief the Combined Chiefs of Staff in a naval uniform borrowed from an admiral! Nevertheless, he won them over, especially Gen. 'Hap' Arnold, head of the US Air Corps, who would have to provide almost all the C-47s for Wingate's venture.

When word got back from Quebec to Stilwell in India, he was a fountain of vinegar. Pounding the table of SEAC's commander, Lord Louis Mountbatten, Stilwell demanded that the Marauders remain under his control. Mountbatten acceded, and Wingate reacted to the switch by inviting Stilwell to march the Marauders into rectal defilade.

The other major Allied players at that time in the ugly SEAC alliance were Chiang, of course, and the British 14th Army CG, William Slim. Both resented Wingate's preemption of air assets, Slim because he felt they would be better used in his, the main theater attack. Chiang resisted any diversion of planes over 'the Hump', his high altitude lifeline between India and China.

So for their own aerial lifeline, the Marauders would have to depend on the influence of Stilwell in a four-cornered hassle umpired by Mountbatten. Into this cauldron must factored the Japanese plans, of course unknown but to themselves, which was a decisive offensive of their own. That is, as the Allies' aim was to clear Northern Burma and open a land line of supplies into China, the Japs were planning to clear Northern India and fold the Allied front! Both campaigns were planned to kick of about the same time, at the end of the monsoon season, a predicable weather break from December 1943 till May 1944.

Ironically, the Japanese CG had based his plan upon what he perceived to be the success of Wingate's first (and actually disastrous) expedition. Gen. Kawabe had been startled that some 2,000 effete British had penetrated so far and caused such mischief. If they could move so effectively then vastly superior Japanese jungle experts, Kawabe was sure, could infiltrate over the same forbidding terrain in the opposite direction. This they were to do and did, not with 2,000 troops, but 20,000—to the total surprise of Gen. Slim.

That was how the future was anticipated on New Years Day, 1944, when the 1688th Casual Detachment was redesignated the 5307th Composite Unit (Provisional), to be commanded by BG Frank D. Merrill, Stilwell's G-3. Merrill was 39 years old, a West Pointer who also had three years of enlisted service. His branch was cavalry—a useful background for the Marauders who eventually were to employ almost as many horses and mules (plus a couple of elephants) as men—but Merrill's most valuable qualification was as a Japanese specialist, one of very few in the pre-war army, who had served as assistant military attache in Tokyo. At the time of Pearl Harbor he was MacArthur's G-2 in Manila, but on a mission in Rangoon he was cut off by the Jap invasion of the Dutch East Indies.

Stilwell took him in because Merrill also knew some of the Chinese languages. He marched out of Burma with his boss, a retreat that ended with Stilwell's famous quote that "We took a helluva licking." A genial, open faced man with a pleasant personality that wore well, Merrill was a shrewd tactician, who knew his enemy, with excellent strategic foresight. His weaknesses were physical: a severe astigmatism and an arrhythmic heart. Merrill had plenty of heart, and it belonged to his men—it just went out on him at times of great stress. There were plenty of such times ahead in 1944.

With Stilwell, Merrill worked out the Marauders campaign plan. It is notable that there was no intervening headquarters between these two generals: the Marauders were a regimental under the personal direction of the theater commander.

The plan was the American variation of Wingate's long range penetration, a series of relatively short hooks rather than a single long jab like the Chindits' first expedition. Their second and far larger expedition, as mentioned, was to be predominantly air mobile with the aim of establishing brigade-size patrol bases in 'strongholds' (Wingate's term) located to force the Japs to attack in order to preserve their lines of communication and supply in Northern Burma.

By contrast, the Merrill did not plan to have any strongholds, but instead expected to attack most of the time, though such attacks would conclude in ambushes and road/trail blocks subject to Japanese reaction. The Marauders should not have to hold much ground for very long because driving down the axis of advance would be two Chinese divisions. They would attack the Japanese frontally while the Marauders cut off retreat and reinforcements.

The closest conventional maneuver to which this campaign can be compared is the pursuit where there is a direct pressure force and an encircling force. The differences were, and would be proven in spades, are that the Japs were not being pursued by the Chinese whose pressure on them was sporadic and extremely cautious; and that the Marauders were not strong enough to encircle any sizable Japanese forces—but rather constantly stood the risk of being encircled themselves.

But these were developments that emerged from reality during the next six months. As 1944 dawned, the Marauders were ready to test long range penetration American style. Organizationally, each of the three battalions had regrouped into two 'combat teams', somewhat of a misnomer because they had no organic artillery, their highest caliber indirect fire weapon being the 81 mm mortar. With multiple routes ahead, Merrill felt that he would need task forces (combat teams) smaller than battalions to advance on more routes.

For mobility, Marauders were introduced to pack animals—360 Missouri mules and 340 Australian horses—but less than a month before the attack jumped off.*[*FN: The reason for the delay was the Germans. A U boat in the Arabian Sea sunk the ship carrying the Marauders first boat load of mules.] Inexperience, especially in care and treatment of quadrupeds would result in many frustrating hardships ahead.

Col. Hunter, now Merrill's XO, described the status of the Marauders on the eve of the Galahad campaign:

"Although not completely equipped or, by any stretch of the imagination, fully organized and trained, it was decided [by Stilwell with Merrill's concurrence] in January that the unit could function, and it was ordered to move to a forward staging area in Assam." Controlled by Stilwell's Northern Combat Area Command, they would march out from Ledo, terminus of both the Indian railroad and a spur (still under construction) of the Burma Road, above the Hukawng valley that debouched into Burma.

It was down that valley, containing headwaters of the Tanai and Mogaung Rivers, that would be the area of operations (AO) for the Marauders and the Chinese 38th Division. Their ultimate objective—scheduled for capture late in 1944 after the next monsoon—was Myitkyna, terminus of the Burmese railroad, and the main fighter base for the Japanese air force in interdicting aerial Hump traffic between India and China. Myitkyna was also an on order objective for the Chindits, who would be operating some fifty miles south of the town across the Irrawaddy.

The British high command, especially Slim and Wingate, gave Galahad little chance of taking Myitkyna at all. Stilwell was determined to prove them wrong, and to grab Myitkyna during the current dry season, not waiting for the next one.

The approach march on the Ledo Road to the end of the spur at Shingbwiyang was a valuable period for shaking down and learning how to handle the pack animals, especially at river crossings. It was really the mules who decided when and where to cross! This was also the Mara 'ders introduction to the jungle, still dripping ! om the receding monsoon and brightly lush. The march of 112 miles crossed two 4,000-ft. passes from where the snow-capped Himalayas could be seen, and lasted nearly two weeks. Initially the Marauders marched at night in hope of concealing their presence, but the Japanese station, Radio Saigon, soon announced that they would be bombed off the road, so they marched by day to make better time.

They closed on Shingbwiyang (site of a cargo airstrip and field hospital) on 16 February. For the next eight days they did pre-combat checks and liaison with US Army advisors to the Chinese 22nd and 38th Infantry Divisions who would be pushing down Kamaing Road, an unimproved extension of the Ledo Road that eventually and circuitously reached Myitkyna 200 road miles away. The Marauders were told not to expect too much from the Chinese: in the last two months they had advanced only 25 miles.

The principal purpose of their commanders was to avoid casualties, for two very valid Oriental reasons. First, though nominally under Stilwell's command, they answered ultimately to Chiang Kai Shek in far away Chunking who favored a war of political not tactical maneuver. Second and most important, a Chinese Division's payroll was based upon its morning report strength. The troops rarely saw any of this money—the commanders kept it—but the longer the roster, the larger the payroll. Because most of the money came from Uncle Sam, American advisors spot checked purported strengths, so Chinese commanders were loathe to lose any men in combat which, against the Japanese had already lasted seven years, and bode to last many more against the Chinese communists.

These long established corrupt practices were not something many Marauders knew about, and for morale purposes not told. They were just assured that the Chinese would fight well if inspired by Marauder aggressiveness.

The enemy intelligence picture had not been developed by the Chinese because patrolling against the Japs was dangerous. Yet Merrill learned a great deal about the Japanese 18th Division commanded by Gen. Tanaka, thanks to the remarkable Kachin Rangers, native tribesmen trained, employed and supplied by the newly formed OSS, the forerunner of the CIA. As a strategic intelligence agency, the Kachin Rangers were not controlled by Merrill, and even Stilwell had only loose reins on them, so coordination was skimpy. Nevertheless, the Kachins, though pretty much conducting their own war, were very cooperative about sharing information about the Japs when contact could be made with the Marauders.

The 18th Division were 7000 veterans of the conquests of Malaya and Singapore as well as Burma, one of the best Jap units on the mainland. Because of their confidence and the advantages of very defensible terrain, they had little respect for the Chinese. Tanaka's strategy was, in his words, to fight an aggressive defense: "Though threatened by enemy envelopment, we will exploit advantages of interior lines, utilizing every opportunity to defeat in detail the slow-moving Chinese whose coordination has proven poor."

He envisioned many little battles, and indeed there would be scores. Of the strategic picture, Ogburn observed,

One is struck by the meagerness of the forces engaged. Battalions here counted for as much as divisions on other fronts, or as whole armies in Europe. It was all either side could do to make a military effort [in Burma] *at all. One had the impression of two Goliaths fully extended and barely able to reach each other in this remote sector, and constrained to grapple—albeit with undiminished ferocity—with their fingers alone.*

Good Missions:

XIV. Walabum, Burma, Feb., 1944

Obviously success or failure for the Marauders would depend on the accuracy and timeliness of intelligence. Once the enemy was located and sized, the jungle offered ample avenues to get around or in back of him, ambush and cut him off. As in all ranger operations, surprise was the key to everything.

Surprise making in the Marauder battalions was the prime responsibility of their Intelligence and Reconnaissance platoons. Consequently, those three platoon leaders were as important, often more so, than even their battalion commanders. 1st Battalion's I&R Platoon was led by 19-year old Lt. Sam Wilson, 2nd Battalion's by Lt. Bill Grissom, and 3rd Battalion's by Lt. Logan Weston. Though their missions were identical, each platoon had an entirely different personality.

Except for certain key NCO's, all members were volunteers for their platoons. Such double volunteers are even harder to explain, but perhaps if they have two common characteristics one is a low boredom threshold, the other is competitiveness. Being 'back with the column' irks such men, even if that column is slinking its way behind enemy lines. They take unusual satisfaction in being the first to learn what's around the next bend in the trail, but routine grates upon them more than others. Their competitiveness is not the kind commonly seen in the army which is "My squad is the best squad in the best platoon in the best company, etc. ... in the army,"; but rather the rare enthrallment of contest with the enemy—the ultimate competition in which the winner lives, the loser dies.

After careful selection of a cadre of profoundly reliable NCO's, Wilson filled out his platoon with men who had disciplinary problems arising from their boredom with routine. He felt they would do anything for the opportunity of independence, and exceed their limits in order to remain out front if that meant being away from organizational discipline. They had one chance to prove themselves; failing that they were back in the column.

Wilson's selection criteria produced superb results, but the 'independence more than anything' appeal was not eternal, not after contact after fear-sweating contact which drained the reservoir of courage till Wilson's men resented him volunteering the I&R Platoon for battalion point. Competitiveness was also attenuated, till it was no longer an anticipated excitement but a survival skill to kill or be killed.

In 2nd Battalion, Grissom looked for very different records in his recruits. He wanted men who valued and welcomed discipline, illustrating the old army saw that if you can't follow orders in garrison you won't on the battlefield. Experience was to prove that assertion very debatable, and there seemed to be a trade off with competitiveness. Grissom's platoon had the highest turnover of the three; intentionally so because he believed in keeping the cutting edge sharp. After spending some time back in the column, Grissom sometimes brought back former I&R men who had reignited the fire in their bellies.

Weston's selection criteria were startling religious. In *The Marauders,* he described them himself:

"About 200 were interested, from which we chose 55. These men were carefully selected and thoroughly checked out, past combat records and all." What Weston wanted most were men who "had been in a hopeless combat situation during which time they realized their own helplessness and their need of God. These men, with very few exceptions, had trusted the Lord to deliver them in combat... What I want to say about them is that when they couldn't advance another step, they went ahead on their knees in prayer."

For the first hook around the 18th

Marauders advance.

Arthur J. Lynnemeyer, Edward A. Wade, Vaughn J. Miller, George E. Harmon, Ervin A. Mathis, Isaac W. Ross, George B. Lowe.

Division's right flank, all three I&R platoons were to go way out, two days ahead of the Marauders' main body, to find the first southbound trail leading to Walawbum. There were three possible routes reported by the Kachins, but they did not know which if any of them were blocked by Japanese.

Wilson would lead because his route recon would be the longest if the eastern-most trail (through the village of Tanja Ga) was the only one where the Japs could be bypassed. Weston's platoon would follow, and after eighteen miles veer south to check out the village of Lanem Ga. Similarly, Grissom's platoon would follow Weston and recon the trail to Nzang Ga, the most direct and preferable route to Walawbum. All three lieutenants were to report their results by radio, but since there would be much masking and jungled terrain between them and Merrill's CP on the Tarung River, they were to use couriers as relay if necessary.

In the absence of any communications with Merrill, Grissom and Weston were to block their trails if they encountered advancing Japs, protecting both Wilson's trail for his withdrawal and the regiment's flank. Even if they had to fight for every mile of what was supposed to be an undetected flanking movement, the Marauders were to press on for Walawbum 90 miles from their IP. Stilwell said so.

Wilson had an additional and vital mission: if he found that Tanja Ga was not defended too strongly by the Japs (Kachins had reported a platoon there), he was to get around them and find a ford for the Marauders across the Tawang River. Tanja Ga was thirty miles away when Wilson hit the trail at 0100 on 24 February.

At first light his men began examining the trail for Jap bootprints, very distinct because of their cloven appearance. The platoon was supported by two mules and a horse, some of the few animals that had been 'devocalized'; i.e, had their vocal cords surgically severed so that they could make no sound in the jungle. The late arrival of the Marauders' other pack animals prevented their devocalization.

Wilson's men made frequent stops to listen but by nightfall, after a seventeen-hour march, they pulled off to sleep with only two OP's to surveil the trail. The jungle was so full of 'noise sentries' no further security was called for.

The platoon sped along the next day, till at 1330 they heard fire from their right rear which they surmised was Weston's platoon encountering the Japs on his trail. Not wishing to bother Weston in the middle of his first action by confirming their surmise with a call on the radio, Wilson pressed on. Every man was a little changed from this point. They had heard enemy fire. Their war had begun, and while they were looking for Japs, the Japs would be looking for them. Only the jungle was neutral—a hostile protector for both sides.

A couple of hours later, fire broke out again: fainter this time, farther to their right and rear. This, Wilson assumed, was Grissom making his first contact. Wilson tried to raise his fellow platoon leaders because this was serious if both alternative routes for the Marauders were blocked. That was the case, but Wilson didn't know it because he could not raise Weston or Grissom. Weston was able to contact regiment, however, and was instructed to strengthen his trail block—the Marauders would force march to get by him on Wilson's route. Grissom received similar orders after his action in which his lead scout was killed, the first Marauder KIA. The distinction of killing the first Jap went to German-born Werner Katz of Weston's platoon.

There was already an SOP for bodies. Enemy dead could be left where they fell unless burial was required for sanitation. American dead were to be buried in the field as close as possible to a trail or other landmark that could be located on a map. US graves were usually marked with a cross of bamboo lashed together with vines. The Japs had an SOP of their own. Their Shinto faith called for homeland burial. As bodies were not easily transportable, body fragments had to suffice. Thus, on occasion Marauders came upon caches of arms and/or legs, sometimes heads, marked with the names of the deceased and their hometowns for delivery in Japan.

Though he had not confirmed it, Wilson felt he was now recon the only route left for his regiment. The independent minded members of his platoon were now beginning to miss their buddies—the excitement of being out front and alone was wearing off!

The 47-man platoon was strung out over 150 yards of twisting trail. The loudest sound was the muffled hoofbeats of the animals that for this reason were placed in the center of the column. Wilson kept at least one Pacific combat veteran out front at all times. Such men said they could smell the fishy odor of Japs yards before they could be seen. The density of the adjoining jungle obviated the need for flank guards whose progress would be too slow for the speed that was now necessary.

It was necessary because the Marauders were closing on Wilson, hustling by Weston's and Grissom's trail blocks so that the Japs could not further interfere. Merrill reinforced those blocks, pushing the Japs south with mortar and machine gun fire that Wilson could hear from five miles away. He had to make agonizing tradeoffs between speed and safety. As Ogburn put it, "From first to last that was probably the worst part of the campaign for those who had to endure it: *what was around the next bend?*""

Finally around a bend was Tanja Ga, a sleepy village of huts on stilts called bashas. The bashas were in a clearing encircled with trenches and bunkers. But there were neither villagers nor Japanese to be seen, though from ganglions of bootprints it was apparent the village had served as a Jap patrol base. Manning the emplacements with half his platoon, Wilson set out with the rest to find a ford for the regiment.

They found one right away. Wilson eagerly cranked up his SCR 284 radio to get this vital news to Merrill. Contact was made but quickly lost because of atmospherics. Darkness was falling fast as it does in the jungle, but Wilson decided he had to become his own courier. The horse he had named Pride and Joy was saddled up, and Wilson cantered off into the night.

Besides branches that could dismount him, Wilson's most realistic fear was being shot like Stonewall Jackson as he approached friendly positions. However, his closest call came from the animal kingdom.

As Wilson's steed struggled up the steep bank of a stream, a pair of big cats—probably tigers— suddenly screamed as they fought or mated in the night. Quite understandably, Pride and Joy spooked and reared. Wilson found himself upside down, being dragged by one foot in a stirrup. He still had a hold on his carbine; by reflex he shot it, hoping to scare off the tigers. He shot high so as not to wound one. This tactic worked, though while Wilson remounted he was sure that any Japs within a mile had been alerted.

But the greatest perils of the night were over

after he rode by Weston's and Grissom's positions. Merrill and Hunter acted at once on his news of the ford. A two-day march brought the regiment to the mile-wide Tawang River which they forded by moonlight. Mysteriously the Japs in Tanja Ga had dissolved into the jungle. They probably got word back to Tanaka that there were some Americans off to his right flank, but Merrill could be pretty confident that his adversary did not have a good idea of what kind or how many.

With the monsoon departed, the Tawang was low, only waist deep. From their newly acquired knowledge of pack animals, the Marauders took no chances even with what should be an easy crossing. Guidelines were strung to keep the beasts within them; each was checked off with its handler like an aircraft manifest to be confirm at departure then again upon arrival. The animals' loads were too precious to be lost if they bolted off into the jungle.

After an uneventful three-day march, the Maruders reached the Tanai River, site of their first scheduled supply drop. Ogburn:

The Dakotas [C-47s] roaring in from Assam provided almost our sole link with the known world, the main evidence we had that it was still there... Grain for the animals was free-dropped in half-filled sacks (full sacks would have exploded like melons on hitting the ground) and these were nearly as dangerous as bombs to heads below. The K rations, ammunition, and medical and signal supplies were floated down on red, white, blue and green rayon parachutes.

There was this colorful air about the drop zone, but this was deceptive. Taking a drop was a strain on us. The battalions, which usually were strung out under cover, each a mile or two in length, had to assemble around an open area. We were like sea serpents which must leave their element to come up on land and expose themselves to danger in order to perform some vital function... We could never tarry where those gay parachutes had flowered in the sky; we had to distribute supplies fast and get out.

South of the Tanai, Weston's platoon was point, and they came upon growing evidence that the Japs were becoming aware of a hostile presence on their flank. Merrill felt he had to move faster before the Japs chopped down more trees to block the trail to Walawbum. He sent out 3rd Battalion before they could completely load up from the resupply drop.

Weston came upon booby traps, and bootprints still oozing water. At Lagang Ga, only a few miles from the Marauders' objective, his platoon came under fire. His lead squad dove for cover and set up a base of fire, while the next two squads hooked up around both sides of the trail. After thirty yards of jungle cutting, they got behind the Jap block but they were a bit too late, only able to pick off the 1st two men of a counter-reconnaissance patrol.

Weston had to keep pressure up the trail or else the Jap patrol would set up another delaying position. Rotating his squads on point, Weston succeeded, driving the Japs back, keeping them on the run. During this action, the lead company of 3rd Battalion had closed on the I&R Platoon. This company turned south for Walawbum while Weston pushed on through

A column of Marauders plods through the jungle toward Walabum.

Lagang Ga to the small Numpyek River, then too turned south to await the rest of 3rd Battalion. Night fell on the Marauders overlooking Walawbum from the swampy east bank clouded with miasmal fog, humming mosquitoes, a breeding ground for hungry leeches. In heavy hanging soaked fatigues, Weston's men settled in for a miserable night. His orders were no fires, noise or movement. Use helmets for latrines.

The next day, after passing through Wesu Ga, 1st and 2nd Battalions echeloned in on the Kamaing Road northwest of Walawbum. Merrill put 1st Battalion in a central position to reinforce either of his other two. Hacking their way as quietly as possible to bluffs above the road, 2nd Battalion was in position before nightfall. Surprise seemed to be with them. Tanaka had apparently lost the Marauders, believing them to be in unknown strength somewhere along the lower Tanai River where it appeared likely they would link up with the Chinese.

In the morning, 3rd Battalion crossed the Numpyek and cut the Kamaing Road with ease. A couple of miles away, 2nd Battalion had been watching the Japs on the road carry on administrative activities as if they were on a camping excursion by Fujiyama. But early in the morning, these REMF's looked more concerned, and soon they were gone, leaving nice rows of foxholes by the road which 2nd Battalion occupied. The reason for the Japs' ill-considered evacuation was that news of 3rd Battalion's cut came up the road and the unit in front of 2nd Battalion had gone south to reopen it. These REMF's were soon disappointed and many of them dead when 3rd Battalion laid into them with well positioned fire.

It looked as though the Marauders had finessed their way to victory in their first battle, but up north two of Tanaka's regiments were coincidentally making a planned disengagement in front of the listless Chinese advance, so these Japs were already on the road and marching south, soon to strike 2nd Battalion's road block.

Tanaka had plenty of artillery in the vicinity. They had only to reposition their guns and salvo on 2nd and 3rd Battalions. Concurrently the Japs mounted hasty, uncoordinated attacks that took large casualties while giving the Marauders small problems. McGee's and Osborne's men were learning something they found hard to believe: Japs would attack, not until the issue was decided, but till the last man fell in a charge. Against these courageous but crazy onslaughts, ammo conservation became a growing concern.

Another paradox developed about the Japs. When they had been outmaneuvered, as Weston had done to them time and again, they would fall back; but when attacking they were indeed kamikazes. It seemed that the presence of officers had much to do with the attitude of the troops. This was not so much a matter of leadership but of presence, as if the officers were personal envoys of the emperor, bidding his soldiers to conquer or die. The higher ranking their leader—and majors waving swords were not uncommon—the more tunnel visioned their assaults. Marauder snipers (some of the best in the army) were particularly alert for waving swords.

A fluid front developed as Jap attacks overlapped each other and also seeped around the Marauder flanks. Consequently there was no 'rear' for them. In its reserve position, 1st Battalion found itself shooting in all directions at groups of leaderless Japanese heading both forward and back. A litter party of seven Japs stumbled into 3rd Battalion's CP, carrying a wounded officer. They all died of their wounds.

For Marauder casualties, Merrill chopped out the jungle from around a paddy for a single engine aircraft to retrieve a maximum of four wounded men per sortie. As the L-5 approached, so did a Jap patrol. The wounded evacuees had to defend the strip, and did, driving off the Japs by adjusting mortar fire.

Merrill's CP at Wesu Ga was hardly any safer. Beach and McGee both got into direct combat with infiltrators or stragglers, it was hard to determine what the Japs were doing.

What was clear was that the Chinese were not doing much of anything as the battles around Walawbum continued for another night and day. Over the radio Merrill protested that if the two Jap regiments that had previously been in front of the Chinese were now identified fighting the Marauders, a Chinese advance should be swift. Hold on, he was told, the allies were bringing up some tanks to support a linkup. Hunter proceeded, luckily with a strong security party, to the proposed linkup point—and ran into fifty Japs infiltrating toward Merrill's CP. Hunter's men broke up the attack but then had to fall back. The Chinese tanks never made it to the rendezvous.

To Stilwell, the reports from the Chinese of their "relentless advance" made it sound as if he had Tanaka in a vice. Tanaka knew different: it was he who had Merrill in a vice. With a couple of battalions the Japs were effectively delaying the Chinese. With the rest of 18th Division, Tanaka was executing his strategy of defeating the Americans in detail.

US air power—the great P-51 Mustangs mostly— saved the second day, pummeling Tanaka's reinforcements coming up from the south. Yet the Japs launched several coordinated attacks that day, 5 March, against Beach but were thrown back by his strong positions on the riverbank. Ammo became a critical shortage that night as the Japs pressed harder, now that they had all the Marauder battalions located. They also had what the Marauders did not, artillery, and they used it plentifully during darkness when there was nothing to fear from dive bombing Mustangs.

Weston's I&R platoon, providing Beach with security on his north flank, was the first Marauder unit to get in desperate trouble. This happened the next morning, beginning the fourth full day of combat, when the Japs assaulted every twenty minutes for three hours, surrounding the platoon on three sides, with a bend in the river at Weston's back. They held out thanks to very close protective fires from 81 mm mortars, by using two tommy gunners as a mobile reserve, and by listening to the Japs' plans for their next attack. The lines were so close that Henry Gosho, Weston's Nisei interpreter, could eavesdrop on Jap officers screaming out their orders. The lines were so close that when the culminating banzai charge swept onto the platoon, attackers from all directions were caught in each others fire. The lines were so close, Weston tried to break contact under a mortar barrage called right on his position.

Beach had managed to swing a squad up to cover the withdrawal where Weston tried to cross the river aided by smoke rounds from the mortars. Raging Japs were close behind but were picked off one by one by one of the Marauders' most successful snipers, a Sioux chief named Norman Janis. Like Janus, he seemed to have four eyes. Covering the crossing, he accounted for six Japs and two machine guns. I&R platoon got away with only two dead and five wounded.

The rest of 3rd Battalion had not been quite so heavily engaged but was in tough fights all along their position overlooking Walawbum. It was 2nd Battalion, several miles west on the Kamaing Road, that was hit the hardest as they were the first to feel the weight of Tanaka's two regiments closing from the north. Five times that day they hit from the north, then once more in a coordinated attack from the south. In between they shelled McGee's positions with direct fire 70 mm and high angle 90 mm. The effect of both was augmented by tree bursts in the low jungle.*[*FN: Double canopy jungle was rare in this part of Burma and triple canopy never seen.] McGee had one dead and four WIA's from artillery against which he had no defense except for occasional air strikes.

But he did have a section of the main telephone line between 18th Division's main CP and the forward regiments. He also had Nisei Roy Masumoto who knew how to tap and eavesdrop on a line. The information he provided was priceless for vectoring air strikes and allowed McGee to confidently move his reserve platoon around, and even redistribute ammo from less threatened sectors.

One intercepted call may have been crucial to the entire outcome of the Walawbum campaign. It located the Jap's forward ammo dump which was hit by dive bombers, destroying one of Tanaka's principal advantages. Artillery vollies tapered off, infantry assaults were less supported by fire, and the Japs seemed to be losing heart. Matsumoto heard their anxiety as the trap they thought they had the Marauders in began to come apart. Late in the day he heard Tanaka's headquarters allow some limited withdrawals because he could get his forward regiments no further help in reducing the Marauder roadblocks.

So the Marauders were saved by the bell. They were down to a couple of clips of ammo per man, and no one had a ration or any sleep for 36 hours. Even drinking water was scarce—the Japs had prevented access to the river. This shortage began the first scourge of dysentery, the all-time biggest casualty producer among the Marauders.

As in any first action, especially one which lasted twelve days and ended in four days of pitched battle, a great deal was learned, but more individually than organizationally.

The combat team organization did not prove to be much of an improvement over having the more familiar relationship of companies to battalions. Air drop procedures, especially to reduce air-ground radio traffic, were less than satisfactory. It was found, however, that airmen on the ground talking to aviators in planes resulted in less antagonism than a soldier talking to aviators, and this procedure was continued whenever possible. The major organizational failing was in artillery support. This was to have come from the Chinese and finally did, but very late in the game that probably would have been lost if it had lasted another half an inning.

Nevertheless, Hunter's opinion (quoted earlier) of Marauder readiness as they set out from Ledo could surely be revised based upon the results of the Walawbum mission: at least 800 Japs had been killed at a cost to the Marauders of eight killed and 37 seriously wounded But non-battle casualties were a pregnant cloud on the horizon—four times the number caused by the Japs.

Strategically, Stilwell had snatched a third of the Hukawng valley from Gen. Kanabe, causing him concern for his north flank just as he

Marauders made the best of what they had.

A friendly shake.

was preparing for his own surprise offensive into Assam. Tactically, Tenaka had given more ground since the Marauders moved against him than he had lost to the Chinese in three months. Clearly any mystique about Japanese invincibility no longer lingered in Marauder minds. Indeed, they had spontaneously mounted a little of their own PSYOP against the enemy: on the second night when Japs were wrapped around all three battalions, the chilling howl of alien wolves began to rise eerily in the Burmese jungle—"Fuck Youooooooo..." The wolf call may not have frightened off many Japs but it gave them a good appreciation of Marauder defiance.

Marauders learned most about themselves and those around them. They learned how important their animals were, and something about how to handle them—getting them to the rear fast when an action started, and into defilade if possible.

Chinese tanks rumbled down the Kaimang Road—they had been of some help in convincing Tanaka to pull back—and the Marauders pulled off into the jungle to lick wound and relate war stories from their virgin combat. They all agreed the army had fulfilled its promise to them for hazardous duty.

600 of the 700 pack animals that had started toward Walawbum were fit to continue. 2,300 of the original 2,600 Marauders were deemed capable to carry out their next mission. It would be another hook to the east, a longer one, to set up blocks on the Kamaing Road at two far away villages, Shaduzup and Inkangatawng. Before heading south, the Marauders had only two days full rest, the only two days with full rations. No one had a bathroom scale but battalions surgeons estimated that, on average, the mens body weights were down about five percent. They would plummet further as wide-spread amebic dysentery took hold during the march.*[*FN: Readers lucky enough to have no acquaintance with amebic dysentery are advised that it is a tropical disease of the digestive tract, the main symptom of which is persistent diarrhea that turns bloody if not treated. Dysentery comes from a microscopic, single-cell animal found in contaminated water. It can be spread further by hands that have come in contact with infected stools, then food.

Boiling water destroys the organisms; so too did the Marauders' iodine tablets. But the tablets must dissolve thoroughly in canteens to be effective. In their constant thirst, Marauders fell into the bad practice of popping the tablets like aspirin, then washing them down with suspect water. Water discipline improved because of dysentery, and also when Marauders found they could get quarts of pure, cool water by cutting sections of thick diameter, green bamboo.]

And another insidious effect, the significance of which could not yet be estimated, was also moving through the ranks as the ranks started out into the jungle again. That factor could be called 'behind the lines time', an experience novel to any US unit (including Rangers) and really known only by veterans of Wingate's first expedition.*[*FN: The longest duration of constant combat by Rangers was that of Darby's during their Salerno campaign, a period of three weeks. The longest behind enemy lines operation was 4th Battalion's six-day penetration into the Mignano valley.]

After six weeks behind Jap lines, only 600 out of the original 3,000 Chindits ever took the field again.

So six weeks had been planned—and even acknowledged by Stilwell— to be the Marauders outside limit. As they departed Walawbum, everyone felt he had no more than a month to go.

Bad Missions:

XIV. Shaduzup-Inkangatawng, Burma, March, 1944

Shaduzup was about fifty trail miles away. It would take two weeks to reach it. Inkangatawng was eighty miles off but a trek of only eleven days, illustrating how terrain and enemy reaction are nearly unpredictable in a jungled wilderness like northern Burma. Shaduzup was 1st Battalion's objective, Inkangatawng, 2nd and 3rd's. Despite their separation (at the narrowest, 10 miles), differences in terrain, and manifest difficulties in coordination, Merrill's forces were ordered to hit the Kamaing Road simultaneously. Each prong would be followed and supported by a Chinese infantry regiment—followed very slowly and supported very skimpily, it turned out—while the remainder of two Chinese divisions once again applied their version of 'pressure' down the road.

This was Stilwell's plan, his order, which Merrill could not convince him to change. It was 1st Battalion's hook that alarmed Merrill—too close an envelopment for long range penetration, so close that Jap infantry on the Kamaing Road could patrol out and discover, then interfere with Osborne's march around their flank. When his Marauders reached Shaduzup they could be hung out on the road as had happened at Walawbum with almost disastrous consequences. Yes, the Chinese were supposed to push down that road again, but the Walawbum experience made Merrill very wary.

Better, he suggested to no avail, that the Marauders stay together and set up a really heavy block at Inkangatawng, forcing Tenaka to pull back farther south to reopen his MSR. Further more, with a battalion to secure a drop zone while two manned the road block, there was less chance the Marauders would run out of ammo again at a critical time.

Stilwell's counter-argument, a legitimate one, was that on two routes to the Kamaing Road the Marauders would have twice the probability of cutting it at least one point. If the whole regiment got bottled up in the jungle way out to the east, the campaign would lose momentum and the Chinese bog down again.

After counting the difference between the number of stars on Stilwell's collar and his own, Merrill saluted and moved out on the mission. He would accompany McGee and Beach.

Swinging far out to the east, 2nd and 3rd Battalions had easy marching and no enemy contact. The only rain on their parade was from the unusually early arrival of monsoon clouds. Half the distance to their objective was along the Tanai River valley, a 'high speed' approach through the jumbled mass of hills called the Jambu Bum. Midway down the valley at the village of Naubum, Merrill picked up a very welcome detachment of 300 Kachin guerrillas organized by the OSS. They provided excellent intelligence for the rest of the route (a major reason why it was covered so quickly) as well as a couple of wonderful prime movers—elephants.

They proved to be versatile as well, especially adept at clearing a drop zone where there had previously been a field of six-foot elephant grass. The only problem was that they had never seen mules and vice versa. Both species of pack animals stampeded if brought together.

Merrill returned Kachin support by operating a medical clinic in their village of Janpan. After that the Marauders began to march the ridges, so high they were above clouds in the valleys. At Auche the two-mile long column turned off the main trail to avoid detection as

Marauders crossing a stream in Burma.

they did a U turn and approached Inkangatawng from the south, a twenty-mile detour. They crossed the Mogaung River twenty times in a space of eight miles. All this effort was worth it in order to achieve surprise.

Slowly, cautiously, Merrill had his Marauders feel their way forward toward the Kamaing Road. He still had time because 1st Battalion was not yet ready to put in their block ten miles north. The Kachin G's, however, perceived that a big fight was imminent and withdrew north. The Chinese regiment behind Merrill was not yet within supporting distance. Instead, it was back plundering Kachin lands around Naubum. He understood why the guerrillas despised the Chinese almost as much as the Japs, and sympathized with the G's return to protect Naubum.

Without the Kachin's intimate knowledge of the terrain, Beach was required to use half his battalion to screen open ground on his southwest flank, which if the weather stayed reasonably dry, was passable by Jap armored cars. Such a situation had developed for Wingate on his first expedition: the map had shown jungle but it was actually a cultivated teak forest in which Jap vehicles chased down the Chindits, nearly getting them all. Merrill could not handle such a vehicular threat as once more he was poised to achieve complete surprise. The other half of 3rd Battalion was able to slip patrols within a few hundred yards of Inkangatawng before they made contact.

Second Battalion, meanwhile, had succeeded in fording the Mogaung undetected and come at Inkangatawng from the northeast. Now, however, with both his battalions poised to cut the road, Merrill faced frustration: 1st Battalion was not yet in position to make their cut up at Shaduzup, and the Chinese main attack had faltered almost from the start.

It did not take long for the Japs around Inkangatawng to counter-patrol and come upon Merrill's waiting force. 2nd Battalion could get no closer than 300 yards from the road before he had to dig in to turn back increasingly powerful probes. 1st Battalion could not seize Inkangtawng itself but effectively interdicted the road.

A Walawbum-type action then ensued, the Japs aroused and initially confused, but soon able to unlimber artillery and pound Marauder positions. Merrill's countered with air strikes that silenced enemy guns during daylight. It was at night that Jap howitzers resumed domination of the battlefield, but fortunately coordination with their ground attacks—sixteen of them on the second day—was poor. Because of the prevalence and profusion of elephant grass, engagements were at extremely close quarters with tiny fields of fire that exposed large groups of attackers very suddenly. In this sort of fighting, the tommygun was the preferred Marauder weapon. Ogburn described a typical action in elephant grass:

It was only the lifting of the enemy's mortar barrage and the sound of shouting and of men rushing through the grass that warned [Marauders] of what was to come—and then the attack was on them. All around the Japanese came running into view. At twenty paces the Marauders' line opened up. The effect was searing. The front ranks of Japanese went down and those behind were piled on top of them...

Pack mule which fell off narrow trail west of Turnok, being pulled up by the neck by muleskinners.

Holding their own, while cutting down or driving off droves of enemy, 2nd and 3rd Battalions could keep up this slugfest so long as any of three possibilites did not occur in combination: 1. Jap artillery did not become significantly heavier. 2. Ammo did not run too short before an airdrop could be received. 3. Jap infantry encircled so effectively that the Marauders were cut off from fall back positions to the north and east.

Merrill learned on the radio that he had to prevent these possibilities from happening for three more days, then his prospects should improve. That was how much longer Osborne reported it would take before 1st Battalion could make their cut at Shaduzup, thereby relieving some of the Japs' pressure from the north.

First Battalion had hit Jap trailblocks and delaying positions from the first day out of Walawbum, one of the reasons Merrill had objected to the hook so close to the Kamaing Road. From that road the Japs sent out reinforcing patrols to skirmish every couple of hundred yards on the single trail Osborne was using. Eight small fights for every mile of trail was typical—and very trying on the point platoons. This was not acceptable progress, so Osborne took the only alternative, to cut cross country and through the trackless jungle.

No map or even natives can provide trustworthy information as to what density of vegetation will be encountered on a compass azimuth through the jungle. Osborne had more bad luck than good, running into stands of bamboo that had to be cut twice, first at ankle height, then again at a level above the top of the mules' loads. Only four mean at a time could hack away with their machetes, rotating with four fresh men about every half hour. The terrain was like a sawtooth, but not shown on the maps countour lines (maps were all pre-war and often wildly inaccurate anyway) because the saw's teeth were less than twenty feet high. Yet that was plenty high, too high for loaded mules to climb, and soon to slippery as well. Their loads were unpacked and passed up slope after steep slope hand to hand.

Then there were the leeches, thin black worms swarming thicker than the Marauders had ever encountered, that seemed clustered by the millions in vast spreads where cattle had once grazed around abandoned slash-and-burn Kachin settlements. To the Marauders Wingate himself had condensed his advice on long range penetration into two most important rules: always speak in whispers in the jungle and never try to pull off a leech.

Ogburn: *On everyone's uniform there were dark, stiff patches. These were dried blood from leech bites . . .The repulsive little rubbery monsters.. did give us the pleasure of going after them with lighted ends of cigarettes. It required nice judgment to bring the heat close enough to persuade them to disengage... One of the platoon leaders went around with a ruby-red eyeball, the result of a leech's fastening onto it during the night. All of us were more or less bloody all the time, and the mules suffered worse than we; their fetlocks were always dark red and slimy with blood.*

With a spoonful per leech, these vampires drew far more Marauder blood than all the Japanese army. Leeches, tangled masses of vegetation, and jagged hillocks kept 1st Battalion fighting nature for twelve hours every day, sometimes with progress of less than two miles.

By such arduous jungle bucking and bashing Osborne did manage to elude Jap patrols from the Kamaing Road.*[*FN: To cover his battalion's disappearance into the jungle, Osborne feinted down the trail with a platoon led by Lt. William Lepore. He and his men fought with extraordinary savagery and were barely able to get away.] After the leech invasion, Marauders were eager to turn upon a human (though inhuman) enemy, and at 0300 on 28 March they took the 300 defenders of Shaduzup by surprise. Osborne struck with six columns, two on the village, two more on the road both north and south, imparting the impression of an attack larger than his battalion. The approach of one column was made easier by following the tracks of a herd of wild elephants. The location of one Jap platoon was indicated by the sound of them fishing in a stream with hand grenades.

The shock the Marauders achieved quickly wore off. At 0700 came the first counterattack; at 0900 Jap artillery began zeroing in. Osborne called in Mustangs for revenge. But he needed more from the air—a supply drop or 1st Battalion would have to ration ammo by dark. The C-47s (Dakotas) delivered in the afternoon, though losing a plane and its brave crew to ground fire.

Resupply was a big boost for 1st Battalion who were able to consolidate and move around positions for complete dominance of the Kamaing Road on which they had under interlocking fire from both sides. Marauder spirits zoomed higher when their Chinese follow and support regiment reached the battle and weighed in with counter-battery fire from their 75mm pack howitzers. This regiment was far more interested in fighting than the one following 2nd and 3rd Battalions. The next day Osborne gladly turned Shaduzup over to the Chinese, and pulled 1st Battalion up the Chengum river out of range of the infernal Japanese artillery, into a reserve position. For the first and only time, Stilwell's hook and linkup scheme of maneuver worked as planned.

Pleased and proud how they had survived their cross country ordeal in the jungle, and praised by Stilwell for their success at Shaduzup, 1st Battalion was taking another supply drop—this time including rations and new uniforms—and bathing in streams with the satisfaction of a tough job well done, when a long coded message came in from Merrill: 2nd and 3rd Battalions were struggling desperately to escape a deadly trap. Osborne was to come to their aid at once. Leaving many of the parachute bundles unopened, 1st Battalion wearily tightened their belts and chinstraps, saddled up and started back-tracking north.

Many unexpected factors had resulted in Merrill's desperation. Strategically, Kanabe had broken into India with a corps that threatened Slim's key bases of Imphal and Kohima where the fighting was touch and go, nip and tuck— a decisive battle that came to hinge on the possession of a tennis court. Till the Japs were turned back at the net, SEAC would spare nothing to save Slim's 14th Army.

But concurrently, Wingate had launched his air mobile attack into northern Burma. The only fight more furious than that on the tennis court raged between Slim and Wingate, the Chindit CG demanding that he retain his air resources rather than divert them to fly in reinforcements (including two battalions of Chindits) for 14th Army as Slim had insisted upon to Mountbatten.*[*FN: While fully engaged fighting Slim and the Japs, Wingate was killed when his US B-25 flew into a mountain during a lightning storm. Even after death, the bizarre surrounded him. The force of the crash commingled beyond identification the remains of the crew and passengers. Pre-established Allied policy for such cases was that all the remains would be buried together in the country to which the crashed aircraft belonged. Churchill—who was probably Wingate's greatest admirer— appealed to the United States for an exception to the policy, and both Ben Gurion and the Emperor of Ethiopia urged that the remains be interred in their countries. Quadrangular diplomacy continued even after World War II, till finally President Truman settled the matter: Wingate would remain where he was, in Arlington Cemetery.]

With such utterly vital matters still unresolved, Stilwell's operations were shoved to the back burner. Air drops to the Marauders dropped in SEAC priority. Chiang Kai Shek wrung his hands about committing two divisions (he had only 300) deeper into Burma while Kanabe appeared about to turn the tables.

And indeed Kanabe had told Tenaka to take the offensive. Tenaka's vigorous execution of that order was the major cause of Merrill's very serious problems. In its previous withdrawal, 18th Division had shortened the distance to Myitkynia, their supply base. Tenaka also now had a more extensive and usable road net, providing him some vehicular mobility. And as Kanabe had been impressed by Wingate's first long range penetration, then imitated it on a larger scale, Tenaka had been impressed by the Marauders penetration, and ordered a two-battalion task force to make a similar hook to the east and cut Merrill off.

When he got word of this serious threat to his rear—the Japs could move north for a long way in trucks— Merrill pulled his Marauders off the Kamaing Road and got them on the trail back to Nhpum Ga, 25 miles north. Orders from Stilwell were to retreat no farther than that village.

But first Merrill had to get there before the Japs. That meant blocking the narrow trails converging before Auche. Re-fording the Moguang and shooting back at Jap infantry emerging from Inkangatawng, 2nd and 3rd Battalions set off on their most grueling forced march. Preceding them at a pace that seemed impossible to sustain were Weston's I&R Platoon and another platoon of 'foot cavalry' led by Lt. Warren R. Smith. Their mission was the most vital ever assigned to Marauders: hold off ten times as many Japs till 2nd and 3rd Battalions slipped by Auche. To do so, the two dauntless platoons would have to soon separate, Weston for Poakum and Smith for Tatbum. At all costs, Hunter told them, delay the Japs for at least two days.

Their job was identical to that of the Japs who had harried 1st Battalion for so long above Shaduzup: shoot and fall back, hit and run, ambush and break off. Engaged the advancing enemy as frequently as possible but not decisively. It took hard-wire nerves, fine judgment, and utter faith in fellow Marauders, but hour by hour of running battle Smith and Weston began to believe that the impossible might be accomplished, that the Japs could be held up.

Smith and Weston were favored by dense vegetation that slowed Japanese flanking maneuvers, and the fact that the two trails were only shoulder wide. The 'noise sentries' of the night also warned them when threats were coming in darkness, and also let them set up positions to the rear. Most remarkably, time after time they were able to draw the Japs close enough to hit them with surprise fire. Two small but bold spoiling attacks drove Jap points back and off balance.

After a day and a half of brilliant delaying actions, Smith and Weston could hear fire from their adjoining engagements. On the single trail to Auche, they joined forces to move back by alternate bounds. In a great feat of arms, hardly recognized by history, they had bought enough time. Exhausted, dehydrated, but with no one killed, the two platoons crossed into 2nd Battalion's perimeter at Auche at dusk on 27 March. On the claustrophobic trails back to Poakum and Tatbum, were at least sixty dead Japs, supposed masters of the jungle who themselves had been mastered by the Marauders' jungle craft and expedient tactics.

Merrill's relief was short lived. 2nd and 3rd Battalions were strung out between Auche and Nhpum Ga, but they were also strung out physically, something that could be called simultaneous burnout (true as well of 1st Battalion which was force marching a hundred hilly miles to join fellow Marauders in their hours of greatest need).

The following grim facts obtained, as enumerated by Ogburn, even before Tenaka poured fresh battalions north to crush the Marauders at Nhpum GA:

All three battalions, just about to a man, was beginning to wear out.

The physical exhaustion resulting from

Brig. Gen. Frank D. Merrill and Lt. Gen. Joseph W. Stilwell talk during a meeting at Naubum, Burma, 5 May 1944.

seven weeks of marching through mountains, mud and water, from insufficient food, and from disease probably made us more susceptible to the nervous strain of being always in the enemy's territory—which was by far the worst part of it.

Not to know from one instant to the next, week after week, when the silence would explode around you created a suspense difficult to describe. It ironed out all individual differences, and everyone in that heterogeneous organization looked alike. All faces—uniform in their emaciation to begin with—wore habitually the same expression of furtive intentness, sharp with the effort to exercise a sixth sense... Worst of all, perhaps, was the way you came to see yourself, as a prey of apprehension, much as if you were pursued by guilt, and this further sapped your resistance to that which you feared.

In the five days before its arrival back in Auche, 2nd Battalion had marched 70 miles and made over 100 river crossings. During one of those days it had fought off continual attacks from point blank range. It was soaking wet on its departure from Auche and it was still wet on its return.

Note that 2nd Battalion, like the rest of the Marauders, already had exceeded the Chindits' time behind the lines limit by a week. And of the immediate future, there was a saying of the times, "You ain't seen nothin' yet."

At first it appeared Tenaka's onslaught might be exaggerated. 2nd Battalion broke contact easily and pulled back from Auche in good order. But when Merrill's CP group started out for Nhpum Ga, Jap artillery began to range in on the trail between two villages five miles apart. At first there were just a few 77 mm pack howitzers but they had spotters in the jungle who knew how to adjust fire. Merrill's column sprawled off the trail. One round hit full on a Marauder, blowing him into bits so small there was nothing to bury.

If there was any good news, it was that Jap infantry was not looking to fight it out with the Marauders. The bad news was they didn't have to because more and bigger artillery pieces were moving north. The worst news was that most of the trail from Auche was along a ridgeline without cover or concealment. And it grew more slippery from hard use and rains, rains so heavy at times that drain holes had to be cut in litters so that the wounded would not drown or wash out. Running the gauntlet to Nhpum Ga, Marauders began to stagger drunkenly from exhaustion, shell shock and snapped nerves.

With his men cracking, Merrill's task was as simple as it was difficult: hold Nhpum Ga so that the Japs could not take Hsamsingyang, his supply base with a rough air strip, under direct fire. 2nd Battalion immediately began preparing a perimeter 400 yards long and a third as wide around a saddle just south of Nhpum Ga where McGee had good observation of the trail and the slopes east and west where the Japs might try to flank it. In case they got around 2nd Battalion, 3rd Battalion set up a 360 degree defense of Hsamsingyang and the airstrip where Merrill dug in the regimental aid station and his CP. Contact patrols ran between the two battalions' positions, but frequently they were turned back by Jap infantry who, as their numbers increased, cut the trail, beginning an eleven-day siege.

Riflemen returning Jap fire from under a basha in Loi Kang Village, attack by B and C companies.

Like most sieges, (an exception was Brest) most of the attackers' damage was done by heavy guns. First to fall were 2nd Battalion's pack animals, unprotected and silhouetted as they were. Their carcasses, a hundred of them, produced the first of a low lying stench that would eventually suffocate McGee's perimeter. To try to bury the animals would expose his Marauders to sniper fire that was abundant from all directions; besides, few men had the strength to dig more than slim foxholes in the rocky ground.

There was not even enough defilade for cat holes; so fumes from excrement all over the ground added to the miasma of rotting flesh and fly-covered viscera. Yet this olfactory hell was no more than an annoyance after the second day when the Japs shrewdly invested and seized key terrain—McGee's only water hole.

It was a seeping spring about 50 yards downhill from the western tip of the perimeter. McGee had posted it with a squad which was able to beat off several small probes, but holding the water hole was not the same as having access to water because canteen runs had to cross fifty yards under accurate fire. Several men were hit going for water so McGee pulled in the squad. From then on, 2nd Battalion relied solely on rain they could catch in their helmets and muddy soup that rose in shallow wells, contaminated by the hecatomb of mules and horses.

From these sorry sources, 2nd Battalion was rationed to a half canteen cup of water per day. The temperature was in the low nineties.

Cpl. Willie H. Castleberry, with a light machine gun crew in Burma, trades pipes with a native Chinese.

The same day 2nd Battalion lost their water hole, all the Marauders lost their commander. Merrill was incapacitated by a heart attack. So as not to aggravate it, news of 2nd Battalion's plight was kept from him. Merrill refused aerial evacuation until all the seriously wounded went out ahead of him; that number grew daily, aircraft were increasingly scarce, so Merrill stayed at the CP, turning over effective command of the Marauders to Hunter.

Even at Hsamsingyang there was not enough water to mold plaster casts for broken bones, and sulfa drugs to fight infection had to be taken dry. McGee was stretched beyond his limits to cover his perimeter which also came under probing as the Japs were able to get around Nhpum Ga. On the third night of the siege, 2nd Battalion was probed all night from every direction. The Japs' purpose seemed to be to force them to fire up their very limited ammo and stay at 100% alert. All the next day, men cowered in their flooded foxholes under a steady drizzle of artillery fire. On the fourth night, when McGee could not keep his men awake, both battalions were hit by attacks: battalion-size against Nhpum Ga, a company against 3rd Battalion to prevent Beach from supporting McGee.

This was the Marauders' longest night, men standing in their foxholes punch drunk and sometimes hallucinating, screaming at the Japs to come on, the Japs screaming back that they were; crisscrossing fire shot by instinct and reflex; insane courage by men driven insane by the accumulation of deprivation and horrors. The most vital aspect of leadership in such stygian combat was fire discipline, for now every bullet was worth its weight in gold.

Lunatics luck seemed to have more to do with the outcome than leadership, however: the Japs were well led and their task force commander, Col. Maruyama, had laid his plans with farsighted care. He had softened up 2nd Battalion for days and nights with accurate artillery, denied them water and isolated them, while rotating his own infantry units so that their battle edge stayed keen. He had felt out McGee's perimeter and learned its hard and soft points. Maruyama's assault platoons knew their objectives and had surveilled their approach routes. They had crept as close as possible to their assault lines on a moonless night, under smothering fire from mortars and artillery, plus machine guns firing from the flanks of the main attack. His men closed in concentrically on the narrow oval of the Marauders perimeter which in most places was only one foxhole thick. For all his excellent preparations and execution, Maruyama deserved victory—but he didn't get it.

Instead he got a lesson like that from the adage that says "Never fight a drunk". You can't anticipate what he will do, and that may be some ability that he couldn't do while sober. Second Battalion that night were in a similar altered state of consciousness (sometimes lapsing into unconsciousness). What they did they can not really remember. Whatever it was most of them came out of it to see the morning and 200 Japs sprawled dead, many of them within the perimeter. Air bursting grenades seemed to have been a main killer, heaved high with only a couple of seconds left on the fuze, the explosions raking anyone above ground but not often wounding a man in his foxhole. A number of Japs also died of groin wounds from bayonets shoved up from foxholes. The ground level defense was also indicated by many Marauder helmets dented by fire of their comrades.

Machine gun crew stands watch.

Dogs attached to 5307th Marauders in fight for city of Myitkyina.

The morning after looked like the nadir of Marauder fortunes but it was also where they bottomed out in their second campaign. McGee had seven dead and dozens wounded, an unusual percentage of them medics—too often they have to get out of their foxholes to do their job. Three of his nine machine guns had been knocked out; the rest were down to a short belt of ammo. All his animals were dead, their unsanitary presence augmented by the 200 Jap corpses. Burial of anything, even the American dead, was impossible because the perimeter was still under indirect fire. 2nd Battalion was on one-third rations, with virtually no medical supplies and less water. At this darkest hour came the glimmer of a faint dawn.

It came from the voice of Hunter on the radio. "You still up there, George?" Incoherent from sleep deprivation, McGee grunted something. "OK," Hunter continued, "we got a supply drop coming. What do you need?"

"Everything. Including a relief force."

"Priorities."

"Water."

In savage fighting, 14th Army had broken the Japs hold on Imphal-Kohima where C-47s had been shuttling in supplies on a huge scale. Also, the Chindits had been inserted, freeing up some air assets for Stilwell's use. Now with a high priority, Hunter called for 500 gallons of water in heavy plastic bags to be dropped in to 2nd Battalion. Braving heavy small arms fire—Maruyama recognized how a successful supply drop could turn his tide—a Dakota was badly damaged and most of its parachute drop of ammo and rations floated off to the Japs. But the water bags fell like bombs, straight into McGee's patched up perimeter. One of the accompanying light aircraft courageously landed at the Hsamsingyang strip, and flew off with some wounded, including Merrill.

As soon as he landed at Ledo, he got off the stretcher, rubbed up the stars on his collar, and demanded that two pack howitzers be flown back an dropped to Hunter that very day. Actually that request had been anticipated, volunteer artillerymen called for and obtained. It was a huge Marauder morale high when that same day two disassembled guns were dropped by double parachutes on the strip, reassembled there and were firing missions for 2nd Battalion within hours.

From then on they fired till their tubes glowed, though without much effect as a counter-battery against dozens of Jap 150 mm guns. Nevertheless, the Marauders seemed to be bottoming out, an impression that grew as Hunter received reports of 1st Battalion's progress toward Hsamsingyang.

But Osborne was still five days march away, and hardly constituted a relief force of fresh troops spoiling to fight. And Maruyama was not about to give up the initiative while he still had superior strength. The day after the air drop he directed a series of cagey attacks to determine the degree McGee had been revitalized. Return fire from 2nd Battalion was still sparse—

not much ammo had reached them by parachute—so Maruyama decided to strike them another blow. He was also prompted by new and aggressive patrols from 3rd Battalion feeling out Jap trailblocks between Nhupam Ga and Hsamsingyang. And he knew that the fortunes of war were beginning to swing toward the Allies; the next few days would tell if this would be true on the ridge along the Tanai River.

So Maruyama went for it. He had a ranger-like appreciation for surprise; in this case surprise came in the form of a mid-day attack, the hottest time of the day when McGee's men were trying to sleep to recover from the previous nightime vigil. Two Jap companies crawled forward as slowly and camouflaged as jungle reptiles. 2nd Battalion's line was like a stressed fault line ready to crack open in an earthquake. The fifty percent of Marauders on 'alert' (a gross misnomer in their condition) looked out from the perimeter in a daze. Men, regardless of the degradation of their previous proficiency, can get up to superhuman levels for an Armageddon, but the equivalent deflation immediately following reaches an equivalent low.

That's where McGee's Marauders were when the surprise daylight attack hit. What prevented the Japanese from overrunning them was a Japanese-American, Roy Masumoto.

Springing out of the grass, Maruyama's men almost took McGee's men in the first rush that crossed fifty meters before it was met by effective fire. Grenading and grappling, the Japs opened a squad-size gap in the perimeter, a gap covered inadequately by a machine gun at McGee's CP bunker. Enfilading Marauder fire killed the attackers in the gap but could not close it. Maruyama's battle commander had anticipated such a result, but did not know where the gap would open. He had a reserve of two platoons to exploit this opportunity. Fortunately McGee had Matsumoto who scurried out into the brush.*[*FN: Lest there be an impression that 2nd Battalion was asleep at the switch, it should be noted that McGee had the foresight to order every foxhole be boobytrapped with an owner-operated 'toe-popper' mine, activated when he pulled out, so when he lost his foxhole, the Japs lost their first man to occupy it.]

What Matsumoto did (preplanned earlier during consultation with McGee's S-2 on the subject of contingencies) was launch the Japs' follow-on attack himself and prematurely. Hiding dangerously close to their LD, the nerveless Nisei shouted *Sumai! Banzai!* and Death to Americans! — in Japanese, sending the Japs when and where McGee wanted them. They never got out of his kill zone, and once more Matsumoto got him out of a tight spot. They left 54 bodies, including two officers, to putrefy on the slope.

That night Hunter briefed 3rd Battalion's leaders:

"Gentlemen," he announced, "in the morning we start an attack that will drive through to 2nd Battalion. It may take two or three days, but we <u>will</u> get through." Hunter then described how economy of force would put every man who could walk in the attack: just Kachins would do the security patrolling around Hsamsingyang. The only muleteers who would not fight were those hauling the two pack howitzers for direct fire. 81 mm mortars would fire smoke and HE every three seconds while 3rd Battalion crossed the LD. The scheme of maneuver would be a direct pressure force up the trail and a flanking force moving where jungle was thinnest. "Ruses, feints, and anything else you can do to fool the Japs are in order. A fake message will be dropped from a plane so as to fall in the Jap lines. This message will be to 2nd Battalion, saying that a battalion of paratroopers will drop south of Auche at 1700 hours tomorrow."

Air evacuation of wounded.

The attack kicked off as a strike by Mustangs bombed and strafed the trail 150 yards ahead; then the mortars and two pack howitzers took up the prep. The Japs between 2nd and 3rd Battalions were back on their heels. McGee sent out combat patrols to intercept any reinforcements Maruyama tried to get around the perimeter.

Beach's force on the trail had tough going, moving up the pack howitzers by slow bounds where they could blast away Jap bunkers that straddled the trail. Initially his flanking force had light contact but entered deep jungle and had to have McGee's 60 mm mortars mark the limit of his perimeter by firing three rounds due west. The flanking force turned toward where they heard the sounds. Providentially, some Mustangs returned to strafe in front of them. The pilots reported that they thought they saw some Japs cutting out of the fight. It looked as though Beach might break through to McGee at that point, but with an hour till dark, 3rd Battalion had to dig in. It was well they did because the Japs filled the night with flares.

During that vicious night the Japs captured a Marauder machine gun and the Marauders captured one of the Japs'. Both were used on their previous owners till ammo ran out. In the morning, Maruyama made one more bid to crack 2nd Battalion's perimeter, and did but only a squad broke through to be wiped out by headquarters personnel and former muleskinners turned grenadiers. Animal carcasses were used as berms to fire on the retreating Japs. Sweating at their close call from the last Jap attack, 2nd Battalion was buoyed somewhat by 3rd Battalion's good progress so far. The next day and the next it bogged down because of enemy tenacity and bad jungle.

Ogburn: *The story of 3rd Battalion's tortured struggle toward Nhpum Ga is of men groping through thickets of bamboo, making desperate dashes from a fallen tree to a concavity in the hillside, hugging the ground with throbbing veins, inching forward in the stifling stillness of stagnant air...and the sight of friends transformed by death into impassive and appalling strangers.*

Now that combat support was more even, the Japs pulled back from the impact areas of fighter-bombers, the two doughty pack howitzers, and mortars, to reoccupy positions when these bombardments lifted. If Beach was ever to reach McGee, 3rd Battalion had to follow closer to their new combat support. This they did first on a thrust up the trail. Dirt from US indirect fire was still in the air as they rushed into it. From the opposite direction came the Japs to regain their foxholes. The race was a draw, ending in hip shots, machete hacks and blows. One Marauder lost his arm, and later his life, to a Samurai sword. But the Marauders took the ground—another fifty yards closer to 2nd Battalion.

Though his siege was being penetrated,

Maruyama refused to surrender the initiative. Suspecting that Hunter had left Hsamsingyang pretty bare, he sent out a platoon out along the Tanai River (most of the fighting had been on the Nampama River side of the ridge) to raid Hunter's airstrip.

On the morning of 7 April, 3rd Battalion was still 500 yards short of McGee's perimeter with an uphill fight the rest of the way. For the rest of the day they could buck ahead only another fifty yards shooting, crawling, grenading against well entrenched highly disciplined Japs who killed three and wounded eight Marauders. Heart breakingly, Beach lost this hard-gained ground when darkness fell and his trail force consolidated into a night defensive position. His flanking force floundered in matted jungle, unable to bring enough fire power forward to blast closer to 2nd Battalion. McGee's aid station that night was accurately mortared with the loss of four medics and half their 28 patients rewounded. Those able to crawl took positions on the perimeter which was down to skeleton strength. At this point, it seemed Maruyama was beating Hunter by attrition. Both sides would know in a few days...

And the next morning, Maruyama's raiding platoon from the Tanai Valley peppered the Hsamsingyang strip with small arms fire, and a few rounds of small caliber (so called 'knee') mortars. It looked as though Hunter's gamble had failed, and the Japs' aggressive persistence would prevail.

But that afternoon, an hour before dark, a newer combat factor was injected into the grinding, eleven-day long, seesaw battle. Down the trail from the north, stumbling with empty stomachs and light packs, came the advance guard of a grimy column. Though they looked as though they needed relief themselves, they were Hunter's relief force—1st Battalion. They had struggled by day and night, from Shaduzup countermarching some 150 miles, reeling from exhaustion the last thirty that they covered non-stop.

Ogburn: *Night marches were always difficult and seldom justified except in extreme circumstances. Unless the moon was better than half full, the darkness of the trail was impenetrable, as if you were walking with your eyes closed. The guides had to grope their way, and in order to keep the column intact, the rest of us had to hang our compasses on the pack of the man or mule in front and follow close enough to keep the feeble glow of the dial in view. Those without compasses substituted pieces of decaying, phosphorescent wood picked up from the ground...*

In our mud-and-blood-encrusted fatigues, our faces drained, we must have resembled the remnants of a Confederate force before Appomatox. A third of our number [now down to 800] *had dysentery and were only draging along. Still, we had crossed the last 30 miles of unrelenting mountains and now, by God, we were here!*

Osborne dropped off 500 men, the ones who were barely ambulatory, at Hsamsingyang. They were to protect the air stip and CP while Cpt. Tom Senff set out to the west with 250 of the hardiest Marauders for a long range penetration to Kauri, a few miles south of Nhpum Ga. This maneuver was hoped to lift the siege—or at lest help 3rd Battalion to do so—and cause Marayuma concern for his rear.

Senff's men threaded their way west of 3rd Battalion's flanking force, crossed overgrown clearings and, with light contact, came onto the Kauri-Nhpum Ga trail after 36 hours, a remarkable feat of cross country movement by anyone, a practical miracle for men in 1st Battalion's physical condition. The stench of death still reached them from McGee's perimeter miles up the trail.

Senff ambushed a squad-size carrying party headed toward Nhpum Ga and killed them all. Aggressively, he pushed a strong combat patrol down toward Kauri on the hunch that they would encounter only REMF's because most of the Jap infantry were probably battling Marauders to the north. Senff was right. His patrol came upon several trailside caches, and pursued by fire some lightly armed Japs fleeing into the jungle, when he received orders on the radio to turn back north and, in coordination with 3rd Battalion, break through to McGee's perimeter.

After calling back his combat patrol, Senff approached Nhpum warily. His point hit a Jap position about a mile out, but the trail block was oriented north and the Japs quickly broke off. That was true too for 3rd Battalion which renewed their attacks: the Japs seemed to have switched from an aggressive defense mode to one of delay—shoot and scoot.

The last half mile of terrain in front of the perimeter was a scene of ghastly devastation. Air strikes and Jap artillery had stripped all trees of branches and lacerated the tilting trunks. The ground was gouged with shell holes, a patchwork of burns from flares, and strewn with Jap bodies that seemed to have been dumped like bags out of airplanes. Where there had been bunkers, there were only crushed logs with arms and legs thrust out between them like sticks. The sound of feasting flies sounded like an orbiting fleet of C-47s The nauseous stench seemed as heavy as liquid—something to wade through—and set men from 1st and 3rd Battalions vomiting so that some could not cross the stinking fields.

But on that Easter Sunday, 1944, the beleaguered Marauders had a three-way reunion at 2nd Battalion's foxholes. The siege had been broken and so was Maruyama. More than 800 of his men—half the number he started with—stayed to be burned and buried with lime chloride. Hunter had lost 52 KIA and 302 WIA in the battle. The rest, to a man, were non-battle casualties. Everyone, from the most junior squad leader to Hunter, agreed that the Marauders were now combat ineffective—surely not a stigma but a badge signifying relentless action behind the lines for an unprecedented duration. Until the Chinese 38th Division finally got down to Nhpum GA, the Marauders had operated alone in enemy territory for over seven solid weeks.

With Maruyama driven off (Tenaka needed him around Myitkynia which was now being threatened by Chindits who had been in the field for three weeks), the Marauders took well deserved relaxation, received their first mail in two months, were issued new uniforms, new mules, and ate their fill. There were no promotions or decorations, except the automatic awarding of the Purple Heart, because Stilwell didn't believe in tangible rewards for soldiers doing their duty.* [*FN: After being broken up, Marauders submitted citations through other headquarters, and received a deluge of decorations, including six DSC's for Myitkynia alone.]

Nonetheless, positive rumors abounded, the most enjoyable being that the regiment would be sent home as a group (like the remnants of Darby's Rangers were that year) for a parade, medals and public appearances. If the 'point' system had been in effect, almost all the Marauders would have qualified because they had received no replacements. Nearly as optimistic a prospect was that the Marauders would be scattered to R&R centers in India, fabulous resorts like those in Kashmir where British troops recuperated after far less time in combat. After all, the Marauders were part of SEAC, entitled to facilities as much as their allies. A month of home leave was certainly the minimum expectation, though Marauders joked that they would go AWOL in the states, eluding MP's as they had the Japs.

Whatever was in the future it would not include mountains, mosquitoes, jungle or Japs because everyone knew and agreed that the Marauders had passed the red line, their fuel lights were glowing, their mental lamps had dimmed, and their reserves of courage and stamina drained to a point beyond replenishment. Count the Marauders out of action till <u>at least</u> after the next monsoon. And if they ever marched behind the lines again it would have to be with a complete changeover of personnel because virtually every veteran had come down with intestinal disease, mostly amebic dysentery that takes a long time to clear up.

The United States Army in World War II —the Pentagon's official history— regarded the Marauders at this point this way:

"The most serious result of Nhpum Ga was the chronic exhaustion of the troops. The fighting edge of the most mobile and most obedient force that Stilwell had was worn dull. From this fact were to flow consequences of great magnitude."

And indeed they did. Stilwell still wanted Myitkyina, and he wanted his forces to take it. That meant the Marauders were to saddle up again for their last ride because Stilwell would ride them till they dropped.

While convalescing from his heart atack, Merrill was consulted about the Myitkyina plans and there is no evidence that he raised his voice in opposition. Perhaps, knowing Stilwell, he knew that any protest would be futile. Hunter and his staff dutifully went to work on the plan handed down to them. The most they could do was give it the best chance to work—a small chance, they agreed.

Marauders, in general, did not give it any chance at all. Two Chinese regiments were to go along, but what good were they? Perhaps from personal guilt, Merrill discharged himself from the hospital and came down to Nhpum Ga to take overall command of the expedition which was to be called the Myitkyina Strike Force (MSF).

In an effort to raise rock bottom morale, he promised the Marauders that upon seizing their objective—the Myitkyina air field—they would be pulled out immediately to be sent on R&R and furlough. Even Stilwell concurred with that.

Down to a thousand men, all much weaker than when they left the Ledo Road three months earlier, the Marauders hit the trail again without much hope but with a promise from two generals, and what Ogburn described as "a pitiful sense of duty."

Bad Missions (The Worst):

XV. Myitkyina, Burma, May 1944

MSF, under overall command of Merrill, moved out on 29 April. For the first time the Marauders were integrated into other units; indeed, the 5307th did not exist in the MSF organization which was made up of H Force (1st Battalion, a Chinese regiment and a pack howitzer battery) commanded by Hunter, M Force (2nd Battalion plus 300 Kachin Rangers commanded by McGee), and K Force (3rd Battalion, a Chinese regiment and a pack howitzer battery) commanded by Col. Henry Kinnison, a Stilwell staffer who had checked out the Marauders and determined them fit to go.

Kinnison was to have a brief period to learn what life—and death— on backwoods trails was like, wildly overgrown trails that had not been used for ten years. He along with 150 other trekkers ran into an infestation of mites whose bites caused scrub typhus. Like most of the men so stricken, Kinnison died of fever. Marauders only wished Stilwell could have joined him.

MSF headquarters was at Naubum, where it would stay, directing its columns by radio, trying to supply them by air. H and K Forces headed east over the mighty Kumong Range with M Force on a parallel trail some 20 miles to the south. The Kumong were the highest and most rugged mountains Marauders ever confronted, and also the most drenched because monsoon rains had returned with fury. Even Stilwell was taken back at what nature was throwing at MSF, so much so that he didn't even tell Mountbatten—his theater commander—that he was going for Myitkyina. Otherwise Stilwell's face would be red when MSF failed.

By May 1st Stilwell was already losing hope, confiding to his diary that the rains were a bad omen, the Japs might be reinforcing Myitkynia, and MSF attempting too remote a penetration. He concluded this entry with wondrous empathy—for himself: "The nervous wear and tear is terrible. Pity [me] the poor commanding officer."

The Marauders didn't have any pity to spare for REMF's as they slogged up into the ranges, with every step their boots either sucking or carrying clay-like mud that doubled their weight. Not even torrents of rain could wash off such concrete overshoes. Switchbacks and traverses were treacherous cat walks. Mules slipped constantly. During one day 3rd Battalion lost twenty of them, and the two tons of equipment they carried. Even Marauders hardened to grim scenes winced and trembled watching mule after mule loose footing, tumbling, thrashing down a mountainside to break their necks or impale on bamboo.

When the sun broke through, the jungle slopes steamed into dense fog. So whether in rain or shine, the Marauders were constantly wet, thier socks abrading, boots rubbing soggy skin into massive abcesses. Bandages would not stick, more water was the only lubricant. So footsore was everyone that it became common to see men hobbling along on crude crutches made from branches.

Ogburn: *For the first time you began to pass men fallen out beside the trail, men who were not just complying with the demands of dysentery—we were all used to that—but were sitting bent over their weapons, waiting for enough strength to return to take them another mile... We had stragglers. Whenver we bivouaced, men who had been incapable of keeping up with the column.. would come plodding in for hours afterward.. There was a feeling in the organization that it was coming apart. And Myitkyina was still 60 miles away.*

Ahead in those 60 miles was the very toughest terrain, culminating in a 6,200- foot pass, reached by a climb impossible for laden men and mules.

All the way up we had to spend as much time resting and giving our hearts and lungs a chance to catch up as we did climbing. One thing that helped you keep going was the thought that this was the worst there could be. It was so bad it was preposterous. At times it was possible to drop to your hands without bending over, and we did so, scrambling up on all fours. We were unable to reach the top the same day we began the ascent of the pass and had to camp halfway up.

Reaching the top of the pass next day, Marauders felt they had climbed right over roof of Burma; indeed at that altitude they had ascended into a new climate of biting wind and chilling cold that produced fuzzy moss and stunty trees rather than jungle vegetation. Then the eastern ranges of the Kumon proved to be more tumultuous than even the western slopes had been: more jagged, precipitous, and more densely packed with vine-matted trees and ripping undergrowth.

Weary arms again swung machetes to widen the serpentine trails for mules. Up, ever up, looking up under the lip of an increasingly heavy helmet till necks cramped, looking for the top of a ridge that was hours in appearing. Then down, ever down, angling ankles so feet could be little platforms digging into the buttery earth. Every step down was a jar to the knee as it accepted all the weight of man and his load.

There was but one mercy found in these pouring hills: they were so inhospitable that they harbored no Japs. The war against them had been suspended pending the outcome of the Marauders' deadliest battle against nature. They would agree that the Kumon won. MSF, as it slithered slowly like a wounded snake toward the valley of the Irrawaddy, was ready for intensive care they would not receive. The Marauders, as their official history states, "did not deteriorate on the battlefield . . .but in the jungle somewhere between Nphum GA and the Myitkyina airfield."

But the airfield was still a faraway objective that no one was yet thinking about. First a 15-man garrison in the village of Ritpong had to be wiped out so completely that word of MSF's presence 35 miles north of the airfield did not reach Tenaka. This total annihilation was for K Force now, with Kinnison debilitated by typhus, effectively commanded by Beach to accomplish.

Well outside Riptong, 3rd Battalion began

Pack train leaving airstrip with food, ammo, weapons and other supplies for American Infantry.

Chinese troops arrive at airstrip to replenish supplies after 112 mile march through jungle with Marauders.

hacking off trail to get around the village on both sides, then ambush the trails south. Beach had the Chinese rush Riptong from the north. A pretty lively battle ensued, the Japs pulling back in good order but right into Beach's ambushes. 146 of the 150 Japs were thus accounted for, and left to rot in the dripping jungle.

But it had to be assumed that Tenaka now knew that some sort of threat was developing in the north, so now Beach's job was to fool him as to the Americans' objective. To do so 3rd Battalion proceeded to Tingkrukawng, a village southeast of Riptong where a British-led force of Kachin guerrillas and Gurkhas had trekked down from Fort Hertz to harass the Japs. 3rd Battalion joined in, but the Japs were unexpectedly strong and also alert because of skirmishes with the G's.

Beach probed for two days, fighting hard against a Jap battalion where intelligence had reported only a platoon. He could make little headway while losing eight dead and 21 wounded. Moreover, because there had been no resupply drops and much of his ammo had been lost when mules perished, he didn't have the firepower to press a coordinated attack. 3rd Battalion puled back, the first time Marauders had failed to seize an objective. But their action was successful in diverting Tenaka's attention from H and M Forces who linked up at Arang, brushed aside a Jap platoon there, and force marched straight for Seingeing less than 20 miles above Myitkyina.

K Force now brought up the rear, lagging behind as scores of men fell from heat prostration and short rations. After two weeks into this most recent mission, the sands were running out faster for 3rd Battalion than for their brothers, but they were not far behind in the descent to utter exhaustion.

It was decision time for Merrill when, on 13 May, Hunter reported that with H Force leading, MSF was at the headwaters of the Namkwi River on which they could guide for the final thirty miles. A Kachin boy felt he could lead Hunter over a concealed route that bypassed Seingneing and led to a point just outside the Myitkynia airfield.

Merill signaled Stilwell: "Can stop this show up till noon tomorrow.. if you think [Hunter's proposed march] is too much of a gamble. Personal opinion is that that we have a fair chance and should try."

This opinion was formed a hundred miles away in the safety of the static MSF CP at Naubum, and Stilwell's reply was very predictable. "Tell 'em to roll in and swing on 'em," he radioed back to Merrill from even farther in the rear.

Leaving about 40 sick Marauders for K Force to pick up, Hunter faithfully followed the Kachin boy who truly knew the Namkwi valley like his own basha, weaving through light jungle and around paddies in daylight and darkness. But on the night of 15 May, only fifteen miles from the airfield, he was bitten on the foot by a deadly snake. He tried to go on but soon his foot swelled to the size of a watermelon, and he fell vomiting and delirious.

Two Marauders slashed the bite mark and sucked out blood and poison for two hours in the darkness. Wobbling and woozy, the brave boy said he could continue. Hunter mounted him on his own horse for the rest of the way.

At daylight H and M Forces were hiding only four miles from the airfield. To prevent detection, they had rounded up all the Burmese in the area so there would be no chance the Japs would be tipped off.*[*FN: In contrast to the Kachins, many lowland Burmese were Japanese sympathizers.] The railroad and telephone line into Myitkyina were nearby and Hunter could have easily cut them but did not because the Japs would probably investigate. The size of their garrison had been reported to be no more than 500—as it turned out, there were triple that number and more would pour in.

To prevent such reinforcement, Hunter's plan was for 1st Battalion to bypass the air field and seize the ferry landing across from Zigyun on the Irrawaddy about two miles away. Enroute Osborne would guide the Chinese 150th Regiment to their attack postion at the corner of the airfield. At 1000 hours on 17 May, the Chinese were to sweep across the runway.

Neither M nor K Force were within supporting distance for this attack. Beach was not much closer than Arang and McGee still north of Seingneing. Hunter had taken the healthiest and most able of MSF for this "'dash", as Stilwell called H Force's bold thrust. The Marauders would live or die, a they always had, from the element of surprise.

It was total, it was complete. "Merchant of Venice" was the signal Hunter flashed to Merill, meaning that the airfield had been seized. In an hour 1st Battalion had the ferry and by noon the Chinese controlled the airfield (no planes were on it). Merill glowed with pride in his men far away, called Stilwell on the radio, and Vinegar Joe hopped, skipped and hollered with joy. Official Army history called that moment the high point of his military career.

Almost dazed by their own success, the Marauders suddenly felt exposed. They had always been jungle fighters. Now they were manning positions on flat, open country with a big town, a railroad and hard top road near by. They adjusted their battle sights for the first time to compensate for long range visibility.

Hunter had delivered. Now it was time for Stilwell to do likewise. The first preplanned lift into the captured airfield was to be ammo, rations and reinforcements, all direly needed.

Instead Hunter got runway engineers and a British anti-aircraft battery—though there was nary a Jap in the sky—neither of which he controlled. They had been sent in by Gen. Stratemeyer, the senior US Air Corps commander who felt that now that Myitkyina was an air base in friendly hands, he would decide what planes went in there and in what order.

Monsoon downpours delayed lifts for another day, a day in which Osborne had to send his I&R platoon into the countryside to forage food for the battalion. The next day Hunter was directed to send his Chinese into the town to determine the strength of the Japanese garrison. They did but got into a fire fight with each other and fell back (perhaps the origin of the expression, 'Chinese fire drill'). He sent another Chinese battalion to relieve Osborne at Zigyun, but the march that had taken Osborne two hours required 48 hours by the hyper-cautious Chinese who told him, "No food, no ammo—no move." Those were probably Hunter's secret thoughts too, and pretty well described his situation.

What was holding up airlifted food and ammo was not just the rains— though they darkened the sky for hours every day—but congratulatory confusion at higher headquarters. No one had expected the airfield to fall out of the blue as it did, and now every military nabob wanted in on the action that had transformed the strategic picture in northern Burma.*[* FN: In strategic importance, the Ranger successes at Omaha Beach no doubt rank first in this history. But second, in the author's opinion, was the Marauders' seizure of the Myitkyina airfield, which in the CBI was as important as the capture of the Remagan bridge that set off the Allied avalanche into Germany. Darby's actions at Salerno would be my third choice.] Except that no one, Stilwell and Merrill included, had a plan to exploit or even consolidate the startling success.

On 18 May, the two generals flew into the airfield to take a look for themselves. They brought neither necessary resupplies nor reinforcements, and not even a frag order to guide Hunter. On his own he urged M and K Forces to proceed to the airfield with all speed. The next day, Merrill suffered a severe heart attack and would play no more part in the campaign. Stilwell replaced him with the MSF deputy, Col. McCammon, who quickly succumbed to fever and was evacuated ten days later. He was replaced by the only man whom Marauders more despised, if possible, than Stilwell: his chief of staff, BG Hayden Boatner. Why Hunter, the proficient, proven man on the scene—and as fine a soldier as any described in this history—was not put in command of MSF is a baffling secret that Stilwell took to his grave.

At least MSF was dissolved and the Marauders reunited under Hunter, the Chinese operating separately. Tenaka got a grip on the situation before Stilwell did, calling in every unit within a 20-mile radius to retake the airfield. Stilwell simply told Hunter to "finish the job"—whatever that job was. What was finished was the Marauders.

The first Jap reinforcements (eventually they would total 4,500) were commanded by the Marauders' respected and ruthless opponent, Col. Maruyama. A patient planner as the battles for Nphum Ga had demonstrated, he extended his lines in front of Myitkyina in both directions, overlapping the Marauder flanks in the next few days so that Hunter had to refuse them by bending around the airfield. Once again, the Marauder line was one foxhole thin.

After probes, Maruyama hit 3rd Battalion on 23 May, taking heavy losses but breaking into the perimeter where the penetration was eliminated only after day-long fighting. This desperate action caused Hunter— whom the emminent historian, Barbara Tuchman, described as a loyal, brave, "dedicated officer who knew the rules"—to present Stilwell an itemized report that charged his staff, beginning with Boatner, with callous indifference and long standing hostility toward the Marauders, culminating in staffers' allegations of malingering. Hunter's scathing report ended with the recommendation that, at the conclusion of the current battle, the Marauders be disbanded.

Actually they were being informally disbanded by medical evacuations, that grew to scores per day forcing Hunter to contract the perimeter to the fringes of the airfield itself. On 25 May, Hunter handed Stilwell the indictment. A Marauder on the beleaguered perimeter recalled wistfully that on this occasion, "I had him [Stilwell] in my sights. I coulda' squeezed one off and no one woulda' known it wasn't a Jap who got that son of a bitch."

How much of a son of a bitch did not become clear till Stilwell's staff roamed through the evacuation hospitals in search of Marauders with temperatures lower than 103 degrees. Such men, they announced, were fit to be flown back to Myitkyina and man the perimeter. Those who were thus shanghaied by 27 May were on hand for the last action by a Marauder battalion.

The remnants of 2nd Battalion had been ordered by Boatner to push a Jap battalion out of Charpate. After a staggering approach march of several miles, a good preliminary recon found a weak sector of the enemy positions, and McGee sent two of his scraggly companies into it. They cleared out the Jap foxholes but then fell into them and fell asleep from sheer exhaustion. McGee himself lost consciousness three times while directing consolidation of the breech. But his men could only hold what they had and for not much longer. They could no more continue the attack than swim the Irawaddy.

McGee dragged himself forward to look them over. What he saw were scarecrows with rifles, all with the seats cut out of their sodden fatigues so they could squirt while they fired. Flesh was bleeding from ulcers as if the men had been hit with buckshot. McGee could watch no more and knew his men could stand no more. He radioed back to be relieved—declaring that his proud battalion was unfit for further combat.

The aftermath for the Marauders was as sad as the ending had been tragic. The IG for CBI began an investigation that amounted to a whitewash of Stilwell, blaming instead the Marauders for "an almost complete breakdown of morale"! Stilwell tried to stop the story there by sending Hunter home on a slow boat, but the story got out and resulted in a Congressional hearing. The *New York Times* observed that

The tough Marauders cracked up when, by somebody's blunder, their convalscents were ordered back to the flooded slit trenches... In the long history of warfare good outfits have disintegrated before, but seldom has their commander [Hunter] *swallowed his pride enough to tell the whole story... Now that the story was out, Army men wondered what would be done. The answer was: very little.*

Except for a Distinguished Unit Citation (it should have been Presidential) labled 'Myitkyina'.

Supreme Allied Commander Southeast Command Admiral Lord Louis Mountbatten visits an airfield captured by Merrills Marauders in 1944. Center, Capt. John M. Jones and Lieutenant General Peck, facing camera. Others are Chinese and American officers.

Preface

In the long history of the American Ranger, a history that spans three centuries of "Leading The Way" into battle by land and water, a new dimension was added during the Korean War. While the 8th Army Rangers were fighting in Korea, a unique force of volunteers were being assembled to train as Airborne Rangers. In 1950 at Fort Benning, Georgia, the Airborne Ranger was born.

We were the first of that Rare Breed to take to the skies. We were the original "Airborne Rangers." We were the first to wear the black-and-gold Ranger Tab, the first to wear the Black Beret.

Of the 18 Ranger Companies formed during the Korean War, 17 were Airborne, and seven fought in the Korean War. Our record on the battlefield was brilliant. At a time when the United Nations forces numbered more than 500,000, less than 700 Airborne Rangers were fighting to the front of each American Division engaged in battle. We set the standards for leadership, sacrifice, courage, motivation, and professionalism throughout the U.S. Army.

We proved once again that Rangers can compensate for lack of numbers, by valor.

RANGERS LEAD THE WAY,

Ranger Rex G. Masters
President 1994-1997
Ranger Inf. Co.(ABN)
Assoc. of the Korean War

Rangers in Korea

(L to R): Elmer Cassatt and Fred Crispino. 8th Army Ranger Company, Korea, 1951.

Introduction

The first volume of this history described ranger operations in World War II. To take up the story from there, we return to the joyously victorious months beginning in the fall of 1945 when President Truman, with pride and pleasure, acceded to the public demand to "Bring the boys home!" The boys, of course, included all Ranger and Marauder veterans, most of whom had exchanged army khaki for civilian clothes by the end of the year.* [*FN: As volunteers (though most had been draftees or reservists), Ranger-Marauders opted to stay in the Army at a higher rate than other soldiers inducted for wartime service.] The colors of their battalions, except for the 6th Rangers, had been retired even before VJ Day as the Army liquidated specialized units (including four of the five airborne divisions) with the haste of a fire sale.

The reasons were economic. The national debt had soared to all-time heights to pay for World War II. For the first time in American history the largest item in the federal budget had been defense; it was also the one most easily cut as the Soviet threat had yet to be widely perceived. Moreover the postwar economy was booming into unprecedented prosperity from pent up consumer demand. Fears of peacetime unemployment were discarded as expanding American industry quickly absorbed the troops coming out of uniform. Additionally, the GI Bill was a wonderfully successful incentive for veterans to personally fulfill the American Dream they had so effectively defended against fascism.

Consequently in the late 1940s it was unusual, almost quaint, for men of talent to remain in the regular Army; even for ranger-type soldiers whose talent was for close combat. Combat was much read about in the reams of books coming out on War World II, not something to anticipate and prepare for in the immediate future. By 1946 there were no ranger units on the Army rolls and the scroll was seen only on the shoulders of veterans marching in Memorial Day parades.

During this down time the heraldic office of the Army sorted out, at the same time muddling, ranger lineage. Merrill's Marauders were orphaned because they had been a provisional unit. Flags of the six Ranger battalions were in mothballs with the possibility of reactivation a low priority in the Pentagon.*[*FN: Even for afficionados of military lineage (the author is not one of them), the paper trail of Rangers is almost laughably convoluted. Here, highly, simplified is its outline:

The history of the eventual parent organization, 75th Ranger Regiment, goes back to the 5307th Composite Unit (Provisional), better known as Merrill's Marauders, activated in the CBI in 1943, to be consolidated in 1944 with the 475th Infantry Regiment which was then inactivated in 1945. In 1954, toward the end of the Korean War, the 475th was redesignated the 75th Infantry Regiment. Till 1956 when it was inactivated in Okinawa, the 75th contained no Ranger units. Its history will be resumed in the Vietnam portion of this volume.

By comparison with the lineage that grew out of the WW II Ranger battalions, the 75th Regiment's history is simple. In many cases the battalions' component companies were split out and do not appear on Pentagon paper until 1960 and were not activated till 1974 or "84. When they did, they appeared in lst, 2nd or 3rd Battalions of the 75th Ranger Regiment, not to be confused (but inevitably were) with the same numbered battalions of WW II. These had altogether different peacetime histories.

Grady Shinpaugh holding burp gun. Otto Grotte, holding ammo drums, was killed Sept. 1952, nearly a year after photo was taken.

Darby's originals, the lst Ranger Infantry Battalion, was disbanded after Cisterna along with 3rd and 4th Battalions. 2nd, 5th and 6th Battalions were inactivated in 1945, the latter after a stint of occupation duty in Japan. Between wars, lst and 2nd Battalions were redesignated and had brief histories as straight infantry in Panama.

In 1950-51, the first four Ranger Battalions were activated by companies which were redesignated by numbers rather than letters (e.g., Company F, 2nd Rangers, became 10th Ranger Company). In 1952, the Ranger battalions were 'restored'—that is, re-redesignated— with all their former elements; but this was only a paper exercise and the numbered companies went on with the war as before.

If the reader is confused at this point, the Gordian knot of lineage weaves thicker between Korea and Vietnam and thereafter.]

But Rangers remained on the wish list of every general who had benefited from their exploits in World War II. Generals like Bradley, Patton (deceased in 1945), MacArthur and Clark had called for a Ranger battalion at field army level. They were now in positions to make that happen, if only on paper. If there was any debate on the merit of this level of employment it was not preserved in Pentagon archives. What is known is that when the need and some resources emerged at the outbreak of the Korean War the Army G-3, LTG Lawton Collins, directed that Rangers be "tested" at divisional level.

The Ranger look. 1950. Merer Simpson, Ted Jewell.

It would seem that World War II had been sufficient test for Ranger battalions (even a regiment) but the army had apparently come around to the view that what was good at field army level might be even better at division. Because this decision shaped the Rangers' Korean experience, it deserves some detailed analysis.

First it must be noted that there was no official Ranger input for this decision. There was no Ranger command to provide it. Second, Collins as a corps commander in Europe had practically no experience with Rangers. He was impressed, however, by the North Korean proficiency as infiltrators and sought a comparable force of raiders in Eighth Army. Bradley, a longtime Ranger proponent, was chief of staff but largely refrained from the issue. MacArthur, the theater commander, and a general of unprecedented influence, did not involve himself in Army doctrine; nor did the Eighth Army CG, Walton Walker, whose background in World War II was in armor. In short, the Rangers had no champion for their point of view. But the question remains, which level of employment—division or higher—was best for Korea?

Eight Ranger companies—the equivalent of almost a regiment—saw significant action in Korea. Their missions, some good, some bad according to the criteria of Volume I were assigned, with the few exceptions noted, by division. To assert that better missions were missed because of the level of employment would be idle speculation. There was no obvious Pointe du Hoc or Cabanatuan that was foregone because a Ranger battalion did not stand ready to perform it. The one strategic mission assigned at army level failed for the same reasons found in the 'bad' mission criteria: faulty intelligence and inadequate support. A Ranger battalion HQ would not have saved it. Yet, after June 1951, many opportunities were present for small-scale amphibious and/or airborne raids up both coasts behind the static Chinese positions. Such short hooks to their rear would have caused serious problems for the Communist commanders strained to keep open lines of supply, but such raids were never undertaken because domestic political clouds in the U.S. darkened any flaming national desire for victory.

The 'bad' missions assigned by division are certainly open to justifiable criticism but must also be examined in light of the situations the divisions confronted. Generally, 'good' missions occurred during UN offensives; 'bad' ones when the Chinese had the initiative. On those latter occasions—with scant intelligence, support, and time for reconnaissance—Ranger companies were set out as early warning posts, 'speed bumps', small breakwaters to deflect an onrushing enemy tidal wave. Yet to fault such employment a critic must answer what he would have done differently had he been the division commander. Which is to say that desperate situations rarely allow optimal employment of available forces.* [*FN: Review of Volume I shows eight 'good' missions, all of which were undertaken when higher HQ was on the offensive. Of the three missions judged to be 'bad', two were offensive (Venafro and Cisterna), and one defensive (Dernia Pass). As a rule, Rangers spend more time and take most of their casualties defending their gains rather than in achieving them—Salerno and Mogadishu being the best examples.]

So though the case can be made that assignment of Ranger companies to divisions in Korea did not work well operationally, it does not perforce follow that assignment at a higher level would have worked better *in so far as operations are concerned.* What would have worked far better was mission development, logistical support, replacement procedures, and staff planning.

Each of these areas can be repeatedly illustrated in the history that follows, but can be boiled down to the absence of a parent Ranger organization. Parents go to bat for their progeny, see that they are sheltered, fed and are not run over by trucks. Parents deal on equal footing with other parents for the general welfare of the neighborhood.

None of these parental functions were available to Ranger companies in Korea. A company commander, typically a captain but often only a lieutenant, worked for a division G-3 who was a lieutenant colonel with a flock of assistants. The Ranger CO was his own staff with no backup. It was a grossly tilted relationship that could work only if not abused by the G-3, and depended almost entirely on his sound judgment, appreciation for Ranger capabilities and limitations, plus a large quantum of generosity. G-3's are obviously officers of high standing but that did not make them consistently good employers of Rangers.

Well employed or not, Rangers took the jobs handed to them and performed to the utmost limits of determination, endurance and courage. The story of Rangers in Korea is not one of ill advised doctrine but of a fiery spirit that took a lot of high ground, held it, and never backed off from a fight no matter what the odds. It is the second chapter of Rangers in the 20th Century, not the chapter of much fame as was the first but one that stirs equal pride and respect in the next and all succeeding generations of Rangers.

Eighth Army Ranger Company

As booty from World War II, the Soviet Union had taken the remainder of huge Sakhalin Island, a mere twenty miles from the northern Japanese coast. Even closer by were the Kurile Islands which had also been added to the Soviet empire. A hundred miles off Japan's south coast was the Korean peninsula which, in Communist hands, would nearly complete the encirclement of the key free world bastion and the foremost US base in Northeast Asia.

Diplomatic miscalculation and unpreparedness in South Korea led to the Communist gamble to seize the lower half of the peninsula at the beginning of the 1950 monsoon season when U.S. air power would be least effective. This would be a collaborative effort by China and the Soviet Union executed by their henchman, North Korea. The Soviets provided modern weapons and equipment with instructors; the Chinese repatriated thousands of Koreans who were veterans of Mao's successful civil war. Thus in the five years North Korean Army (NKA) became a rugged, well-trained force of 127,000 ground troops, seasoned with combat veterans and animated by genuine zeal to reunite their homeland peninsula under proletarian, totalitarian rule.

Across the 38th Parallel was the Republic of Korea Army (ROKA) numbering about 98,000 men in eight infantry divisions, none of which had tanks or heavy anti-tank weapons. Their indirect fire capability was lighter, vastly outnumbered and outranged by the NKA. Air and naval forces on both sides were negligible. That lack in the south was to be filled by the U.S. in the unlikely event of invasion. In the north that lack was deemed irrelevant because the invasion was to be a blitzkrieg, over before the U.S. could effectively react..

The U.S. presence in South Korea was the Korean Military Advisory Group (KMAG), some 500 specialists who worked for the ambassador rather than the theater commander. As the principal intelligence agency on the ground, KMAG did not foresee an invasion but predicted that the ROKA could repel one. At 0400, 25 June, it thundered down across the border with armored spearheads and numerous guerrilla landings along the east coast. The next night, the UN Security Council called on its members to give South Korea military aid.*[*FN: At the time, the Soviet Union had walked out of the Security Council so was not able to veto this action, a remarkable example of uncoordination among the communist regimes on the eve of their joint venture. Without a United Nations mandate to defend ROK, the American effort to do so would have been much more difficult. Third country U.N. members contributed more than the equivalent of a division of troops and of some of the best fighting formations in the war.]

In three days the ROK capital of Seoul fell and the ROKA was routed all along the front. Truman authorized MacArthur to commit U.S. ground troops (air and naval forces were already in action) to hold an enclave around the southeastern port of Pusan. The advance party was from the 24th ID, Task Force Smith, a battalion-size combat team which was chewed up and overrun on 5 July. The CG of the 24th was captured when the rest of his division tried to make a stand at Taejon. On 18 July, the 1st Cavalry Division and the 25th ID came ashore, followed by the 29th RCT. These units comprised the initial field forces of Eighth Army under LTG Walton Walker, a dogged bulldog of a commander, just the sort of man needed for the do-or-die defense of the Pusan perimeter.

To make matters still more desperate, the Eighth was a hollow army. All regiments had been stripped of one infantry battalion; all artillery battalions reduced from three firing batteries to two. The divisions had come from Japan where they performed sedentary occupational duties and little worthwhile training. Their equipment was outmoded and worn, with serious shortages in ammunition and spare parts. Once again the United States had entered a war (though Truman called it "a police action") quite unprepared, and this time there were no allies to buy time. Instead the currency would be American blood.

With too few troops to hold too much terrain, divisions had to defend more than ten times their normal frontage; so of course positions were thin and porous, lending themselves to NKA infiltration, especially at night, which became the enemy tactic most dreaded by the Americans. This tactic also was the spark for a counter-capability for infiltration—and that meant the Pentagon would begin thinking about Rangers again.

The first glimmerings came from Eighth Army G-3 which ordered a study of the feasibility of friendly guerrilla operations in North Korea, hopefully to take some pressure off the Pusan perimeter. That was a pipe dream, but it got the ball rolling toward creation of a provisional company of Rangers in Japan. Their first job would not be to raise friendly 'G's' but combat unfriendly and pesky ones in some wild terrain at the north end of the Pusan perimeter.

The company would be commanded by 2nd

LT Ralph Puckett Jr. who can be considered sort of the Bill Darby of Korea because both officers started Ranger units in-theater from scratch. Though Ranger companies were being formed in the States, time was critical so Eighth Army decided to take a company 'out of its hide', a very thin hide because replacements for the embattled front line units were extremely scarce. Nevertheless, Puckett was authorized to sweep up how many volunteers he could from service support units in Japan to a total of three officers and 74 EM.

Here again was historical analogy: Puckett, like Darby, answered to a general officer's staff; Puckett, like Darby, was to raise his force from volunteers, most of which were not in combat arms. But Puckett had to stand up his force faster because combat missions were already waiting for him. Thus, in late August 1950, the Eighth Army Ranger Company was provisionally created on paper, later to add the designation 8213th Army Unit.*[*FN: In 1942, the Army tasked its scientists to develop a 'Sub-intelligent Numbers Selector' (SINS), the only characteristic of which was that the numbers it selected be utterly random so that the enemy would know nothing about units from the numbers they were assigned. SINS' (the nature of the selector is still classified TS/SCI) first product was the numbers 1688, 5307 and 475 for Merrill's Marauders. After WW II, SINS was deactivated because it was too successful, baffling Americans more than the Japanese by its lack of reason and logic. Reactivated in Korea, SINS gave Puckett's company the number 8213!]

Puckett had seven weeks to take a company from paper to Pusan, so his selection process was necessarily informal and hurried. At the repo depot at Camp Drake he found two of his West Point classmates awaiting assignment and quickly recruited them, thereby filling his officer quota. For the EM, word of activating a Ranger company spread by official channels and word of mouth, bringing in several hundred volunteer statements but few mentioning combat experience. Screening records, the three lieutenants looked for solid duty performance, weapons qualifications, and athletic ability, though there was insufficient time in initial screening to administer PFT's. Single men under 26 were preferred.

Genuine motivation was determined by interviews. The first quick cut was made after Puckett announced to all the volunteers assembled that his only guarantee was for hazardous duty without hazardous duty pay, so anyone not drawn by those conditions should feel free to leave—at which point about a third of his audience walked out! Those remaining were given individual interviews.

Those selected were poured into the TO&E of a World War II Ranger company, the only relevant table that happened to be available at G-3. There were two 36-man platoons and a five-man headquarters element. The platoons were broken down into two assault sections and a special weapons section containing a 60 mm mortar, two 3.5" rocket launchers, and a BAR. There was one sniper with the platoon CP. Everyone else was armed with an M-1 or carbine. Puckett's company wore the Eighth Army shoulder patch.

8th Army Ranger Co., Korea.

Elmer Cassatt, Frances Bell, Harvey Olson, Fred Crispino. 8th Army Ranger Co., Korea, 1951.

Three days after the company assembled for their first reveille, they sailed for Pusan. G-3 had reserved a training site for them called Ranger Hill in an area infested by communist guerrillas who were perfect 'opposing forces' for the most realistic sort of training. Puckett's foremost objectives were to attain proficiency in recon and combat patrols, trail blocks and motorized scouting (though the Rangers had no vehicles, not even a mess truck). Intrinsic in this training was Puckett's goal that "Each Ranger will achieve the best physical condition of his life." Underlying the unit training was individual competence in land navigation, camouflage and concealment, and the old Ranger equalizer, adjustment of indirect fire. Live fire and night exercises in the surrounding rice paddies and hills were the norm. Officers and NCO's held classes in raids, ambushes, and infiltration, then conducted field exercises in these subjects.

Puckett's particular training emphasis could be called 'review and redo', meaning that after every exercise he not only conducted a thorough critique but drew out individual impressions from each participant, a process that could become rather personal at times because not even leaders were spared criticism from those who had followed their orders, resulting in somewhat of a breakdown in the conventional officer/NCO-EM relationship, and resembling the communist practice of self-criticism. Whatever its origins and temporary discomfort for its participants, Puckett's system worked, and that's all that mattered to him and his company.

Hurt feelings, he believed, were far better than getting people hurt later in combat. He also inculcated the truth that field manual doctrine was initial guidance, not a rigid mould for on-the-ground tactics which ultimately depended on flexible thinking and initiative. The 'redo' that followed such exhaustive reviews meant repeating the exercise till mistakes were understood by all and corrected. Though makeshift and informal, training around Ranger Hill accomplished a great deal and bonded the company in a unique way: the men had experienced unprecedented stress and utter exhaustion together, conditions that caused them to know themselves and each other to a degree that can not be duplicated short of combat.*[*FN: The author experienced MAJ Puckett's brand of training in 10th Special Forces and can attest to its effectivness. I believe that most great trainers—in contrast to just good ones—are or appear to be a little crazy until the student realizes that they are indeed crazed by the passion to make him the best soldier he can be. Puckett was such a model, wild-eyed at times, disconcerting to his training cadre that he lashed mercilessly, driving them harder than even the students. The method in his madness was not evident till the effect of Puckett's training settled in at a deep level, often much later, to emerge at times of great stress when the former student realized that he had done this before and prevailed.] Another of Puckett's training innovations was in rotating leadership through the ranks (simulating the affect of leader casualties in combat) during exercises so that every man had to keep aware of the general situation and work out in his mind what he would do if command rotated to him in the next few minutes. In this way, all of Puckett's Rangers were leaders.

Eighth Army Ranger Company had completed but five of Puckett's seven-week training program that he felt necessary before his men were combat ready. He was cut short because of MacArthur's great counter-stroke, the daring landings at Inchon that turned the tables on the NKA and sent them streaming north. Concurrently, Walker broke out of the Pusan perimeter and by the first week of October (a period of perfect flying weather for air power that further punished the NKA retreat) two UN corps had crossed north of the 38th Parallel to attempt their own reunification of the Korean peninsula. By the end of the month their patrols were approaching the Yalu River, the international border with Red China.

In the train of the UN advance , however, were many pockets of bypassed NKA troops who quickly converted to guerrillas, and good ones. These were a threat to Eighth Army's extended lines of communications and had to be dealt with. Puckett's company was 'chopped' to 25th ID who was most involved in counter-guerrilla actions in a 6,500 square mile area of mountains east and south of Taejon. Through ambushes and saturation patrolling, the division had killed or captured more than 1500 G's by the middle of October.

Puckett and his quartering party joined the 25th ID at Taegu and received a mission to clear the village of Poun 32 miles to the northeast. Eighth Army Ranger Company was twelve men short because of dropouts and training injuries, but Puckett picked up ten ROK interpreters.

The Rangers' first operation in Korea more resembled search and destroy missions of Vietnam than what they did in World War II or later in the Korean War. But this was because the initial enemy were guerrillas or NKA stay-behind units which operated like guerrillas. In field manual terms, Puckett's was a denial operation, continuously combing the high hills, setting up trail blocks and ambushing defiles through which G's might travel. Contacts were few, scattered and no larger than squad size. Some excellent ambushes were carried out in which entire G parties were anihilated. Except for one instance of friendly fire from ROK constabulary that illustrated language and coordination difficulties, the cherry missions of Puckett's Rangers were an excellent field exercise and sort of a graduation from training to combat. As he had on Ranger Hill, Puckett conducted detailed critiques after debriefing each patrol that made contact. In this way, training never really ceased. Particularly fined tuned were immediate action drills, the life saving reaction in guerrilla warfare.

Also honed was the ability to scrounge, a function for which Korean Rangers would become famous or notorious, depending upon one's perspective. For Puckett's men, scrounging was a necessity not just a passtime. With a single corporal to provide all logistics, the Rangers had to live off the land no matter who owned the land. All classes of supply were dependent upon the unit to which the Rangers were attached. Mobility was entirely in brown leather personnel carriers until a jeep found its way into Ranger hands and soon lost its bumper markings and previous ownership. It was an all-purpose vehicle, used to evacuate the injured, visit outposts, haul supplies and evacuate EPW's. Almost as useful was rice from 'friendly' paddies which became a constant supplement to C rations. Like the guerrillas they were fighting, the Rangers had to live off indigenous and UN supplies, whether or not it was donated.

Twenty-fifth ID, with Puckett's Rangers attached to their Recon Company, adequately cleared their AO of guerrillas during the next month and by November went to work again on the same sort of mission around Uijongbu some twenty miles north of Seoul. The only novelty for the Rangers here was some mine clearing and flnal security for armored columns heading north across the 38th Parallel.

That was the direction of MacArthur's entire UN command as total victory over the NKA seemed in sight, raising visions of "Home for Christmas!" in everyone's mind. Holiday plans were changed, however, by some party poopers who called themselves the 'Chinese Peoples Volunteer Force', but were actually the cream of Mao's Army, uninvited guests whom MacArthur brashly assumed would stay north of the Yalu. Even verified contacts and EPW reports could not persuade him otherwise. Indeed, the American strategic intelligence failure was probably the worst in history when in November 300,000 well trained troops of the Chinese Communist Forces (CCF) poured into North Korea fully prepared for winter warfare. The UN forces were prepared only for Christmas.

Yet MacArthur blithely resumed the offensive on the belief that Mao was bluffing. As part of Eighth Army's general advance, 25th ID was ordered to attack toward Unsan to develop the situation. Their spearhead would be armor heavy but included the Eighth Army Ranger Company, perhaps on the recollection of similar missions performed by Rangers years before in France.

As part of TF Dolvin, built around a medium tank battalion, the Rangers first scouted ahead about 5,000 meters without contact but uncovered a number of escaped POW's from the lst Cav. Division. The Rangers then mounted tanks to continue the reconnaissance in force. They dismounted when taken under fire, losing their first KIA. Two more Rangers were killed by fire from Dolvin's tanks, an inauspicious beginning to combined arms operations. Nevertheless, Puckett cleared a hill that commanded the axis of advance and dug in for the night during which he lost two men to frostbite. The next day when the TF moved out, the Ranger Company was down to 51 effectives.

This time a platon of tanks were attached to Puckett so that there was no question of who worked for whom. By mid-morning the entire task force was in a serious fire fight that ended with the little Ranger combined arms team in the lead to seize Hill 205, a thinly wooded crest of high ground that fell to the assault supported by air strikes. Six Rangers and three attached ROK's were casualties in this action that ended with the withdrawal of the CCF a few hours before dark. Puckett knew he had not seen the last of them and had his men dig in as deep as the rocky terrain allowed. When he studied the TF fire support plan for the night defense, Pucket noticed a gap of several kilometers between Hill 205 and the adjoining infantry company. Rather than expose a flank, he had the Rangers readjust their positions to an all around defense, a wise precaution because by 2100 the Chinese knocked off the nearest friendly platoon.

An hour later, a preparatory mortar barrage fell on Puckett's position, signalling a succession of infantry assaults that went on all night—a very cold night. The Rangers were introduced to the CCF control measures of whistles and bugles. Good overlapping fields of fire and adequate artillery support, including illumination, turned back the first onslaught that ended by midnight. But several Rangers were killed, and Puckett wounded by a mortar fragment in the arm. In the next two hours, more waves of Chinese tried to retake Hill 205. Some reached grenade range of the Rangers positions but all were repulsed with heavy Chinese losses—losses which would have been even greater if U.S. tanks at the base of the hill had participated but did not because of no experience nor training in night fire.

The first five CCF attempts, the last one by about two companies, amounted to increasing strong probes. This was the Chinese method of developing a situation: first send a platoon, then a company, then incrementally whatever it took to seize an objective. With good intelligence about Puckett's position from their previous failures, the Chinese launched a reinforced battalion for a sixth assault at 0300—the 'dead man's hour'. They struck at the Rangers' flank at the widest point in the gap between it and the adjoining infantry company. The CCF also hit that company with a supporting attack that caused Puckett to lose priority for artillery fire. Under too much mortar fire to move to supplemental positions that would have thrown off Chinese intelligence, the Rangers had to meet the Chinese where they had before, with every foxhole zeroed in by automatic fire and showers of stick grenades.

Under the sheer mass of men and metal, one after another Ranger foxhole went under, opening a breach in Puckett's perimeter that he had no men in reserve to close. Two mortar round straddling his CP rewounded him badly and killed LT Cummings. Chinese silhouettes filled the horizon: Hill 205 was being overrun.

The sound of American weapons diminished as Rangers ran out of ammo. Friendly artillery fire dribbled as the entire task force was fighting for its life. Through his artillery FO, Puckett reported that he would try to bring survivors off the hill. Dolvin could not help, but only wish him luck.

Self-extraction under extreme pressure was not an exercise the Rangers had ever rehearsed, but they seemed to know what to do instinctively, first gathering the wounded and moving them toward the best route out. Individuals and small groups of Rangers moved back toward each other, steadily contracting the perimeter, cutting down Chinese who were penetrating, saving ammo as much a possible by firing only at definite targets and only when there was a high hit probability— whenever possible using bayonets to do the work of bullets. Shouting was about the only means of identification in the swirling, close-range fight in the darkness. Without being told, volunteers stayed on the hill till the last second to provide covering fire for the withdrawal. At the base of Hill 205, First Sergeant Pitts created a rally point to provide maximum fire support for Rangers still coming down. To read about such horrific danger and confusion is by no means to experience them; in the deadly deluge on the hill, each Ranger's fate was determined by reflex and blind luck.

Puckett and one platoon leader survived along with only 21 EM in the company. The Korean Rangers' baptism by fire was almost their last rites. Yet their near anihilation was barely noticed because all UN forces were sent reeling and bleeding by the massive Chinese offensive all across the front that collapsed the ROKA and exposed every US division to defeat in detail and encirclement.

Now commanded by First Sgt. Pitts, the Ranger remnants provided security for the 25th ID command post as it hastily withdrew south. On 5 December, with Puckett evacuated and indefinitely out of action from his multiple wounds, CPT John Paul Vann from division G-4 took command to refit and reorganize under the TO&E of Airborne Ranger companies that had been formed stateside. He was an exceptionally energetic and ingenious man who had demonstrated his courage and resourcefulness with single-engine air drops to cutoff forces in earlier fighting— personal qualities that would bring him much notice, controversy and high positions in Vietnam which made him the posthumous subject of a Pulitzer Prize-winning biography.

With a new strength of five officers and a hundred EM, most of them cherries, Vann divided his company into three platoons of three ten-man squads. There was a 57 mm recoiless rifle and a heavy machinegun with each platoon plus two jeeps and a deuce-and-a half at company HQ. By 15 December, filled out from the replacement pool at Camp Drake and with volunteers from 25th ID, Eighth Army Ranger Company was operational again and attached to the Turkish Brigade with a mission to screen the extreme western flank of the UN command on a large island at the mouth of the Imjin River.

Relentlessly, the CCF armies forced that river line and pressed UN forces south of Seoul which changed hands for a third time. Initially with the Turks, then back again with 25th ID, the Rangers were busy in counter-infilatration operations, one of which on 12 January was significant.

With two platoons reconning a huge mountainous area that was sort of a no man's land, the patrols moved out at dawn on a typically bitter cold morning. A 90-minute contact occurred when they converged on a small village around noon. In column formation, both platoons retraced their steps to a company assembly area, thereby violating a Ranger principle of never taking the same route out that was used gong in. This lesson was re-learned when, with a river blocking one flank and a defile closing maneuver room to the rear, a force of about a hundred communist troops ambushed the column just after it had taken a five-minute break. A ChiComm machine gun opened a hot action that continued for two hours, when with the arrival of darkness and by fire and movement the Rangers were able to break out, but at a cost of two killed (whose bodies could not be recovered) and 19 MIA, including both platoon leaders.

By late January, the new Eighth Army CG, Matthew Ridgway, had revitalized his demoralized command in perhaps the greatest feat of American arms in the 20th Century. The most difficult task in leadership is to convert defeat and retreat into an offensive spirit, but this Ridgway was able to accomplish, mainly by his frequent presence with the troops, showing them, convincing them, that now the United Nations had what was needed to reverse its fortunes again. Blunting the CCF offensive, he started grinding north, using supporting fire and air power extensively, concentrating armor wherever the terrain allowed, in a rolling succession of limited-objective attacks, not so much oriented on terrain but for killing large numbers of communist troops and preempting any enemy notions to resume a major offensive.

Defied by Ridgway's tactics and over extended logistically, the CCF was badly in need of rehabilitation, so the bulk of their armies pulled back to a line just south of Seoul, leaving light screens of infantry to hinder Eighth Army's slow advance much the way Kesselring had done against Fifth Army in Italy.

Like Darby's Rangers, Vann's company proved quite effective in penetrating such screens because their cross-country mobility was among the best in the UN command. Forward reconnaisance was an appropriate mission for the Rangers who could best do it at night.

A typical 'good' mission was on 2 February 1951 near Suwon when Vann broke down the company into squad-size recon patrols which moved out at 2000 with the mission of locating enemy positions and taking a prisoner. For eight hours until dawn the patrols combed forward of friendly lines, finding no Chinese which was a major bit of intelligence that enabled Rigway's left flank to advance ahead of schedule.

From 9 to 14 February, along with 25th ID's Reconnaissance Company, the Rangers screened the division's west flank and conducted a zone recon that featured a lot of mine clearing around the port of Inchon. After this action, Vann had to return to the States on emergency leave, turning over command to CPT Charles Ross, Eighth Army Ranger Company's last CO. Like his predecessors, Ross was the man for the job. Though he had no previous Ranger training, Ross had led many patrols on Bougainville and Leyte, as well as running a patrolling school at battalion level. During the inter-war period he had commanded basic training, cavalry and airborne companies. Another important personnel change was the promotion of First Sergeant Pitts to lieutenant and his reassignment as leader of 2nd Platoon.

Further good employment of the company came about during Eighth Army's Operation Killer.*[*FN: As the name Operation Killer suggests, Ridgway never lost sight of Eighth Army's ultimate objective which was the destruction of the opposing armies. Thus he treated politically important objectives, like the recapture of Seoul, as secondary and kept focus on the later much maligned term, 'body count', which actually in every war is the prime determinant of military victory.] In order to keep the CCF off balance, the Rangers were ordered to conduct raids into their lines, take prisoners and generally raise hell—the sorts of missions for which every Ranger is adept and enthusiastic.

They are hardly less hazardous. For example, on a raid on railway tunnels across the Han River a platoon of Rangers was caught on the ice (no white camouflage was available)

LeRoy May setting up gun. Harvey Olson laying down. Frances Bell in back. 8th Army Ranger Co., Korea.

under fire from five machine guns. Only poor NKA marksmanship prevented casualties.

On 28 February, 2nd Platoon got its chance at crossing the Han which by now was flowing with huge chunks of ice. The mission was to raid the village of Yangsuri, destroy enemy positions around it and snatch a prisoner. For two days Ross devoted the company to recon for a suitable crossing site (he would be crossing himself) and set up overwatch positions. At 2200 hours five two-man boats ventured into the rushing river. Three capsized and returned for a second try, though the crews were virtually paralyzed by the cold immersion. Ross, with only five other Rangers, felt very lonely on the enemy bank. But in the next hour he was joined by the rest of 2nd Platoon which used eight-man boats, far more suitable for fighting the current. The Rangers formed up and moved on to Yangsuri, found it deserted, and patroled into enemy terrain for another mile, still without contact. They made it back to friendly lines with this valuable negative information.

On 1 March the Rangers went back again, this time to check out the railroad tunnels north of Yangsuri. 1st Platoon battled the ice this time but at mid-stream it had congealed into a sheet that stopped progress. First the ice was attacked with oars and rifle butts but with no success; then when several Rangers crawled out on the ice to try and pull the boats, it cracked, dunking the men who were barely saved. The mission was then abandoned. During the next few days, Ross persisted in finding a way to get across but to no avail except for valuable data acquired about currents, banks and depth, important information for 25th ID's later crossing when Ridgway kicked off his next meat grinding operation called Ripper. Here Rangers again led the way, scouting as much as six miles ahead of 25th ID's attack.

Twenty-fifth ID's progress during Operation Ripper carried across the 38th Parallel and provided the opportunity to cut off retreating CCF using the northbound road and railway which channeled through a defile. Ross got the mission to infiltrate seven miles to ambush the defile, supported by long range artillery and TAC air. Two platoons broke down into squad-size infiltration elements and by different routes moved out on the evening of 18 March, using maximum concealment which made the routes longer and more circuitous. By 1530 the next day, all squads had successfully rendezvoused at a hidden assembly area for the ambush and had the site under surveillance.

During the night Ross installed a road block with secondary ambushes on the flanks but nothing came down the road. He abandoned his positions and took a direct exfiltration route back to friendly lines which he crossed at 0500. Despite its disappointing results, this operation behind enemy lines was a good mission and a well thought out employment of Rangers. Like so many of their combat patrols, information obtained enroute in and out made them more valuable as recon patrols.

For the next couple fo weeks, Ross's company acted as flank guard for 25th ID's slow drive north, the Rangers advancing from blocking position to blocking position as a neighboring regiment ground forward.

Though unknown to them at the time, Eighth Army Ranger Company's last mission would be its next one, a company raid, preceded by infiltration some six miles ahead of friendly lines, to sweep through the small village of Changgo-ri believed to by held by a CCF delaying force of unknown size. If they could be significantly damaged before they set up as a rear guard, 25th ID's offensive would accelerate once a break through was accomplished.

That all Ranger raids to date penetrated to a maximum distance of about seven miles was no coincidence, because this was the range of supporting artillery. Ross plotted concentrations along his infiltration route. A different and more direct route was selected for exfiltration. The raiding party passed through friendly outposts at 2200 on 27 March. Climbing carefully through hilly, wooded terrain the Rangers were able to guide on the road to Changgo-ri, arriving at the western edge of the village at 0100. The plan was to sweep from north to south, but upon examining the terrain at the objective Ross decided to make his LD a creek in front of his position, thereby saving some time and time was short.

To enhance surprise, he relied on stealth, rather than call for preparatory fires and told his assault platoon to infiltrate the village, firing only when fired upon. They gobbled up an enemy outpost, but fire broke out and spread rapidly. The CCF were present in strength though taken by surprise. The initial assault made it through the village, destroyed a rice cache, cut down a high score of enemy, but ran into a trench line on the far side of Chonggo-ri where the CCF were able to regroup and even the fire fight. However for the next two and a half hours the Rangers still held the initiative and upper hand, when Ross decided that maximum damage had been inflicted at no loss.

At this point he wisely signalled for the assault to break off and rally between the fire support elements. This withdrawal was accomplished just as the Chinese began to counterattack, met by the full weight of the Ranger company's concentrated fire. As the counterattack sputtered out, Ross pulled back quickly from the village and onto his exfiltration route under protection from artillery fire that neutralized CCF pursuit. At 0500, friendly lines were recrossed without incident, completing a text book raid which had the additional benefit of determining the strength of the enemy rear guard (about a battalion), information that greatly aided the success of 25th ID's subsequent advance.

After a debriefing and several hours rest, the Rangers' high fell to the ultimate low. Ross called the company together to tell them that they were to be disbanded within the next week. Members had several alternatives for reassignment but none sounded at all appealing for men who had indeed become a unique band of brothers.*[*FN: CG, 25th ID had authority to reassign the former Eighth Army Rangers within his command, but those jump qualified could volunteer for the 187th RCT (Airborne) or any of the Airborne Ranger companies coming into country. Except for his company officers, most of Ross's men were not jump qualified so they remained with 25th ID.] The reason for the Eighth Army Rangers disappearance will be presently explained, but first some questions about their existence and effectivness.

First, in Korea was it a good idea to revive Rangers with a TDA company? Apparently the Eighth Army CG thought so, and with Pentagon concurrence this was done. It is hard to criticize a decision made by those with the most appreciation of realities on the ground and approved as well by those responsible for the big picture. Both authorities had to make do with what was available which was very little— an undersize company commanded by a 2nd LT.

Envisioned for specialty missions in the category of short range penetration, the company was employed not much differently than a conventional infantry unit of comparable size. This was not because of perverse conventional commanders or ignorance but because of necessity. Like all Eighth Army forces in the desperate days and nights of 1950-51, the Rangers were a reaction force. Their reactions were not quite routine infantry stuff of that time but pretty close: counter-guerrilla operations, behind the lines ambushes and raids, infiltrations, specialized patrolling, and some tank-infantry teamwork as part of reconnaissance in force. The point is that such operations were necessary or conventional commanders would not have required them; and though straight infantry could have done them, the Rangers did, preserving the commander's conventional assets. All told, in the most general sense, the Eighth Army Ranger Company was an excellent economy of force.

But were they better at what they did than straight infantry equivalents? What they did better than other U.N. units was operate effectively at night. With this advantage, they were properly employed, a credit to 25th ID, commanded by MG Kean. Almost alone among U.N. ground forces, the Rangers made nightime miserable for the enemy.

Preserving (economizing) conventional assets and staking out claims to the night are the two principal benefits Eighth Army Ranger Company provided. But at what cost? That is, would Puckett's Rangers, before they were Rangers, have been more cost effective in other roles? The answer is an emphatic—and somewhat surprising—NO!

The raw material for the company were men from service support units on occupational duty in Japan. As David R. Gray, one of Puckett's originals, described it: "[We] trained soldiers who might never have surfaced as combat leaders in the first place... These personnel served [previously] as mechanics, bakers, cooks and radio repairmen, not as infantrymen; very few were noncommissioned officers... in no way a leader drain on Eighth Army when viewed in the total context of personnel then available in Korea."

Though small by every measure (virtually nothing logistically), Eighth Army's investment in its Rangers was well rewarded no matter what cost/benefit analysis is used. This is true even though the company's name was something of a misnomer because it operated exclusively (except for a brief attachment with the Turkish Brigade) for the 25th ID, not at field army level. Testing Ranger employment at division level was one of the rationales for the company's creation. The test was a valid one, and Eighth Army Ranger Company's experience was pretty much duplicated by the numbered Airborne Ranger companies that began arriving in Korea during the winter of 1950, just before the CCF changed everyone's plans for Christmas.

Ranger Training Command (Center)

We return to the opening months of the Korean War, a period of great alarm in the Pentagon about the NKA success at infiltration, especially at night.*[*FN: Such alarm was understandable, but successful night infiltration by the NKA was hardly surprising, given the vast frontages US divisions had to cover and the absence of night sight devices.] To provide a counter-capability, the Army Chief of Staff, J. Lawton Collins, directed his G-3 to develop small 'marauder' companies which could regularly infiltrate enemy lines and raid rear area installations. Collins' guidance was that the companies be all volunteers "of high intelligence", lean adminstratively, contain scout jeeps mounted with a machine gun, and that one company be available with each division in Korea.

In August, G-3 began to flesh out this concept. The result was that the machine gun jeeps were stripped out, the companies were to be airborne, and named Rangers rather than marauders. To implement the rebirth, Ranger Training Center (Airborne) became operational at the end of September 1950 at Ft. Benning, GA. Two key elements of the unfolding concept shaped the training and operations of Rangers for the rest of the war and thereafter: the divisional level of employment and the airborne requirement.

A concept that is hazy at its inception is not likely to get better as it is implemented by real world plans. The concept of airborne rangers was contradictory to Collins' guidance that the marauders be used in short range penetrations ("infiltrate through enemy lines and attack command posts, artillery, tank parks, and key communications centers or facilities" were Collins' words). Infiltration implies short range while airborne means long range.

In addition to the implicit contrast in missions between infiltrators and paratroopers, the divisional level of employment also points to shallow rather than deep penetrations. Infantry division commanders are not concerned with strategic targets for paradrops but instead are focused on what is right in front of them, capable of being reached by foot-mobile, short range combat patrols. The airborne requirement for Rangers in Korea was, therefore, an inappropriate and unnecessary capability (even though it was used once): something like requiring cops to be lawyers—useful on rare occassions but extremely uneconomical in training and mission accomplishment.

But that was the requirement Ranger Training Center (RTC) was to carry out. The pros and cons of the other crucial requirement, division level of employment, has been discussed earlier.

Collins personally selected COL John G. Van Houten to stand up the RTC. An ROTC graduate from the University of Georgia in 1926, Van Houten's military background included wartime service with the 9th ID as division chief of staff and regimental commander. When the Korean War broke out he was on the faculty of the Command and General Staff College at Ft. Leavenworth, KS. His experience in ranger operations was nil, but his executive officer (appointed by Department of the Army) had had more than plenty.

He was COL Edwin A. Walker, a former regimental commander of the combined 1st Special Service Force who had seen many forms of ranger action in the Aleutian Islands and Italy. Walker was, and remained, a masterful trainer in the essentials of combat, with open contempt for spit and polish, a deep suspicion of 'show' but ardent advocacy of 'go'—hallmarks forever after of what would became known as the Ranger School, offspring of the RTC.*[*FN: The author was additionally fortunate to have undergone Walker's mode of training while he commanded the 24th ID in Bavaria. Above all other features of training, Walker emphasised realism. Thus in his division's Army Training Tests (ATT's, an earlier name for ARTEP's), when the real world situation called for battle groups to fight outnumbered, Walker ganged up three or more against the tested battle group at Hohenfels. Nothing was simulated except for ball ammunition. So, when as a 2nd LT I manned an outpost line with my platoon, I reported what I actually saw—a division worth of armor thundering down on my battle group, not a few jeeps whose number I was to multiply and translate into enemy tanks.

For Walker, realism definitely included the fatigue factor. Battle group ATT's were open ended, so that a commander who had resolved to carry on indefinitely was reduced to a hollow-eyed ghost, too groggy to notice when an umpire put him to bed because of exhaustion, and subordinates ran the latter stages of the ATT. Similarly, a staff that didn't have a realistic system of rotation fell apart after five days and often flunked.

Like many men of narrow genius, Walker was nearly incompetent outside his speciality. In 1962, he was relieved from command of 24th ID for trying to censor a tabloid, *The Overseas Weekly*. Walker had to resign, and in civilian life is remembered as a right wing fanatic.]

While Puckett's first volunteers were mostly service support troops, Walker had the pick of airborne combat arms. Former WW II Rangers and Marauders were sought, mainly for cadre, but the bulk of the first Ranger classes at RTC came from the 82nd Airborne Division, already jump qualified. The 82nd was nicknamed "America's guard of honor", implying that they pulled a lot of duty as ceremonial troops—understandable because of Ft. Bragg's proximity to Washington, D.C.. "All Americans" who had tired of parades, starched fatigues and chrome helmets flocked to the call for Ranger volunteers. Nearly 5,000 eventually applied.

At Harmony Church, RTC headquarters at Ft. Benning, those selected certainly entered a different environment. Their mirror-polished Corcorans were exchanged for combat boots to be oiled to a dull finish, soon to be scuffed white by rocks, lacerated by thorns, and immersed in the mud of Georgia bottom lands. Show, even in barracks, was out and go was in.

Go meant a whole lot of running in formation, followed by many more miles of forced marching. Van Houten's goal was for each company "to move from forty to fifty miles cross-country in twelve to eighteen hours, depending upon terrain." Classes were almost entirely outdoors regardless of weather. The training schedule was so tight it allowed for six hours sleep between the last field exercise and reveille.

But only rarely was an exercise completed on time. It could be if performed perfectly but the Rangers were learning and mistakes were paid for in lost sleep. The next day's training began at 0730, no matter if students had trudged in at 0430 from a night recon patrol that should have returned at 2300. And no one ever hit the sack till his weapon was cleaned and inspected. The scheduled training week listed sixty hours, not including maintenance; so actual time spent learning to be a Ranger was closer to 140 hours per week.

After instruction in heat injuries and frostbite, the elements were virtually ignored by the cadre who sneered when they spoke of 'creature comforts'. The only part of a Ranger's body that needed to be comfortable were his feet—his prime movers.

Indifference to comfort became known as 'misery practice', and has been debated ever since. Detractors of such Ranger training practices feel that misery in combat situations must be endured because it has to be, there being no alternative; therefore, experiencing previous misery is of no value in a combat situation. Proponents of misery make the same point they do about the notorious lack of sleep in Ranger training: if you've done it before, you have more confidence that you can carry on during the next similar experience. During the first couple of weeks in the field, a new Ranger feels himself enormously renewed after three hours on an air mattress where before he was exhausted if he did not have at least seven hours in bed. Thereafter he knows he can go on and on with three hours sleep each night because he did so for extended periods in Ranger School.

In building such confidence, Ranger training somewhat 'psyched' its participants; that is, it convinced them about their proficiency by deceiving them in a beneficial way. PFT scores administered at the beginning of Ranger training were uniformly higher than at the end, a surprise to trainees who felt they were in the best condition of their lives when they graduated. They were, not in terms of strength but in endurance. Their chin-up scores were probably less but their five mile times faster. Ranger work depends much more on endurance than strength—except for mental strength.

Headquarters, RTC, contained 22 officers, 34 EM. In their first cycle, they trained 19 officers and 314 EM organized into four provisional companies. One of them was entirely black, the one and only unit of its kind, the result of the integration of the army by President Truman. This first black company had to show a lot of people they were Ranger quality and they did.

The companies' TO&E was the same as

Eighth Army Ranger Company after it was decimated on Hill 205 and reorganized under command of CPT Vann. Crew-served weapons were usually pooled at company headquarters to be issued to the platoons depending on the mission. The new companies were high on fire-power—higher than a standard infantry company half again larger, but short on transportation—one jeep and a deuce and a half— to haul the weapons and necessary ammo.

In four weeks, and after a 30% dropout rate (including injuries), the first four companies received their new guidons. Though slated for Europe and stateside as well as Korea, the CCF offensive changed orders so that all headed for Eighth Army. The very next day, (Thanksgiving) RTC received the trainees for 5th, 6th, 7th and 8th Companies. It was this second cycle which, upon graduation just before Christmas 1950, that first donned the new Ranger insigne, an oval of black trimmed in gold.

Red, white and black had been the traditional Ranger colors. Why gold was substituted for red and white remains a mystery; so too is why an oval replaced the scroll. In time the oval evolved into an arc identical in size and shape to that of the airborne divisions, and became known as the Ranger tab, a qualification badge rather than a unit designator but worn with commensurate pride.*[*FN: In nature, any combination of red, black and gold signifies danger. For example, all poisonous reptiles and plants display one or more of those colors, as do most predatory felines, fowl and fish. Benign flora and fauna rarely combine those colors, and when they do scientists have determined it is for warning. Thus to link ranger and danger not only rhymes but is fitting: Ranger graduates may be familiar with the jingle, "Danger, stranger, I'm a bad-ass Ranger."]

Daniel Gollock, William Knudsen, Fred Crispino, Carl Lillard.

Harland Morrisey, Marvin Vickrey, John Summers, Harvey Olson.

Elmer Cassatt and Earl Doyle. (All photos, this page, of 8th Army Ranger Co., Korea.)

The day after Christmas, 1950, the RTC—though no one asked them— finally had a say in the employment of their graduates. Not surprisingly, a major conclusion of RTC's staff study opposed the employment of Ranger companies at division level. The principal arguments cited in the study for a Ranger battalion HQ at corps level or higher were that there would be better mission planning and necessary administrative support. But with a war going on, and the infanty divisions in Korea scratching for any resources, Department of the Army would not relent. So five of the first seven Airborne Ranger companies would be attached to line divisions in Korea (6th Company went to Germany; 7th stayed at Ft. Benning as RTC troops).

At the dawn of 1951, RTC knew they would be around for the long haul, not just a wartime ad hoc operation that would shut down when a temporary emergency passed. This was because the end of the Korean War, after the entry of China, looked to be open ended—really, no end in sight. The U.S. commitment to the U.N. cause deepened, as exemplified by calling up reserves, notably National Guard divisions. With the fixed DA policy of one Ranger company per division, RTC set about their third cycle to train volunteers from the 45h ID (Oklahoma) and the 40th ID (California).

This meant activating the 10th and 11th Airborne Ranger companies (9th Company was cadred to remain at Ft. Benning), and starting up a jump school because few national guardsmen were qualified parachutists. At first volunteers among them hardly needed to be recruited. For 10th Company, 600 volunteers applied for fewer than 200 slots.

Meanwhile 3rd, 5th and 8th companies received a 'bonus'—cold weather and mountain training at Ft. Carson, CO—before they shipped out for the cold weather and mountains of Korea. The real bonus of this month in the frigid Rockies was issuance of the arctic gear that would be so necessary in 'frozen Chosen' (i.e., Korea), gear in very short supply over there. An overlooked asset were mules that would have been extremely useful in the Korean mountains. Rangers at Ft. Carson were trained in mule handling, but for reasons unknown never saw the animals in Korea.

By March 1951, Rangers had proved their worth sufficiently in Korea that a classic debate broke out at DA concerning their future employment. This was not the old question of 'division or higher', but something more fundamental: whether to have Ranger units at all or spread the wealth throughout the Army. The RTC raised the following points in defense of Ranger units:

1. *Rangers could not be mass produced.* The technical skills might be taught throughout the Army, given enough time, but the Ranger spirit could not. Such spirit was present only in what was to be called a cohort; that is, a group that from its first days trains together then stays together in combat, thereby understanding each other to the point of 'battlefield telepathy'—knowing what the other guys are going to do in a tight spot without even talking with them. Battlefield telepathy can not be developed except in a cohort.

2. *A seasoning of Rangers in a conventional unit would not produce the quality of an all-Ranger unit.* Though obviously true, this point has a down side because the alternative is all-Ranger units which concentrate too many shock troops in one place; and, as at Cisterna or Hill 205, they can be wiped out in a single disaster. More commonly all-Ranger units are ground down as in the pre-Anzio Italian campaigns, a process called the 'selection-destruction cycle'. Ranger purists argue that this wouldn't happen if the missions are appropriate, but in two wars it has been proven that there will always be a percentage of bad missions, sometimes with disastrous results.

3. *In conventional units, Rangers would be held down by the existing rank structure.* This point is somewhat dubious too because in all-Ranger units ability is less likely to be rewarded because ability is so abundant.

4. *In conventional units, individual Rangers would feel compelled to show rash courage, justifying the Ranger reputation and setting a Ranger example.* This indeed happened with some Rangers who, unable to join Ranger units in combat, served extremely heroically with line units. Actually this is the sort of effect proponents of Ranger dispersal hoped for so, it is not a good argument for Ranger units.

5. *Led by hard charging Rangers, conventional units might attempt missions too difficult for their level of training and proficiency.* Another weak argument because conventional commanders would be the ones who assigned missions.

So the strongest argument for concentration of Rangers in Ranger units was one not mentioned: if an important and really important and difficult mission came up (e.g., Pointe du Hoc), it could be accomplished with a high probability of success only if a Ranger unit were on tap to carry it out. Ranger-type missions of lesser difficulty could be accomplished by line units with some Ranger-trained personnel but success probability was reduced. The question really boiled down to whether a small number of extraordinarily proficient units were preferable to a large number of better than average units.

The question remained rather academic while the Korean War sucked up Ranger assets to the point where few graduates could be dispersed into the Army at large. This situation was illustrated when, in March 1951, for the first time a volunteer quota fell short of its goal. 246 volunteers were sought for the 28th ID (Maryland-Virginia NG) but only 197 men stepped up. As in World War II, the yen for combat tends to dry up after six months of heavy casualties reports.

In April, the Ranger Training Command became the Ranger Training Center and took a look at itself, finding that its average student age was under 22, with an educational level higher than the requirements for OCS. DA was so pleased with the RTC's output that Van Houten was asked how his school could raise the standard of training army-wide. This tasking seemed to be the Ranger dispersal argument returning through the back door, but Van Houten neatly converted that subterfuge into a proposal to 'Rangerize' the army without breaking up Ranger units.

He proposed several courses of action:

1. *Send key personnel from the line divisions to become Ranger trained, then return them as training cadre for their divisions.* This sounded plausible but Van Houten and Walker might have had their tongues in their cheeks because there would surely be significant dilution of expertise, not to mention the elixir of spirit, in transmitting rangercraft from the RTC to the cadre, then from the cadre to the line. Training standards and quality control of instruction would certainly degrade.

2. *Whole infantry battalions would be run through the RTC in three-month courses to weed out the unfit and identify natural leaders.* This made good sense, and the concept is present today in the National and Joint Training Centers at Ft. Irwin and Ft. Polk respectively. However during the Korean War no infantry battalions anywhere in the world could be spared for Ranger training. Moreover, to conduct such battalion-level training would have required a huge build up of the RTC and, one would suspect, draw frowns from the Infantry School (the RTC's landlord at Ft. Benning) which quite rightly held the teaching of battalion tactics to be their province.*[*FN: Does the Ranger tail wag the USAIS dog a bit at Ft. Benning? In 1960, as a new lieutenant, the author was surprised to find his first instruction in platoon tactics taught by the Ranger School.]

3. *An indoctrination course, including practical exercises, in Ranger ops would be conducted for battalion, regimental and division commanders.* This didn't happen—probably because not many O-6's and above were eager to take even a recon patrol beyond Custer Terrace.

By June, RTC received feedback from Airborne Ranger companies in combat and brought in veterans from these companies as trainers. New or newly emphasized subjects were night fire control measures, village fighting, and use of foreign and/or silent weapons. Fourth cycle Rangers were also expected to be better in the use of assault boats and know how to operate a tank, friendly or captured. The average age of students in the fourth cycle had crept up over 22 years.

In September, 1951—exactly a year after it stood up—RTC graduated its last cycle. A circle also was completed when in the first week of September, Ralph Puckett received his DSC for Hill 205 at a ceremony on French Field. Soon the Airborne Ranger companies were furling their guidons, and one by one followed the World War II battalions into deactivation and history. The RTC would become a department of the Infantry School and award the qualification badge.

So Rangers, though not Ranger units, would continue. The dispersal proponents at DA finally won, but the Army was a winner for the preservation of the best school ever attended by American soldiers. It is time now to return to the grim histories of the first RTC graduates, 160 of whom never saw Harmony Church again.

First Airborne Ranger Company

First Company's lineage traced back to A Company, lst Battalion of Darby's Rangers, a tremendous heritage that began at Dieppe. lst Company was activated on 29 September 1950 at Ft. Benning, composed mostly of volunteers from the 82nd but also 30 from the llth Airborne Division at Ft. Campbell KY.

The fledgling Rangers (with a morning report strength of five officers and 120 EM) had volunteered principally to experience combat, but two weeks prior to their activation the Inchon landings seemed to be bringing the Korean War to a swift close. Motivation for the rugged Ranger training was diminished because of these expectations of peace, making the jobs of both the RTC cadre and the company leadership uniquely difficult. It was not until November when the lst Company was at sea that the entry of the CCF opened a new ball game in Korea, and the Rangers knew they would be playing hard ball very soon. Nonetheless the last four weeks of training at Ft. Benning were intently serious, the biggest problems arising from the start up of RTC itself, training as they were the first cycle while learning their jobs themselves.

On 15 November 1950—with the Ft. Benning band playing, "Now Is the Hour"— lst Company entrained for Camp Stoneman CA, coincidentally the same port of embarkation of Merrill's Marauders seven years before. There had been no pre-deployment leave, so when the train pulled into a Texas station for a two hour stop, the CO allowed his men a brief R&R, but with the caution that there must be no AWOL's. This was the first non-training opportunity the company had to test their faith in each other—and they kept the faith: everyone was back at the train on time.

Briefly billeted at Ft. Stoneman (to receive innoculations and unmarked fatigues), the Rangers heard some disrespectful comments from troops in adjoining barracks. lst Platoon was detailed to teach the 'legs' how Rangers should be addressed, a lesson in protocol that left more bloused lips than bloused boots among the victims of the first Ranger raid outside Ft. Benning.

They sailed on 25 November, crossing the stormy North Pacific in 15 days. The close confines and constant roll of the ship allowed for little training except PT and weapons study. On board were a number of medical personnel, straight from civilian life. They sought instruction in battlefield survival from the Rangers and this was willingly provided. Some saw each other later—during the war, both Ranger coaches and medical pupils would benefit from each others expertise.

After a week at Camp Zama, Japan, checking out cold weather gear, lst Company set sail again, this time on a ferry to Pusan. There they flew up to Kimpo Airport outside Seoul, then by truck to the 2nd ID's sector on 23 December. 2nd ID had three U.S. regiments fleshed out with two battalions from France and Holland. All had been hard hit by the CCF surprise offensive. Indeed Korea was 2nd ID's bloodiest war, in which they lost more men than in either of the World Wars where they had gone through some very heavy fighting. In Eighth Army reserve, refitting and training replacements when joined by the Rangers, the division was commanded by MG Ruffner, fortunately a general with a good appreciation for Ranger capabilities, limitations and employment.

Further attached to 2nd ID's 23rd Infantry Regiment, the Rangers soon felt their lack of organic service support, particularly in the form of chow. Satellited on other units for messing, hard feelings developed on both sides as the line battalions had no surplus of chow themselves. Nor was even potable water plentiful. lst Company's original CO, CPT John Striegel, came down with debilitating hepatitis and had to be evacuated. The Rangers received no replacements, so LT Alfred Herman took command on 6 January 1951.

First Company had seen some action and suffered casualties by then. On the day after Christmas they began patrolling—began 'taking back the night' from communist forces. On New Years's Eve they learned that this would not be a task without cost: a four-man patrol was ambushed, trapped and captured by the CCF. One Ranger, Polish born, was handed over to the Russians, never to be seen again. The other three men died in the horribly brutal communist prison camps. To be captured was almost a certain death sentence, so much so that a tentative jump behind CCF lines by lst Company was scrubbed early because of high level fears about the fate of POWs.

Korea was that kind of war. Though no less resourceful and adept at E&E than their forefathers in World War II, no Rangers escaped from CCF prison camps (nor did anyone else), and years later this was true of prison camps in North Vietnam as well. Korea was the first time Americans were pitted against communism. From very early on it became clear that this was a merciless enemy, and therefore deserved no mercy.

By January, 2nd ID was back in action around Wonju, a rail and road hub that had to be held or Eighth Army's center would bend dangerously. On the west coast, Seoul and Inchon had already fallen—for the second time— to the communists and now Wonju was the objective of the CCF-NKA main attack. To the east, a ROK corps had caved and 2nd ID looked to be flanked.*[*FN: The NKA's ability to infiltrate had hardly diminished, even after they were smashed in the fall of 1950. Rehabilitated by the CCF, in January an entire NKA division actually infiltrated through the ROK's east of 2nd ID and penetrated fifty miles behind Ridgway's front. He had to commit Eighth Army reserve, the US lst Marine Division, to drive this NKA division back. They exfiltrated over a hundred miles in the dead of winter and reentered CCF lines near the east coast. So when it came to infiltrating, the Airborne Ranger companies were up against masters.]

First Company's job was security patrolling and identification of the main CCF-NKA effort to take Wonju. Generally, the line infantry patrolled by day, the Rangers by night. The enemy was not hard to find. On 7 January three NKA divisions attacked the two forward 2nd ID regiments which had to fall back through Wonju. lst Company helped in the delay by blowing bridges. A 2nd ID counterattack to retake Wonju failed, but the NKA were contained on the high ground about ten miles south of the city after suffering extremely high loses (200,000 in the whole communist offensive). Just south of Wonju would be about the high water mark for the communists for the rest of the war.

It was also gut check time for the Rangers who endured cold as deep as the snow. Mortar tubes froze, contracting so much they would not take a round. Hot coffee froze just as hard, even before it reached the lips. A punctured boot meant rapid frostbite, and faces had to be inspected constantly for the telltale gray patches of dying skin.

Sources of shelter were as important for the communists as for U.N. forces, so Eighth Army directed that the enemy be denied any havens from the paralyzing cold. This meant wiping out villages (their inhabitants had long since fled), usually by torching them. This amounted to a 'scorched earth' policy but the press, who associated the term with Soviet tactics in World War II, were told the principal purpose of structure destruction was to clear fields of fire. In fact, fields of fire were already excellent in the bleak, snow-whipped mountains.

Lack of concealment and high reflection from snow fields did not make Ranger patrols any easier except for land navigation. It was therefore a good practice not to move around much at night but rather penetrate quickly to places of good observation and stay there, sniping or calling artillery on NKA infiltrators. Thus, in the initial months of war, lst Rangers were largely a counter-infiltration force, using patrols for defensive purposes.

Ambushes were frequent, raids rare, during this period. When patrols had to penetrate much distance, the leader designated a 'get away man', someone oriented on the route back rather than the route ahead. More than once he was required by himself to make it back to company for help. Meanwhile his pinned down patrol relied on indirect fire to neutralize the opposing force. When a Ranger relief patrol was dispatched it was used to distract the enemy so that the engaged patrol could break contact and exfiltrate.*[*FN: This 'distraction' tactic is an interesting variation of the communist tactic so common in Vietnam of 'bait and ambush', where the bait was pinned down and called for help, then the relief force was ambushed.] These tactics had not been taught at RTC (though they would be), but Ranger initiative once again came up with what worked and used it.

On 16 January, the lst Rangers began their first operation as a company. Ridgway had blunted the CCF offensive all along the front and was looking north again. The initial step was to feel out what was in front of Eighth Army; to do so in the 2nd ID sector meant sending out Rangers, the best explorers on the peninsula. "Utilize them to the fullest", Division G-3 directed the 9th Infantry Regiment to whom lst Company was attached.

Among the many well founded stories of Ranger misuse and abuse, is this not one of them. If a line company had been singled out by G-3 to be "utilized to the fullest", the CO might well have groaned and his troops felt discriminated against to the point of going to the IG. In line units, rotation of dangerous missions is a mark of fairness and concern by the commander. Line soldiers expect to be called upon to do their jobs—jobs sometimes as dangerous as Ranger missions—but such jobs are considered occupational hazards, not plums.

Rangers have a quite different attitude, one that hankers for the dangerous, and this may be the most significant temperamental difference with their line counterparts. Explaining this difference requires examination of American characteristcs.

The United States has not—never in this century— been a warlike nation, and fights only when forced to. Accordingly, American soliders are not expected to like war but to do well at it from a sense of duty and rightness. As a general characterization, this is true, and in no way diminishes the quality of the American soldier for whom Sgt. York of World War I fame might be a prime example.

A conscripted turkey shooter from Tennessee, York single-handedly (by breathtaking marksmanship) took out the best part of a first rate German company from their trenches that could have withstood a well supported battalion attack. For this epic individual feat, York became the ultimate symbol of the reluctant warrior—doing his job when he had to with superhuman proficiency but not enjoying what he was required by duty to do.

America is a land of infinite diversity. Sgt. York may epitomize admirable American traits, even an American ideal, but he does not by any means represent all American soldiers. Among them is the ranger-type, born fighters under control, who, like lst Company were dismayed that they would miss the combat they yearned for like sex. Bob Black, a member of lst Company, noted that when news of the Inchon coup was relayed through his troop ship headed for Japan, the non-Ranger soldiers aboard cheered because they hoped to be spared the ordeal of lethal combat. By contrast, the Rangers gnashed their teeth because they would be *denied* the experience of combat for which they had so ardently prepared.

The Ranger attitude is as natural as Sgt. York's, just more uncommon, and by current fashion, politcally incorrect. In my introduction to another book, *Lightning in the Storm,* the author described this conundrum as simply as I knew how:

A man joins the army recognizing that someday he may be in combat; a man volunteers for the elite of the army anticipating the prospect of combat. Such an attitude does not denote a war lover or Rambo, but rather a willingness, even eagerness, to test himself with the ultimate risk. It is in the makeup of many young men to do so. Across all cultures and millenia this has been true.

This truth may explain why 2nd ID G-3's admonition to exploit the Rangers to the fullest drew no special comment or reaction. The Rangers of lst Company wanted to be out front, so there they were sent.

In fleeing toward Wonju, 1st Company worked a mile or more in front of friendly lines, probing, stopping to set up ambushes, then probing forward again. Second ID expected them to run into trouble at any moment and so put a l05 battery in direct support.

For three days they pushed out, lengthening their patrols and on l9 January took a prisoner, returning him to G-2. Contact was so light, lst Company was ordered to scout aggressively till they came up against significant resistance. To develop the situation more quickly, machine gun jeeps were employed (the same sort of jeeps that had been stripped out of the Airborne Ranger TO&E) that lst Company had acquired in Pusan.

The first such motorized Ranger patrol soon came upon a roadblock manned by a couple of NKA platoons. A sudden and violent attack by dismounted Rangers, overwatched by the jeeps' machine guns, overwhelmed the NKA by fire and dispersed them. The Rangers could sense from this engagement that morale and momentum had shifted in favor of Eighth Army, and that the fight might be oozing out of the communists. Boldly the motorized patrol drove on through the ruins of the next town.

Conspicuous in the rubble was a massive metal vault, all that remained of the town bank. The vault was intact and still locked. Soon the CO, LT Herman, was called forward to determine what to do. He directed that the bank vault be cracked Ranger style—with a 3.5" rocket and a demo charge. As the smoke cleared, currency notes settled like autumn leaves after a windstorm. First Rangers had liberated a huge quantity of Korean Won, the legal tender of ROK.

The company fund was about to become multiplied a thousandfold. Herman could imagine *buying* the vehicles he needed off the black market. Hell, he could now hire a Korean mobile catering service—who needs 2nd ID's mess trucks!

His fellow Rangers kidded him about early retirement as he told them to "Put *my* money in my jeep." Alas, returning to friendly lines the jeep was ambushed—the driver captured, the shotgun wounded—and the money reverted back to the North Koreans.

Embittered to have gone from rags to riches then back to rags again, the Rangers continued their reconnaissance in force northward around Wonju and then east of Chipyong-Ni—a town they would be seeing again. The critical feature in this area were twin railroad tunnels that had changed hands several times. The tunnels were especially important for the communists because they were one of the few places safe from mauling air attacks. lst Company was to keep those tunnels clear, which meant many eerie patrols through utter darkness and over bodies from earlier actions as well as rockslides where enemy could be lurking.

First Company pulled back to Wonju after more than two weeks of continuous patroling. In early February they were out front again, searching for contact. One way they found it was by sending small elements where they were sure to be sighted from enemy OP's, then await reaction. On one occasion, a single Ranger platoon put themselves in a position where the NKA became convinced they could maneuver against it, with high hopes of cutting it off. The platoon, however, turned out to be both bait and hook, drawing a ground attack that fell into an ambush by the rest of lst Company.

On 3 February, lst Company received their first replacements from RTC, one lieutenant and six EM. They arrived just in time for the most celebrated Ranger raid of the war. At last, for the first time since WW II, Rangers were tasked to do what they do best—strike deep with stealth and surprise against an important enemy command and control center.

By now, with a less fluid front across the peninsula, U.S. intelligence had been able to locate some large enemy headquarters by radio intercepts, agent reports, and aerial photography. One of the headquarters suspected was that of the 12th NKA Infantry Division, a tough outfit that had been in action since the beginning of the war. It was believed to be deployed north of Hoengsong in 2nd ID's sector, building up strength for a spoiling attack against the ROKA in the central front. There were strong indications that a regimental HQ, maybe even that of 12th NKA Division itself was located some ten miles west of Hoengsong, in or around the town of Changmal. lst Company had done the most to determine that the enemy screen in front of the l2th Division was porous. Gen. Ruffner therefore gave LT Herman the OK to see if the Rangers could break through the screen all the way to an NKA headquarters. Besides the possible disruption of enemy command and control, the raid, even if only partially successful, could have the effect of spoiling the communists' spoiling attack.

Could a heavy air raid have achieved the same effect? No, because intelligence was not so precise that a headquarters had been pin pointed. So one of the purposes of the Ranger raid was to find out just what was in Changmal. Air raids take no prisoners. And even though an air raid could be more destructive, it would not impart the sense of vulnerability on the enemy as a limited objective ground attack. The communists were already fearful of air power and had developed some countermeasures. What Ruffner wanted was to give them something new to fear, and focus their attention on security rather than offensive operations.

A ground raid on a headquarters is something personal: a statement of 'in your face' from one commander to his counterpart. This attitude, though unreported, has probably been an underlying reason for many a raid that would have been vetoed for other considerations.

Changmal was nine straight distance miles, beyond friendly artillery range, behind enemy lines. To get there undetected, lst Company would have to travel many miles more. Following good Ranger doctrine, Herman determined first what forces he wanted in and around the objective. That was a difficult decision because enemy strength in the vicinity of Changmal was not known—much the same enemy situation CPT Ross, CO of Eighth Army Ranger Company, had to consider at Chonggo-ri six weeks later.

The CO's plan of action on the objective were remarkably similar: both felt their primary strength should be in the fire support elements so that no matter what the assault element ran into it would have heavy support from the flank. Consequently both plans called for two platoons in fire support, one in assault. This plan allowed the assaulting platoon to back out quickly if the objective was too strongly held.

Back planning from actions on the objective, Herman soon realized that his approach would have to be in two stages in order to take maximum advantage of the sparse concealment. The same would be true of the withdrawal—two days each way so that the final approach, assault, and initial getaway took place at night. An added planning factor was to have different ingress and egress routes.

In coordination with Division G-3 and the regiment responsible for passage of lines, Herman had less than three days (but a lot of time by RTC standards) to develop his plan. Communication with 2nd ID would be by aerial relay with two contacts scheduled, one in the evening of 6 February, the other on the morning of 7 February. The aircraft, of course, was not to fly in a way that would compromise the Rangers location.

If the most difficult terrain was traversed, Herman felt his company could move out undetected during the early afternoon of 6 February. This was something of a gamble but was indeed accomplished.*[*FN: For his timetable to work, Herman had to move out in daylight or else add an extra night to the schedule. 'Laying dog' in the frozen hills for an additional day was too cold an experience to contemplate. The alternative Herman chose, however, was frighteningly dangerous; that is, he never *knew* if the company's daylight movement had been detected and the NKA alerted. A daytime snowstorm would have been an answer to his prayers.] After thorough briefings on every phase of the raid, lst Company headed for the hills, carrying extra ammo and grenades but traveling light (no bed rolls), wearing white camouflage, navigating first by terrain features, then by compass point to point after darkness settled in. Additional navigational aid came in the form of artillery marking rounds (while the Rangers were still in range) mixed into H&I missions.

There were no contacts during the afternoon, an encouraging sign for Herman. He knew, however, that with some daylight left the company would have to slip through a danger area when they passed an abandoned village in a valley that was probably outposted by NKA. He dropped off two squads on a ridgeline to cover this passage. First Company made it through undetected, but then the sound of fire from the rear meant their covering positions were engaged.

What to do about that was a very tough call for Herman. Mission accomplishment was foremost, but leaving two squads to fend for themselves was unpalatable. They were good men with good squad leaders who knew the route back to friendly lines if they were cut off from the company. But if they could not, death in the snow was by far the best thing that could happen to them.

If things had turned out differently, Herman

would have been faulted for what he did—which was turn back his rear platoon to flank the force attacking his OP's. The rest of the company provided fire support for this maneuver which was tactically successful: the NKA were driven off with about a dozen killed, the OP squads rejoined lst Company without suffering casualties, and the trek toward Changmal continued.

But had the all-important element of surprise been lost? Herman had to gamble that it had not. He was right. Apparently the NKA judged that they had turned back a probe, not suspecting that the force they had engaged was continuing deeper behind communist lines. The sound of the fire fight itself was no cause of alarm for the communists. All across the front sporadic gunfire echoed back and forth through the hills, making it difficut for anyone to determine exactly who was shooting at whom, where and why. And now night fell, obscuring whatever suspicions might have arisen about the clash.

The 'impossible' terrain no doubt helped in fooling the enemy.*[*FN: NKA communications were so austere that radios were a rarity below battalion level. Thus it is not surprising that the engagement with lst Company was not reported up NKA channels because probably no more than an NKA company was involved. A company would have reported with a foot messenger, and even if one were dispatched he could not have reached Changmal before the Rangers.] Herman's preplanned route took lst Company along razorback ridges and slopes so steep that for descent the Rangers slid down as 'body-toboggans'. Frequently crossing ice-cold rushing streams added misery but also ensured the avoidance of NKA. The snow was deep, muffling the pants and labored breathing of toiling Rangers. The 'cold soaked' snow did not melt under their tread, and in the morning many tracks would be visible, but that was one of the reasons this would not be the return route.

Around 0100, lst Company struggled up to their objective, Changmal. There the best early warning for the NKA were village dogs, but they were not out in the bone-chilling night. Sound transmitted well in the clear cold. The Rangers could hear slight, undetermined activity around the town. The NKA were probably making preparations to go underground before morning brought the U.S. Air Force overhead, when movement would be impossible. There were a few lights on in town, a sign that the inhabitants felt secure. The NKA were in for a pre-dawn wake up—and it wouldn't come from the air.

Drowsy NKA voices could be heard from as close as five yards away as the Rangers crept up on the ragged cluster of buildings that were what was left of Changmal after it had experienced a great deal of war. In the faint light reflection off the snow, there was nothing to be seen of the usual trappings of a major headquarters—antennas, vehicles, barb wire, etc.—but the NKA were light infanty, so there could be much in the huts, especially underground, that would not attract notice. If indeed there were headquarters personnel in town they would soon experience some close combat.

Squad by squad the fire support platoons slipped into their positions and checked their fields of fire, while the assault platoon swung around to attack from a flank. Fire from the support element would be safest if they could see the muzzle flashes from the Rangers assaulting on line as they swept forward. Eventually the assault would mask supporting fires but by then the raid should be complete.

Some buildings had stone foundations, but most were wooden. A few had a second floor. Rangers crouched to throw grenades into the upper windows. Around 0300, the boom of grenades going off inside signalled the opening of the raid.

The effect was devastating. Every bullet of thousands fired seemed propelled by pent up emotion. The raiders were doing what they had trained for so long and always hoped to do. The buildings under their fire seemed to shred as if a typhon was ripping through the streets. Screams were the only reply for the first minute, then NKA popped out of the houses like grilled corn kernels. Before they could get off many rounds, they were cut down as the assault platoon quickly cleared building after building, leaving only the dead. In front of them small groups of NKA, probably emerging from underground shelters, recoiled and broke away but into the flanking fire of the support elelment. Carnage and chaos reigned for about five minutes (a very long period for intense fire), during which the Rangers' heavy loads of ammo and grenades were expended with fearsome results.

Buildings were burning, scores of NKA rolling on the ground, when Herman signalled his assault platoon to break off and rejoin. Their rampage had almost reached the far side of town, cutting down everything in front of it, grabbing the highest ranking prisoner they could find, decommissioning five machine guns, and absconding with dozens of lesser caliber weapons which might be useful in the withdrawal when American arms ran short of ammo.

Quick reports from his platoons informed Herman that he had only three casualties, WIAs, and only one of them seriously, wounded in the leg. A litter made of two rifles and two field jackets would suffice for now. That was good news; another bit was that the EPW was a lieutenant, a regimental REMF attached to a battalion quartered in town. His division was the 12th NKA Infantry. Maybe lst Company had hit a headquarters after all!

The bad news was that the EPW's regiment was beginning to react. From low hills outside town, some machine guns opened up in a wild search for the targets. Herman knew it was high time to evacuate Changmal, a name that would be added to the long history of great Ranger raids. With two hours of darkness remaining, lst Company broke contact and left town.

Herman had planned for some WIAs and felt very fortunate to have so few (KIAs probably could not have been carried back, depending on their number). But because his egress route was different from the ingress, Herman had no pre-planned hides where he could cache his wounded till they could be brought into friendly lines by another patrol. He had also laid on a helicopter for possible medevac. Choppers were almost experimental in the early 1950s, but had proven very valuable in some cases when weather and terrain conditions were not too bad. Maximum capacity was two litter patients.

A notable feature of lst Company's exfiltration was that the route initially went deeper into enemy territory. The additional distance was well worth it to throw off any pursuit. Even in daylight and across open snow fields, the Rangers were able to avoid contact. However they did come across some enemy commo wire apparently leading to OP's that overlooked Hoengson. These were snipped, and the OP's plotted for air strikes.

One serious problem was that Herman did not have commo himself. The aerial relay plane could not find his column because of considerable fog.

In a hut at the abandoned village of Su-Dong, Herman cached his wounded with a medic and seven men for security. Nearby was a bit of flat ground that had good flight paths in and out for a small helicopter.

These stay behinds were provided with rations, extra ammo and told to expect extraction within 24 hours. The rest of lst Company continued south, intent to fulfill the promise to those remaining at the cache.

Without further event, the Rangers crossed into friendly lines at 1730 on 7 February. Herman immediately set about dispatching the medevac chopper but it was grounded for mechanical reasons. Calling for volunteers, he sent out a patrol with a litter party the next morning to relieve the cache force and bring back the wounded. This patrol did not reach their objective but came upon results of surprising success.

The point man got the drop on seven NKA who came off the ground with their hands up. The reason they were so close to U.S. lines anyway, they said, was to surrender, and held up safe conduct leaflets that psyop had strewn by air all over the front. A field interrogation revealed that they were members of lst Regiment, l2th NKA Division. They said their regimental headquarters had been hit by the Rangers at Changmal, then pulled back northwest to safer ground. During this displacement the seven EPWs had taken the opportunity to desert.

This intelligence, if true, was too important to wait until the patrol extracted the Ranger stay behinds, so the patrol returned with the EPWs. This was surely the right decision, but it did no good for the wounded at the cache. The lst Company First Sergeant, Romeo Castonguay, decided to take matters into his own hands. With a small detachment he set out for the cache to bring in the wounded.

He did not find them at the hut where they were last seen. During the intervening hours, the cache security force had surveilled the approaches to Su-Dong. While he was a one-man OP, CPL Glen Hall was to observe and report any enemy activity. When an single NKA messenger came along on a bike, Hall could not resist the temptation to drop him. With one shot he did so, frisked the body, and returned to his OP.

Hall's comrades came up and scolded him, sternly advising that today he was not a sniper, just an observer. Hall apololgized, said he understood and wouldn't do it again, but in the next hour another messenger came along and fell to Hall's marksmanship. More valuable documents were obtained, but the Rangers at the cache felt they could trust Hall no longer to hold fire. They displaced, so SGT Castonguay arrived at an empty hut.

But he was hailed from nearby high ground where his litter party was warmly welcomed. They loaded up the wounded, and Castonguay led them back over very rugged terrain to friendly lines. Though delayed, the finale to the Changmal raid was as succcessful as the rest of it.

More than any other U.S. foreign war, Korea was one of ebb and flow. One side never had the initiative for long before the other side tried to wrest it away. By February 1951, the communists realized that their best chance of resuming the offensive was by concentrating blows on the ROKs whose combat capabilities were far less than US forces. They also suffered somewhat from psychological intimidation by the Chinese. On 11 February, only four days after the Changmal raid, the CCF took over the central front from the NKA and barreled into the 8th ROK Corps which had been advancing through Hoengsong.

CCF success soon rolled up three ROK divisions and exposed 2nd ID. To refuse his flank, Gen. Ruffner had to hold the crossroads at Chipyong-Ni, and to do so he committed his 23rd RCT reinforced by a French battalion and lst Ranger Company, some 4,000 men against three CCF divisions. COL Paul L. Freeman would command the defense in a three-day battle that resembled Bastogne, the last occasion when so many Americans fought succesfully while completely encircled.

Freeman chose his ground well, eschewing outlying high ridges where his line would be too thin, entrenching instead the rim of a bowl formed by lower hills where he could reinforce against threatened penetrations. Delaying actions north of the town allowed him some 48 hours to dig in and stockpile ammo, especially artillery,for the imminent CCF onslaught.

The first mission for the Rangers was to patrol from the perimeter to locate the main avenue of enemy attack. Accompanied by a French war correspondent, lst Company moved out at 1900 on 11 February to reconnoiter the village of Miryong-Ni a few miles northwest of Chipyong-Ni. They were to determine the CCF strength there and, if possible, take a prisoner.

Cautiously slipping around the town they investigated it without contact, but at 2200 a platoon clashed with Chinese on a ridge to the west where for an hour lst Company got into a full blown fire fight that brought down on them so much fire from 82mm mortars and heavy machine guns that the Herman could consider one part of his mission accomplished: the CCF was around Miryong-Ni in regimental strength. Closing quickly, the Rangers fought their way into enemy positions to complete their mission. Two prisoners were snatched at the cost of two KIA and two WIA, one of which was the French correspondent who died of his wounds. The patrol re-entered the 23rd RCT perimeter around 0400 bearing his body on a door torn from a building in Miryong-Ni.

With an infantry company, the Rangers then became RCT reserve. Within the tight perimter they were still subject to indirect fire, and one Ranger was killed by a mortar round. All the next day the encircled RCT was bombarded, then waves of infantry struck on the night of 13 February. They were preceded by bugle calls, ringing cymbals and pyrotechniques that lit up the sky. From midnight till dawn, the CCF main attack shifted around the perimeter, probing for a soft spot. This was chest-to-chest infantry action, continuous attacks that Bob Black described as "beating like waves against an island shore."

With every battalion's flanks tied to another, the RCT hurled back each thrust, sometimes resorting to 'mad minutes' that left no attackers standing. Leaving piles of dead and bodies in pieces, the Chinese pulled back before daylight exposed them to air strikes and observed artillery. Napalm got down into the CCF positions where nothing else could. Most of their wounded died because they could not be evacuated under the hawk eyes of U.S. fighter-bombers.

But the relentless attackers came back at dark, driven by commanders who realized that if Chiyong-Ni went under Eighth Army would be psychologically rocked and shaken. COL Freeman spared nothing to ensure that wouldn't happen. His worst problem was dwindling ammo; though a huge quantity had been stockpiled, it would not last if the fighting tempo continued as it had the night before.

The night of 14 February was still worse. The CCF concentrated their battering on the boundary of two battalions so as to grind them thin enough to rupture. After midnight, several of the hardest hit sectors were penetrated; the postions were restored by counterattacks, but this depleted Freeman's reserve until lst Company was all that remained.

Then at 0315 a CCF surge ripped open a gap, and the defending US company had to fall back in tatters. Before the Chinese could flood in, the Rangers headed off further exploitation with a platoon led by LT Herman. He joined up with a rifle company pulled out of another sector, and though Herman tried to have his whole company committed under his command, he instead led his platoon (33 men) to retake the right half of a hill that was the counterattack's objective.

The Rangers would have no preparatory fire except from their own puny 60mm's. With daylight approaching, they charged uphill as their machine guns supported from the flanks. The Chinese had had time to prepare, and their interlocking fire took down nearly a quarter of the Rangers. Friendly fire in the confused darkness added casualties. As the Rangers closed in, the Chinese dropped protective barrages of white phosphorous from 4.2's captured earlier in the war.

Barely a squad of Rangers reached the hill crest on which stacks of bodies smouldered and smoked. LT Herman had been knocked out of action, the platoon leader, LT Heath, and lst Sgt. Castonguay killed, and a score seriouly wounded. The high ground had yet to be taken, but as long as there was a Ranger left standing, he would fight to secure the objective.

CPL Glen Hall—tha man who couldn't resist sniping— was one of those few. Originally a machine gunner in the fire support element, he had burned up all his belts before the gun malfunctioned. Grabbing a fallen carbine, he charged up hill to join what was left of the assault. He found few friends but a number of stunned Chinese who were not ready for the hand-to-hand combat that ensued.

"Get some men up here!" was shouted down to the wounded downhill, who relayed the desperate request back into the perimeter. But there was no immediate reinforcement, partly because the RCT S-3 believed that the counteratack had succeeded and secondly because there were penetrations elsewhere. The snafu meant that the handful of Rangers on the hill would have to hold what they had by themselves.

Fortunately the Chinese were in not much better shape. But there were many more of them, and the Rangers were running out of blood and ammo. For two hours they struggled to control the hill crest, then they were forced off.

Very soon the balance of lst Company, plus a line company, were ordered to take back the lost positions. By then friendly artillery was able

Ranger Emmett Fikes, Im Jin River, Korea, 1951.

to interdict CCF follow-up forces and isolate those on the hill. And fortunately, the new Ranger attack was supported by a quad-50 halftrack, sort of the B-52 of Korea, a high rate of fire weapon that cut a swath of death and destruction wherever it was focused. A couple of tanks added their fire power, and 155's weighed in with direct fire of white phosphorus. Bob Black noted, "the sound of the [155's] firing and resultant explosion came almost as one."

Thus, the Chinese were literally blasted and burned off the hill, but on the reverse slope they stopped the counterattack. So the situation on 16 February was about the same as it had been some twelve hours earlier, but now thousands more men were dead, and lst Company shattered after grisly warfare reminiscent of World War I. Many Rangers could have been saved if Herman had been allowed, as he requested to be, to keep his company together for the night counterattack.

There was no time for hand wringing. Day broke on the perimeter intact but very short on all types of ammo. An airdrop was the most urgent priority, and securing the bundles was the foremost mission. Here the Rangers airborne training proved invaluable. Until they set up good ground-air commo, the C-119's were overshooting the DZ's and bundles were lost to the enemy. Soon the Ranger ground controllers had the 'chutes floating down into the narow perimeter, but bundle retrieval was still extremely hazardous because of CCF mortars.

Even with much badly needed ammo, Freeman was in no shape to repel the Chinese as he had the night before, so a relief force from the 5th Cav Regiment cranked up to break through to his perimeter. This counterattack sputtered at first—the CCF had too much at stake to forfeit ther fortunes at Chipyong-Ni— but preceded by waves of napalm-dropping tac air, the 5th Cav ground ahead in the afternoon against dogged opposition.

First Company held a portion of the perimeter where the 5th Cav was headed. The last possibility the Chinese expected or were prepared for was a break out from the perimeter. But this is exactly what the Rangers did, charging from their foxholes, routing a CCF battalion in front of them and driving them into the 5th Cav who cut down Chinese in droves.*[*FN: lst Company's surprise attack seems to have been a surprise to everyone, for higher HQ had not ordered an attack at all. Apparently it arose from instinct and initiative when squad leaders (there were few officers left) saw the opportunity to hit the enemy while he was looking the other way at the 5th Cav coming up from his rear. A more stirring example of Ranger aggressiveness is not to be found in the history of any war.] Tac air also had a field day, strafing swarms of fleeing targets. The 5th Cav CO rolled into Freeman's CP at 1730, and the blood soaked three-day siege of Chiyong-Ni was lifted—the first utter defeat of a large Chinese force (40,000 troops) in the war.

From the hard hitting patrol that opened the action at Miryong-Ni till the surprise attack to open the way for the 5th Cav, lst Company had made major and conspicuous contributions to 23rd RCT's valliant and notably successful defense. Comparisons of Chiyong-Ni to Bastogne are not far fetched at all.

The cost to the Rangers was ten killed and 31 seriously wounded. lst Company would be recognized by a distinguished unit citation, and acknowledgement throughout Korea that there was no finer fighting force on the peninsula in any uniform. They gained fame at Changmal by doing what they were specialized to do; they won immortality at Chiyong-Ni doing what they should not have been required to do.*[*FN: The French battalion held the high ground alongside lst Company when the Rangers launched their surprise attack toward the 5th Cav. Frenchmen, more than most nationalities, are admirers of espirit and elan. As they watched it being displayed so abundantly in the Rangers ferocious charge, they called across the battlefield, "Ah, Rangers.. *Tres bien!"*]

Down to platoon-size, lst Company slowly recovered strength as 28 replacments dribbled in from Ft. Benning. There was a change of command when CPT Charles Carrrier, from 7th Company and briefly with Eighth Army Ranger Company, assumed command from LT Herman who remained XO.*[*FN: Why a lieutenant with Herman's record was not instantly given a battlefield promotion is a question that baffles the author. The promotion would have not only ben richly deserved, it would have saved Herman's life.]

March 1951 was a month of counter-guerrilla patrolling and rounding up stay-behinds as Eighth Army battled north to the 38th Parallel. Spring thaws slowed Ridgway down considerably but Seoul was once more recaptured. As lst Company renewed proficiency, a new sort of mission was added to their repertoire: mine laying behind enemy lines.

In April, Eighth Army (now commanded by Gen. James Van Fleet as Ridgway moved up to CINC after President Truman's relief of MacArthur) shoved forward beyond the 38th Parallel and aproached a three-sided plateau about ten miles across that contained important transportation junctions vital for the enemy's logistics and assembly for their anticipated spring offensive. This known a the Iron Triangle, was too vital for the CCF to abandon.

On May Day—the 4th of July for communists worldwide—the CCF struck back to protect the Iron Triangle, launching their main attack on the west coast but supporting it with another at the boundary between the U.S. Ninth and Tenth Corps. At this junction 2nd ID was to hold and repel any attempts to split the division from the adjacent ROK's. G-3 formed Task Force Zebra for this purpose, composed of a composite infantry battalion and a tank battalion. Initially tasked to run contact patrols with the French on the left and ROK's on the right, on 5 May lst Company was then designated to be TF reserve.

At first the situation seemed so well under control that scheduled R&R's to Japan continued, but on 16 May the CCF bore down on 2nd ID's neighbor, the 5th ROKA Division, and tore them open. This began what was called the 'second impulse' of the communists' spring offensive, shifting their main attack from the west (where they had once more threatened Seoul) to the eastern front, centered at Inje. To thwart them, 2nd ID had to hold the shoulder of a sixty-mile-wide penetration that extended all the way to the Sea of Japan and had pushed the ROKA back from ten to thirty miles.

Tactically, the communists had nothing new except that Russia and China had bolstered the offensive with more tanks. The CCF attacked as before, freely expending manpower and exploiting their expertise in cross country movement and infiltration, striking at night and going to ground by day.

Second ID had the benefit of extremely rugged terrain to defend (though this also favored CCF infiltration) and did so by deploying on 'No Name Line', a wide angle that hinged on Hill 1051, held by the 38th Infantry Regiment to whom lst Company was soon attached. The brunt of the CCF attack fell on 38th Regiment, Hill 1051 being the objective of 30,000 veteran Chinese.

Guided by stones marked with luminous paint, utterly disregarding horrendous loses, a Chinese division was able to storm Hill 1051 and fracture 38th Regiment on the night of 17 May. This created a situation for the Rangers that was Chipyong-Ni all over again: counterattacking in the darkness and chaos, without time for planning, to seize an objective (Hill 710 to the rear of 1051) they had never reconned. The two major Ranger strengths— stealth and surprise— could not be exploited but guts and aggressiveness took up the slack: Hill 710 was seized and secured against repeated Chinese counterattacks. At dawn's early light, Rangers still held the hill and 2nd ID had time to readjust forces for the next days battle.

Where the day before there had been three CCF divisions for 38th Regiment to contend with, by morning on 18 May there were but two, the 45th and 181st Infantry.*[*FN: During the course of the entire battle along No Name Line, the Chinese Third Army hurled a total of twelve divisons—120,000 troops at 2nd ID] The 44th Division had been virtually anihilated (5,000 casualties) and was now out of the fight. It took another day for the Chinese commander to move several more divisions against 2nd ID. One of them was so cut up by long range artillery and air strikes that it never reached the battlefield. Van Fleet had located the center of the communist offensive and had shifted reserves and artillery assets east to meet it. 2nd ID had done the most to buy him the time to make these vital redeployments.

Occupying bunkers built by friendly forces that had been kicked off Hill 710, daylight found lst Company overwatched by Chinese on higher ground. The Rangers were neither strong enough to interdict CCF infiltration through the gullies to the rear, nor could they do any more than exchange fire with CCF overwatching forces. CPT Carrier knew that the next night would be one where he would have to defend in all directions.

This indeed became the situation but one he prepared for by bringing up all the ammo available and tightening his perimeter for self-sufficiency. One problem he did not foresee was strafing by U.S. aircraft. In the fluid situation, hills were misidentified and two Navy Corsairs swooped down on Hill 710, 50 calibers blazing. Ranger Bob Morgan's bunker was blown apart.

Stunned but still thinking, he crawled out with an identification panel. This drew heavy CCF fire, causing him to tumble downhill into a pile of Chinese dead where he remained, imitating them. Morgan could not move back uphill till dark.

For all their preparations, the Rangers had to evacuate Hill 710 without a fight. This was because 1/38th Infantry had been doubly enveloped by CCF infiltration, and were pulling back several miles—if they could. G-3 ordered Carrier to form a rear guard for 1/38th. This was a euphemism for getting the hell out any way you can. At nighfall after ruefully destroying a lot of equipment to prevent its capture, lst Comany began exfiltrating by squads. Running a gauntlet of fire, squads eventually broke down into sections then were separated as individuals, an ordeal Bob Black described as "for each man the ensuing hours became a private hell... Men of 2nd ID were scattered about the hills in groups that would flow together then break apart under fire. As the surviving Americans tried to reach friendly positions, the Chinese[ahead of them] killed or captured large numbers."

Among those killed were now famous names: LT Herman, lst Company's original CO, and CPL Hall, the sniper of the Changmal raid and winner of the DSC at Chipyong-Ni. Captured were seven veterans and CPT Carrier who heroically defied eleven months of torture before the communists executed him.

Such losses were extremely bitter for the Rangers, caused as they were by higher HQ ordering an evacuation of a position lst Company had seized with blood and bravery, then relinquished because of the collapse of a sister unit. Lives lost for temporarily needed terrain was a dark hallmark of the Korean War.

The martyrdom of so many Rangers was like losses in a flash flood. Suddenly they were gone forever, but the CCF onrush soon dashed into droplets against 2nd ID's fall back positions. In the division sector, 35,000 communists would never see China again, and by its end their spring offensive became known as the 'May Massacre'. That term could apply as well to lst Company which by 20 May were down once more to half strength.

Their only consolation was a second distinguished unit citation.

Plus the satisfaction of becoming combat ready, and back in action, within a week.

The rest of the war for lst Company was an anti-climax. May was devoted to security missions, mostly rear area patrolling, and also some training of ROK Rangers. June brought the company back to full strength, and a new commander, CPT Charles Ross, most recently CO of the now disbanded Eighth Army Ranger Company. He would have the unwanted distinction of closing out the another company, for lst Ranger Infantry Company (Airborne) was inactivated on 1 August 1951.

Thus ended a chapter as proud as any in Ranger history, sealed with names cited in the award of 110 Purple Hearts (including 15 oak leaf clusters). Among them were 75 KIA. Another 16 were MIA, never seen again. Eight died in captivity; six more were POW's who survived to be released in September 1953.

Why did lst Company leave Eighth Army rolls when it did? For reasons that are laughable when considering that since VE Day all Ranger campaigns have been against non-Caucasions. Said MG Carl Ruffner (an otherwise sensible officer) in his letter of appreciation to lst Company as it broke up:

Undoubdtedly many of you are wondering why your company is being deactivated. A thorough study of the past actions here in Korea indicates that deep patrol missions by small units, for which the Rangers are intended, are made most difficult in the Korean Theater by reason of the racial differences between the Oriental and the Caucasian. For this reason it has been decided that the utilization of Ranger units in Korea would be discontinued.

Second Airborne Ranger Co.

If, as Gen. Ruffner alleged, race was a factor in Ranger operations against the enemy, race too was signficant in U.S. Army attitudes toward the Rangers. This was because President Truman's order to integrate the armed forces had been in effect for only a few years, and the Army had dragged its feet in implementation. Previously, except for a single squadron in the Air Corps, 'colored' (i.e., black) troops had been consigned to service support units because of the unofficial army position that blacks were not reliable fighters. This in spite of the sterling combat performance of blacks in the Civil and Indian Wars, and even service under George Washington in 1776.

The RTC was no more progressive in integrating black Rangers than anyone else. 2nd Company was formed all black, in a sense integrating RTC by admitting black Ranger candidates but consolidating them in one company.*[*FN: Until 1954, "separate but equal" was also the US educational policy approved by the Supreme Court.] This was a step forward but the leap to full equality would have to be made by 2nd Company itself, the springboard being the record of their performance, the last word to be written with blood.

The first unit to approve that record was their white sister company, the 4th, as both underwent training together at Ft. Benning. Mud and snow have a way of obliterating skin color.

Both companies, each with about five officers and 110 EM, entrained for Ft. Stoneman on 3 December 1950. Packed on what seemed to be cattle cars, they took up the practice of lowing like a herd of steers to indicate collective displeasure. On New Year's Eve, they reached Korea, 4th Company to join the lst Cavalry Division, and 2nd Company to 7th ID.

2nd Company's first fight was not with communists but to prevent the Rangers from becoming a casual company for all blacks assigned to 7th ID.

Division policy, prescribed by the corps commander, LTG Ned Almond, was that:

Colored troops who arrive in this division, if they have the proper MOS's, will be assigned to the 2nd Ranger Company. Others who are assigned by error will be returned to the Eighth Army Replacement Battalion. Personnel will not be mixed within units. If Ranger Company becomes greatly overstrength, another company will be formed.

In other words, Almond required that *any* black solider coming into 7th ID with an infantry MOS—the most common in the army—would be assigned to 2nd Company with no consideration for his lack of Ranger qualification.*[*FN: Almond also had an inflexible policy that under no circumstances would a black officer command a white one. Naturally 2nd Company preferred service in IX Corps whose CG was not a racist.] Even more important, he would not even have to be a volunteer for Rangers—in fact he might want no part of Rangers at all—any more than every white soldier wanted to be a Ranger.

And the phrase, "assigned by error" is a clear message to G-l that if a black replacement did not have an infantry MOS (and therefore could not be assigned to 2nd Company) he was not wanted in 7th ID because "Personnel will not be mixed within units"— that is, mixed into white units.

2nd Company's CO was LT Warren E. Allen. He was in no position to tell Gen. Almond to stuff his non-Ranger, non-volunteer replacements. So 2nd Company would unwillingly be a tool for back door segregation, because they had no other choice. Their job in the Korean War would just be doubly difficult—to concurrently fight communists and train replacements, as diverse as cooks and bandsmen, in ranger craft. But things had always been twice as tough for black Rangers, and they would just drive on to do what had to be done. In so doing, these original black Rangers did in fact lead the way, both to victory and equality.

7th ID had been Eighth Army's amphibious division since Inchon when it came ashore, linked up with U.S. forces breaking out of the Pusan perimeter, and helped liberate Seoul for the first time in September 1950.*[*FN: Contrary to popular belief, amphibious operations are hardly the exclusive province of marines. In WW II, there were far more army amphibious operations, landing far many more troops, than the USMC ever did.] Back to sea again in October, 7th ID sailed around to the east coast, landing at Iwon and advancing to within a few miles of China before the CCF boiled over.

Near Chosin Reservoir, elements of 7th ID were overrun but the rest of the division managed a successful amphibious withdrawal from the port of Hungnam with moderate casualties and getting out most of their heavy equipment (in contrast to the rout of U.N. forces on the west coast). Though short of everything, 7th ID was probably in as good shape as any of the U.S. divisions that had seen considerable combat.

In early January 1951 when 2nd Company joined them, 7th ID had completed their third voyage through Korean waters and were deployed on the central front, preparing for a continuation of the CCF winter offensive. To police up his rear area before focusing on the forces in front of him, the CG, MG Ferenbaugh committed his newly attached Rangers to counter-guerrilla operations. Like Eighth Army Ranger Company and lst Company before them, 2nd Company would see their first action in such operations—a good sort of 'pre-season' tune up to learn about Korean terrain and conditions.

They weren't long making contact because 2nd Company's AO was an ant hill of communist activity—night time activity mostly until the Rangers moved in. In frigid darkness around 0200 on 7 January, their first kill was a major one when twenty NKA's were ambushed at a roadblock. Four hours later an NKA company hit a nearby regimental aid station that was within the Ranger perimeter and soon fighting spilled over, threatening 2nd Company's CP and 7th ID's MSR back to Taegu.

LT Allen split his force to take on both engagements simultaneously. Both enemy attacks were broken up as the Rangers defended and maneuvered at the same time. With daylight approaching, the NKA found they were pinned down while being flanked. Unexpectedly they were not able to break contact and disperse as they had against conventional units. The Rangers could move cross country as fast as they could and were intent on hunting down the survivors of 2nd Company's first battle. About fifty guerrillas were accounted for at a loss of SFC Isaac Baker killed and three Rangers wounded.

To pacify this mountainous and defoliated area, 2nd Company imposed a nightime curfew, turning their AO into a free fire zone after dark. Thereafter, a number of communists fell in small meeting engagements with Rangers positioned where they could detect movement even at night by skylining ridge crests. Individual guerrillas had to move along or over these crests—messengers mostly—to provide communications. Guerrilla supplies too had to be carried on foot, and though the Rangers did not intercept any columns, they forced the G's to 'convoy' their supplies with combat guerrillas and divert supply routes over worse terrain where going was slower. Losing the initiative and many men to LT Allen's tactics, the guerrilla battalion commander decided his force would be better off if it drifted out of the Ranger AO to less dangerous areas. The G-3 was impressed but now he had to contend with G's elsewhere.

They became a lesser concern as the CCF offensive spilled south, bending the ROKA line on 7th ID's right flank. To their rear was the deep valley of the Han river through which ran a road that was the division's MSR. ROK forces covering the pass were shakey. G-3 moved 2nd Company back to the pass to secure it in case the CCF broke through.

Enemy infiltrators did approach the pass. On 9 January while patrolling, 3rd Platoon of 2nd Company came upon an equal size force of infiltrators which seemed to be an advance guard, digging in to await arrival of the main body. Though the NKA were on good defensive terrain, the Rangers attacked using a high volume of fire. Infiltrators can not carry much ammo, so their return fire was less.

With this advantage, the Rangers slowly worked around their foe in a fire fight that lasted from 0900 till 1445 hours. Rooted out of their positions, the communists were able to take their wounded with them, leaving eleven dead but losing no prisoners, a disappointment for the Rangers in this otherwise good demonstration of brains and aggressiveness.

Neither quality was of any use against the unconquerable enemies: frost biting cold, penetrating wind and hard-granuled snow. Through it all, in knee-length white parka shells, the Rangers constantly patrolled. Occasional scattered contacts would not be remembered as well as the elements of nature that had to be fought every day and night. For example, just an exchange of fire meant hitting the ground, not moving for several minutes as the leader decided on a scheme of maneuver. He had to maneuver— to be pinned down started the clock for frostbite— and even in those few minutes of decision and tactical instructions, limbs began to stiffen and facial skin turn pale.

During the next week of January, 7th ID's situation turned grim as the CCF infiltration to their east had erupted from a trickle into a flood. 2nd Company was rushed around to fill gaps as the CCF turned from their penetration of the ROK's to take 7th ID from the rear. So the G-3's decision to commit 2nd Company (attached to l/32nd Inf) to security of the pass turned out to be a wise one.

LT Allen went out to meet the threat rather than wait for the CCF to cut off the pass. At 0730, 14 January, he snaked the company down from icy slopes to a cross trails village which he knew would be vital to the CCF advance through the hills east of 7th ID's MSR. A U.S. infantry company was to go around and beyond the village (Majori-ri) in sort of a spoiling attack.

About a battalion of CCF had already reached the village, just before 2nd Company arrived. Opening the fire fight, Allen sensed that the enemy was in hasty defensive positions and that a hard thrust might push them out into open country where air strikes (though very scarce at that time) could reduce the possibility of a successful counterattack. Grabbing for the initiative, 2nd Company went on line and closed in on Majori-ri. With flanking fire from their machine guns, they hit the objective ferociously, carrying the edge of the village in the first rush. But then a great deal of return fire broke out as the weary Chinese battalion regrouped and fought back desperately.

That battalion had trekked for five days on short rations, little rest over high mountains and through cold, snow and bitter winds. They could not stand up for long against the determined Rangers who drove them from what little shelter the village provided. The Chinese broke, fleeing Majori-ri without a hundred infiltrators who had entered it that morning. For their bold success in attacking a superior force, 2nd Company lost five killed and the same number seriously wounded.

Regretably there were no air strikes available to finish off the CCF battalion. Its parent regiment heard the fire fight for Majori-ri and so were prepared for the U.S. infantry company that had bypassed the village. The Americans ran into a buzz saw and were so chopped up that 2nd Company was ordered to move up and take pressure off. This time it was a Ranger company against an alerted, dug-in regiment and Allen could find no vulnerble spots to penetrate. One Ranger was killed, two badly wounded trying, and the probe was stopped.

Matters got worse as the enemy realized they had the Americans vastly outnumbered and began to counterattack. Low on ammo, Allen told the other company to pull back through the Rangers who would conduct a delay. Warily the Chinese pushed forward, then broke through Allen's thin covering force and some Rangers were cut off.

CPL Bill Rhodes—known as 'The Thin Man'— was one of them. Three of his few comrades left to cover the withdrawal were killed nearby him when his last clip pinged out of his M-l and he realized he was completely out of ammo. The Chinese were over his position in a minute. He dropped as if shot. His 'body' was rolled over and his watch and wallet removed before the Chinese moved on. After thirty minutes and nearly freezing, Rhodes stirred and slipped away toward friendly lines. In the open he was fired upon, and went into his dead man's act again. This time his 'body' was stripped of boots, socks, parka and cap.*[*FN: Ranger Bill "The Ghost" Washington had a nearly identical hair raising experience while escaping and evading back to 2nd Company the night before Rhodes' return.]

Frostbitten and virtually naked to the cold, Rhodes eventually found shelter with a kindly Korean couple. On the third night of evasion, he made contact with a friendly patrol and returned to 2nd Company. Amazingly, the company had suffered only three more killed and five more wounded, including Allen,—and still more amazing, none captured.*[*FN: When the lst Platoon leader, LT Bernard Pryor went down with a wound along the center of his head, his platoon sergeant, MSG James Freeman took over for the next 35 days and received a battlefield promotion.]

Their shrewdly aggressive actions—always taking the fight to the enemy— had been instrumental in deflecting some very serious threats to 7th ID's MSR. By 24 January, the steam was gone from the CCF offensive and Ridgway began looking north again. As 7th ID headed in that direction, 2nd Company was busy mopping up bypassed communist groups.

On 6 February, a spot check on Ranger performance requested by the Pentagon reached Eighth Army and was passed down to the appropriate division CG's. Gen Ferenbaugh's comments regarding 2nd Company were these:

In view of mission assigned this division requiring its employment by battalion combat teams in torturous mountain terrain over extended distances and widely separated sectors, it has been difficult to employ the 2nd Ranger Company entirely in its prescribed specialist role.

On every occasion, performance of 2nd Ranger Company has been highly praised by commanders who have assigned missions to the organization. Their specialized training and high espirit due to their volunteer status makes their assignment desirable to augment any fighting unit.

This unit is presently at 61 percent strength. No forecast indicating increase to authorized strength has been given this headquarters. In order to utilize this highly trained unit efficiently and maintain morale, qualified replacements should be made available through normal channels with the same frequency as for other units of a division.

In view of outstanding performance of this unit, it is believed that one Ranger company for each infantry regiment with a battalion headquarters and service company for administration and logistical support would be desirable.

What a difference a month of war makes! From directing that 2nd Company become a repo depot for all blacks coming into 7th ID— Almond's policy, it should be emphasized, not Ferenbaugh's— Ferenbaugh was doing flips about black Rangers combat performance, and suggesting a Ranger company at regimental level with a Ranger battalion HQ, presumably at division level.

Some other changes can be read between the lines:

Ferenbaugh credited 2nd Company's "volunteer status" for its "high espirit", so evidently he revised Almond's previous policy that 2nd Company need not be all volunteers (but they still needed to be all black).

As for replacements, Ferenbaugh requested that those in the future be "qualified", which can be inferred to mean Ranger qualified. So apparently he no longer expected replacements to become qualified through OJT with 2nd Company. Instead he wanted a steady flow of Ranger fillers through the regular pipeline. Ferenbaugh could not have assumed that such qualified replacements would be all black, so either he envisaged 2nd Company becoming a racially mixed unit, or, as he recommended, additional Ranger companies would be added till there was one per regiment. Given the army-wide personnel shortages of early 1951, he probably meant the first alternative which meant he accepted that 2nd Company would become racially mixed.

To get ahead of the story, this is indeed what happened, and that outcome came about solely because of 2nd Company's performance in combat. Acceptance of integration can be considered to be a struggle for attitudinal change on the part of whites. In 7th ID, that change was rapid and the attitude which resulted was respect—that unique respect of one soldier for the competence of another.

2nd Ranger Company demonstrated that competence once again on 20 February in an action that brought wide notice in U.S. newspapers. For the first time there were reporters around to record the 'Buffalo' Rangers at work. This nickname was derived from the black cavalry in the Indian Wars whom their foes dubbed 'Buffalo Soldiers', presumably from their coarse hair, dark color and the fact that many wore beards. "Buffalo!" became 2nd Company's war cry, and also an emergency pass word for evaders and any patrol that returned so late for its regular pass word to be current.

On the central front the CCF held an annoying salient that needed to be reduced. In coordination with a neighboring division, 7th ID was tasked to take down the western shoulder of the salient within which was the village of Chuchon, eight miles behind the enemy front lines. This range of penetration was Ranger territory, so G-3 assigned Chuchon as 2nd Company's objective.

An accompanying correspondent from International News Service reported 2nd Company's AO to be a "white mountain wilderness". Camouflaged in parka shells, the Rangers moved out with platoons abreast to infiltrate to their objective. Enroute, a platoon detected movement of another party headed in the same direction. There was not time to maneuver for an ambush, so the Rangers trailed the communists who were headed for Chuchon to sack it. To scare away the villagers the raiders opened fire on the village. This was the distraction the Rangers needed to close in.

The reporter was awed when 2nd Company turned the strange encounter into a rout, in full cry racing after their quarry right through the village, an attack the reporter described as 'banzai'. 2nd Company's mortars sealed off exit from the village, and the communists were brought to bay and bayonets. It wasn't till then that the number of enemy proved to be many more than had originally been pursued— others had come at the village from different directions. 2nd Company evened the odds by killing a great many in house to house fighting. Almost all the houses were wooden, and at the end of the battle shredded or burning. The communists were either dead or cleaned out. The reporter headlined his piece, "Hard-hitting Rangers Make Life Miserable for Surprised Enemy."

With their publicized victory at Chuchon, 2nd Company seemed to be on a roll, a feeling that increased when they were attached to the Rakkasans of the 187th RCT (Airborne) on 28 February. Airborne, after all, meant miltary parachuting, something Rangers had not done since they left Ft. Benning. Rejoined with 4th Company, the Buffalos and Rakkasans conducted unit training jumps leading into tactical exercises. Something seemed to be in the wind— something like a lot of parachutes descending behind enemy lines.

An original paratrooper himself, Ridgway planned that March would be the month for Operation Ripper, another meat grinder for the CCF and NKA, and an opportunity to give them some more worries about their MSR. To do so, Ridgway had his G-3 consider a regimental-size drop to cut off supplies, and possibly an enemy retreat, north of Seoul. For three weeks then in March, TF 187 (the RCT plus 2nd and 4th Ranger Companies) trained up for Operation Tomahawk.

While in the marshalling area, LT Antonio Anthony joined 2nd Company with a platoon of black airborne Rangers fresh from RTC. Allen had also opened the faucet for his own replacement pipeline by recruiting the best of the black trainees foisted upon 2nd Company by Gen. Almond's segregationist policy. These volunteers got a three-day airborne course right in the marshalling area.

The DZ's for Tomahawk would be centered on Munsan-ni, a town some 24 air miles northwest of Seoul where the paratroopers were to hold an airhead that would interfere with the NKA-CCF's logistical network. Within 18 hours linkup was to be made by a combined arms ground force built around the 6th Medium Tank Battalion. D Day for Tomahawk was 23 March 1951.

Compared to airborne operations in World War II, the Tomahawk jump is most similar to that of the 101st A/B Division as part of Operation Market Garden in Holland where the Screaming Eagles seized a long stretch of highway for exploitation by an armored column with the purpose of deeply cleaving enemy lines. Operation Tomahawk, however, was not designed so much for penetration but to land an airborne anvil against which an armored hammer would shatter the enemy rear area.

Tomahawk launched from an airfield at Taegu, with 150 transports carrying 3500 paratroopers and twelve 105mm howitzers in five serials. 2nd Company was in the second serial. It flew east out over the Yellow Sea, then inland from the North Korean coast. History was in the air, for this was the first time Rangers ever parachuted into combat.

Morning fog had lifted (the main reason for the relatively late drop time) when at 0915 2nd Company spilled into the arctic air. Aerial navigation had been good, and all Buffalo Rangers landed within their designated DZ. As combat loads go, they were traveling pretty light at about 70 lbs per man. This proved to be an initial advantage as CPT Allen—promoted while in the marshalling area— got his men off the DZ in a hurry. They had to be because 2/187 jumped on the same DZ only minutes later. By that time 2nd Company was formed up in an orchard, ready to move out for their first objective. Despite the rock-hard frozen ground, the Buffalo Rangers had only two jump injuries which were efficiently treated by members of the 60th Indian Field Ambulance Unit whose very first jump was their entry into combat that day.

The communists had been taken pretty much by surprise, but soon a couple of their machine guns began hosing 2nd Company's assembly area from long range. Even before his company was fully assembled, First Sgt. West grabbed some early arrivals and went after those threatening positions. His aggressiveness paid off: the enemy abandoned the guns which West's force captured before they could be turned on the descending troopers of later serials. 2nd Company also took the first prisoners of the operation, two members of the 19th NKA Infantry Division.

Allen's initial objective was Hill 151, terrain that dominated a DZ as well as the planned linkup point, and about a mile from his assembly area in the orchard. Sgt. West's prompt action protected the first phase of the Buffalos' approach march toward Hill 151, but in front of a small village they had to deploy and clear it out in order to continue for their objective.

Expecting alert and organized resistance in the village, Allen was as surprised as he was pleased when his men were able to sweep through, killing six and capturing twenty enemy soldiers. Hardly breaking stride, the Rangers regrouped for the assault on Hill 151 as it was bombed and strafed by tac air. As the P-51's pulled off, 2nd Company's 60 mm mortars began their preparation of the bleak, treeless hill.

Stunned and confused by an unexpected battle that had been brought to them by air, the NKA defenders nonetheless had good defensive terrain that they would not give up easily.*[*FN: All the EPW's taken in this action had no idea that there were even such things as paratroopers. Some said they just gawked instead of shooting at them.] Hill 151 was a not yet fully occupied blocking position prepared in case of an armored attack from the south. 2nd Com-

Ranger Jump.

pany would have to fight for it by maneuvering over open terrain and uphill.

Betting on the expected linkup, the Rangers expended ammo freely in fire support for the assaulting elements. They could not completely encircle the hill but worked around to three sides before moving for the top. Though the three assaulting groups could not see each other, they stayed pretty much abreast and advanced close behind LT Anthony's mortar barrages. By noon, Allen could report his objective secured with eleven enemy dead counted and one EPW. The rest of an estimated two platoons had managed to escape down the back side of the hill. Ranger losses were one KIA and two WIA, an amazingly low cost for any wartime airborne operation, much less the first combat jump.

Ready for further missions, Allen was told to sit tight, dig in, and prepare for linkup as the armored task force was reporting good progress. Darkness fell, but at 1800 the Rangers had friendly tanks in sight and established commo with them. They turned out to be from the 64th Tank Battalion from 3rd ID. War produces some strange coincidences and reunions. Many of these tankers were from the same part of Ft. Bragg where most of 2nd Company volunteered for the Rangers. Thus D Day for the first air assault in Ranger history came to a nearly picture perfect close (4th Company's part in it will be described in their history).

Some hard fighting went on in other Tomahawk sectors but for 2nd Company the next five days were largely pursuit and mopping up disorganized enemy resistance. The Buffalos fell in with the tankers in a drive that continued for twenty miles deeper into the ruptured communist front.

Tactically an overwhelming success, strategically Operation Tomahawk turned out to be a disappointment. This was because the CCF— unbeknownst to Eighth Army G-2— was in the process of taking a long step backwards, leaving only delay forces in the area where the paratroopers dropped. The CCF had already displaced the logistical facilities that Tomahawk had hoped to hit. In a sense it had struck at air. A week earlier the drop would have hit pay dirt. Though the paratroopers were ready then, through no fault of their own Operation Tomahawk's D Day came seven days too late.

In April, 2nd Company was back with 7th ID training black replacements who reported in with a variety of MOS's. Their number would reach 400 EM before they were returned to their TO&E units in early May. Among the less brutal practices the CCF inflicted on POWs was constant propaganda to exaggerate racial differences, especially to turn blacks against whites. But ironically it was the war the communists had started that did much to unite the races on the battlefield, where the only colors that mattered were red, white and blue. The first white Buffalo was Joe Russo, a medic SD to 2nd Company.

By May Allen had 123 men for duty at the start of the CCF's spring (and last) major offensive. Attached to the 31st RCT, 2nd Company protected their east flank where it tied in with a Marine regiment. From a fortified outpost on Hill 258, the Rangers operated a reconnaissance and surveillance line between the two U.S. regiments. Chinese were so thick in the hills that most Ranger patrols had to be platoon size just to handle the many contacts along the outpost line which was only about a mile north of 2/31st Infantry's CP. Allen warned them that he could hardly stop all CCF infiltration along a shallow river, and before long a lot of Chinamen could be expected in the rear, though he would try to block them around Hill 258.

Everywhere the front was similarly porous. Eighth Army was just spread too thin to seal itself tight. Fighting in all directions was to be expected, fall back positions were being prepared, and furthermore Van Fleet was much more concerned about the reliability of the ROKA on the east flank than the ability of US divisions to hold ground. For its part, 7th ID had no one as good as 2nd Company to operate an early warning line. Allen was provided with plenty of ammo to stockpile on Hill 258, but he could forget about reinforcements if the CCF hit all along the division front, an event he felt sure was coming after all the activity reported on his outpost line.

During the first two weeks of May the Rangers strengthened Hill 258 by mining daylight approaches while leaving narrow night routes open for small patrols to 'hand check' the Chinese—to use the basketball term—so they could not build up into a siege force. Allen also scrounged extra medical supplies, preregistered artillery fire and made sure he had alternate commo to the redlegs, vitally important in the mountains where radios were often masked.

Then came a lull, and the threat in 7th ID's sector seemed to subside. R&R's were to get back on schedule, whch meant that Allen, LT's Pryor and Freeman plus eleven EM departed for Japan, leaving LT Queen in command of a foxhole strength of one other lieutenant—Albert Cliette— and some 80 EM.

Unfortunately the Chinese weren't taking any R&R's! Infiltrators began a rising tide that in days washed through the hills on all sides. Where they hit small breakwaters like Hill 258 the communists diverted to softer points and sought out seams between units. Experienced units responded with indirect fire so as not to reveal positions, but with all the infiltrators and recon elements the Chinese had in the hills, Queen realized that if he did not react soon 2nd Company would be pinned down before they could do anything to disrupt the enemy's growing momentum. On the nights of 15 and 16 May his night patrols reported that CCF forces were approaching Hill 581, an elevation though several miles away would dominate the Rangers' outposts.

Boldly he decided to strike for Hill 581 before the CCF could assume a dominating position. This meant almost abandoning Hill 258 because neary every Ranger would be needed for the spoiling attack. It also meant the probability of a meeting engagement on the peak of Hill 581 because the Chinese and Rangers were coming at it from opposite sides. Obviously time was of the essence.

Even to reach Hill 581 meant a nightlong approach march through Indian country and several skirmishes enroute. The objective was at the point of the CCF's deepest penetration so far, a factor that led Queen to hope that the hill was manned by tired, poorly supplied troops (similar to the situation 2nd Company used to advantage on 9 January at the Han River pass).

No matter how many Chinese might be on the hill, he felt most of them would be concentrated on the slope facing generally southwest toward 7th ID's line. Consequently, the Rangers would come at the hill from the same direction as the Chinese were taking; i.e., from the reverse slope. Possibly the enemy might mistake 2nd Company to be CCF reinforcements. This was the sort of trick often played by the communists on the Americans.

To succeed with their surprise, the Rangers needed much stealth to get in position behind the reverse slope of Hill 581. To keep the enemy's attention in the other direction, Queen called H&I artillery fire on the forward slope. All this positioning and preparation took up most of 17 May. Then the Rangers attacked, driving uphill until taken under fire that increased till they were stopped around 1500.

Surprise had gotten them more than half way to their objective by not firing until fired upon. The Chinese on the reverse slope had been unprepared for an attack from the rear, but soon those on the forward slope turned their weapons around to meet the unexpected threat. Queen kept them fixed with heavy and close fire support from preplanned artillery concentrations. The Rangers came around at them from both sides, killing 50 and wounding 90. By 1700, HIll 581 was in Ranger hands.

Queen's decision to continue the advance in the afternoon paid off. With more than an hour of daylight to prepare for counterattacks, the Rangers feverishly dug in, coordinated interlocking fires, and redistributed ammo. On top of the hill commo, for once, was good back to division; but because of the sharp drop off of the peak, registration rounds for defensive fires constantly dropped short.

Grenades were laid in rows for rapid throwing, then the Rangers settled in for a few hours rest. The veterans made bets on when the Chinese would be back. Most thought the first probes would start around 2100. A machine gun section from 7th ID sent up to strengthen Queen's position soon fell wounded to CCF mortars and snipers.

Most bettors were wrong: contrary to CCF SOP, the Chinese struck first— full force— at 2300. Apparently they already knew the hill well enough to skip a reconnaissance. Nonetheless the Rangers were prepared for the sudden onslaught. As much as possible, they let artillery break up the attack before engaging with small arms that would reveal positions. When they were revealed, it was at such close range few Chinese got back down the hill to provide information to the CCF commanders.

LT Queen handled artillery adjustments. He knew his job would be vital—the great equalizer against overwhelming odds. Closer and closer he walked the 105mm concentrations, following the progress of Chinese waves up the hill. Huge gaps were blown in their assault lines, but on they came until Queen had to call fire on the very brink of Ranger fox holes. 2nd Company positions were raked with friendly shrapnel, but the few U.S. casualties were an acceptable trade off for the decimation of Chinese attackers.

Frugal with ammunition, the defenders rose to fire on the backwash of human waves before the next ones formed and came on. Marksmanship was more accurate in that situation than when Rangers could barely peek over the lip of a foxhole as an attack was in progress.

By 0200, U.S. grenades were all expended, but there were plenty of 'stick' grenades to be picked up from enemy bodies. By 0300, after four all-out attempts by the Chinese, BAR men were demanding M-1 clips to refill magazines. Riflemen then began looking around for AK-47s among the Chinese dead that littered the slopes in front of the foxholes. Artillery still whistled overhead, following enemy survivors downhill. Fortunately, the CCF had not had time nor transportation to bring big mortars up to support their attack. They thought an infantry battalion was enough for the job. The first one was now combat ineffective but a second battalion renewed the attack and managed to drive the ammo poor Rangers back from the peak. Around 0500 Queen counterattacked and controlled the peak.

By 0645, the CCF had been shattered, with only enough men walking to remove half their casualties that numbered in the hundreds. Though not strong enough to police the entire hill, Rangers could see sixty mangled bodies in and around the foxholes of their original foxholes. Thanks to their tight organization and interlocking fire, no Rangers had been killed and only ten seriously wounded. Their evacuation, however, proved to be a problem because Korean litter bearers were not available. Consequently Queen lost more of his meager foxhole strength in moving his wounded off the hill.

Delighted by the outcome on Hill 581, 7th ID sent a company out to push the Chinese further back. To do so, the CO led his company into the Rangers' perimeter, finding them cleaning their weapons matter of factly, and oblivious to the carnage they had caused on the hill.

The CO looked down the hill at the scene, then turned to his men and said, "I want all of you to look around. This is what happens to the enemy when soldiers like these Rangers don't panic." With a high ball to 2nd Company, he added, "Gentlemen, I salute you."

For the CCF, the defeat on Hill 581 was part of the undertow that reversed the tide of their last offensive which eventually became known as the 'May Massacre'. Much hard fighting in the last two weeks of that month produced the massacre from huge fire fights that blazed one after the other all the way along the front, especially in the east. In such a running battle, 2nd Company had little time to remain kings of the hill on 581 before they were ordered to push east to take pressure off the US Marines.

In this role the Rangers were sort of a spearhead and combat patrol, using somewhat the CCF tactic of seeking out soft spots in enemy resistance, bypassing and pressing on. Follow up forces were often diverted to higher priorities, however, so Allen had to weigh the risks of 'getting in too deep'. But aggressively he fought his company deeper into the hills, forcing the CCF's attention toward him and away from the embattled 7th Marines.

In three days of intermittent fighting, 2nd Company had a constant but changing objective: whatever happened by daylight, high ground had to be secured for a night defensive position (an NDP, to use the Vietnam term). On 24 May they were approaching Hill 545 from where the Marines had been repelled in an attack and were taking considerable fire. Allen made the hill 2nd Company's objective of the day, and they knew they had to have it by nightfall for an NDP.

Infiltrating to the slope away from the Marines, Allen did pretty much a repeat of 2nd Company's successful attack on Hill 581, storming up from the enemy's rear rear, killing 15 Chinese on the way to the objective. At the summit of Hill 545 the CCF (an estimated company) turned furiously on the Ranger attack, but short of ammo they could not stop 2nd Company's momentum which carried the hill with the loss of only two Rangers seriously wounded.

The successful attack had been unexpectedly swift. The Marines were asked to lift their artillery fire for which they had piled up stacks of 105 rounds for the many barrages they expected would be needed to support 2nd Company. As the Rangers mopped up the forward slopes of Hill 545, waves were exchanged with the Marines on the next terrain feature who now had a secure flank.

By June 1951 it was clear to both sides that U.N. forces had the upper hand. Why else would the Soviet Union's delegate have appealed to the Security Council for a cease fire? Which he did on 24 June. At the Ranger level, the sea change meant more offensive operations for which they had originally trained.

2nd Company sought such missions, often requiring shifting attachment along the front as opportunities came up. Back and forth between battalions of the 7th ID they went, a travelogue for the Korean mountains just north of the 38th Parallel. Finally, on 8 June Allen, nailed down a solid mission: scrape off whatever enemy was to be found on Hill 772, a hill that deserved the title of mountain.

Night time is prime time for Rangers, but in the long hours of daylight during late spring in Korea, U.S. artillery and tac air ruled so strongly that Allen preferred to take the high ground while he could best control the plentiful fire support. In broad morning light, 2nd Company followed a creeping barrage uphill while P-51's worked on the summit with the weapons the Chinese most dreaded, napalm. The fight was gone from the CCF. By dusk, 2nd Company had Hill 772 with a body count of eight enemy KIA and 30 wounded. Seven Rangers were lightly wounded.

Backed by an RCT, the Buffalos continued the attack on 11 June, seizing critical high ground near Sanying-ni while losing one KIA and eight WIA. This action forced another CCF withdrawal.

Action tapered off by early July—good fortune for 2nd Company because Van Fleet was ordered to let up and the CCF had time to catch their breath and dig in. Indeed they honeycombed the hills in front of 7th ID so that any objective was strongly fortified and deeply entrenched. Under these circumstances, raids were too costly for the benefits they might produce.

Then, on 10 July, the DA message came down ordering the deactivation of all Ranger companies in Korea. The ridiculous reason was the same as Gen. Ruffner had proposed to lst Company—that racial differences prevented deep penetrations. That must have been a laugh for 2nd Company.

Third Airborne Ranger Co.

The pipeline from Ft. Benning was turning out Ranger companies for Korea on schedule. With practical feedback from Eighth Army Ranger Company and lst and 2nd Companies, who were already operating in theater, three more companies were ready to set sail by March 1951. Because of lessons learned in Korea, and returned to Harmony Church, the second cycle was probably better prepared. For example, they had the advantage of such combat-tested refinements like tracer designation of targets during night attacks and introduction to standard communist offensive tactics.

Eager to get into the fight, 3rd, 5th and 8th Companies found their first ones against disrespectful Marines on the troop ship. Interservice relations were better with Air Force cherries on board who had been issued fully automatic M-2 carbines. 3rd Company's carbines were semiautomatic M-l's. Sales, swaps and thefts provided many Rangers with the ideal light-weight, high-volume weapon they sought for close-range fire. It may have been this acquisition that first led to the innovation of taping two 'bananas' together (though perhaps WW II Ragners had previously discovered the method) so that bullet feed was hardly interrupted while changing magazines.

3rd Company's transport made port at Yokohama, then stopped at Kobe where the Rangers had a few hours for their first shore leave. Short of money, they stuffed their apparel with anything on board they thought they might barter with the Japanese—especially the women— for quid pro quo. Japan was still recovering from World War II's austerity, so even blankets and linen from the troop ship were valuable commodities.

Trades went well, but time was short. When all Rangers were accounted for and the gangplank went up, there were still several hours before the ship sailed, so members of 3rd Company practiced rope techniques by descending the mooring hawsers to the dock and the enticements of Kobe. The return climb was much more difficult, especially to get around the metal rat guard, and for several Rangers it resulted in a drunken slide for life that ended in the water.

On 24 March the sight of Pusan was enough to sober anyone, one Ranger describing the port as a "mud pile surrounded by anti-aircraft guns." By LST, the three Rangers companies sailed around the peninsula to Inchon. There on 31 March they separated for attachment to their respective infantry divisions.

3rd Company, appropriately, was attached to 3rd ID which at the time was fully involved in rolling back the CCF on the Seoul-Uijongbu axis along the Imjin River through a series of limited objective attacks. On 11 April, 3rd Company, CPT Jesse Tidwell commanding, got into the action as part of a tank-infantry task force whose mission was to recon in force by penetrating up a wide valley. The tank and Ranger companies had never worked with each other and it showed.

3rd Company had some features new to Rangers in Korea: their XO, Bob Channon, was a captain, and they had a 57mm recoilless rifle section. For this mission they would also be their own artillery FO's, and ride to work on the hulls of heavy tanks.

Daylight had barely broken when the task force punched through to the first village in the valley where they received light automatic fire. With the 'reckless rifles' and 3rd Platoon remaining with the tanks, lst and 2nd Platoons, under Channon, dismounted for their first engagement and to clear the uninhabited village. With direct fire support from the tanks, this was no problem—two Chinese were killed with no U.S. casualties—but Channon could not raise the artillery on the frequency he thought they were on.

Moving on rapidly with two tanks overwatching, Channon's platoons reconned high ground that gave them a good view of the valley ahead. Though apparently unoccupied, a trench line of bunkers at the base of a long ridge protruding from the east into the valley could block the task force unless the ridge was given wide berth to the west. Channon radioed this recommendation to Tidwell who, with five tanks, had already agreed with his armor counterpart, CPT Boone, to swing the TF (-) in that direction. This movement put the two tank-infantry teams in defilade from each other.

As they often do, things began to go wrong at the same time. Channon's force moved out to skirt the ridge across broad rice paddies while Tidwell and Boone were moving farther west, thinking they were following Channon who soon received accurate morter fire from the vicinity of the ridge, fire that wounded him and and his RTO. Soon his platoons were also exchanging direct fire with concealed Chinese on the ridge.

Channon reported these developments to Tidwell who radioed back instructions to move out under cover of the tanks. Over the armor net, Boone ordered his tanks to maneuver west. The two tanks with Channon thereupon headed off at high speed to rejoin their company without advising him, thus leaving his two Ranger platoons alone and under fire with their wounded.

Before long Tidwell and Boone's force were also taken under pretty heavy fire, and the order was repeated for Channon to rejoin the TF. A Korean was now his RTO and did not relay the message, probably because he did not understand it. Without effective commo with anyone, Channon realized he either had to withdraw back to the hill under increasingly accurate mortar barrages or try to rush the ridge. The Ranger rule was when in doubt, attack.

It was fire and movement across frozen paddy land—movement being a belly crawl to stay within the slight protection of the low dikes. Spreading his Rangers to the maximum to avoid mortars, Channon slowly progressed through the paddies that terraced up toward the ridge. Where the terraces stepped up, Rangers could sometimes find a good line for providing fire support for further movement.

At about 600 meters from the base of the ridge, two Chinese machine guns opened up from previously silent bunkers. Rangers replied with automatic fire from their bootleg M-2 carbines which effectively neutralized the machine guns. Then a third opened up from atop the ridge. Channon designated it to be the center of the Ranger objective, and his force swept forward.

So far they had only three wounded, but a the base of the ridge the Rangers were hit with preregistered mortar concentrations they could only escape by fast movement through the killing zone. Ten Rangers were cut by fragments, including Channon who was wounded a second and third time. With jagged metal in his face and leg, he then lost his left calf to automatic fire. His second RTO was also wounded, thereby losing commo with the remainder of the task force which, in its own fire fight, did not know of Channon's predicament. Though greivously wounded, he would not relinquish command of his force and directed the two platoons on up the ridge till enemy mortar fire was masked.

Direct fire still riddled them but the Rangers were now able to give as well as they took. Finally able to reach Tidwell by radio, Channon asked for his two tanks back. They rumbled across the paddies and took out two machine gun nests at the base of the ridge. Encouraged, Channon's Rangers were able to seize their objective, the third nest on the ridge crest.

Guided by Channon and lst Platoon Leader, Pete Hamilton (also multiply wounded), the tanks ground down the CCF's lower defenses while the Rangers on the ridge turned to the nearby trenches and bunkers that still held many Chinese. SGT Hal Barber took charge of this sweep, with bayonets rooting out defenders who fought back with grenades. In the swirling melee, concussions knocked Rangers Shafer an Cisneros down the slope. They stopped their descent by rolling into each other. Struggling to their feet, they dusted each other off and climbed back into the fight where SGT Mas

Najako had caught another grenade and threw it back to the owner.

Confronted by the bayonets of relentless Rangers, the Chinese began to abandon their strong position. CPL Bill Osbourne stabbed two of them as they fled the trenches. Their screams brought out cringing comrades who were gunned down as they tried to escape. Barber now had the original objective, and turned his force west to clear the nose of the ridge which was the major natural obstacle to the progress of the entire task force. Cornered by the two tanks below and Rangers closing in from the ridge, the CCF company pulled out.

This amazing action that became known as 'Bloody Nose Ridge' showed how a confused and critical situation can sometimes be rectified by sheer guts and aggressiveness (at no time did Channon have artillery or tac air support). The badly started battle ended in an inspiring but very costly victory: four Rangers died, thirty-two were seriously wounded. But 3rd Company had made a blazing mark for itself in its very first engagement, a mark high on the wall that Ranger replacements knew they would have to reach because cherries just like themselves had set it in the most difficult circumstances.

And action was not over for the task force. During the battle for the ridge the TF (-) had advanced north up the valley. With the ridge behind them now cleared by Channon, they felt the TF could attack to the head of the valley before circling back to friendly lines.

Channon was ordered to leave the ridge and link up for this attack. Burdened with wounded, his battered force struggled up the valley, sometimes under fire from mortars and snipers. When the TF reunited, they cleared the head of the valley which was being used as a hub for enemy supplies and communicatons. While tearing up caches, the TF took a prisoner, adding to a body count of over a hundred for the day's work. In 3rd ID's journal, the G-3 gave the operation a glowing "mission accomplished."

3rd Company was down below half strength after their first action, and the replacement stream was barely a trickle. Yet less than two weeks later the first impulse of the CCF final offensive began to take form. Gen. Van Fleet saw it coming and had heartburn about Eighth Army's MSR where it crossed the Hantan River. He made its security a 3rd ID responsibility; consequently, 3rd Company was given the mission to protect the bridge at all costs.

After a few days of security patrolling, the Rangers were pulled off the bridge to join 3rd ID's Recon Company, as well as engineers and FO's, forming TF Rogers to range north and feel out any enemy buildups. To the surprise of the CG, TF Rogers did not encounter much determined resistance. This was because the first impulse of the CCF would concentrate farther east.

However, they did hit the 3rd ID sector with a supporting attack on 19 April, striking in late afternoon and continuing till early morning so as to avoid retribution from tac air. Their first wave disabled and isolated five 3rd ID tanks about eight miles within enemy territory. Before the tanks could be destroyed, 3rd Company was sent out with the recon company, engineers, a retriever and a quad-50 half-track to bring in the tanks that were strewn beside a dirt road in steeply mountainous terrain.

Cautious for ambush, the U.S. column came instead to deep ditches recently dug as tank traps. Scouting showed they were covered by fire from the hills. 3rd Company crossed the ditches to take out the well sited machine guns, a very difficult assignment till the quad-50 moved up to provide covering fire. This was effective though some Rangers, climbing faster than expected, almost moved into the quad-50's devastating kill zone.

With CCF covering fire suppressed, a tankdozer piled dirt into the ditches, enabling the column to proceed. Advancing another half-mile, it came under mortar attack. The CCF knew the Americans would come back for the tanks, and did all they could to prevent retrieval. Crossing about a mile of rugged terrain, the Rangers forced the mortars to displace. As darkness descended, target identificaion became troublesome for the quad-50, and once more Rangers were almost hit by this most destructive form of friendly fire.

Gradually the enemy grip on the disabled tanks was lost as the Rangers and Recon Company worked their way into the surrounding hills. CCF infiltration was adept at taking terrain but not much good at holding it unless they had time to dig in. The fast work of the tank retrieval task force did not give them that time. After dark, the tanks were secured, inspected for booby traps, then slowly hauled back, the Rangers pulling flank guard on the dangerous route that was the same as the one taken to the objective. This hairy mission was completed with the loss of only two Ranger WIA.

More than any other Ranger company, 3rd seemed to draw the most combined arms missions. This was perhaps because by the time they entered the war, more tanks were available in the infantry divisions. In any case, on 22 April, 3rd Company was once again 'riding to work' with 3rd ID's Recon Company to reduce a deep objective for another probe of the CCF front. This sort of pressure delayed the CCF from building up in front of the ROKA.

The objective this time were two Chinese companies dug-in on a hillside situated in a way that it could be taken from only one direction. Tidwell did not like the inability to maneuver, but with Recon Company's tanks and two Ranger platoons providing a base of fire, 3rd Platoon used the limited frontage as best they could for an early morning assault that initially achieved surprise.

But soon the Rangers moving uphill on line were in a blazing fire fight, in which three dropped wounded. SGT Homer Simpson, a WW II veteran of the Pacific and one of the most popular men in the company, was leading an assault time against a machine gun nest that threatened the entire attack. Preceded by a shower of grenades from his men, he charged it with M-1 and a phosphorous grenade in hand. The white smoke had its effect, ejecting enemy from their holes. But more jumped out than Simpson could bring down with his rifle. One emptied a drum of bullets in his direction, killing him with a slug in the eye.

Simpson's sudden fall set off a wave of shock and rage among the Rangers who then overran the CCF position like banshees, grenading, shooting and bayoneting Chinese as

Fikes' bunker, August 1952. Note skull.

if many enemy deaths could bring back Simpson from the dead. The frenzy of revenge was observed by a light aircraft from division that hastily dropped a note to 3rd Platoon to bring back a prisoner. Fortunately one was saved, and he provided valuable information about CCF plans.

Before long those plans began to unfold. The CCF would try to drive a wedge between 3rd ID and the neighboring ROKA division. At that boundary the British 29th Brigade was OPCON to the Americans. The heavy blow fell on the ROK's to their left who folded, leaving the British cut off while 3rd ID refused its flanks and fell back to block penetrations. Fire of every caliber roared all across this front for the next two days, the battle heaving to and fro, till 3rd ID regained control of the situation sufficiently to release 3rd Company for the purpose of locating and, if possible, aiding the isolated 29th Brigade, its most endangered battalion being the lst Gloucestershires ('The Gloucesters'), one of the most famous units in all British history.

During the first winter of the war, British units had been left in the lurch when the CCF poured through Eighth Army. If this were to happen again in the spring of l951, a great deal of good will would be lost between the two U.N. allies. 3rd Company's mission was to assist in ensuring that it didn't happen by making contact with the Gloucesters.

From the actions earlier in April, Tidwell was down to about 67% strength, but from their security mission earlier on the Han Tan bridge, the Rangers knew something about the terrain on this tributary of the Imjin River where the Gloucesters were believed to be holding out. Marching day and night, paralleling and sometimes wading in the tributary, 3rd Company swung far to the north so as to come at the CCF from an unexpected direction.

What turned out to be unexepected was the strength of the CCF. Five divisions had been launched to anihilate the British, and they nearly succeeded. Tens of thousands of Chinese were deployed between the two Ranger platoons and the Gloucestershire Battalion who were fighting for their lives after shrinking their perimeter from l7,000 meters to less than a mile. Tidwell could see the their tac air support miles away, swarming like bees over a hive.*[*FN: Barely fifty men out of the original 500 Gloucesters survived by breaking out to the *north,* then west and finally south to U.S. lines. The three-day pitched battle by the encircled British ranks with Chipyong-ni among the most stirring defense in any modern war. Though tac air support was greater for the Brits, so too were the odds against them.]

With daylight the Rangers could see that the beautiful Korean spring had been fertilized by death: skulls and fragmented skeletons from the winter's fighting had been uncovered by melted snow and lay around like mounds of chalk among the blooming mountain flowers. A season had passed but the barbaric war continued. This fact soon became evident as 3rd Company began to be peppered by fire from what seemed to be every hill in the vicinity. Engaging one enemy position only brought more attention from others.

LT Eaton, commanding the advance guard of the far too small expedition, quickly realized 3rd Company was in too deep, outgunned and enormously outmanned. He began a fire and maneuver withdrawal to break contact as Tidwell looked for a different route to extract his force; but there were too few ravines through the steep mountains and he had to pull out pretty much the way the Rangers had come in, while keeping a reaction force ready to hold the rear open.

During the morning, five Rangers were wounded while fighting the delay. By noon the sun was bright but Tidwell already began thinking about an NDP because there were still miles of torturous terrain between the Rangers and friendly lines. A hill among the mountains looked to be the best bet. Though it would be dominated by surrounding heights, at least the Chinese would have to climb to attack it at night. Eaton's platoon, which had converted from advance guard to rear guard, now led the company again and went ahead to prepare this hill.

The ground was slate; when mortars struck it added to the fragments. During the afternoon, one Ranger was killed and six more wounded. Fortunately they were not pressed hard by the CCF that night who seemed to have orders to bypass any resistance in order to penetrate as deeply as possible.

All night the horizon bulged with fire balls as the Air Force flew around the clock to keep Chinese hordes off the Gloucesters. When the morning fog cleared the Rangers could see a 3rd ID tank-infantry team in a valley 2000 meters away trying to break through in another attempt to relieve the British.

The team quickly came to grief. Exposed to heavy fire from the mountains, the tanks were knocked out one by one. There was not enough U.S. infantry to protect them from squads of Chinese sent out to permanently destroy the tanks with satchel charges, which they did.

Flush with that success, several companies of Chinese then went after the Rangers, but they had improved their position on the NDP, established commo with Div Arty, and thus were able to rake the CCF attack with barrages across a mile of fairly open ground. Only three-quarters as many Chinese pressed the attack to the base of the hill.

By chance and by good appreciation of terrain, 3rd Company had placed themselves like a boulder in a flash flooded stream; that is, they would surely be rolled away but till that happened they were constricting a bottle neck. Calling concentrations on three sides of the Ranger perimeter, Eaton could see the Chinese recoiling but coming on again, time after time, each wave picking up the remnants of the last.

The friendly artillery slackened because 3rd ID was being hit again with multiple attacks. So far as Tidwell knew his mission was still to weather the storm, then continue to march, somehow find a way to the trapped Gloucesters. Unknown to him at the time, the British had begun their own attack to the rear and would eventually link up with a 3rd ID relief force that would not turn back. 3rd Company did not want to either, but G-3's orders were now to pull into the tightening 3rd ID lines. Protected by time-on-target salvos called by LT Eaton, the Rangers evacuated their position just as the Chinese came up the other side of the hill.

The stand of the 29th Brigade had done much to break the CCF momentum on the western front. With the time the British bought, 3rd ID could brace for the Chinese breakers, which when they came were not as strong as expected. The CCF strength had shifted elsewhere, to the central front where it hinged to break the ROKA farther east.

What this relative slack meant for 3rd ID and 3rd Company was the opportunity to hit instead of being hit. By indirect fire and tac air. In the defense, the CCF submerged into the ground—no matter how rocky or steep— but they had to surface to attack. Though darkness was their preferred hour for attack, they had to move around to prepare by day, and that's when they became vulnerable. 3rd Company spent the first week of May on 'target acquisition' patrols, infiltrating to high ground where they could see signs of buildup and activity, reporting it, then adjusting the artillery and air strikes.

The effectiveness of such target acquisition was demonstrated by the disappearance of CCF artillery supporting their attacks against 3rd ID. Because their fire support methods were so primitive, CCF guns had to be laid in daylight. On small recon patrols, 3rd Company Rangers observed several attempts by CCF redlegs to lay batteries secretely. The secret was soon passed on to the Air Force who hit the sites, reporting secondary explosions.

Such patrols were good employment of Rangers, exploiting their nearly unique ability to move unobserved across terrain where the enemy did not look for them, terrain too difficult for most other friendly infantry units to negotiate. 3rd Company's contribution to the May Massacre was therefore optimal, inflicting (by adjusting fire) many enemy casualties while suffering very few of their own.

By 11 May, the CCF spring offensive had been subdued to the point where Eighth Army called for 3rd Company to be released for parachute proficiency training at Kimpo Airport. Understandably, 3rd ID G-3 took a dim view of losing their best 'target acquisitors' for what evidence and logic indicated was an irrelevance—training for airborne missions in a war where victory was secondary to containment, implying that no audacious efforts (e.g., jumping behind enemy lines) were even being contemplated by the Pentagon.

This was certainly the case. In the American Army—because it is the army of a democracy— national policy is quickly realized, then restated in the foxholes. By June of 1951, the restatement echoed a justifiably cynical realization: the Pentagon was asking US forces to 'die for a tie'.*[*FN: Similar to 2nd Company's war cry of "Buffalo!", 3rd Company adopted one from Bloody Nose Ridge—"Die bastard, die!"]

3rd Company spent early June in division reserve where they received a miscellany of missions like convoy and CP security, surveillance and visiting patrols. Replacements began to refill Tidwell's ranks, and some Rangers wounded in earlier actions returned, lifting the spirits of both the company and returnees. There is probably no better mark for the quality of a unit than the efforts of its WIA's to get back to their old outfit, often having to overcome REMF policies and red tape to do so. The company is just as gratified as the returnees, maybe more so because casualties usually happen so fast that a man is gone before anyone can wish him well. Then, weeks or months later, the familiar face reappears—a face that may have been bloody when his buddies last saw it—and everyone realizes there is life after medevac.

The opposite sort of feeling settles on a unit when a man is listed as missing, a feeling of guilt and uneasiness that keeps men subconsciously looking for the MIA whenever they are out on operations. In June such an event befell 3rd Company.

With the U.N. front largely stabilized, clearing of the no mans land between friendly and CCF lines became a routine practice. So one day a 3rd Company went out to evacuate some civilians from a small shot up village that was in the line of defensive fire. The civilians were in bad shape, diseased and malnurished, some of the children vomiting foot-long tapeworms. As the Rangers escorted them to ambulances parked back in defilade, CPL Dave Rauls noticed two Koreans fade into background, as if to remain in the village. He went to get them. Alone among the shattered huts, three Chinese soldiers appeared out of hiding and got the drop on him.

Rauls was quickly whisked away before his patrol noticed his absence. The Chinese seemed to be crowing that they had accomplished a snatch—and they had. Later they were joined by an English-speaking officer who tried to interrogate Rauls who was desperately depressed by his fate, and said nothing.

Carefully moving alongside a river bed, the group traveleld southwest to rendevous with a vechicle where another U.S. captive was being held. Slowly and with many stops the vehicle moved along, constantly alert for U.S. aircraft.

Upon reaching a dirt road, two more vehicles appeared, forming a convoy with about a dozen POW's (their hands tied but not blindfolded) and fifteen guards. Rauls felt even more ashamed than frightened.

For some reason the Chinese seemed to feel that ridgelines offered more concealment than valleys. They may have been right because roads were in the valleys and drew the most aerial attention. Staying just below the skyline, the three vehicles pitched along, hugging the hills as they wormed deeper into enemy terrain. Rauls felt his chances of escape ebbing with every mile.

Just before the vehicles reached safety in a long, deep ravine, two fighters streaked down and blew them up, pulling off when they saw all three vehicles explode and burn. Captives and captors were doused with 50 caliber fire as they tried to flee. The strafing was over in less than a minute, then the Chinese apparently forgot their mission and began to take vengeance on the POW's. All were clubbed, bayoneted or shot. Knocked unconscious by a burp gun barrel, Rauls was left among the dead. Sometime during the next eight hours he was found by a friendly patrol, sent to follow up the air strike. He woke up in a field hospital, with nightmarish memories and an indelible lesson in the importance of personal security while on patrol. He had been MIA for less than a day.

On 17 June 3rd Company, along with the Recon Company, an artillery battery and a tac air forward control party, became Task Force Ferret manning an R&S line some six miles north of 3rd ID's main positions. The mission once again was target acquistion— and early warning if necessary.

By now, however, the mole qualities of the Chinese had become an advantage in this static stage of the war. Large CCF formations would not surface because they were no longer on the offensive. So pickings for TF Ferret were lean.

Just as trench warfare neutralized United Nations supremacy in the air and artillery, it also took the premium off 3rd Company's ability to penetrate. Field fortifications had sprouted like weeds. There was no longer any terrain that was merely difficult to approach the enemy: now all approaches were covered by fixed positions. Surprise (except for time and size) was pretty much impossible. All patrols could expect to cross enemy minefields as well as their own.

In this stalemate situation, recon patrols were sure to have contact, while combat patrols required the kind of heavy support that earlier in the war would have been provided to a regimental assault. Raids were sure to be costly, and limited objective attacks (e.g., Heartbreak Ridge, Porkchop Hill) were planned at field army level. Eighth Army's primary mission was now to hold what they had while holding down friendly casualties. This was a no win attitude antithetical to the aggressive, offensive-minded spirit of Rangers.

But as someone said in Vietnam, "This is the only war we've got", and 3rd Company continued to fight it no matter how chafing and dispiriting were the rules.

As the war faded into positioning and politics, 3rd Company approached its last act before the Pentagon dispersed all the Rangers in Korea. It turned out to be a meaningful and exciting finale for 3rd Company, one that left them with a better feeling than its sister companies would experience.

In July, 3rd ID was holding the base of the strategic Iron Triangle, a plateau surrounded by commanding hills that was the most important terrain in the war at that time. Neither side could control the triangle though there were many thrusts and parries to do so. To strengthen forces at two corners, 3rd ID thinned out its line at the southern base of the triangle by evacuating Hills 682 and 717. To deceive the Chinese about this repositioning, 3rd Company was to man the two abandoned hills and give the impression that they were still defended by full strength battalions.

Employing plenty of H&I fire, broadcasting some dummy transmissions, and frequently moving unseen between positions then showing themseves to Chinese observation, the Rangers rather enjoyed their deceptive show of force. But to make it thoroughly convincing, they had to show that they had the initiative.

Carefully surveiling their front by daylight, 3rd Company used a platoon to check out night routes to a few potential ambush sites on the enemy side of no mans land. This platoon would be guides for combat patrols. After eight days of thorough familiarization with their AO, Tidwell was ready to add some fear to the respect he had been instilling in the Chinese across the valley. His purpose was counter-reconnaissance. If he could cause CCF recon patrols to worry about their own security, others would be more cautious about investigating American defenses on Hills 682 and 717.

On three successive nights, Rangers were guided out to five of the most promising ambush sites. Early the first night, one site had contact with an outbound Chinese patrol with unknown results but turning it back with casualties. All sites were evacuated before dawn, with indications of occupation only at the site that had the action.

The next night, another site saw action, again apparently with an outbound patrol, this one larger than the one the night before and later in the evening. Contact was broken by preplanned concentrations of artillery which the Rangers enjoyed in abundance.

The third night two ambushes were laid and one paid off. From the previous actions, Tidwell had a sense of the Chinese patrol pattern and he wanted to hit one inbound to create the maximum demoralization. That's when a patrol is just about home, relieved that they have made it, and losing their edge of alertness. Just before BMNT on 25 July, two CCF recon patrols (totaling about a platoon) started to merge in a hollow right outside their minefield. As they started re-entering their lines in file, a squad of Rangers raked them with fire. The firefight was fast, furious and one-sided. When it ended, there was enough light for the ambushers to see five enemy dead. Picking up twenty enemy grenades, the 3rd Company Rangers pulled off their last action with one WIA.

As with all the Ranger companies in Korea, August 1951 was 'end x'—end of the exercise. It was hard to remember that the whole thing had been an exercise directed by the Pentagon—a 'test' of Ranger employment at division level was the way it had all begun. Perhaps it was better to end as it did, with the maneuver phase of the war over. 'Die for a tie' is hardly the motto for any fighting man, and alien to every Ranger instinct and tradition. 3rd Company had added new honors to that tradition, and it may have been just as well that they had to quit when they did with pride and prestige.

Fourth Airborne Ranger Co.

As previously noted, 4th Company was 2nd Company's twin, having trained together in an RTC cycle, taken the same train across country and the same ship to the Far East. They even shared the same battle cry—Buffalo—but for 4th Company it derived from getting to know cattle as their troop train rolled endlessly across Texas.

Buffalo was also apprорpiate for the division to which 4th Company was attached, the lst Cavalry whose lineage began on the high plains.*[*FN: The lst Cavalry has the unique distinction of having fought in four major wars, as a different type division in three, but not as cavalry in any of them. In WW II and Korea they were infantry, in Vietnam airmobile, and in the Gulf armor.] When they joined the Cav in January 1951, the division was in reserve near

Lt. Ralph Puckett, Lt. Charles Bunn, Lt. Barney Cummings on Ranger Hill near KiJang, South Korea.

Seoul. 4th Company (CPT Dorsey Anderson commanding) got familiar with Korean terrain and weather by securing the divsion CP.

When the CCF resumed their drive south, the CP hurriedly displaced to the rear in vehicles but there was none for 4th Company. This was an all too typical problem for Rangers in Korea: no organic transportation, no mobility except in leather personnel carriers. This situation was bad enough for Anderson but an even more acute problem was his lack of a mess. When the Cav CP pulled out they built bonfires, burning all excess Class I supplies to prevent capture by the enemy. The Rangers had to scavenge in the embers and ashes for charred cans of chipped beef and sweet potatos. A chow line heating rack was improvised from a discarded urinal trough!

Despite such serious scrounging, Anderson's food was running low while the promised trucks from division headquarters never returned to pick up the Rangers. The only thing moving toward them was the CCF offensive, so it began to look as though 4th Company's first action in the war might be its last. In the lonely cold this impression intensified when the corps covering force, a battalion from 24th ID, started to displace through the Ranger position. Anderson hailed them to make an offer: 4th Company, 115 rugged Rangers ready to fight, was at their service—in exchange for some chow and vehicles to move to the rear after the delaying action was fought!

The battalion commander eagerly agreed to this deal, but when he radioed corps with news of his unexpected reinforcements, they advised lst Cav who came up with some trucks at the last moment. This incident got relations between division and 4th Company off on a sour note, relations which did not improve much in the next couple of months.

Indeed when asked by the Pentagon in February to evaluate 4th Company's performance, MG Bryant Moore was rather tepid in his comments, alleging that the Rangers had not been in sufficiently good physical condition to promptly begin operations.*[*FN: 4th Company's physical condition was every bit as good as 2nd Company's—both were outstanding— because they had done the same amount of PT together on the troop ship. Of all the shortcomings any Ranger unit can be accused of, poor physical conditioning is least likely to be true. Almost by definition, Rangers thrive on exercise and exertion, so a CO who allowed his company to get out of shape would have to actively stop his men from doing what they do naturally, almost compulsively. Were there such a CO—and Dorsey Anderson certainly was not one—he would soon be out of a job and probably out of the Rangers. For this reason alone, MG Moore's comments about 4th Company's physical conditioning were obviously way off base.] No doubt covering his curvature for almost abandoning the Rangers in an unintended stay behind position, Moore mentioned plans for leaving 4th Company "in various ambush positions when the division withdraws to successive positions."

Better understanding between cavalrymen and Rangers would develop as the war went on, and 4th Company would excell whenever given a meaningful mission. But in late January there was minunderstanding again when 4th Company was detached for an airborne raid on a suspected guerrilla headquarters. Refresher training and mission preparation near Taegu took several weeks and reached a point where 4th Company was on the airstrip waiting to board planes. A last minute intelligence fumble lost the location of the elusive enemy and the mission was scrubbed. Disgusted, the Rangers returned to the lst Cav who had not missed them. Soon they were pulling REMF duties like securing division trains. The men were moody and morale was drooping.

Then in March, another airborne operation was in the wind—Operation Tomahawk—a drop with 2nd Company and the 187th RCT behind CCF lines.

This jump came off on the morning of 23 March, 4th Company in the second serial of four C-46's. Air speed and ambient cold almost produced frostbite before opening shock. Heavy with extra ammo and two days assault rations, the Rangers were light on winter clothing—with nothing more than a blanket wrapped in a poncho for sleeping.

There was much to do before anyone thought about sleeping: 4th Company had drawn the toughest objective in Operation Tomahawk, Hill 205, described by Bob Black as "a hill upon a hill". To get there from the DZ, where fortunately there were no serious jump casualties, required crossing a mile of frozen rice paddies divided by several tree lines.

Before there was much serious reaction from the North Koreans, Anderson had his platoons formed up and advancing, separated into three prongs to spear Hill 205 like a trident. The first tree line was unoccupied; there Anderson set up his CP. Crossing the next belt of paddies, the center platoon received some mortar fire that they double timed through without casualties. The left platoon flushed out two NKA who were probably mortar FO's and killed them with long range rifle fire.

The next belt of paddies was crossed without incident, and Anderson displaced forward to the next tree line. One more paddy and a tree line remained before the Rangers reached the base of the hill. The platoons spread out on line before enemy mortar fire resumed, this time with more accuracy. It seemed to be adjusted from the military crest of Hill 205 (as Black implied, the hilltop was considerably behind the military crest). Tac air was busy elsewhere around the AO but Anderson got one mission to work over the military crest. The mortar fire let up as 4th Company took the last tree line and a number of abandoned enemy trenches which became the company CP and aid station for the next phase of the attack—assaulting the military crest.

Anderson had to chose beween sending all of 4th Company up the forward slope or flanking the hill with two platoons while sending the center platoon up the middle. The air strike seemed pretty effective; flanking maneuvers would take more time and expose the flankers to more mortar fire, so Anderson ordered the assault to proceed straight ahead and straight uphill. That turned out to be the right decision though the assault ran up against strong resistance—but not at first.

A company or more of NKA had slid back off the military crest, so when the Rangers came over the crest there was a deafening exchange of close range fire. This benefitted the Rangers by removing NKA mortars from the combat equation, but it benefitted the NKA because Anderson at the base of the hill could not determine the scope of the battle at the military crest and affect its outcome with tac air.

There were only four pack howitzers in the airhead; priority for their fires was elsewhere so tac air, though also heavily committed, was Anderson's primary method of influencing the battle. With his RTO he began climbing the slope to get a close hand view of what was happening and adjust an air strike.

By the time he reached the military crest 4th Company had prevailed with light casualties. This was their first fight and, like 3rd Company at Bloody Nose Ridge, every man was setting a standard. They took on the NKA like savages. The enemy got off the first bursts but the Rangers would not be pinned down or turned back. Shooting, moving, diving and throwing, they pushed into the NKA positions that had been hit by the air strike. The NKA melted away to the flanks, many to circle around the hill and join their comrades on the top.

By dark 4th Company was still looking up from the military crest at an enemy company holding commanding terrain. Between them was a bowl that was a shooting galley for the defenders. The enemy side of the bowl rose very steeply to the hill top. Anderson decided to massage his final objective during the remaining hours of daylight and try to take it that night. As fighting elsewhere around the Munsan-ni airhead diminished, he got more artillery and air support. With time to look around, 23 enemy dead were counted, two machine guns and a light mortar captured. NKA who evacuated the military crest had been seen helping many of their wounded.

Action during the day had cost 4th Company five seriously WIA. More significant was a shortage of ammo that required a carrying party to make a long round trip to the DZ. Numbing cold left the Rangers shivering as they awaited this necessary resupply. The night was moonless so the time made available recon for the night attack was not very useful beyond the line of departure.

This time Anderson would go at his objective from both flanks and the middle. Muzzle flashes would be about the only control method as the platoons moved uphill. Preparatory fire (the pack howitzers were very low on ammo) had to be confined to the hilltop. The Rangers moved out around 2300.

The action that night was a rarity where both sides attacked, and nearly simultaneously. Hill 205 was the last and most important high ground the NKA held in the vicinity. Apparently they planned it as a hub for arriving reinforcements dispatched by the CCF. So the NKA's missions were to hold the hilltop and counterattack the military crest to reoccupy it—not a very difficult objective, it would appear, because the NKA would be attacking downhill over terrain with which they were very familiar.

In three distinct engagements, the two forces ran into each other in total darkness. Perhaps the more alert, Rangers got the drop each time but confused fighting broke out all over the hillsides. Anderson's center platoon made the most progress, nearly reaching the top as the NKA were distracted by fighting on the flanks. The NKA had not tried to descend straight down into the bowl because the wall was so steep. They retained plenty of fire power on the top to support their two flanking attacks. Coming up the wall, Anderson's center platoon was stopped and had to hold on by their finger nails while enduring a shower of grenades. In an untenable position, they climbed down around a knoll where they kept a deadly fussilade on the NKA, preventing them from reinforcing their attacks.

Anderson's flank platoons were able to reorganize themselves after the initial clash. With several wounded, they consolidated into all around positions and picked off NKA who tried to get around them. Anderson had to decide whether, when daylight returned, they would be better off where they were fighting or back at the LD where there was concealment and some cover, perhaps the best place for 4th Company to be for a morning attack to seize the hilltop.

He left the decision to his platoon leders: if they wanted to pull back and reorganize they could do so, but if they did not want to fight to take ground they already controlled, they also had that option. Both platoon leaders on the flanks opted to stay in their exposed positions, even if coordination for the next day's attack would be more difficult. The center platoon returned to the LD which became a firing line to support the flank platoons.

The NKA had broken off their attack in plenty of time to take cover before daylight. A dozen or so bodies marked the boundaries of their advance the night before. It was good luck and hard fighting that had stopped them halfway down the hill before the NKA could envelope the military crest.

But now at dawn, they held all the advantages of terrain and numbers. It was going to be a long day for 4th Company unless they got a lot of big time fire support. Fortunately tac air was all theirs that day. Fighter-bombers were loaded with napalm, the air weapon the communists most feared. Fireball after fireball blossomed along the hilltop, with the Rangers so close behind that their faces flushed from the heat.

Longer range air interdiction prevented CCF reinforcements from reaching Hill 205. Bunker by bunker, trench by trench, the NKA (perhaps as many as a hundred) went up in flames or were shot down at close range. Bodies were plentiful but often too crisp to recognize. 4th Company's grim work went on into mid-afternoon when a three-Ranger fire team reached the top and shot their way to the head of a narrow draw, a concealed route which the enemy apparently had used to evacuate the scorched hill.

With the draw under Ranger control, fighting on the hilltop quickly petered out and several NKA surrendered. 4th Company had their objective secured after this very tough two-day fight, really their first of the war. It was a job to be proud of but cost two killed and nine wounded.

When U.S. armor broke through to the airhead, 4th Company first mounted up on the tanks then, when the terrain steepened, scouted for them in a four-day push north to the 38th Parallel. There was no stopping there, as Eighth Army wanted as much defensible terrain as possible for a buffer. The Tomahawk task force (187th RCT and a tank company+) had the delicate mission of maintaining signifcant pressure while not getting in too deep, especially in mountains where strong bands of infiltrators could cut the task force's tenuous supply line.

To prevent this and clear overwatching hills on the flanks of the armored column, 4th Company had to push CCF off ridgelines. This meant very arduous ascending and descending between ridges: slow, tough climbing that only troops in superior physical condition could accomplish. Regretably, MG Moore was not around to note this.

By and large, bands of enemy formed on hills but melted away if hit with an organized, supported attack. On 28 March two 4th Company platoons had to veer off to eliminate such a force that threatened the RCT's left flank from Hill 279, an action recorded by photo journalists from *Life Magazine*.

The Rangers were successful—and very timely. Anderson's platoons went up two sides of Hill 279, with tac air and direct fire support from task force tanks. Against such a combined arms attack the CCF resistance was unexpectedly tough. They seemed to be fighting under orders to hold at all costs. Rooting them out, the Rangers consolidated and began to understand why the CCF had made a stand. Approaching across the barren hills were columns of infiltrators who would have made Hill 279 their base to attack the task force's line of communication. Instead they ran into the 4th Company night defensive position. The Rangers turned them back, and the threat to the task force dissipated.

On 29 March, the RCT received orders to turn back too. Their deep reconnaissance in force had been most effective but Ridgway did not have sufficient strength to follow-up. For the Rangers the Munsan-ni operation ended where it had begun, at Taegu. Total casualties for 4th Company numbered 22.

Earlier in March, about the time 4th Company started training for Operation Tomahawk, Eighth Army G-3 sent out a call seeking volunteers for a really deep airborne raid. Rangers from all over the country (at that time lst, 2nd , 3rd and 4th Companies) signed up. Only four Americans, corporals Ed Pucel, Bill Miles, Marty Watson and PFC Ray Baker—recently busted from sergeant for punching out a straightleg lieutenant— all from 4th Company, were selected.

Their mission was to lead a force of nineteen ROK's to blow up a railway tunnel that had proven invulnerable to bombing and was now well protected by anti-aircraft batteries. This target was thirty miles inland from the port of Wonsan, a hundred miles behind the lines. Entry to the objective area would be by a night drop—the cherry jump for most of the ROK's. Destruction of the objective was to be attained by a pressure activated charge, thereby adding train wreckage to the problem of reopening the tunnel, but requiring that the tracks be mined secretly.

Egress would be by exfiltration for amphibious extraction on the coast, across some of Korea's most God-awful mountains, then through an NKA division bivouacked on the coastal plain. This trek—along with the drop, infiltration, recon and mission accomplishment—was to be accomplished in 72 hours in arctic conditions! All together the mission was a fantasy of optimism, illustrated by the notion that the ROK's could somehow disperse if necessary into the rural population. In this case aerial or maybe amphibious rescue of the Americans might be attempted, but no one at G-3 had much of an idea as to how it could be done.*[*FN: At Eighth Army, the tunnel mission was labeled Virginia One, the brainchild of G-3 (Miscellaneous), a joint-combined office roughly equivalent to what later would be considered a special operations task force (JSOTF).

Much of what went on at G-3 (Miscellaneous) is well described in Ed Evanhoe's *Partisan and Special Operations in the Korean War, 1950-53,* (Naval Institute Press).

The author ranks Virginia One as the most ill-conceived and desperate operation described in this two volume history. The principal reason for its kamikaze nature, I believe, is that though the raid was Ranger led, it was neither Ranger planned nor directed. How could it have been when the senior Rangers involved were corporals! Instead Virginia One was a mutant form of partisan warfare conceived by field grade officers without experience or appreciation in what they were doing, but with high zeal for demonstrating the value of G-3 (Miscellaneous).

The second reason Virginia One was a one way mission even before it started was that it was to be executed mostly by foreign forces. Whether the ROK's involved—they were OCS cadets—should be called guerrillas, saboteurs, irregulars or something else is an unimportant question of semantics. What mattered was that they were foreign nationals whose lives, implicitly, were discounted in value. A same size force of Americans would never have been permitted so deep in enemy territory during the Korean War because casualties, especially captives, were an extremely sensitive factor in U.S. politics. Later in the war several Ranger operations, far more feasible and promising than Virginia One, were scrubbed because of the likelihood of unnecessary casualties during what the press was describing as "The wrong war, at the wrong place against the wrong enemy."

This attitude was already jelling in March 1951; indeed the commander of Virgina One, a U.S. Army captain, was not allowed to jump in with his force for fear of his capture. Only the four Rangers were deemed expendable. Though conducted under the aegis of Eighth Army, the plan was, as one observer noted, "conceived by spooks to be carried out by gooks". Most charitably, it could be described as a small gamble for high stakes.]

In volunteering, the Rangers had been warned that the job would be covert and dangerous, but this tunnel raid amounted to a one way mission. There was no backup, contingencies, or margin of error in any phase of the plan, and unproven ground to air commo would be relied upon utterly.

COL John McGee, who had first stood up Eighth Army Ranger Company, was the action officer for the operation. Though the plan he proposed and bird dogged to approval was outrageously hazardous, he demonstrated an obligation to the Americans he inserted so deeply into the jaws of almost certain death or capture. No doubt the raiding force was written off even before they boarded their plane but McGee never gave up hope of retrieving the four Rangers.

They were chosen for their backgrounds and but mostly for their specialties, somewhat the way special forces A teams would be in the future. All except Miles were World War II veterans. Pucel had parachuted into the Balkans with the OSS to work with partisans. Watson was an expert at E&E, having gotten away from the Germans on six occasions. Baker, a former marine, knew about beaches from several combat landings in the Pacific. Miles was highly qualified both as a radio and demo man.

All, of course, were paratroopers who keenly appreciated the value of an accurate drop. Consequently they requested that a small pathfinder team jump in on D minus 1 to mark the DZ. The Rangers' request was denied by G-3 (Miscellaneous) for the puzzling reason that if the pathfinders were spotted the DZ might be compromised and raid would have to be aborted. For the planners apparently, mission failure because of an inaccurate drop was less of a concern than mission abortion. But not to worry, the Air Force liaison officer assured the worried Rangers, the C-47 would put everyone right where they wanted to be.*[*FN: But the AFLO did not mention that the C-47's crew were reservists newly arrived in country, and the Virginia One drop would be their first incursion into North Korea!] And by the way, for deception after the jump, the plane would make several leaflet drops as if that were the reason it was flying over North Korea.

For a raid that required everything to go right, nothing did. The night jump missed the unmarked DZ by eight miles, dispersing the raiders into four groups divided by snow clad mountainous terrain where they could not see each other, much less at night. Watson came down over a village, landing on the (fortunately sturdy) roof of one of the huts— and since this was North Korea no one inside thought the noise was a late visit from Santa Claus. With dogs barking all around, Watson still had the presence of mind to pack his 'chute in a GP bag before double timing out of town. For some warmth at night, that parachute silk would come in very handy.

But his aerial arrival pretty much blew the deception plan which was that the ROK's, in NKA uniforms, were to appear as guards for the Americans who would pretend to be POW's. That fanciful ploy was to have been reserved for action on the objective. For now the first problem was to get the raiders assembled. Each carried a portion of the 300 lb of C-3 explosive required to mine up the tunnel.

Miraculously there were no jump injuries, and during the next three days the raiders were not further compromised as they searched for each other; but Chinese forces were obviously aware that there was either a downed U.N. pilot or airborne infiltrators somewhere in the area. For either possibility, the CCF began intense patrolling. Meanwhile, with the raiders finally all together, Watson led them off toward the tunnel about ten cross country miles away.*[*FN: Another incredible aspect of Virginia One was the absence of an officially designated on the ground commander for the mission. The four Rangers sort of worked this out among themselves with Watson, though he had a later date of rank than Pucel and Baker, emerging as the leader.] They were soon halted by a howling blizzard that covered their tracks but forced them to take shelter till mid-morning. Already Virginia One was eight hours behind schedule.

For Watson, the raid was shaping up to be the hairiest of his career, and he had already survived the worst fight in Ranger history. As a young PFC in lst Battalion he had fought till out of ammo while cut off at Cisterna, then been captured and endured two years imprisonment in a German *Stalag*. Now he was far deeper behind enemy lines with even less support, less hope of relief, ass deep in snow with only three other Rangers. At least he had had real life experience in the deception plan because he really knew how to act like a POW!

While looking for each other the raiders had used up rations that were planned for movement to the objective, so for food they were now down to frozen chunks of chocolate and bullion cubes. Snowshoes or skis were the way to travel in mountains but because the local communist soldiers didn't have any, the raiders couldn't either or else they would blow their cover as NKA. Their best maneuver was to impersonate the CCF search parties who were looking for them!

Wallowing through snow drifts, struggling up steep mountains enroute to the tunnel, the raiders frequently crossed the tracks of search parties. The intelligence for the operation had been as inaccurate as the drop: the objective area was teeming with CCF where the plan had assumed there were only some REMF guards. And now three days into the operation, Virginia One was two full days behind schedule with prospects looking even worse.

Watson could not even report the predicament because of damage to the SCR 300 radio and the effect on the batteries of intense cold. A resupply drop of rations had been scheduled for the third night but without commo, and cloud cover preventing visual signals, the drop could not be confirmed. Though the ROK's, masquerading as NKA, were able to recon in the vicinity of the tunnel, the intelligence they brought back proved that penetration of the objective area was unthinkable: there were too many CCF alert and dug into strong positions at both ends of the tunnel. This news was as bitter as the cold. The mission was now demonstrably impossible; to attempt it suicidal. The Rangers forced themselves to turn to another problem equally grim and discouraging—how to get out.

For the next few nights they could hear McGee's C-47 flying in the vicinity but couldn't raise him, a maddening frustration because the radio was a huge weight to carry up and down the mountains. Then the flights stopped, and after five days behind the lines the Rangers assumed they had been written off. They decided to E&E to the coast but would try to take out a secondary tunnel along the way.

To reach it took another two days of struggling through snow and slipping on icy rocks. ROK scouts reported that this tunnel was also well defended and impregnable. The Rangers then stashed the demo, booby trapped it, and moved on with lighter loads. Ten days into the mission they were hit by another Siberian blizzard that kept them holed up for 48 hours. The next day the wind shifted, bringing warm air from the south. Every possible trail was now a raging torrent. Heavily soaked, the raiders trudged on.

At last, after two weeks of starvation survival, they got a break. Atop a mountain in sunlight, Miles tried once more to contact base station. If the long dead batteries warmed up sufficiently, he would make a blind transmission, then destroy the radio.

By some strange atmospheric fluke, the balky SCR 300 raised a friendly station, that of an aerial FAC with 7th ID eighty miles away. Of course the FAC had no awareness of the raid, and suspected Miles' transmission was some kind of Chinese trick; nevertheless, he relayed Mile's message to Eighth Army G-3.*[*FN: After days of silence from the raiders, McGee had alerted all U.N. units on the eastern front to flash any word they heard from the call sign Virginia One.] Four hours later CPT Perry, OIC for the raid, flew up, made contact with Miles and arranged a suppy drop. Close air support was also put on call.

It was soon needed. CCF were starting up the raiders' mountain; a shoot out was expected any minute. Three Navy Corsairs expended their bombs and 50 mm's on the plentiful targets; a little later Air Force jets followed with napalm and more bombs. Though they had not yet located the raiders, the CCF knew someone was adjusting effective air power on them so the intensity of the search was redoubled.

If air support could hold up the enemy till nightfall, the raiders might have a chance. A developing snowstorm would provide more concealment for getting away, but the most important factor they had going for them was the US Navy who had a cruiser task force on another mission steaming off Wonsan Bay. The TF commander was asked to commit four light helicopters to attempt a rescue. Three were available, and their pilots volunteered despite fading light and heavy, snowy winds. Their lift capability was only about 450 lbs each—two Rangers or three ROK's.

Commo between the rescuers and raiders was spotty: only the approximate location for a pick up was worked out. At 0800 on 30 March, Navy LT John Thornton went in first but was hit by ground fire and crashed. Uninjured, Thornton bravely directed the second chopper, piloted by LT Bob Felton, into a position to winch out CPL Baker. Up went Baker, as ground fire searched for him in the ice fog—but then the winch jammed, swinging him fifty feet from safety. For fifteen freezing minutes, Felton towed him out of danger to a beach 5,000 feet below. The chopper landed beside Baker, and with ice-caked limbs he was hauled aboard for delivery to an LST.

Back on the mountaintop, Thornton conducted similar winch extractions for Pucel and Miles. The latter Ranger was hit on the face by a bullet as he rose through the air. Blood gushed from his wound but then instantly froze—a very effective thermal tourniquet. His chopper was also hit, and only made it back to the cruiser *St. Paul* because the skipper sailed his ship dangerously close to shore.

With no more operating helicopters, Watson, Thornton and a squad of ROK's (the rest had been ordered to disperse and evade) were stranded on the mountaintop with CCF closing in. With a thermal grenade, Thornton destroyed the classified components of his downed chopper, then followed Watson off the mountain into a blizzard that let them pass within feet of Chinese soldiers.*[*FN: If there should ever be a hall of fame for 'Honorary Rangers' from other services, John W. Thornton, USN, is an eminent candidate. Besides his heroic effort to lift out the Rangers in his own chopper, Thornton insisted that Pucel and Miles go out ahead of him onto the next winches, a selfless act that resulted in their salvation but his captivity. He did this in the Navy tradition that the captain is the last to abandon ship.]

Without food, shelter, commo or any help whatever, Watson and Thornton (accompanied by five ROK's) reeled into the frozen wilderness. During the first night, the ROk's went their own way as chances for successful E&E were higher in NKA uniforms and without two occidentals. Two ROK's made it to friendly lines after a two week trek. Their debriefing, however, raised some suspicions about their story. Further interrogation revealed that they had been caught and 'turned' by the CCF. After this confession the two ROK's were summarily executed.

For another week Watson and Thornton tried to trudge toward the coast but mostly they had to hide out from search parties. Thornton's flight boots soon came apart and his feet became frostbitten. Realizing he was a liability, he waited till Watson was away searching for food; then Thornton hobbled down a mountain and gave himself up. That heroic self sacrifice did Watson no good— he was captured the next day. To addd to the bitter irony, he was only a few miles from friendly lines.

As they had shown during their ordeal in the mountains, Watson and Thornton were extremely tough, thereby surviving three years of brutal captivity and intermittent torture. Watson was the last American to be exchanged at the end of the war. His total time in two wars as a POW added up to more than five years. No Ranger in history ever proved more indominable.

Early spring of 1951 saw some cross-leveling among the six Ranger companies which were now in country. In this way a least the Rangers were 'regimental', drawing on a common replacement pool and occasionally transferring some men between companies. 1st Company, which was understrength after the longest period of hard fighting, gained a squad from 4th Company.

In Korea, April is the time of floods. During its long occupation by Japan, the principal rivers of the peninsula had been dammed to control flooding and provide electricity. Now in time of war these dams were critical: the side who controlled them could use directed floods for denial and obstacles. In April, U.N. forces were on the move north, movement which could be greatly hampered if certain dams were breached, allowing billions of cubic feet of water into valleys that otherwise would be U.N. axes of advance, and washing away bridges necessary for lateral communications.

The most tactically important of these dams was near Hwachon where it held back more than thirteen square miles of a deep reservoir. 275 feet high, 825 feet across, the Hwachon dam was also a source for vast hydro-electric capacity that before the war had powered the entire city of Seoul. With its spillway gates closed, the dam was designed to hold the springtime water level 32 feet above the main structure.

The CCF held this dam like the Sword of Damocles over IX Corps' plans. Twelve feet above the normal flood crest on the Han River was predicted if the dam was sufficiently ruptured at the right time. That time was fast approaching as snow melted in April.

It would take a corps attack to wrest the dam from CCF control, an attack that could be thwarted by blowing the dam.*[*FN: Total destruction of the dam was beyond CCF capabilities. That would have taken a stupendous quantitity of explosives which could be emplaced only by expert engineers which the Chinese didn't have. Such emplacement, with the extensive drilling it would have required, was virtually impossible because of U.S. air superiority.] The threat of uncontrolled flooding might be precluded, however, by a raid to destroy the flood gates and the spillway machinery—sort of a preemptive strike, causing a minor flood in order to prevent a major one.

The IX Corps CG, MG Hoge, obtained Ridgway's endorsement for such a strike. The 1st Cav was about to go into Eighth Army reserve, thereby freeing up 4th Company for the raid that would be the most strategically significant Ranger operation since Pointe du Hoc.

Indeed the raid would be something of a throwback to Pointe du Hoc, the last Ranger amphibious operation. But this time, rather than LCI's with grapelling hooks, the Rangers' vessels would be small rubber assault boats with paddles.

Planning time would also be quite different: on 8 April the CCF began opening the floodgates. The resulting torrent prevented use of five bridges in U.N. hands downstream. Before matters got any worse, the Rangers had to strike. Because the CCF could only crank up the floodgates manually (it took ten hours to lift each one), they had managed to open only six of eighteen.

Though 4th Company was earmarked to execute the mission, responsibility for its accomplishment was with 1st Cav who was in the process of being relieved in place by the 1st Marine Division. Division G-3, preoccupied with the relief, delegated OPCON of the raid to the 7th Cavalry Regiment who passed the mission down to its 2nd Battalion. Somewhere along this chain of command and amid other concerns, the raid lost some priority and focus.

But not of course for 4th Company who were alerted for the mission on 7 April. As 2/7 Cav was going over fire support overlays and range fans with the marines, CPT Anderson grabbed a combat engineer and rushed over to a similar dam in friendly hands to study the floodgate mechanisms. Then on 9 April Anderson went up in in a Piper Cub for a personal reconnaissance of the Hwachon dam. Analyzing the twisted, extremely precipitous terrain, he saw that a short amphibious sortie across 1100 yards of open water would bring the raiders to their objective in minimum time with maximum surprise.*[*FN: There is no better illustration of the necessity for a Ranger planning cell (i.e., a Ranger battalion HQ) than the Hwachon dam raid. Such a cell would have seen the mission coming, done the esssential recons and coordination so that the Ranger CO for the mission would have been conducting rehearsals rather than these other vital functions.]

Anderson was still aloft, musing on his alternatives, when on the radio he heard that 2/7 had kicked off in a premature attempt to grab the spillway. Landing immediately, he had to brief his platoon leaders on an operation he had no part in planning. The Cav's attack had gone off half cocked because Gen. Hoge had come down to see what was developing to neutralize the flooding. To his anger, he perceived that the dam raid was on the back burner. With a few well understood X-rated phrases, Hoge revised some priorities and 2/7 launched a hasty attack that called for 4th Company to pass through the cavalrymen and seize the floodgates.

In war, complete surprise can be achieved only by an act that surprises even its participants. This is what happened to 2/7 Cav who were suddenly redeployed while turning over positions to the marines—and thrown across an LD toward objectives pointed out to them on the run. They were surprised and confused but not so much as the CCF who had been watching a orderly rotation of enemy divisions. Surprise carried the cavalrymen's attack nearly to the dam. But favored by dominating heights, the CCF regrouped and resisted, holding the only access road as night approached on 9 April.

Hoge stayed around to make sure his message had been understood. To reassure him, the next morning 1/7 Cav fell out for a supporting attack to take pressure off its sister battalion. The time gained allowed 4th Company to request assault boats and appraise the unfolding situation. The worst part of it—apart from the improvisation—was the weather which denied air support.

This same bad weather (poor visibility) would be turned to the Americans advantage when 4th Company paddled across the reservoir to amphibiously flank the CCF that had stopped 2/7 Cav. That was now the new plan—entirely oral—dashed off by elements of G-3 which were scattered twenty miles between a forward CP and a reserve position. In general, the division resembled Keystone Kops, rushing about in all directions. That was their fault for misunderstanding their missions, but it is to the great credit of the 7th Cav that they could make something of an operation that had become hideously hilarious—for someone with a very morbid sense of humor.

10 April produced a break in flying weather. That was good news for air support but bad for concealing 4th Company's flotilla, now tentatively scheduled to embark after midnight—if enough boats could be rounded up by the Division Engineer. With 1st Cav strung out to hell and back, his efforts were frantic to get boat-carrying trucks pulled out of convoys and turned around on treacherous single lane mountain roads. Smoke generators and life preservers were almost as difficult to return to the attack assembly area. Amphibious aircraft were called in from all over the country to haul power launches to 4th Company's embarkation site.

Anderson watched all these efforts with dismayed anticipation. An attack that was so confused and rushed in its preparation could have few expectations for success. An additional discouragement was when he learned that only slow-firing 155 mm howitzers were available for artillery support. The division's 105's were all in march order, and due to restrictive terrain could not be returned to locations within supporting range of the raid.

The final briefing at 2200 did nothing to reassure Anderson. 7th Cav's S-2 estimated that a CCF regiment was dispersed along high ground where they could influence the Ranger landings.* [*FN: Enemy forces were underestimated by a factor of two. Actually the entire 115th CCF Division had dug in to protect the dam.] The S-3 announced H hour for the first three Ranger boats (one platoon) would be 0230, the rest of the company following at 0330. 2/7 Cav would kick off its land attack at 0400, the same time 1/7 began a diversionary supporting attack. 3/7 was initially in reserve with the mission to exploit whatever success was achieved by the attacking forces.

Before the briefing 4th Company had been preparing demolitions and rehearsing as much as they could their actions at the floodgate. Anderson had task organized his three platoons. Third Platoon (LT Mike Healy) would lead with one squad boat carrying men armed with carbines and pistols—backup for silent weapons like knives and hatchets. They were to secure the opposite shore. Next came another 12-man boat carrying most of the demo. There were not enough explosives for redundancy. If that boat went down, so did the mission.

The third boat in the spearhead carried fire support: an M-1 with a night sniperscope, BAR's, a light machine gun, a 57 mm recoiles rifle, and Healy's 'artillery'—60 mm mortar rounds rigged to be fired from rifle grenade launchers.

An hour later, in the second wave of six boats would be company HQ, 2nd Platoon and most of 4th Company's attachments: a 20-man heavy machine gun section, 81, 4.2, and artillery FO parties, all from 2/7 Cav. These boats would have to return to pick up 1st Platoon and the XO, LT John Warren, who would then become beachmaster for resupply, reinforcements and evacuation.

For the resources, time and intelligence available, Anderson's plan was as good as could be expected. The success of its execution would now depend on how well the Rangers adjusted to inevitable changes brought on by mechanical and coordination failures. The first of the latter occurred while moving through 2/7 Cav to the embarkation site. The route Warren had reconned was no longer feasible because the cavalrymen had repositioned for their own attack. Consequently the launch got off an hour and fifteen minutes late.

All of Healy's boats (each with an empty weight of 410 lbs) had outboard motors which were not to be used unless the flotilla was taken under fire while on the water. The night was moonless. Because of the late start, the darkness was already shattered by the sounds of 2/7 Cav's attack while the Rangers silently paddled, navigating by compass azimuth only. The crossing took a half hour, and though reveille for the Chinese had sounded by a bombardment of 155's, the Ranger spearhead slipped onto the rocky shore undetected. So far, so good, and 2nd Platoon was signalled to come on.

Despite the thunder of 7th Cav's preparatory fires, Anderson decided not to use the outboard motors. That was just as well because, as Warren in the third wave discovered, only two of the six motors worked. Even before all the Rangers were ashore unscathed, 2nd and 3rd Platoons began climbing for an intermediate objective, a bowl-shaped knoll named Objective 79. It was occupied by a platoon of Chinese trying to peer through the pre-dawn fog.

The total surprise of the amphibious attack became apparent when a CCF squad stood and shouted at the shadows approaching them, apparently in the belief that they were reinforcements. The Rangers silently waved greetings as they closed in. They were within a hundred meters when the CCF realized that the approaching silhouettes were too tall to be Chinese. One of their two machine guns cut loose, killing two Americans.

Now surprise was gone and the fight was on. Maneuvering over steep and coverless terrain, the Rangers knocked out one, then the other machine gun with grenades and small caliber assault fire. The Chinese broke, leaving fourteen of their dead, but as dawn broke 4th Company had seized the hill, yet they were still six hundred yards short of the dam.

From the hill, Anderson listened for the report that 1st Platoon (LT James Johnson leading) had landed, and he also monitored the progress of 7th Cav's other attacks. Faced with well prepared CCF positions that included reinforced concrete pillboxes, 2/7 Cav had stalled shortly after crossing their LD. 1/7 Cav's supporting attack had fared no better. To make matters worse, early morning light had revealed 1st Platoon's crossing to CCF mortar FO's. Soon water geysers were erupting among the boats. They had to change course and frantically paddle for cover in some of the reservoir's coves. Twenty additional boats were at the embarkation site for resupply. Ten had motors; none of them worked.

One piece of high ground (Objective 77) lay between Anderson and the dam. To attack and move forward would expose 2nd and 3rd Platoons to flanking fire from an adjacent hill (Objective 80) where CCF machineguns were already blazing. Such an advance would also extend the Rangers to where they could be cut off from 1st Platoon once it landed. To make matters worse, four of 1st Platoon's boats had now been damaged by fire, and could make headway only by hugging the shore.

Undaunted, 1st Platoon finally came ashore around 0700, then were guided up to Anderson's position. 2nd Platoon had already begun to probe toward Objective 77, so Anderson told Johnson to go for Objective 80 while the rest of the company supported by fire. With their attached supporting arms, this fire was considerably greater—especially with three heavy machine guns—than the Ranger company had ever put out. Indirect fire support was not as good because of lost commo with the 81 mm's.

1st Platoon had the toughest going. 2nd Platoon kept Objective 77 too busy to interfere, but even so the CCF had enough firepower on three other positions to keep 1st Platoon in helmet defilade. By mid-morning their attempt to take Objective 80 from the front became an obvious non-starter. Johnson then backed his squads out from under the worst of the enemy fire that was pinning them down and guided them toward a flanking route as Anderson burned the air waves for more indirect fire support.

However, it was the CCF mortars that picked up the tempo and cut short Johnson's flanking maneuver. 3rd Platoon tried to apply frontal pressure but had to return when lst Platoon ran into a sizable counterattack. With two squads of 3rd Platoon diverting to hit a shoulder of the CCF attack, Johnson was able to turn back the enemy with many casualties. The action so far had also been costly for the Rangers. Two had been killed by mid-afternoon, seven wounded.

With lst and 3rd Platoons repulsed and bleeding, the CCF concentrated their attention on 2nd Platoon's probe toward Objective 77 which also was going nowhere. Anderson pulled them back to Objective 79, a good decision because medics evacuating wounded to the boats reported Chinese getting around to the Rangers' rear. Anderson therefore organized 4th Company into a goose egg with all-around defense. This was completed with no time to spare: at 1415 a full battalion of Chinese swarmed toward the Rangers, intent on expelling them from Anderson's goose egg where he could keep the floodgates under efffective fire.

If he could hold this position the Chinese could not cause a flood themselves—that was the next best result to causing a pre-emptive flood which was the objective of the entire operation. So the Rangers knew there could be no pulling back, even if they were cut off from their boats. Shortly they werecut off by groups of encircling Chinese. Hundreds had been scurrying across the dam as Anderson tried to stop them with 155 howitzer fire. Low cloud coverage in high terrain prevented USAF bombing and strafing runs.

The CCF between 4th Company and the beach tried to feel out Anderson's strong position by bobbing up to draw automatic fire. The only fire they drew were single shots from Ranger snipers who drilled a decoy, then the others gave up on such sucker plays.

The major threat developed to Anderson's front where the CCF battalion deployed professionally, covered by abundant mortar fire the Rangers could not neutralize. Chinese mortars would be the greatest casualty producer that long day (12 April) because Anderson's position was sheer rock with no overhead cover. CCF infantry, though brave and resolute, was not so much of a problem. At critical positions along the perimeter, Rangers were spaced five yards apart with automatic weapons well sited for overlapping coverage of enemy avenues of approach.

Fire on both sides built up during the afternoon. The Rangers were grateful to be no longer attacking but rather defending on such barren terrain. They worried most about what would happen that night when the Chinese began the sort of stealthy infiltration at which they were so accomplished.

But that CCF battalion must have had orders to push back the Americans from the dam before nightfall. Maybe the Chinese high command wanted to blow the remaining floodgates under cover of darkness. For whatever reason, in broad daylight the battalion suddenly rose, blowing bugles, screaming warcries, and swarming toward Objective 79. Their charge was met by a mad minute of defensive fire that opened huge gaps in the first wave of attackers.

But more waves followed and some reached defilade so that the Rangers had to stand to take them under fire. Exposed on the skyline, U.S. casualties began to increase. Luckily the CCF did not employ supporting fire except from their mortars; otherwise Anderson would have been in too close combat. As it was, the Chinese were being engaged between 300 and 100 yards from Objective 79 and taking a hellacious volume of effective fire.

Losing too many men on this killing ground, the CCF began to work toward Anderson's flanks For some reason, the CCF between Objective 79 and the beach did not take part in the initial assault, and Anderson had been able to shift some Rangers from the back to the front of his goose egg. But now as their comrades tried to infiltrate around the flanks, the Chinese on the beach side supported by fire.

To interdict the enemy maneuver, Rangers ventured out from the goose egg looking for moving groups of Chinese. BAR men Forbes and Sanchez wriggled down a drainage ditch to get better firing positions. Peeping over a knoll they spotted a platoon of CCF bunched together around their leader who was making his own visual recon. At thirty yards range Forbes and Sanchez opened up with both barrels, firing six magazines, hitting a dozen Chinese and dispersing the rest.

Pumped up from blasting such a lucrative target, their ears ringing from the pounding of their weapons, the two Rangers at first did not hear the shouting from their buddies on the perimeter. When Sanchez and Forbes looked up, they were being frantically waved back, waves that were soon accompanied by screams to return. What they could not see from the drainage ditch was that while interdicting they had become interdicted themselves by a squad of infiltrators.

In but a few moments they understood their predicament. With no more delay than it took to stuff empty magazines in their pockets, (BAR magazines were precious—without them the BAR was a hand-loaded, single-shot weapon) they began to fast crawl back toward safety. Friendly covering fire was snapping over their heads, so Forbes and Sanchez realized that some unfriendlies must be very close by. On the final sprint back up hill, Forbes learned how close when he felt two sharp blows below the ribs. He looked up amazed to see a Ranger firing at him. Forbes had not been shot by the Ranger but by a Chinaman a few feet behind him whom the Ranger killed in the next second.

At a tremendous cost, by dark the CCF completed a strong encirclement of Objective 79. The nature of the fighting that afternoon is well described by Bob Black:

... The artillery FO from 7th Cav, and the 4.2 mortar observer called down accurate fire within 75 meters of the Rangers, but the Chinese came on into machine gun, rifle and carbine [fire] until, finally, some Rangers were using pistols. CPT Anderson, SGT Schroeder (the comunications sergeant), and six men in the company headquarters position fired for fifteen minutes, stopping only to reload.

Elsewhere on the line, CPL Angarano was hit, his arm almost severed. [Medic] Goolsby pulled him to cover, treated for shock, and tried to inject albumen, but Angarano's veins had collapsed and he died from loss of blood. PVT Young, a BAR man, was hit in the head, convulsed, and died. Ammunition was taken from the dead and redistributed. Squad leader Ken Robinson rationed out five rounds per man, saying, "Don't use 'em all for just one kill."

The Chinese were closing in, disregarding their horrific losses. It was not enough to repulse them.*[*FN: More than a hundred bodies were counted within grenade distance of Objective 79. Because in the end the Rangers relinquished the position, most of the battleground fell to the CCF so the most accurate body count was for them to perform; but if a hundred had fallen on the edge of the Ranger perimeter, at least triple that number were cut down in the outlying area where the Ranger fields of fire were the most devastating. The estimate of four to five hundred CCF dead was corroborated by the disappearance of the CCF regiment from subsequent orders of battle.] They had to be drastically winnowed before dark or the depleted and ammo-short Rangers could not make it through the night. Anderson was already wondering how he could exfiltrate his survivors back to the beach. Grimly, he contemplated the necessity of leaving the wounded on the position (as the British had done on Gloucester Hill),arming every one of them who could still pull a trigger, to cover the escape.

2nd Platoon began an 'attack to the rear' to open a route to the beach. Firing off their last 57 mm rounds, the Rangers then destroyed the recoiless rifle and slowly pulled back from their position. With only one box of ammo for their light machine gun, they used it sparingly while withdrawing to the beach.

In his shrinking perimeter, Anderson had half a box of ammo for each machine gun, about sixteen rounds per M-1, and no more grenades. With 2nd Platoon now holding the landing site, Anderson radioed 7th Cav for permission to consolidate his Rangers in a small, tight beachhead. To remain on Objective 79 overnight was to invite wholesale capture. Drizzly rain and dense fog would aid him in exfiltrating to the landing site.

Then came a series of jolting surprises for the Rangers. First, 7th Cav denied Anderson permission to withdraw from Objective 79. Instead he was to hold it for reinforcement by a Cav company who would also bring him a basic load of ammo. This company was already in assault boats; 2nd Platoon should prepare for its arrival and provide guides to Objective 79.The second surprise—really a shock—were 7th Cav's orders for the two US companies, once they linked up, to seize Objective 80 that night and be prepared to continue the attack the next morning. The S-3 apparently didn't know about the CCF regiment camped out on Objective 80 and controlling all its approaches. The Rangers half joked that nuclear weapons were about to be introduced to the war because there sure wasn't any other way take Objective 80.

Nevertheless Anderson began making his plans for an attack that would kick of about two hours before dark. Then the third surprise came over the radio: Attack cancelled. Pull back through the Cav company, re-embark and return to the friendly side of the reservoir. This was done with only a few more casualties though the Cav company had a rousing fight that night while holding the beachhead.

Shaking their heads, at 0130 4th Company walked away from the assault boats at the point on shore where the raid had begun some 24 hours earlier. Gen. Hoge had aborted the operation himself when he learned how strong were the CCF defenses of the dam and how many troops the lst Cav would have to commit to control the floodgates.

So the surprise achieved by the daring amphibious raid was frittered away before sufficient reinforcements arrived to support and exploit 4th Company's initial success. It was a bitter pill for the Rangers, to have had the floodgates under fire then having to abort so close to the objective. Many good men had lost their lives and their blood (total Ranger casualties were more than a quarter of the company's strength) because higher headquarters was indecisive—then finally having made a decision, could not stick with it.

In a way, 4th Company was a boy sent to do a man's job. Willing and able enough, the Rangers were too few, too lightly supported to accomplish the mission by themselves. Maybe if the weather had cleared and tac air support had weighed in heavily, they could have reached the floodgates and no doubt disabled them. But that was not to be, and Hwachon will always remain in 4th Companay's memories as "that damned dam."

As afterword to the operation, the CCF did complete opening all the floodgates and the resultant erosion downstream hampered U.N. movements for about a week.

With the lst Cav in a relatively quiet sector during the 'May Massacre', 4th Company performed pretty routine patrolling; but 'routine' for Rangers can be highly hazardous duty. Often called upon to hide behind enemy lines for a day or more, 4th Company men plotted enemy locations and movements for artillery and air strikes. Two Rangers were captured by CCF on one such patrol that began with their becoming separated from their rubber boat, carrying their weapons and all equipment, in the raging Imjin River.

This was more than enough reason to abort and get back to friendly lies, but Sergeants McPherson and Ward still had trench knives, a map— and a mission. Rangers simply do not quit, so they decided to continue deep into enemy territory where they had planned to rely on stealth, rather than their weapons, anyway.

Shivering from cold water immersion, Ward and McPherson infiltrated their objective area which they found to be teeming with Chinese. This would have been enough and valuable information with which to return, but their mission called for a day of clandestine surveillance. So they stayed, not beginning their return till nightfall. They slipped by position after CCF position, but their AO had become a Chinese assembly area.

Shortly before first light they were fired upon. Pursuit and evasion went on well into the morning when they were finally tracked down and surrounded. With fire arms they would have probably gone down fighting, but with only knives they had to surrender. Both survived three years of inhuman captivity.

Two large morals can be found in this tiny operation. The first is that the rationale for disbanding the airborne Ranger companies—because non-Orientals were too conspicuous behind the lines— is ludicrous when men like McPherson and Ward had to conceal size and skin color any time they patrolled. 'Correct' pigmentation and bone structure would have hardly prevented their capture!

The second moral is never send Rangers on a mission unless you are prepared to have it carried to finality, even to a bitter end. In McPherson and Ward's case, a well-trained, well-led recon patrol of experienced conventional troops would have turned back when unable to reach their objective area. And rightly so—very few operations, especially patrols, are designated as being so important that mission accomplishment is to be pursued till the last man falls.

But that was exactly Ward and McPherson's attitude from the time they moved out. Their battlefield judgment was clouded by the unique Ranger determination to carry on whatever the circumstances.*[*FN: McPherson and Ward began the mission as part of a larger Ranger patrol. After the mishap in the river, the patrol leader, using good judgment, aborted the mission, a decision for which no one found fault.] Which is to say that the adage, 'discretion is the better part of valor', is often not found in the Ranger psyche.

Over time, and through interviews with Rangers of the Korean War, the author has come to believe that this pit bull mentality may have been the unstated reason—rather than the announced 'non-Caucasion' reason—that the airborne Ranger companies were disbanded when they were. It was a time when US public opinion was fuming with impatience to end the war and with no more casualties.

In their previous operations, the Ranger companies had taken disproportionate casualties, partially because of the pit bull factor (the predominant reason was probably because they were most often committed where the fighting was already or about to be the fiercest). The US high command, therefore, knew that if Rangers stayed together in companies, where the pit bull mentality tends to be reinforced, that every patrol had a good chance of ending with casualties like Ward and McPherson.

Pit bulls neither back up nor back off. "Cry havoc! and let slip the dogs of war," is a famous line by Edward DeVere from which the titles of two war books was taken. When havoc is the purpose, a pit bull is the war dog of choice; but when fighting has diminished to snaps and growls, the pit bull is caged. This was the state of the Korean War when the airborne Ranger companies, including 4th, were disbanded.

Fifth Airborne Ranger Co.

In November 1950, the second RTC cycle produced the next wave of airborne Ranger companies, numbered 5 thru 8. 6th Company was assigned to Germany while 7th became the holding company and training cadre at Ft. Benning. Many hard-charging Rangers in these two companies were frustrated not to be going into combat, so members of 5th Company counted themselves among the lucky.

They were also lucky to be commanded by CPT John Scagnelli, an officer so hard bitten his nickname was 'Nails'. Built and hard like a fire hydrant, he had been a World War II paratrooper, and his face looked like he had landed on it in every jump. 'Scag', as he was also called, was profanely fluent in Italian as well as English. To distinguish his company, he authorized Mohawk haircuts though to be nearly bald during winter training meant loss of a great deal of body heat. And 5th Company was to get an extra four weeks of winter training at Ft. Carson CO in January. First they needed to warm up: the send-off party from Ft. Benning resulted in property damage of $1,200!

Rangers are individualistic and their companies separate; so more so than conventional units, they reflect the personality of the company commander. And the CO—as this history has shown—is very nearly permanent, exempt from the usual command rotation practiced elsewhere in the army, removed only if he becomes a casualty or liability.

Scagnelli's personality (indelibly imprinted on 5th Company) emphasized the 'band of brothers' characteristic of Rangers. Other CO's— Puckett may be the best example—focused most on efficiency; that is, assuring instinctive reactions to changing situations, fostered by repetitive training and thorough critiques. LT Herman of lst Company seemed to call for individual initiative more than anything else, as demonstrated so stirringly by the spontaneous Ranger attack at Chipyong-ni that cracked the CCF's choke hold on the perimeter.

CPT Tidwell of 3rd Company found collective strength in discipline. Not the appearance of discipline like dismounted drill or footlockers squared away, but discipline which meant that in everyone's mind was the essential conviction that doing his job as required would produce the synergy for overall mission accomplishment. It was this kind of discipline that carried Bloody Nose Ridge. Perhaps the foremost characteristic of CPT Anderson's command personality was ensuring good planning, and this was featured all down the line in 4th Company. Amid all the confusion and garbles of the Hwachon dam operation, Anderson kept his head and came up with a daring but not reckless plan that would have succeeded if adequately supported. All his Rangers knew what they had to do and did it—because he made sure they knew the plan.

Command personality, of course, does not mean a fixation on one quality to the exclusion of others equally important. So it would distort the point by specious comparisons to allege, for example, that Eighth Army Ranger Company was less disciplined than 3rd Company or that 4th Company displayed less initiative than lst Company (remember Ward and McPherson!). The best commanders are the most flexible, concentrating on the qualities most essential—which may change— for the situation at hand.

Nevertheless the command personality of the CO, especially in the Rangers, is reflected from the bedrock of his company. For Scagnelli this bedrock was the 'band of brothers' attitude expressed as 5th Company against the world which included the 'straight leg' army as well as the communists. At Ft. Benning, as example, 5th Rangers closed the barber shop where they received their Mohawks—by hurling empty beer cans at any non-Ranger patrons who dared to enter! Except for 2nd Company , unique because of its racial composition, 5th Company made the most of strength through separateness.

After a month of cold weather training in the Rockies, 5th Company was well prepared for winter in Korean mountains.*[*FN: This training at Ft. Carson included considerable mule handling, an ideal vehicle for ranger cross-country mobility. But for reasons research had been unable to uncover, mules were never employed in Korea. Perhaps it was because of KATUSA's—Korean Augmentation to the US Army—who were virtually coolies, each man carrying on an A frame almost as much as a mule— and eating less.] The down side to the training at Ft. Carson was that it had kept 5th Company from a month (January 1951) of some of the heaviest fighting in the war. It was not until 5 March that Scagnelli had his men embarked for Yokohama.

Spoiling for action, they found it aboard ship when marines were given preference for seats in the movie room. Knuckle sandwiches convinced them that Rangers should be accomodated. The skipper of the ship noted wryly that the presence of only one Ranger was enough to produce a riot. A favorite article passed around on the ship was one by the noted columnist, Drew Pearson, which stated that, "If captured, the Rangers knew they would be killed." 5th Company was delighted and, though none would ever acknowledge the possibility of being captured, adopted the war cry, "We'll all be killed!"

On 1 April, 5th Company was attached to the 25th ID, a hard fighting division that had been in action since the earliest days of the war. It was now deployed on the western front, almost on the 38th Parallel in mountainous terrain. The CCF had taken a long step backwards in order to refit and resupply. Ridgway was intent on preventing them from stablizing and digging in, and to do so he insisted that IX Corps keep maneuvering and hit with limited objective attacks designed to kill great numbers of enemy and keep the rest off balance. If that were successful, the CCF would have trouble mounting an offensive of their own. Spring flooding made life miserable on both sides.

In the fluid situation that had obtained since the beginning of the year, there were many stray infiltrators and pockets of bypassed enemy troops in IX Corps' rear. Like the Rangers in Korea before them, 5th Company cut their teeth against such miscellaneous enemy. But their first mission was a very odd one—a denial operation against friendly forces!

25th ID had recently recaptured an airstrip, one of the very few flat areas in the mountains, and therefore coveted by all the support units following up the advance. G-3 wanted that strip for light aircraft, and told Scagnelli to secure it for that purpose alone—repelling or evicting any trespassing unit, no matter what rank its commander. Never known for his diplomatic language, Scagnelli accomplished that mission, at one point dissuading an artillery recon party that had already begun surveying the runway.

Friendly neighbors soon became convinced that 5th Comnpany's 'Keep Out!' sign was serious, and the Rangers had their first chance at in-country training in the surrounding hills. On its first day of patrolling, the Rangers captured a lost CCF infiltrator. Though the terrain of their AO reminded them of the barrens of Colorado, there was a bleak grimness peculiar to Korea that impressed the new arrivals. It may have been the occasional body parts protruding from the ground of previous battles. From patrols, lst Platoon had plenty of locations to refer to graves registration.

Scagnelli's initial company-size combat patrol brought it the distinction of being the first U.N. force on the western front to venture north of the 38th Parallel since the CCF onslaught the year before. 25th ID was cautiously advancing with two regiments forward, one back. The division had been the foster parent unit for the original Rangers in Korea, the Eighth Army Ranger Company (since disbanded), so G-3 had a high regard for Rangers in general, even unblooded ones like 5th Company. Consequently G-3 put them out front of the entire division in Scagnelli's cherry operation.

His mission was to cross the Yong-Pyong River and set up ambush positions about a thousand yards into enemy terrritory. The targets were CCF mine-laying parties. Anti-tank mines had been slowing 25th ID's armor; so rather than have engineers taking time to remove mines, G-3 decided to have Rangers remove the mine layers. On a night in early April, 5th Company passed through the frontline foxholes of the 27th Infantry Regiment for a first look behind enemy lines. It was a good feeling everyone remembered: the company was intact (actually eight EM overstrength) on an independent mission, the kind they had prepared and longed for.

Fording the river was done in text book fashion, thanks to training in the Rockies. Scagnelli felt the crossing had to be made undetected in order to achieve surprise. It was achieved because two platoon ambushes (the third platoon was held as a reaction force) slid into place without the Chinese knowing that there had been a penetration at all.

One dirt road led into, and a secondary track ran beside the ruins of an abandoned village. The next day U.S. armor was to advance through and around this village. 5th Company was to prevent any delay. Scagnelli's main ambush was on the road, a smaller one on the track. The night was moonless, and both ambushes were in place with OP's out by 0100. 5th Company was cold but warmed by expectation.*[*FN: 5th Company's unofficial history, a monograph by Ranger Walt Eckhardt, is titled, 'Travel Light and Freeze at Night', a lament of every soldier who has been to Ranger School. The origin of this lament can probably never be precisely traced, but because 5th Company used it when coming out of the second cycle of the RTC, they are credited in this book.] If intelligence was good, there should be some good hunting before morning.

There was good intelligence—the essential ingredient for successful Ranger operations. No sooner were the ambushes in place than the OP's to the north saw movement on the road. Shortly, six Chinese scurried south, each carrying two AT mines. The OP's did not have radios, so they had to report the enemy movements in person. Unfortunately, they were spotted by the mine layers who turned around and fled north. Scagnelli fumed with frustration that 5th Company would have no kills that night, but the mission was accomplished: there were no mines to delay the 25th ID's advance the next day.

Pulling off the ambushes, the disappointed Rangers had a chance to vent their frustration. As dawn broke, the Chinese came down from the hills to see what had scared off their mine layers. The Rangers were hungry as well as pissed—they had missed chow the night before, and were cracking open C rations when the CCF began to probe. They were met by a flurry of furious fire—ammo did not need to be conserved because linkup with the 27th Infantry was imminent—and Rangers went out to meet them. Two Chinese machine guns were knocked out, and the force that accompanied them dispersed farther uphill than where they had started, leaving seven bodies behind.

Soon US tanks rumbled by the village, shooting into the hills, and G-3 was quite pleased by their uninterrupted progress. The Rangers were less than pleased: their first significant action had been successful but action at the objective had been less than the best possible. At breakfast back across the river, Scagnelli counducted a harsh critique and everyone resolved to do better though they had done well.

From the river the company ranged out with recon patrols, one of which resulted in their first death, that of PFC Eddy who died of wounds received when he stepped on an unmarked mine. Another Ranger lost a fingertip in an accident. Evacuated to a MASH, he recovered from anesthesia and decided to hike back to 5th Company without medical clearance. His buddies were glad to have him back, but two weeks later word came down that he was being carried as AWOL from the hosiptal!

'Hooches' were sometimes bunkers and sometimes above ground structures, but either were highly prized for protection against the elements. In their patrol base by the river, 5th Company had found and improved a number of underground hooches. One, however, turned out to be a source of danger instead of protection.

Bob Wilson had been heating C rations when his small burner ignited the straw roof of his hooch. This conflagration soon set off his basic load of ammo as well as that of two buddies in the hooch. Rounds cooked off rapidly until 5th Company was pinned down by its own fire—adding a new meaning to the term, fire discipline!

25th ID at that time was digging in about seven miles from the southern base of the Iron Triangle where 5th Company became employed in securing positions for artillery raids by 155 batteries firing into Chorwon. On 11 April, Scagnelli obtained a more aggressive mission: to penetrate farther north and seize Hill 383 which the CCF were preparing, apparently for defense. As the Chinese progressed with their work, Scagnelli felt that if 5th Company could infiltrate the hill at night, they could use the

CCF's own positions to defend the Rangers from counterattack the next morning.

The idea had a certain amusing appeal. The commies were great diggers, so let them dig the Rangers' positions. What made that plan more ironic was that the CCF work parties pulled off Hill 383 at dark to return in the morning. 5th Company would take over the positions, almost as if they were the night shift.

But to do so required an infiltration of the enemy's front lines, passage through their rear, then ascent of the hill, all without compromising the mission. As division increased its interest in Scagnelli's daring plan, they decided to reinforce it with a strong reaction force: the 24th Infantry Regiment would stand by to either launch a diversionary attack or rescue in case of trouble; or, if the Rangers were successful, attack and relieve them on the hill.

The scheduled pattern of division's H&I fires was changed so as not to interfere with 5th Company's infiltration routes. Priority of fires would go to the Rangers if they became engaged. With hopes as high as their caution, they moved out at EENT for Hill 383, an attack very similar to that 2nd Ranger Battalion's on Hill 400 in December 1944.

To travel light and ensure commo security, the platoon radios were not to go on the mission. Only an SCR 300 for coordination with 24th Infantry Regiment would be carried by the command group, plus fire support radios. Apparently Scagnelli felt that non-electric control measures would be enough. For this purpose his initial formation was a company column. Scouts had already pre-reconned the thinnest sector of the enemy front—the most rugged stretch of rocky terrain with good concealment— before the main body crossed the LD. The order of march was 3rd, 2nd, and lst Platoons, with the command group with 3rd Platoon.

The route of march proved to be an excellent one for undetected travel—Scagnelli had studied and selected the route from aerial photos— that skirted several abandoned villages and forded the headwaters of the Han Tan River. But the difficult terrain and the requirement for utter stealth reduced the rate of march far below his expectations, really to a crawl.*[*FN: Despite the vital need for stealth, there was one unbelievable incident when a ROK interpreter reported that an NKA was holed up in a hut in one of the villages. Instead of bypassing the hut, a Ranger tossed in a WP grenade, setting it aflame. The counseling he received was no doubt as fiery.] Not until 0435—after seven and a half hours of travel— did 5th Company coil into their assault position at the foot of Hill 383.

By this time the communist work parties had long departed; indeed, they could be expected to return and start climbing back up at dawn, less than two hours away. So Scagnelli laagered the company, hoping to remain undetected until artillery softened up the hill for a daylight attack. This was done. Howitzers pounded Hill 383 till the Rangers rose up and began to climb at 1000 hours. The artillery preparation had kept enemy work parties off the hill, so now Scagnelli's attack was more a race to see who would get to the top first.

The Rangers won. But as expected, they soon engaged swarms of CCF coming up the other sides of the hill. The enemy now perceived what Scagnelli was trying to do and launched wave after wave of troops to prevent him from securing the hilltop. The positions already dug were not enough to protect his entire company. 3rd Platoon was able to occupy them and riddle the counterattacking Chinese but Scagnelli could see that if 5th Company were encircled on the hilltop, overhead cover was insufficient against mortar fire which had started to drum down on his men. He probably could not prevent encirclement so he ordered 3rd Platoon back and withdrew the company to its original laager. With CCF exposed all over the hill, he called for an air strike with plenty of napalm.

The strike, another that soon followed it, and heavy support from DIVARTY pretty much scraped the CCF off the hill. Before they could recover or reinforce, Scagnelli quickly ordered 5th Company uphill again with platoons in a 'V'. This second assault carried the objective after close combat with the splintered enemy still reeling from air strikes and artillery.

An EPW was taken, and the Rangers consolidated the objective in a goose egg that took advantage of what the Chinese had already dug.

Unfortunately most of the diggings faced American lines, not the direction the Rangers would be defending because the most likely counterattack would come from the other side. But on the southern face of 5th Company's position was a sheer cliff that dropped down hundreds of feet to streams that fed the Han Tan River. Scagnelli put the bulk of his strength on the other faces where they dug like hell. He then radioed to 24th Regiment that Hill 383 was secure and 5th Company ready to support a break through.

However, during the afternoon the 24th was unable to break through. By dark they had punched a salient into CCF lines but not far enough to protect the Rangers by direct fire. With only about half their ammo remaining, the Rangers would have to hold out by themselves with help from only the artillery. Scagnelli registered concentrations all around the hill as signs of a CCF build up could be detected even before dark. Now the lack of platoon radios might be serious, but Scagnelli placed himself in an OP where he had almost complete coverage of the perimeter.

Preceded by a twenty minute mortar prep, the CCF struck on two sides at 2000 hours. The Rangers held their fire as they heard Chinese leaders yell at their men to get up the hill and through the defensive concentrations that Scagnelli adjusted by sound. In less than an hour, the Chinese fired two red rockets, apparently a signal that the first wave had reached its assault positions. Then bugles blew—still the Rangers held fire. It was not until they heard labored breathing of enemy struggling up the hill that they tossed grenades high in order to have air bursts (allowing one second to elapse before releasing the handle). "Here they come!" was the shout when riflemen opened up on targets. BAR's and machine guns stayed quiet until the attack got more serious.

The CCF closed in on 2nd and lst Platoons but were turned back after about forty minutes of fighting. Another mortar barrage followed as a second and third wave took up the bloody attack. During lulls, Chinese casualties could be heard rolling down the hill. All Rangers except half of 3rd Platoon were engaged, and automatic weapons spoke decisively this time. But the night was young and the next wave struck three sides of the perimeter but, fortunately, not at the same time: U.S. artillery apparently was breaking up formations and preventing coordination between CCF commanders.

Scagnelli could foresee any number of more waves lapping up the hill, so it would be but a matter of time till his company was fighting with only bayonets and empty weapons. Now was the time to do something while the Rangers were still intact and had good control.

The precipitous south face of the perimeter had not yet been used by the CCF. Calling off artillery fire in that direction, Scagnelli sent a few scouts down the cliff to see if the company could egress off the hill. In an hour they reported back that the way was extremely steep but clear of Chinese. Scagnelli began thining the line by squads, those departing leaving ammo for those that remained. The platoon radios were now sorely missed.

The CCF were still peppering and probing the hill, concentrating on 2nd Platoon's sector, the last one Scagnelli evacuated. SGT McAbee's squad was the rear guard. Spreading out to cover more frontage, they became scattered, and as the CCF came on once again, Rangers were cut off. They knew the route of withdrawal, however, and began shooting their way out.

Ron Mott's narrative typifies the sort of individual actions that occurred as squads started to E&E:

The order came to move out from SGT McAbee. Moving back along the crest there was firing everywhere. I kept running into Chinese, dead and alive. I remember to this day the stench of the garlic soaked bodies. When we got down the hill a ways, I found there were only four Rangers left [from my squad].

The rest of the company had gone down ahead, and it was now perfectly quiet. Red Eisman, Louie Goldbaum, one other Ranger and I slid about halfway down the hill and found a deep depression in the rock. We decided to spend the night there, as it was pitch black and we didn't have any idea where we were. At first light we saw we were close to the Han Tan River. When we got to the other bank there was a large explosion at the cave we had been in. When we got back to our lnes, and explained who we were and what had happened, they told us how lucky we were because they had seen some movement at the cave mouth and fired a 57 at it. That was the explosion. We were just lucky that the men on the 57 were slow.

More than luck was involved in 5th Company's withdrawal. Badly wounded by mortar fire early in the fight, SGT Bill Kirshfield and CPL Walt Maziarz knew they could not be carried down the cliff. They volunteered to be the sacrificial rear guard. They were last seen firing BAR's, holding up the CCF, fighting to the death. Their bodies were later recovered.

Even in the pitch black night that Mott described, the survivors of 5th Company barreled down the hill, sometimes so fast that they passed within a few feet of astonished CCF patrols. At the racing waters of the Han Tan River, darkness made crossing extremely dangerous but not so much so as being caught by first light in midstream. Many Rangers slipped and were swept downstream into deep pools and rapids, but miraculously none drowned.

Squad and platoon leaders roamed the far bank, gathering up their men. When Scagnelli felt the bulk of his company was accounted for, he marched them toward friendly lines. Single and small groups of Rangers would dribble back later. It took two more days before the last men returned. The final toll for Hill 383 numbered eight killed, 24 wounded and two MIA.

The bitter battle up and down Hill 383 raised doubts in 5th Company about the fighting qualities of some 25th ID units. This perception increased as the division worked its way closer to the Iron Triangle. During the second week of April, the Rangers were tasked to screen across a wide valley that separated two Tropical Lightning regiments. Contact patrols found some units not to be where they said they were; visiting patrols at night often found everyone at LP's to be asleep. One morning 5th Company looked out from their perimeter to learn that the neighboring regiment had evacuated without telling anyone.

These relations did not bode well for what Scagnelli knew was coming— the kick off of the enemy's big spring offensive. It initially hit, and hit hardest, the U.N. forces to the southeast on the Imjin River (see 3rd Ranger Company's history), requiring 25th ID to backpedal and refuse its flank. Ruefully, 5th Company turned around to trudge south past Hill 383 and its memories, past the positions they had secured for artillery raids, past the site of their very first action. They paused on the original defensive positions at the 38th Parallel but soon pulled out and back some more.

5th Company was part of the division rear guard, at times supported by tanks, and Scagnelli kept plenty of room between his Rangers and the CCF who could not cover ground with the same speed. When smoke from Chinese cooking fires was spotted, Scagnelli usually displaced. He just didn't have reason to be confident that 25th ID would bail him out if Rangers became heavily engaged.

On the morning of 23 April, it appeared that the long retrogradation might be ending when 5th Company was ordered to break contact, and take a long step back on the hulls of tanks with 25th Recon Company. Moving down a long valley, at first they were pecked by small arms from the hills, evidence that CCF infiltrators were abreast working their way south. But after several more miles, hostile fire ceased; the Rangers studied the high ground on their flanks where they could see GI's digging in and mortars registering. 5th Company was entering the lines—the main line of resistance—of 25th ID and the long, sleepless delaying action was over.

To recoup strength after shooting and falling back for so long, the Rangers were initially placed in reserve with the mission of blocking any penetration between 27th and 35th Infantry Regiments, whose flank companies that loosely tied in with each other were both lettered 'G'. Having two 'Georges' side by side turned out to be a confusing coincidence in the battle royal that followed.

Each company had a platoon forward on the division combat outpost line (COPL). Behind these forces, who were improving good natural defensive positions, 5th Company felt relatively safe. But during the previous days and nights as rear guard they had been feeling the growing weight of the CCF offensive as it approached the COPL. Scagnelli advised everyone to scrounge up as much additional ammo as they could—he had a feeling it would be needed. This hunch was reinforced when the division armored screening force came hightailing down the valley with reports of heavy contact with "millions of gooks". The Rangers asked the tankers to drop off any ammo they could spare. It was enough to almost double each man's basic load.

Rather than have to try to blunt an enemy penetration under fire at night, Scagnelli obtained permission to move his company up close to the regimental boundary and add a small Ranger OP to the COPL. The OP was approved, but Scagnelli was then required to detach a platoon to fill in between F and G Companies of the 27th, with the balance of 5th Company given the contingency mission of backing up a gap between two other companies of 2/27th Infantry. The final configuration of his company did not please Scagnelli (who had a three-way mission and coordination with four CP's), but when all was said and done, he had 2nd Platoon forward and separated by about a quarter mile from the other two who were echeloned to the right rear.

All this shuffling had to take place in early evening while CCF patrols began to probe the COPL. Scagnelli's principal concern was that 2nd Platoon would become engaged first and lose contact with the rest of the company which would have to remain poised for blocking and counterattack missions. His fear was realized when constant enemy mortar fire interdicted liaison between 2nd and 3rd Platoons. On a larger scale, this was also happening between the two George Companies as they pulled their platoons off the COPL. Scagnelli received another on order mission to fill that gap if it developed.

Around 2020 hours, scores of enemy flares began a green glare, answered by parachute flares from DIVARTY, as both sides tried to determine what they were up against. The Ranger OP came back on the run—they knew it was something bigger than 5th Company had ever seen before.

Eighth Army searchlights flicked on, causing indirect illumination from low cloud cover. The sight they revealed was indeed unprecedented: Chinese formations up to battalion size, advancing in march order, column after column, for miles to the north. There was a breathless moment for every American along the valley, then artillery RTO's shouted "Fire for effect!" almost simultaneously. The valley shook and earth geysered as the CCF parade entered a caldron of explosions and smoke.

But they came on relentlessly, even under the non-stop fire of fifteen artillery batteries. Hundreds of Chinese dropped or disintegrated but thousands more pushed on like lemmings toward a cliff. The Americans' long range infantry weapons now reached out for them, seemingly ignored by the CCF who crowded forward to bugle and whistle signals.

Scagnelli now received orders to send a platoon into the gap between G and F Companies of 2/27. 3rd Platoon moved out on this mission but was immobilized by a torrent of mortar fire. It became obvious that everyone would pretty much have to fight where they were because it would be impossible to move reserves. The first units placed in this 'die in place' situation were the two George Companies.

2nd Platoon was soon engulfed in the attack that isolated the Georges. Like ocean breakers, waves of Chinese hit and dissolved against strong U.S. positions. But then like incoming tide, the CCF surged around the breakwaters and rolled on, leaving the first tough positions for follow up troops. Any US position on low ground was swept away like drift wood. Meanwhile 5th Company (-) became a blocking force against the breakers battering down the valley. In front of them were two platoons of G Company that fought like the furies but were soon overrun. A quarter-mile of foxholes was ripped out of the U.S. lines. American survivors streamed to the rear, many stumbling with wounds. In the flashing darkness Rangers could not determine if figures coming toward them were friendly or enemy. They were both as forces became intermingled in the roar and screams of the night.

To open some ground and sort things out, Scagnelli pulled back while heavily engaged—the most difficult feat in war. Among the many Rangers wounded in this withdrawal was the lst Platoon Leader, LT Fred Lang. He was hit in both legs but refused to be carried because that would slow and endanger more Rangers. He demanded grenades and extra ammo. He would be a one-man blocking force till the Chinese swamped him.

The attack bludgeoned its way through the Americans' MLR. Station after station went silent on land line, and many on the radio net. Mortars fired as fast as rounds could be dropped but then too went silent as the Chinese poured over their positions. To slow the CCF advance U.S. artillery had to work over former U.S. infantry positions, killing whoever was there. The battle finally turned, and the attack recoiled, when K Company, 3/27 Infantry, held the regiment's last and deepest blocking position.

Around 2130 hours, 5th Company's XO, LT Joe Ulatoski, was orderd to the 2/27 CP to form a security perimeter there because squads of Chinese were now running all through the battalion's rear. Ulatoski assembled about twenty men—cooks, supply personnel, commo people, plus three Rangers—that he distributed among four strong points. He led an eight-man reserve that grew in number as stragglers rallied on the sounds of U.S. fire.

Ranger Walt Eckhardt recalled how the media were covering this last ditch action:

[Two reporters] *were with the 2nd Battalion, 27th Infantry that night. They took refuge in a large well that had a shelf above the water and covered it with tarp. Occasionally they would stick their heads up and ask passers by what was happening. Then they would go back to their portable typewriters and, by candlelight, write about the war.*

Their story was that, like a tidal wave, two CCF divisions had crashed against the 27th Infantry's positions and engulfed them. But not many Chinese could surge through with following waves—there're weren't any because artillery and tank fire had chewed them up and gnawed them down to nothing. Within the U.S. lines the surviving attackers fought on savagely but with diminishing ammunition and dwindling numbers as dawn approached.

At first light American heads were down from plentiful sniper fire; suppressing it was the first job for the Rangers. Out in the valley where the Chinese hordes had appeared the night before, tac air now swooped like gulls over a fishing boat. The sky was overcast but P-51's kept poking through to bomb and strafe anything that moved on the enemy side of the penetration.

That penetration was several miles deep and almost a half mile wide at its neck. But it had been blocked (well to the rear of the battalion CP) and a reserve company closed the neck by mid morning. The CCF—only about a battalion of them were fighting within U.S. lines by this time—were bagged and about to be broken up completely. Scagnelli was given the mission of sweeping back up the east side of the valley where the most desperate early fighting had occurred. It was pay back time for the Rangers who moved out on line as if they were conducting police call.

Everyone hoped to come upon some MIA's from the night's fighting. Revenge became the order of the day when they found LT Lang's body unrecognizably shredded from multiple bayonet cuts, soggy with blood. There were already plenty of CCF prisoners so no more were taken. Moving forward as if on a pop-up target range, 5th Company shot everything that moved—and just for good measure, shot all the 'corpses' because the Chinese were good at feigning death. It was grisly but satisfying work. Over 700 enemy bodies were counted on this sweep.

This battle had done much to dash the CCF attack but the offensive kept coming. So as not to be stuck out alone out in the Han Tan Valley, 25th ID continued to retrograde by successive positions. This maneuvering allowed some slack for the infantry regiments, but the Rangers remained in constant contact, attached each day to whatever battalion was pulling rear guard while the others leap frogged south.

In order to conserve the depleted ranks of his men (company strength was down to below sixty), Scagnelli became a foremost expert in punishing the enemy while they moved forward but disengaging before 5th Company took much punishment themselves. But one day in late April, an adjacent company was hit so hard and fast that it could not disengage as the Rangers were doing. Scagnelli attacked to take off some pressure off, enabling the other company to extricate itself. In so doing, 5th Company was almost surrounded. Only outstanding soldiering by SFC Willis Powell prevented a complete encirclement, as he led an attack to the rear, walking artillery right behind the last Ranger out.

After neary forty miles of withdrawal, I Corps could tell that the CCF offensive was running out of steam. The Chinese could advance only in the dark; under artillery observation and air strikes they had to go to ground by day. Darkness was also the favorite Ranger hours, so 5th Company was constantly setting night ambushes while 25th ID began to gather strength on the last high ground about ten miles northwest of Seoul.

Here they settled on the Lincoln Line: a deep, extended system of trenches and bunkers, minefields and switch positions, that had been prepared in anticipation of the CCF spring offensive. As I Corps delayed back toward it, the Lincoln Line was feverishly improved with fortifications that could stop all the Chinamen in Asia. By May, 5th Company was finally released into inactive reserve where they recouped, and refitted, but mostly the bleary eyed veterans of a month of constant fighting just rested. Not until 11 May was the company fit enough to truck down to Kimpo airport for a training jump.

In mid-May the most serious fighting of the CCF spring offensive shifted over to the eastern half of the peninsula where the Chinese made headway against the ROK's. On 14 May, I Corps launched some limited objective attacks to take pressure off the eastern front. The first of these was a reconnaissance in force by Task Force Hamilton comprised of 5th Company and a beefed up tank company. It was to be a day trip. They were to strike north up the Hanchon valley (the lower Hantan River) to Uijongbu, with the mission of determining enemy strength, taking prisoners, and inflicting damage.

This was ground the Rangers had seen before during the long delaying action. Though still much understrength from the battles of April, their 'will to kill' had grown stronger and they had some scores to settle. The tankers also had more enthusiasm for being advance rather than rear guard for the division. So TF Hamilton slashed north with a vengeance, howitzers pounding suspected enemy positions and tanks shooting up any sign of movement in the hills. The only instrument missing from the combat support orchestra was tac air, hampered by low visibility.

The task force ran into mixed degrees of resistance. Apparently an NKA battalion was screening the valley as the CCF regrouped. Hamilton brushed them aside and kept driving north, intent on clearing Uijongbu before noon. All the noise of the meeting engagement must of drawn the concern of CCF farther up the valley where they were assembling for a probe of the Lincoln Line. About a mile northeast of Uijongbu, they joined the running battle to stop Hamilton's incursion and protect their supply dumps.

The Rangers dismounted from the tanks several times to clear resistance from adjoining ridgelines. This was hard physical work, rapidly climbing toward the enemy, rapidly sweeping parallel to the advancing tanks, rapidly getting back aboard, then doing it again at the next blockage. Around 1000 hours when about a battalion of Chinese entered the fight contesting a defile, 5th Company was a little winded from the earlier actions but took on the new threat without a pause.

The tanks spread out to break through the defile, but soon several were immobilized by mines. Hamilton ordered the Rangers to take some high ground to protect a mine clearing team. 3rd Platoon, led on this occasion by LT Ulatoski, drew the mission to seize a long hill and deny observation for enemy weapons covering the minefield. The Rangers hopped to the job, sweeping up the gradual slope against scattered opposition.

But nearing the crest, resistance stiffened. Rangers shouted that there were tunnels cut through the hilltop. Chinese began to pour out of them to protect the crest. At this critical juncture, commo with the task force was lost so supporting fires could not be adjusted.

To make matters still worse, upon crossing the crest 3rd Platoon found that a CCF company was set up in a reverse slope defense where they could rake the hilltop. Others could be seen gathering to probe the flanks. All Ulatoski could hope to do was prevent encirclement and hold what he had, because to retreat down hill would expose his men's backs. Some wild fighting went on as Chinese mortars zeroed in on the hill where their troops had cover in holes while the Rangers were fighting in the open. The life or death priority was to kill the Chinks in the holes so Rangers could get in them!

And the CCF company on the reverse slope seemed prepared to counterattack. Clearly 3rd Platoon's situation had grown desperate.

Ulatoski:

Right in front of me I saw four Chinese occupy a position to set up a machine gun which would bring us under point blank fire. Being the first Ranger to see this, I grabbed a hand grenade, pulled the pin and was cocking my arm when a mortar round exploded immediately to my rear, seriously wounding me.

I apparently blacked out momentarily because the next thing I recall was a voice screaming, "The grenade, lieutenant, the grenade!" Suddenly someone fell on me, locking my hand to the spoon on the grenade. It was PFC John Hammond, telling me that I was holding a grenade with the pin pulled. He pried my hand loose, threw the grenade and destroyed the machine gun with its crew.

Hammond was one of the many heros of this action, dragging Ulatoski to cover, providing first aid, then rejoining the back and forth fire fight that went on for the rest of the morning.*[*FN: TF Hamilton's reconnaissance in force and subsequent larger unit actions determined that 3rd Platoon's hill was actually a strong point on the enemy's MLR. Days later it required a well supported U.S. battalion to seize the hill that 3rd Platoon almost took in a rush.] The raging fight resembled 3rd Company's dire plight on Bloody Nose Ridge—both were part of an armored reconnaissance in force where faulty commo almost resulted in the anihilation of a Ranger platoon.

3rd Platoon took heavy casualties but were not anihilated, primarily because of their aggressively hard fighting, but also through the initiative of squad leaders when Ulatoski went down. They knew what had to be done: it was obvious to every Ranger on the hill that once more they would have to make a fighting withdrawal or this time they would go under. Every Ranger automatically began that maneuver, squad leaders controlling it as if they had mental telepathy with each other.

Scagnelli, of course, had his heart in his throat from the time commo went out and he could see a big fight erupt on the hill. Most of TF Hamilton had passed through the defile when 3rd Platoon's action removed enemy fire on the minefield. Scagnelli persuaded a tank platoon to break off and move toward 3rd Platoon's hill. Just when the Rangers needed it the most, these tanks were able to provide some covering fire— 50 caliber mostly— for the 3rd Platoon to extricate themselves. Using fire and maneuver to the rear—a tactic well learned during the delaying actions of April— they were able to get all the wounded into defilade at the foot of the hill before the able bodied Rangers pulled off. They did so behind a departing shower of grenades. Throwing their grenades high, they were already running or tumbling down the hill when the 'air bursts' exploded.

Maybe the most remarkable feature of this withdrawal under great pressure was that only one MIA remained on the hill. Under overwatching tank fire, the remnants of 3rd Platoon—some carried piggy back down the hill—rejoined TF Hamilton which had already achieved what it came for : the CCF MLR was located, prisoners taken, and a great many communist soldiers killed. 5th Company's losses were eight dead, eleven wounded so seriously that they never rejoined the company, and twelve less seriously. Loss of sergeants and lieutenants was disproportionately high. On 23 May, 5th Company could muster only 78 men for duty, many of them convalescing.

There would be little rest for the Rangers as 25th ID was now assigned objectives for their part in Operation Detonate, an I Corps attack to push the front farther away from Seoul. The spearhead would be Task Force Dovlin, a name familiar to readers of Ranger history. The previous year a task force of the same name, led by the same armor officer, included Eighth Army Ranger Company, taking it to its biggest and most deadly fight. No doubt Scagnelli wondered if history would repeat as 5th Company jumped off at 0625 with 25th ID Recon Company and two tank companies from the 89th Tank Battalion.

They were headed once more toward the Iron Triangle but the division attack would be a long time getting there. On the first day TF Dovlin had pretty easy going up a valley parallel to the Han Tan. Small enemy concentrations were found spotted around knolls and lower elevations in the valley, but the commanding high ground did not seem to be occupied in strength. That was deceiving, however, because a great many Chinese could be concealed in the rugged terrain where they found shelter from tac air.

About ten miles and five hours into the attack, a long ridge paralleled the road, and it was here the enemy took advantage of this good flanking position; i.e., the Amercans could not continue to advance without first turning and driving the CCF off this long ridge. Diverting a recon platoon to attack up the long axis of the ridge, Hamilton pressed on with the rest of his task force which supported by fire from the road.

Past the ridge, the valley sloped down to a good sized stream that cut diagonally across the axis of advance. A partially destroyed bridge spanned the stream. Hamilton knew this foot crossing would be defended and it was, not by a solid defense but by scattered clumps of mutually supporting trenches, holes and low bunkers that forced the Americans to fight constantly in all directions. No single objective was critical to this kind of 'checkerboard' defense, so the task force felt like they were swatting at hornets. The Rangers liked this sort of work, however, going after small groups with small groups of their own.

Dovlin did not want to spread his force out too much in this kind of bushwhacking. He had reached his phase line for the day; the bridge was beyond it, so he pulled back into a laager before dark.

His success brought reinforcements, an infantry battalion and an engineer company, for the next day's push. But overnight too the CCF determined to oppose the stream crossing in strength. Organizing the task force into three tank-infantry teams, Dovlin decided to ford at as many crossing sites simultaneously. They were all taken under heavy mortar and anti-tank fire from ridges on the northern, eastern and western sides of the valley.

The task force would have to clear at least two of these ridges before it could proceed. Dovlin sent the Ranger heavy team north after the farthest objective. Under fire from two sides, two platoons of 5th Company rode to work, knocking out two 57 mm recoiless rifles (apparently captured from Americans) enroute. As they climbed the ridge, tanks overwatching, Rangers could hear CCF mortars on the reverse slope thumping at maximum rate of fire as if they knew the Rangers were coming and the battery would soon have to go out of action. The communists were right because after knocking out two machine guns in short order, the Rangers surged over the ridge and took the displacing mortar crews under effective fire. As the CCF broke and fled, the Rangers took two prisoners and counted 14 dead.

Fighting for the other two ridges took longer, but in a couple of hours TF Dovlin was across the deep stream and underway again. Division ordered him to exploit and advance as far north as possible without becoming over extended. Dovlin was ready and willing but his trains had not yet completed the stream crossing, so he waited till they had. This turned out to be a wise decision because after fording, the thin skinned trains vehciles came under attack from a couple of CCF platoons who had infiltrated into the center of the valley after the U.S. tank-infantry teams had moved on.

The Chinese sprang their ambush of Dovlin's trains at a shallow defile. He ordered both Ranger teams to turn around and come to the rescue. Within minutes the Rangers were attacking to the rear. For once they outnumbered the enemy and could use tanks at point blank range. The teams rolled over the CCF platoons and virtually anhilitated them by sheer shock. Dovlin's trains were able to continue with small losses.

These sharp and speedily successful actions during the first two days indicated to the Rangers that there may have been a sea change in the fighting. Clearly momentum in the war was shifting, and the CCF did not seem to be as fanatical though they still fought tenaciously. TF Dovlin went into its third day of operations with high spirits and determination.

Their mission now was to hook deep, bypass as much as possible, in order to get into enemy supply areas and cut off retreat as a regiment of 25th ID applied direct pressure from the south. Clinging to whatever they could hold on to on the bucking tanks, 5th Company could only crouch closer to the hulls when Dovlin's columns were taken under fire. On the move, a tank commander might reply with his 50 caliber. Rangers could spot for him, but this was an uncomfortable form of travel where they were fired on but could not fight back.

The task force pulled up about ten miles behind enemy lnes. They were temporarily outside range of supporting artillery but felt they could rely on themselves. Two tank-infantry strong points were established as road blocks to bottle up enemy withdrawal. The third team became a mobile reserve as the Rangers gratefully dismounted and fanned out recon the valley floor.

There was good hunting and easy pickings here. Chinese REMF's were gunned down wherever they appeared. Several NKA trucks were discovered and destroyed. Rangers uncovered a ration break down point, taking what they wanted, grinding the rest into the dirt.*[*FN: Always short of rations themselves, most Rangers developed a tolerance, if not a taste, for rice, the portable staple of the CCF. Boiled and mixed with a wet ration, rice proved to be a filling supplement for stomachs that were often empty.] Next they located an ammo dump that the engineers were happy to blow up. The Rangers were having a field day, but the commotion they raised drew notice from communist combat units who had been focused to their front. Now their rear was threatened and they turned to do something about it. An enemy infantry company was spotted coming out of the hills to investigate. Its general direction of movement was toward one of Dovlin's strong points that the Chinese didn't seem to realize was there (the U.S. tanks were pretty well concealed.) Staying concealed themselves, Rangers slinked up the flank of the unwary Chinese. When the Rangers opened fire in unison, the CCF veered directly into the trail block ambush. Recoiling with heavy losses, they came back into the Rangers' fields of fire with more losses.

The survivors dug in where they were pounded by air strikes which finished them off. There were no U.S. casualties in this wonderfully one sided action that ended the CCF's counter attacks for the day. Free to range the valley, a team blew away a truck park and another ration dump. Rangers took 26 prisoners.

The next day TF Dovlin sealed the valley, acting as an anvil for the hammer of 25th ID's advance. After link up, the victorious task force was dissolved and 5th Company attached to the 35th Infantry Regiment which put the Rangers in the usual place, out front on patrol. The division advance had carried north for nearly forty miles. It now approached the 38th Parallel where, fittingly, the first American across was Ranger Fred Hoy.

Triumphant but tired, 5th Company was the

first unit to determine that there was still plenty of fight left in the CCF. In a jumble of hills just over the parallel, they got into a very sharp fight after hitting a withdrawing column, probably a company whose parent battalion had already reached its assembly area out of sight from the Rangers. The CCF quickly came to the column's aid. Apparently their assembly area was also an ammo dump (confirmed by secondary explosions from air strikes) because uncharacteristically they expended ammo like the war would be over the next day.

By contrast, 5th Company soon ran short of ammo and the CCF had closed into a semicircle around them. Scagnelli's choice was to pull back at dark or dig in. Holding good defensive terrain, he decided to stay. The next morning the Chinese who had nearly encircled him were gone, leaving a trail of bloody bandages.

This brisk action ended the battles of May which had cost the Rangers four dead and eleven evacuated wounded. The lst of June thereby marked the low point in 5th Company's foxhole strength—less than fifty present for duty. Fifteen replacements arrived, and twenty Rangers returned (from wounds, sickness and leave), before the company was attached to the 27th Infantry for the next action which kicked off on 4 June.

The days of TF Dovlin and long, dashing advances were over. I Corps had slashed its way up to the base of the Iron Triangle but that's as far as they were going.*[*FN: The USMA Department of Military Art and Engineering explains the break off—which Gen. Van Fleet stoutly opposed— of U.N. offensive this way: "The JCS, always concerned that Russia might seek to take advantage of the concentration of so large a portion of American forces in Korea, was opposed to another drive north." This reason does not make much sense because the stalemate which soon ensued drew down U.S. forces in Korea hardly at all. The more likely reason was that continuation of the offensive would have caused more casualties than stalemate. During this period, casualties had more political than military significance. The hard boiled truth is that well conceived and executed attacks, though initially more costly, ultimately reduce the number of casualties.] The CCF had dug in, their offensive ambitions shattered, but their defensive positions grew almost as strong as the Lincoln Line. For forces in contact, Korea now began to resemble World War I's trench warfare. And it seems that whenever God sees trenches, He fills them with water.

Under these circumstances, 25th ID set about to straighten its front lines, pinching out a salient here, seizing a hill or two there, to obtain full observation and integrated fires all across the front. These were minor maneuvers in the big picture but often the toughest sort of infantry work.

On an operation typical of this static period of the war, 5th Company was assigned Hill 722, named Objective Sugar, in a limited attack by the 27th Infantry Wolfhounds. Surprise would be unobtainable in a night attack against the alert and well prepared Chinese, so Scagnelli decided to hit them in daylight (and rain) when supporting fires would be most effective.

After an exhausting, soaking approach march, the Rangers reached 3rd Battalion's forward foxholes where they grimly prepared to do their duty. With double loads of ammo and bayonets already fixed, 5th Company was guided through friendly mine fields, then spread out to maneuver toward the flank of the strong CCF position on Hill 722. Enemy outposts were drummed by artillery but the Rangers did not engage in order to conceal the strength and location of the upcoming assault. Apparently few Chinese got away from the OP's because on the lower slopes of the hill, a number of enemy troops stayed huddled in defilade, using spider holes that provided maximum cover from artillery.

The spider holes were also well concealed, so much so that they might be unnoticed and bypassed. This happened to Pat Brady who turned and saw a camouflage mat come off like a lid, a Chinaman pop out like a jack-in-the-box to swing an assault rifle in Brady's direction. But Brady pulled the trigger first—only to hear the firing pin thud in a misfire. A buddy saved him with a BAR burst. Brady had to continue the attack with grenades and a 45.

Belts of spider holes were just an introduction to the CCF's defenses. The hill steepened, the heavily loaded Rangers slipping and falling on the rain-slicked rock. Fire from heavy machine guns began to crisscross their clamoring advance. Under such fire, Rangers needed the CCF's spider holes. Chinese were evicted with bayonets, Rangers often firing to free the blade after it had sunk in. Another technique much used was the belly crawl, feeling ahead with one hand for the lip of a spider hole; after finding it, slipping in a grenade. From this prone position, manhandling the body from the hole was literally heavy lifting.

The advance itself became a crawl with casualties climbing much faster. 5th Company had no room to maneuver: it was platoons on line with squads as skirmishers. The Rangers were a wavey line, inching up hill as friendly artillery silenced enemy mortars and churned the hillside a few dozen yards ahead of the attackers.

A platoon leader was hit and rolled down where he joined ten other wounded Rangers who had been able to evacuate themselves off the hill. Above them, three other WIA's were under too much fire to be pulled back; they were repeatedly hit and died. CPL Walt Eckhardt, when his right arm was shattered, crawled over to a buddy with his BAR so the weapon would remain in the fight. Every automatic weapon was vital because the Rangers had been unable to gain fire superiority. The base of Hill 722 was also under fire, so the WIA's who could walk retraced the approach march back to 3rd Battalion's positions.

Meanwhile 5th Company's assault on the hilltop had sputtered as if doused by the drizzly rain. Interlocking fire from the CCF was too hot on the upper half of the hill where they made good use of many large boulders. Scagnelli passed the word to pull back to the base of the hill to reorganize and find a different approach. Interrupting a difficult attack, then resuming it successfully is not often accomplished in combat but Scagnelli had done so to seize Hill 383 in March. If anyone could do it again it was the Rangers, so Scagnelli called for a council of war even in the heat of battle.

2nd Platoon had discovered a narrow, extremely steep ravine that led into an lightly defended sector of the CCF positions. Scagnelli ordered LT McAbee to try this approach while lst and 3rd Platoons infiltrated back up hill to gain good assault positions for a coordinated attack with 2nd Platoon. In the ravine, McAbee would probably lose radio contact so the signal for the coordinated attack would be when he fired a red flare.

With the longest and slowest way to go, 2nd Platoon set off on their mission while Scagnelli called back for some help from 3rd Battalion. Their support had been less than superb up to this point. Short rounds from their 4.2's had caused several Ranger casualties; they had provided no litter bearers for Ranger WIA's who had made it on their own back to 3rd Battalion lines. So far they had been not much more than spectators for 5th Company's increasingly desperate fight (19 Ranger casualties with more to come).

Now Scagnelli announced that if the 27th Infantry really wanted Hill 722, they had better send up two infantry platoons as reinforcements. They were to answer to whatever Ranger was in charge, even if the Woldfhound platoon leaders ranked him. Abashed, the 3rd Battalion CO complied. The two Wolfhound platoons provided a base of fire for lst and 3rd Platoons who moved back uphill at half strength.

Besides the red flare assault signal, Scagnelli needed phase lines to control the operation. They were very valuable as the second attack moved methodically toward its objective. The Wolfhound platoons performed well, and generally coordination was excellent.

Then Scagnelli was wounded by mortar fire. His life was not in danger but he clearly could not continue in command. The XO, LT Ulatoski, was in a hospital from previous wounds. The senior Ranger officer at hand was 2nd LT McAbee; he was called back from 2nd Platoon (turning it over to MSGT Bartholemew), and though ranked by both Wolfhound LT's, McAbee ran the attack from then on.

The second climb up 722 was no easier than the first—probably even more dangerous because the CCF had brought up reinforcements plus numerous mortars that slid into positions that had defilade from even the considerable weight of the 105 battalion firing in direct support of the 27th Infantry.

Scagnelli's oral order had called for a coordinated attack, launched by 2nd Platoon's red flare. But he was no longer present to supervise the execution his concept. What happened in his absence was that the four-platoon frontal assault never materialized because of withering direct and indirect fire from the hilltop. All they could do—and to do so required great fortitude, determination and dirt level infantry guts—was creep up to the higher levels of Hill 722 where a base of fire could cover the hilltop. This relentless attack took the lives of two Rangers, the wounding of fifteen more. Their sacrifice allowed 2nd Platoon to position themselves at the top of the ravine to carry the objective.

Rangers and flares seem to have a fateful

relationship. Flare failure at Arzew, Algeria, almost undid Darby during the Rangers' cherry operation in World War II. This time 2nd Platoon's red flare was the signal that never happened. Rain and ruin prevented its ignition. The attack stalled. The frontal force was pinned down but ready to charge if the flare went up. But it didn't. It took a messenger a half hour to get that information back to McAbee (as Scagnelli had anticipated, 2nd Platoon lost radio contact in the ravine). McAbee knew he had to get the order out himself—the order for everyone to assault.

He attempted to by going forward (uphill) to use arm and hand signals, supplementing radio commo with the frontal force. McAbee's movement, with his RTO and company runner, quickly drew attention from the CCF who showered his little command group with grenades and peppered it with fire. All three men were badly hit, along with the radio. So the final assault remained coiled, like the strike of a rattlesnake whose head had suddenly been severed.

Thus what momentum and potential 5th Company's attack had developed on its second surge was lost. McAbee, though reeling from loss of blood, was able to send a message to MSG Bartholemew, ordering him to withdraw 2nd Platoon which was losing in the exchange of grenades with the CCF who were tossing dozens of stick grenades like confetti at the Rangers clinging to the top of the ravine. Orders were orders, so 2nd Platoon crept back down hill. Before the medics led him away from the hill, McAbee told the frontal force to also break contact and reoccupy a rearward phase line. His last message was a request for instructions from 27th Regiment.

No instructions were forthcoming. This anti-climax ended the attempt for an objective not worth a minute fraction of the bravery and blood expended to take it. Hill 722 was the Ranger equivalent of more familiar, but equally unworthy, objectives with names like Porkchop Hill and Heartbreak Ridge which were symbolic of the phase of the war that spawned the motto, "Die for a tie." It was no wonder that not many soldiers, even Rangers, were willing to adopt that motto.

A little more of such fruitless combat was in store before 5th Company's Korean history—along with that of their sister companies— ended in deactivation. The first operation was a screening mission between two infantry regiments attacking to establish the Wyoming Line which would be the final front for Eighth Army. On the second day of the push, 5th Company was sent ahead in an area recon to find fording sites for continuation of the attack. No one had to tell the Rangers to hold down casualties— they already knew how to do so with long range fires and artillery.

These operations were successful and lasted from 10 till 23 June at which time 5th Company went into division reserve for a much needed three-week rest. Not much effort was made to bring them up to strength because rumors of deactivation were already being taken seriously. Almost the only replacements were Rangers returning from hospitalization. Yet 5th Company trained itself as if more combat was imminent. Eighth Army's fighting edge might becoming dull but Rangers would never stop sharpening theirs.

5th Company's final action was again with the Wolfhounds, ten days during July, patroling across the Imjin River in a sector recently taken over from the 1st Cav. Fighting almost exclusively with artillery fire, the Rangers inflicted many casualties while suffering none themselves. If only all of the war could have been fought that way!

But for 5th Company the war was over on 1 August. Orders were cut for airborne qualified members to transfer into the 187th RCT; most of the rest readily found homes in 25th ID where 5th Company had an unsurpassed reputation. Before the companies broke up, some sort of swan song was called for—plotted actually. There had not been a Ranger battalion operation during the entire war. Now as the 3rd, 4th, 5th, and 8th Companies collected on a train for Pusan, a battalion and an operation plan was formed.

Though unnamed, it could have been called OPLAN Carnage-Revenge. Already sulking from deactivation, the battle grizzled Rangers were further insulted by their current designation as replacements, a status that officially leveled them with cherries fresh from the States. Company officers counseled for everyone to hold their peace and assemble dociley in what, since Puckett's day, was known as the Ranger Compound outside Pusan. The OPLAN would be executed at night—when Rangers are at their best (or worst, depending on your point of view). An added incentive—if any more was needed—were orders from the REMF headquarters restricting all Rangers to their compound.

On the evening of 4 August, darkness descended like the gloom of the Rangers spirits. Then conversations between companies reinforced a growing rage at their collective fate. Well lubricated with strong Korean beer, talk turned to action and execution of the OPLAN.*[*FN: 8th Company veterans claim to have supplied the beverage that night after hauling a water buffalo to an abandoned brewery where they found huge shattered kegs with about a foot of reeking slurry at the bottom. The fumes from these dregs were so powerful that rodents had died from inhalation and were floating bloated on the surface. Nevertheless this was some form of alcohol and was sloshed into the buffalo which returned for the Ranger sendoff.]Pyrotechniques and demolitions appeared and were put to use.

What followed was part party, part bacchanal, part mayhem, part debacle— and all Ranger. The compound fences were demolished, Pusan illuminated by flares while gunfire erupted in tracers that caused REMF's to believe that they were under an air raid. Rage, frustration and betrayal were equally eruptive. With no communists to fight, fisticuffs broke out among Rangers, including officers. A sergeant had served CPT Elkenberry well at Ft. Benning but when this sergeant was assigned to 5th Company, Scagnelli busted him. This led to a personnel matter becoming personal, and in the Pusan Ranger compound they came to blows. MP's sent to quell the Ranger riot, were brushed aside like so many flies.

There was, of course, official reprisal the next morning when the Rangers companies were formed up in the shambles of the compound to hear a REMF officer accuse them of indiscipline and bestial behavior. Many of his audience were about to ETS or rotate, so he was answered by rumbling growls and salutes that did not include the whole hand. Then, like EPW's, the Rangers were loaded onto flat bed trucks and dispersed to other camps. The finest American fighting forces in Korea were no more, gone with only their memories and the scrolls on their right shoulders as badges of incomparable service and sacrifice.

Eighth Airborne Ranger Co.

This book has taken the regular army Ranger companies of the Korean War in numerical order, consequently ending with 8th Company. By coincidence, in 8th Company was the Rangers' foremost participant-historian, Robert W. Black. Not only did he assemble the raw material that resulted in two classic books, *Rangers in World War II* and *Rangers in Korea,* it is to Black that all generations owe gratitude for sketching the origins of most ranger training experiences. Whether or not a soldier was ever assigned to a Ranger unit, the Ranger School provided him the total environment for the most uncompromising training he ever received, and the closest approximation to the realities of war—war as expressed in infantry combat.

So it is in a monograph history of 8th Company, titled *Devil* by Bob Black, that we turn to consider important traits and training methods that were rewarded by the most coveted of all skill qualifications, the Ranger tab. Paradoxically, in 1950 the first tabs were more scorned than prized by their recipients. That's because the red, black and white scroll was the hard won coat of arms for Rangers from World War II. The tab of black and gold—a combination allegedly selected by a graduate of West Point whose school colors are grey, black and gold—did not make that vital historical connection.

But for once history smiled upon the Rangers: the tab did not replace the scroll but supplemented it because the scroll returned (at first unofficially) to Rangers' shoulders as a combat unit patch. The tab and the scroll then became related like the Expert Infantryman and the Combat Infantryman badges.

Whether in mountains, desert or jungle, war for those who fight it means nearly constant discomfort, fatigue and fear. From the beginning and thereafter, Ranger School provided these three factors in abundance. Weather and terrain along the Chatahoochie valley afford all that is needed for discomfort plus some elements of physical danger to produce fear.*[*FN: Ranger School graduates after the first cycle appreciate how difficulties of terrain were enhanced by dividing advanced training into the 'Florida' (swamp) and 'mountain' phases, thus ensuring experience with extreme terrain.]

Short of arming Ranger candidates and opposing forces with ball ammunition, a fear factor of death could not be introduced into training. But a psychological equivalent could. That was the fear of letting close comrades down. The first comrad in that status was the 'Ranger buddy'. He could have been a previous acquaintance but might be someone assigned at random. However they came together, Ranger buddies were to become practically Siamese twins, each experiencing what the other did, never leaving the other no matter what the circumstances. If one was patrol leader, his buddy was assistant. If one became lost or even captured his buddy was too. In ideal combinations their respective strengths compensated for respective weaknesses. No matter how individualistic or selfish a Ranger candidate might be, the buddy system required that he live outside himself and form one of the strongest partnerships of his life.

The next bond was to the squad, then the platoon, then company. Finally there was the ultimate allegiance to Rangers as distinguished from all others in uniform. In 1950 Rangers were required to be airborne. Later, during the initial period when DA policy was to diffuse Ranger skills throughout the army, the airborne requirement was dropped but most Rangers were parachutists anyway.*[*FN: Presently all Rangers in Ranger units are airborne qualified, so the 1950 policy has been resumed.] Of the two qualifications, there was never any contest as to which was more highly valued: learning to become a paratrooper took five days, to become a Ranger six weeks. The comparative washout rates were roughly five and thirty percent.

There is nothing new in unit pride as a combat factor (the vogue term is 'force multiplier'), an advantage sought by commanders going back at least to Greek phalanxes. In most settings instilling such pride requires artifice by commanders because the testing mechanisms are not present in a real way. Battalions within the same division in peacetime, for example, have very few meaningful ways to demonstrate superiority in combat performance. Ingenious formulas combining garrison and training indicators are the substitute. They create competition but hardly indices of combat proficiency.

Even less convincing is similar competition between companies at a basic training center. There is simply no plausible way to determine the ultimate question, who would do better in combat? But that's an irrelevant question because a basic training company will never go into combat, and it is is dubious to raise the question of 'best' in that context. But proponents of such competition would answer that being 'best' is an attitude which should be instilled.

Coming into Ranger School, candidates already had been exposed to such artificial competition designed to instill a 'best' attitude. Then, like immersion in ice water, they found themselves competing for survival with the elements of war: discomfort, fatigue and fear. If they did not meet these challenges they were not the soldiers they thought themselves to be. Which is to say that what they believed about themselves was tested by what they came to *know* about themselves in Ranger School. That's the way it all started in 1950—and has changed remarkably little in over forty years.

Whether it was COL Van Houten, Walker or Singlaub, or Ranger veterans from World War II, or some combination of inputs, the virgin RTC did exactly what it set out to do— convince its graduates that they *knew* their capabilities, not just what they had hopefully believed. Black's descriptions of the training undergone by 8th Company sounds very familiar to any Ranger graduate. The training worked so well that the essentials remained constant despite enormous changes, over the years, in technology and tactics elsewhere in the army. And the essential of the essentials was self-knowledge through demonstrated ability.

Almost all the first Rangers to be trained at Ft. Benning were from the two airborne divisions extant in 1950, the 82nd and 11th. When the Korean War broke out, both divisions had geared up for imminent deployment but their high alert status never went beyond "stand in the door". The airborne were to remain in strategic reserve in case Korea spilled over into fighting between super powers in more important theaters. Confrontation with the Soviets in Europe was still the main show for which the airborne divisions stood in the wings.

Soldiers have varying appetites for combat, and most of the 'troopers who volunteered for Rangers were the kind who had wearied of the spit and polish garrison life of Ft. Campbell or Ft. Bragg—and fidgeted as they read of the fighting in Korea. More than any other motivation was that to get into the fight. This attitude was to continue as the prime charcteristic of Ranger volunteers thereafter, that and the desire to be your best among the best.*[*FN: Perhaps this attitude attenuated in succeeding decades as the Ranger tab came to be looked upon as a 'merit badge', an expected item on the resumes of regular army officers. Be that as it may, the RTC never relaxed its standards.]

It was the combat knowledge of the original RTC cadre that determined what being the best required. In retrospect several branches of war craft might have been followed. For example, soldiers are shooters so (it could have been argued) the best soldiers should be the best shots. But marksmanship was not practiced in Ranger School. Moving in formation is vital for maneuver, but battle drill—much less dismounted drill— was absent from the Ranger curriculum except for immediate reactions in patrol situations. Many other texts from the school of the soldier could have been taken as the model for Ranger training. What was chosen was development of the individual's self-knowledge of his abilities from the view of how they would contribute to Ranger missions.

And the first component of the individual was his physical condition, addressing the fatigue factor in the three characteristics of combat. Newly assembled, 8th Company at once began a steepening regimen of fatigue.

Black: *Five mile runs greeted the dawn, speed marches in full field equipment were artfully blended with formal calisthenics, log and pit drills. Young men who prided hemselves on being at the peak of physical condition heard from muscles that had taken previous training in stride.*

"What is an Airborne Ranger?" asked a letter from home.

"Tired," was the response.

For nearly every man in 8th Company and others in the second RTC cycle, the spirit was willing but sometimes the flesh incapable. In contrast to basic and other training schools in the army, men were not driven to meet standards. Instead they could opt to be driven away in a jeep that followed every morning run and speed march.

If a man felt training was beyond his capability, he had only to go sit in the jeep. No questions would be asked, no comments made— he would just be driven away and his belongings moved from Ranger barracks before the company returned. No one was forced to be a Ranger, the desire had to come from within. Those who failed were returned to the airborne divisions without prejudice.

Of course there was tremendous incentive to remain, even if that meant exceeding previous physical limitations. A mother would sooner have her first-born torn from her breast than would a Ranger have his buddy fall out and depart Harmony Church forever. Yet the tempting option remained. No one had to continue the sometimes brutal training, and no formal penalties attached if he didn't.*[*FN: In writing previous books, the author became acquainted with soldiers who had left Ranger units for reasons that were never questioned. All were admirable soldiers who contributed greatly to their conventional units. So it would seem, from the 'outside' army, that to have been an ex-Ranger is more a credential than a stigma.]

The concept of physical rigors inducing self-examination can not be credited to the original RTC—though the American adaptation of it can— because, while starting up, RTC more closely resembled foreign schools for elite forces than it did sister U.S. training institutions. World War II Rangers, in significant ways, were offspring of the British Commandos. Darby studied their training closely before setting up his own.

The feature he carried over directly was the principle of test by stress. For instance he noted (but did not adopt) the Commando practice of interviewing candidates naked—a shortcut to observing how a man conducts himself under unique stress. Many of the RTC cadre were Commando trained, so the test by stress philosophy was very prominent in the training experiences of 8th Company's cherries.

But there were as many differences as similiarities between Ranger and Commando training, differences based primarily on the sort of combat in store for the graduates.*[*FN: During the winter of 1950 LTC Singlaub visited Korea to brief the divisions on the Rangers in the pipeline, but also to gauge what sort of operational environment the Ranger companies should expect. He found that because of the do or die fighting going on during that winter, Eighth Army was distracted by more pressing matters than how to best employ Rangers when they arrived. That attitude, plus the constant turnover of key operational personnel in the divisions, convinced him that the Rangers would largely have to make their own way, often on unenlightened missions. His judgement proved to be very accurate, and underlined the self-reliance required in Ranger training.] Commandos were much more amphibious creatures than the 1950's Rangers needed to be. Eighth Army was already ashore, on line and in contact. Though amphibious operations were included in the Ranger curriculum, crossing water to reach the enemy was not an overriding priority.

Commandos were designed to be raiders,

more so than Rangers; their respective slogans could be 'hit and run' versus 'patrol and penetrate.' Thus the duration of Commando operations were expectedly shorter than what 8th Company trained for. Wilderness survival was more emphasized by Rangers than Commandos whose theater could be considered civilized. E&E for Commandos meant making immediate contact with the local resistance; E&E for Rangers meant avoiding civilians altogether while trekking for days through moonlike terrain.

Facing a sophisticated enemy in the Nazis, Commando talents were also required to be more sophisticated (e.g., electronics, sabotage, interface with an underground). Though weapons specialists were at a premium for RTC in recruiting their first candidates, Rangers emphasized more basic soldiering skills than Commandos.

Cultures also differentiated Commando and Ranger training. The foremost example would be use of live fire. Except for the standard infiltration course, Ranger trainees did not hear as much live fire; whereas Commandos used snipers to plunk rounds near trainees during most advanced exercises. This was certainly an advantage for realism but was curtailed by RTC because of safety. Far more than any other U.S. school, the RTC pushed back safety requirements to the absolute minimum but not close to lethal hazards approved by other nations for their elite forces like the Commandos.

Another important cultural difference was the use of hazing and harrassment to induce stress. Commandos condoned personal and sometimes degrading treatment for this purpose. Though not so extreme, other U.S. training institutions also saw acceptable reasons for tearing down before building up a trainee—exemplified by the 1950's image of the overbearing drill sergeant and his epithets like 'scumbag' and 'maggot' for recruits.

The RTC eschewed this derogatory form of stress. Rangers were called Rangers from the first day of their candidacy (though for the secondary purpose of extinguishing rank in Ranger School). GI parties and Saturday morning inspections were out. The RTC wanted the training itself to provide all the stress needed. Black testifies that they certainly succeeded— and probably exceeded their stress goals!

Perhaps the most remarkable aspect of the RTC's training philosophy concerned the meaning of discipline. Conventionally, discipline could be summed up in the proposition that if a soldier can not obey in garrison, he will fall short in combat. Staffed by combat veterans, the RTC knew better. They knew instead, from experience, that men crave orders in combat; that is, they long for their leaders to tell them what to do in order to break the paralysis of fear.

The Prussian purpose of discipline, watered down in varying degrees in the U.S. military, was to cause the soldier to fear his chain of command more than the enemy's bullets. Orders were to be followed blindly and without question, thereby breaking paralysis in combat. In contrast, as Black noted, for Rangers "the desire had to come from within." In other words, the Prussian model inculcated disciplined habits through enforced obedience to orders, while the starting point of Ranger philosophy was self-discipline from which obedience to orders naturally followed.*[*FN: These contrasts in discipline were also operationally driven. In combat, conventional units normally operate in formation and in close concert with other units; whereas Ranger operations are generally more independent and self-reliant.]

It is not surprising then that outward evidence of discipline was not a hallmark of the 1950 Rangers. In passing, Black mentions destructive raids into Phoenix City, rowdyism in barber shops, binge drinking of beer, and other instances of behavior not often found on the record of soldiers of the month. It seems the RTC acted as something of a heat shield for its training companies, and it can be imagined that the post commander stoked up considerable fire under Van Houten to keep his student warriors in line.

It also seems that Van Houten knew the best solution to the problem: keep the Rangers exhausted, in the field, and challenged. One of the most absorbing challenges was learning how to use communist weapons as well as everything in the U.S. infantry arsenal. However, combined arms training seems to have been down played despite after action information from Eighth Army Ranger Company's participation with TF Dovlin. No doubt tanks were pretty scarce at Ft. Benning during this period of the war.

Patroling is a standard and important skill practiced throughout the infantry; but for Rangers in 1950, and ever since, patroling was the training vehicle, the trademark of what Rangers are all about. A patrol, by definition, is a detachment that goes out from friendly lines, penetrates, and returns. Whether to recon or for combat, the RTC foresaw that patroling would be the Ranger companies' stock in trade in Korea—even more so than for the World War II Rangers (with the possible exception of 6th Battalion).

Increasingly larger and longer distance patrols were the perfect mechanisms for test by stress. An early RTC rule was that a well executed patrol could be conducted in the time allowed in the operations order, and the patrol members would then have about four hours of rest before receiving another mission. A less than well executed patrol—or one that ran into extraordinarily bad weather— would take more time, time that came out of the following rest period.

Early patrols by cherry Rangers, of course, often returned late so everyone was soon into sleep debt. Men who never believed they could get by on four hours of sleep out of 24 found that they were gratefully refreshed if they could get two hours. Thus men exceeded their limits, limits that the RTC subtly advanced like moving goal posts. Because as the Rangers became more proficient, their missions became more difficult.

But by the end of the course Rangers felt they could do anything because they done what they would have previously thought impossible. With the luxury of six hours sleep and only 60 instead of 100 lbs loads to carry, a Ranger felt that with one resupply drop he was ready to head up the Appalachian Trail (which ran by the mountain camp) where his company could knock out a well defended objective in Maine.*[*FN: This was a benign delusion the RTC did not try to dispell, but for their own curiosity they tested it by comparing physical fitness scores before and after trainees went through Ranger School. In the author's class—which, incidentally, graduated exactly thirty years after Bob Black's— over 90% of us scored <u>lower</u> on our post-graduate PFT's than we did on our entry PFT. We felt stronger but were not, yet it was how we felt that mattered.]

CPT James A. Herbert was with the Airborne Committee at Ft. Benning when the call first went out for a new generation of Rangers. Pulling strings and whatever levers he could, Herbert managed to join the RTC during its earliest days. As four companies began the first training cycle, he was put to work as a recruiter of volunteers for the second cycle, interviewing them at Ft. Bragg. He noted the names of candidates who most impressed him; then when good fortune put him in command of 8th Company at zero strength, Herbert got out his notebook and had orders cut in record time. Not surprisingly, 8th Company had the highest classification scores in the RTC's brief history!

Christmas, 1950, approached through a sleepless swirl of patrols without end. 8th Company not only learned about themselves as they never had before but became intimately acquainted with the strength and weaknesses of everyone around them—priceless knowledge when locked in combat in Korea's mountains.

After nine days leave, 8th Company was back training again, polishing hard earned capabilities for the last time under the vigilance of the RTC. Black: *With their state of training and cohesion ever improving, the company moved on to exercises of larger scope. Constantly increasing was the distance traveled by foot, the number of obstacles encountered, and the use of live fire. .. The men knew the song of the bullet well before they reached the battlefields of Korea.*

Like Jackson's famous 'foot cavalry', we marched long and fast. The task was to move fully equipped cross country a distance of 50 miles in 12 to 18 hours depending on terrain. By comparison the famed WW II British Commandos considered 42 miles in 19 hours the rule.

The men worked as a team. If a Ranger developed foot problems, others carried his gear. Men dozed, even slept while marching.

The graduation exercise began with a low level night jump, so low there was not enough altitude to use the reserve 'chute. Ron Sullivan was killed when his main 'chute malfunctioned. Twenty-two other Rangers were injured. Sullivan was not found for five days while the exercise continued. Other search parties had scoured the countryside looking for him, but it was not until 8th Company returned and, without rest, took up the search themselves that his body was found.

Grimly all of Herbert's men prepared a new issue of weapons for post-graduate cold weather training at Camp Carson. Their TO&E would be heavy on fire power by the standards of those days. In addition to two BAR's per squad, there were 36 submachine guns, nine 3.5" rocket launchers, nine 60mm mortars, three 30 cal. machine guns, and three 57mm recoiless rifles in the company.*[*FN According to GEN Herbert, 8th Company sought still more firepower after reaching Korea. They obtained it by trading, scrounging, or salvaging till they replaced nearly all their M1's and carbines with submachine guns or BAR's.] As Black observed, *Such armament produced a volume of fire that line infantry units would later marvel at. The men learned early that this firepower required a prodigious quantity of ammunition, all of which had to be carried on ones person.*

The Rockies behind them, 8th Company set sail from San Francisco on 5 March 1951 on a voyage of sea sickness already described in the histories of 3rd and 5th Companies. The last leg was by LST into Inchon, a scarred port Black remembered this way:

For most of the men, it was their first glimpse of the devastation of war. We saw a land pock marked by shell holes, twisted, broken buildings and downed power lines. Among the wreckage an orderly flow of troops and equipment moved toward the front.

The front for 8th Company would be that of the 24th ID, the U.S. division that had been in nearly continuous combat for the longest period. Their original CG had won the Medal of Honor but had been captured during the first dark month of the war. On 1 April when 8th Company was attached, the 24th was commanded by MG Blackshear Bryan, a former intercollegiate football lineman who was later mistaken on a TV show to be a profesional wrestler.

Bryan was wrestling with the rugged terrain and bitter resistance just south of the 38th Parallel when his new company of shock troops reported in to G-3. Impressed by their assortment of knives, facial camouflage and satanic appearance, a stafff officer said Herbert's men looked like devils (or maybe that they looked like the devil!), and decided 'Devil' would be the designator of their station on the division phone net. Devils thus became 8th Company's nickname.

As part of I Corps' grinding advance to take pressure off Seoul, 24th ID slogged ahead with two regiments on line and one following. By 7 April, at Phase Line Kansas Bryan paused to bring up his logistical tail. It was then 8th Company went into position on the regimental COPL where they got their first look at veteran enemies, the CCF.

Herbert's first mission was an ambitious one. Three miles beyond the COPL was a cluster of a half-dozen huts named Topyong-Ni which was believed to be a Chinese feeding station. The hamlet lay in a flat-bottomed funnel with hills on each side and a ridge behind it. Herbert planned to approach Topyong-Ni at night from the open end of the funnel, the side facing American lines.

8th Company pulled off the COPL on 8 April to prepare for the raid that night. The objective would be to determine CCF strength around the hamlet and destroy any facilities within it. Prisoners were to be taken. Two platoons were to block enemy interference from the flanks while the other platoon assaulted. A CCF mortar position was known to be within range of the raiders; it was to be suppressed by artillery fire.

The operation could have been an advanced exercise at the RTC. There was sufficient time for a thorough patrol order and leaders visual recon of the initial route in. A different route would be used coming out, so coordination for passage of lines with two friendly units was required. This was done, as was coordination of signals to lift supporting artillery fire. Slipping out into barren no man's land for the first time, the Devils felt the attainment of a long sought goal and culmination of an enormous amount of training. RTC instructors were remembered, and the Ranger tradition seemed to be following 8th Company like lane graders.

In addition to stealth, their first test was when the plan went wrong. This happened when the first three rounds of supporting artillery fire fell short. Several Rangers were concussed—the rest were cussing. The artillery prep also seemed to be a wake up call to the CCF who began probing fire in the Rangers' direction.

This fire actually turned out to be a benefit, revealing an enemy OP that guarded 2nd Platoon's route to the objective. There was no time in the operations schedule to take it out or divert forces for a coordinated attack. Instead the OP would have to be eliminated quickly and, hopefully, silently. BAR man Jim Reynolds drew the job. With two grenadiers backing him up, he double timed off to locate the OP, snuck up on it, and cut down the two Chinese with a short burst. Phantom lane graders jotted down a 'good' spot report for Reynolds.

But now 8th Company was among an alerted enemy. At nightfall when the patrol moved out, the CCF had also used the darkness to rotate troops from their positions to the chow down in Topyong-Ni. So Rangers and enemy were both moving in the same direction. Whoever moved undetected would reach their objective.

Despite the Chinese advantage of familiarity wth the terrain, it was the Rangers who closed in on the KP's in Topyong-Ni. With his blocking forces in place, Herbert gave 2nd Platoon the signal to open fire. Tracers ripped through the huts. Silhouettes rose in the flames but soon fell. A snatch team dashed through the smoke but found only Chinese dead—and vats of rice spilled on the ground, hissing with steam.

Counterfire was sporadic at first and seemed disoriented, but a lone machine gun threatened 2nd Platoon's withdrawal. Six BAR men went after it with such fury that within a few minutes the crew was killed and the gun destroyed. Phantom lane graders could almost be heard applauding. It was that quiet as the raiders pulled off the objective. Jim Teague recalled that, "the only sound to be heard was the crackling of fires" as every hut in the hamlet burned.

But now the most difficult phase of the raid, egress, began as 8th Company had to back out through aroused CCF forces. First the farthest blocking platoon was able to pull off without contact. The assault platoon also disengaged easily, but the blocking platoon nearest to friendly lines had to hold while the others hot footed away from searching fire. This platoon soon broke contact by exfiltrating squads despite constant illumination from CCF flares. As they withdrew, the other platoons called in artillery on where flares were fired. Though separated, 8th Company continued to support and coordinate.

There was no company rally point (Herbert could imagine a lane grader asking him why not). The reason was Herbert had foreseen that speed would be of the essence in the exfiltration. Though it was planned for all three platoons to use the same route out, time would be saved if they did not assemble first. Additionally, the smaller the exfiltrating elements, the less likely they would be detected. This proved to be the case as every Devil made full use of his evasion skills.

Through the night the 24th ID troops on the COPL had witnessed what sounded like a pretty nasty fight around Topyong-Ni. The battalion S-2 moved up to the passage point to learn how the cherry Rangers had fared. Finally, two hours before dawn, figures were challenged in front of the COPL. "Rangers!" came back as the password. As they entered friendly lines, they were immediately asked how many of their original number were coming back through. Everyone who went out, was the proud reply.

Black remembers there was a moment of silent amazement before the 24th ID veteran muttered, "Well I'll be damned..."

Like training at the RTC, after Topyong-Ni combat for 8th Company became a blurred swirl of one action after another. The Rangers had shown what they could do; now 24th ID provided endless opportunities to do more.

Black: *No matter how well a unit trains, war has its surprises. In the black arena between the lines, a misunderstood signal, the unexpected use of searchlights by a friendly unit, short rounds from mortars and artillery, a wrong turn— these and a thousand other pitfalls lie in wait. War is unforgiving, mistakes are paid for in blood.*

As with Roger's Rangers and the Ranger battalions of World War II, not all our operations went well. To achieve constant sucess at infiltration against an enemy with such vast numbers of men was not to be expected.

Some missions failed when successive lines of Chinese were encountered. Hours would be spent carefully bypassing enemy positions only to be discovered by others. The babble of a Chinese challenge, shots exchanged, then a rapid withdrawal through the night sometimes with four BAR's abreast in the lead and others in the rear...

The care and protection of wounded men was especially difficult when operating forward of friendly lines. Helicopter evacuation was in its infancy.. It would have been a suicide mission to attempt to bring out Ranger casualties by that means... [Yet] *it was unconscionable to leave a wounded friend. Torture and murder by the enemy—especially the North Koreans— was a frequent occurrence. So it became an article of faith that our wounded would be returned to friendly lines. This often required feats of dedication and strength.*

Base camp during this period was a busy place as men prepared for patrols. Company administration was hurriedly processed. LT Giacherine, the XO, lst SGT Craig, and Hale, the company clerk, worked long hours in order to have the opportunity to participate in missions.

These missions in April were what someone called "Chinking away at the Chinks". Like crowbars striking concrete over and over, Eighth Army slowly began to crumble CCF defenses. In 24th ID, the Rangers were on the night shift of this demolition, pecking away at strong points identified by line infantry companies during the day. Though their feeling of Ranger superiority never diminished—indeed, it increased with time and by comparison with the regiments to which they were attached— 8th Company developed high regard and respect for the Oriental enemy who rarely avoided a fight.

Black: *The Chinese were worthy opponents. Their version of advanced individual training was to keep men on line till they became proficient or died. They used their numerical advantage well and were equipped with a wide variety of automatic weapons, light and heavy mortars that they employed with skill. Though outclassed in artillery and air power, they were an agrarian people inured to hardship. They excelled in the infantry arts, were resourceful and brave.*

On 15 April, 8th Company changed over from night fighters for a daylight operation. Moving out at 0800 from the lines of the 19th Infantry Regiment, Herbert's platoons had independent but identical missions: each was to zone recon a different half-mile strip of no man's land, and locate enemy positions which could hold up an attack the next day by the 19th Infantry. Additionally, the Rangers were to search for and remove any mines in their zone.

The recon would be favored by rising temperatures that, after a cold night, left haze over the rolling terrain so that visibility was only about one mile, a factor that helped infiltrators from both sides. Warily, 1st and 2nd Platoons slipped into their zones, checked the trails for mines, finding none. Key terrain was marked on maps for use by the regimental S-2. Silently, covertly, 1st and 2nd Platoons completed their work without making contact, re-entering friendly lines before 1300.

After a traditional war dance in the assembly area, 3rd Platoon had similarly slipped into their zone with scouts out. Pete Torres was one of them. He kept an eye to the flank where the 19th Infantry's I&R Platoon was conducting a recon patrol toward Hill 299, the dominating feature in front of the regimental sector. If trouble was coming, it would probably be from Hill 299.

On multiple trails, 3rd Platoon's squads wound down into the valley floor with its low ridges, the reverse slopes of which would have to be checked out. Several potential enemy strong points were investigated but the enemy wasn't there that day. Through good use of masking terrain, 3rd Platoon sneaked about doing their work like American Indians, satisfying LT Strong who had been concerned how much his men remembered from RTC about daylight reconning.

It had gone very well, and Strong signalled for the platoon to reassemble to begin a different route back to friendly lines. Torres was walking drag as the squads in column moved up a gradual ridge that was sprinkled with low trees and scrub that provided pretty good concealment. Then somewhere far off to the flank in the I&R Platoon's sector, fire broke out, but to veteran ears it did not sound very threatening. 3rd Platoon simply took their weapons off safe and paid more attention to that flank.

Soon Torres saw what the shooting had been about: the I&R patrol had made and then immediately broken contact with a distant CCF outpost. He could now see about a squad of Americans break out of a woodline and dodging back toward friendly lines. Torres pointed this action out to his squad leader, then Strong halted the platoon to observe.

To their surprise—then delight—they had ringside seats to a pursuit of the I&R patrol by a growing group of Chinese who were so intent on their prey that they had no suspicion that the Rangers were on flanking high ground less than 200 yards away. Neither did the I&R troops seem to know that the CCF were after them and gaining. The Chinese were moving swiftly and silently, something the Rangers had a moment to admire.

"Berk Strong is not the kind of guy to conduct a staff study before he makes a decision," Herbert recalled jokingly about the only officer among his three platon leaders. Strong immediately ordered a hasty ambush. It went off with an initial sweep of fire that cut down the first group of Chinese like hay. More enemy poured out of a shallow ravine, squad after squad entering the Rangers' field of fire. From a supporting position, a Chinese company that had been following the pursuit engaged the ambush with counterfire that built up rapidly. 3rd Platoon's ridge had pretty good concealment but not much cover. Dirt began to spurt like a pond in a hail storm. Rangers were hit, Torres among the first, struck below the eye, the bullet passing through his mouth to lodge in his shoulder.

Black: *Prior to this engagement our casualties had been at night. In darkness it had been difficult for another soldier to guage the extent of a wound. Now the sight of Peter Torres' torn and bloody face burned itself in the memory of his friends.*

The outbreak of a roaring firefight quickly drew the attention of the nearest 19th Infantry battalion. Strong's RTO handed him the handset: battalion was calling him.

"You talk to 'em," Strong replied over the din of battle. "I'm busy."

He was busy deciding whether to break off what had been a very successful action—later called the '299 turkey shoot'—quit while he was ahead, in other words, or slug it out with the CCF some more. Youthful in appearance though a World War II veteran with a CIB, Strong chose to stay. Not just on the ridge where the CCF were zeroing in with mortars: Strong deployed his platoon for a sharp spoiling attack downhill to where the Rangers could shower trapped Chinese with grenades.

"When I throw my grenade, everybody does!" he yelled, and off they went.

Black: *In a series of charges, constantly firing, the Rangers drove the Chinese force back. From recollections of the hectic battle, vignettes remain: Antonio Velo and E.C. Rivera killing a Chinese soldier in the act of reloading a drum in his captured American Thompson sub-machine gun.*

Nick Tisak changing magazines in his BAR when Leroy Williamson yelled a warning,

"There's a Chink!"

"Well kill him!" Tisak replied.

Williamson fired a snap shot from his M1, killing the Chinese soldier in the act of throwing a grenade. As the Communist fell rearward from the impact, he released the grenade which soared up to explode above the Chinese position.

"I would have laughed if I'd seen that in a John Wayne movie, but that's what happened," Tisak recalled 44 years later.

As the platoon continued forward, Medic Gill Gregory returned to the wounded Torres. Blinded by blood, Torres waited holding his rifle with bayonet fixed.

At the sound of approaching footsteps, he struggled to defend himself. Gregory's reassurance was most welcome. Setting to work, he used a bayonet and fingernail clippers to remove the bullet, rigged bandages for the wounds, then set out to guide Torres back to friendly lines.

But toward friendly lines was not the direction for the rest of 3rd Platoon. They continued to attack, Strong so caught up in the rush of battle that he he did not seek artillery support. Nevertheless his Rangers were in full cry and drove the shocked Chinese back toward fixed positions on their front lines. A second CCF company came out to help the first. The volume of return fire from the Rangers revealed to the Chinese that they were facing a platoon, not a battalion attack. The second company struck Strong's right flank. He decided to break off while the Rangers still had access to a withdrawal route.

Now it was the Chinese who were in full cry. They were held off in large measure by BAR man Velo, whose weapon grew steaming hot as he fired up magazine after magazine, feeding them so fast his BAR sounded like a light machine gun. He knocked back many a Chinaman but drew the deadly attention of more automatic weapons. He was severely hit in the upper thigh, turning over his BAR to another man, then two buddies took turns carrying him to the rear.

Two other Rangers were also so seriously hit that they had to be manhandled back. This reduced 3rd Platoon's firepower in the fight that had definitely shifted against them. Strong would now have to get them out of a kill bag like the one from which they had saved the I&R patrol.

Fighting even harder than they had in the attack, the Rangers covered each other as they fell back by squads. To the 19th Infantry troops, whose lines were Strong's destination, the approaching firefight sounded more like full scale CCF attack coming toward them. Herbert tried to get some air support up for Strong but the lieutenant's radio had been inoperably damaged. 1st and 2nd Platoons rushed over from other sectors to be available for possible aid in extracting 3rd Platoon.

But haze in the last valley Strong had to cross built up with the growing heat of the afternoon; from the American lines not enough of the storming battle could be seen to influence it.

Black: *On the valley floor, the temperature seemed oppressive. Exhaustion was total—even the sight of bullets striking near ones feet was insufficient to inspire faster movement. There was only heat, thirst and the dusty path stained with the blood of our wounded.*

A Chinese machine gun covered the last open space, a rice paddy that had to be crossed... Tony Velo died out there from shock and loss of blood.

Fortunately this paddy could be observed from friendly lines which provided some covering fire for the retreating Rangers. 19th Infantry troops went out to assist but Strong waved them off as he brought in his platoon intact, carrying two dead and three wounded. None was missing. When the haze cleared, 55 Chinese bodies were counted near where the action began; at least twenty more dead were estimated closer to CCF lines.

The '299 turkey shoot' was an early example of a characteristic that has been both praised and criticised about Rangers—they go for the kill at every opportunity; and if not pulled off like a fighting dog, they won't leave until they can put a foot on the bodies of every foe they have been fighting. Maturity, and the nature of missions modified that characteristic so that it does not inevitably appear, but in the heart of a combat Ranger remains the desire, sometimes rash, to close with and destroy what's in front of him, no matter how much of a threat it is to his own life. And if the life of another Ranger is imperiled, nothing but death itself is going to stop him.

By the last week of April, the Korean War was entering a new and decisive phase. I and IX Corps pressed north in an effort to dominate the Iron Triangle so as to deny it to the CCF as an assembly area and advanced logistical base. The CCF were even more determined to hold the Triangle which was vital to their forthcoming spring offensive.

24th ID's objective was Kumha, at the southeast apex of the Triangle. 8th Company was part of the advance guard, operating between 1000-2000 yards ahead of the two leading regiments. On 21 April the Rangers ambushed enemy counter-reconnaissance patrols, killing twelve.

EPW's revealed that Eight Army's advance was soon to end all across its front. Throughout the afternoon of 22 April, CCF artillery thundered its new presence in the war while the Americans dug in, brought stockpiles of ammo forward and braced for a showdown. 8th Company was scheduled for combat patrols that night but at the last minute they were cancelled: it did not make any sense to go out looking for communists—their artillery prep meant that soon they would be coming to the Rangers.

Come they did, in numbers that 24th ID veterans said were larger than anything they'd seen during the whole war, including the initial invasion. The entire I Corps front was ablaze, positions being lost then retaken in an ebb and flow of attacks and counterattacks. 8th Company was in the thick of it, tied in well with the 21st Infantry and losing no ground—thanks in great part to thousands of rounds of howitzer fire in direct support. The situation was one of tense turmoil, but 24th ID was holding its own.

The situation on their right flank, however, took a very bad turn that night when the neighboring 6th ROK Division turned and ran. Like creeks after a spring rain the CCF poured into an eight mile gap, then broke up into company-size infiltration groups to work down into 24th ID's rear. This development caused MG Bryant to immediately refuse his right flank, and try to regain contact with IX Corps where the ROK's had bugged out. For the latter purpose, 8th Company saddled up as a contact patrol with instructions that the nearest elements of the 6th Division were believed to be about 13,000 yards to the notheast (though not even AM radio contact had been made with them). "Captain Herbert, go find them and report back on their situation."

If COL Van Houten could have overheard that tasking, he would have been justified in a moment of pride before gasping. Obviously 24th ID had become convinced that their attached Rangers were Paul Bunyans who marched in seven-league boots, because 13,000 yards—all through enemy infested cross-comparmented mountains— was almost half the distance across the entire IX Corps front! This mission was more an expedition than a patrol: the Rangers might as well have been told to cross Niagra Falls, then come back to report what the Canadians were doing.

The mission— all the more difficult because much of it would have to be accomplished in daylight while 24th ID was bending its line, moving it farther away— began with tragedy. While he was standing in line for the last hot chow 8th Company would taste for many days, the cotter pin on a Ranger's grenade snagged on something and pulled out. Unable to free the grenade from his belt, he screamed for his comrades to get back, then fell on the grenade and was killed.

Black: *A numbing chill swept through the company. Almost as one, the men threw away their food, mechanically filed by the immersion heaters to wash their mess kits, then loaded on the trucks. There was little conversation. The waste of this good man's life in a tragic accident left a feeling of utter desolation. This was the low point in the existence of 8th Ranger Company.*

As they began their perilous patrol to find the ROKs, 8th Company mustered 87 men and three officers.*[*FN: Unbeknownst to 24th ID, the 6th ROK Division had already retreated 21 miles, losing most of their artillery, all their trucks, and 84 machine guns. If the extent of their rout had been suspected by 24th ID, 8th Company no doubt would have never been sent out.] Crushingly loaded with extra ammo (some men carried twelve grenades), two-days rations and mountain sleeping bags, at 2000 on 23 April the Rangers trudged off to mount the high ground east of 24th ID's dangling flank. For commo they had borrowed two SCR 300 radios from the 21st Infantry. Herbert was also accompanied by a three-man FO team from DIVARTY.

Accompanied by KATUSA liter bearers, 8th Company slipped off into the night, heading northeast to infiltrate the Chinese who were themselves infiltrating to the southwest. Contact was to be avoided until the Rangers reached some high ground where they could interdict the flow of CCF. With a lot of artillery fire being exchanged, it was a noisy night so silent movement was less important than preventing visual detection. Staying off the trails (though this made climbing more difficult), around 0230 Rangers spotted a large CCF column silhouetted on a ridgeline trail and called artillery on it without revealing their presence. Screams from the Chinese indicated that the fire did have an effect.

Herbert had to get his men onto high ground by daybreak, and prayed they wouldn't have to fight before then. Bent under enormous loads, they could hardly climb, much less maneuver without dumping their equipment which would likely roll downhill and never be recovered. Everything they carried was vital, including water which would be much needed after such exertion and the heat expected the following day.

Hill 628 was the high ground Herbert wanted. 8th Company mounted it very slowly, the climb so steep and the loads so heavy that men had to push on their knees to straighten their legs. A hundred yards from the top, his lead platoon cached their loads and spread out to investigate what was ahead.

They detected movement near the summit. Cautiously they closed in, then crept closer. Something about the sounds did not sound like Chinese. Then a silhouette was spotted, shaped like a GI helmet rather than CCF headgear. The Rangers whispered a challenge. The frightened response was in Korean— from a half-dozen cut off ROK's holed up on Hill 628.

This news was a great relief to Herbert who quickly, and before daylight, occupied the hilltop and established control of the ridge leading west toward friendly lines. Technically he had accomplished his mission: made contact with the 6th ROK Division— a scared and stranded squad of stragglers! But he knew of course this wasn't the contact 8th Company had been sent to make.

Observation was good in all directions from Hill 628. To the east there was nothing to be seen of ROK's. If there were any out in that direction they would be in hiding because the CCF had completed their sweep through 6th Division's positions. Herbert felt that 8th Company would have a long way to go before they came across an organized force of ROK's who, he assumed correctly, were probably now well to the south and still retreating.*[*FN: In defense of the 6th ROK Division it should be noted that they were the target of the CCF's main attack and struck by an onslaught of four of the best Chinese divisions.]

The contact he realistically hoped to make was with 24th ID. After a couple of hours rest, Herbert sent a patrol toward Hill 830, the highest ground between 8th Company and the new lines of the division, now several miles away. His first priority was to have a way back to friendly lines if 8th Company became in danger of being cut off by vastly superior forces. Signs of such forces were plentiful to the east and south.

The contact patrol on 24 April had good luck, meeting a patrol from the 21st Infantry on Hill 830. Herbert's patrol brought back some good news: 24th ID was extending their flank toward 8th Company. The bad news was that 8th Company was to continue out itself, so the vain hope of finding the ROK 6th Division had not been abandoned by Gen. Bryant.*[*FN: Actually there was no higher priority in Eighth

Army at that time. From the other direction, the U.S. lst Marine Division was also probing into the yawning gap created by the ROK's in an effort by Van Fleet to restore the United Nations front. When this was done (not until May), the 'first impulse' of the CCF's spring offensive was repelled.]

Yet the linking probes by 8th Company and 21st Infantry were a reassurance for Herbert, stronger than the sense of folly induced by the order to continue the mission. He realized that so long as the Rangers did not become the 'lost' (i.e., a forgotten) patrol, their mission could be pursued in good faith. Faith that 24th ID was still aware the Rangers were out there—if only a speck of rock in Niagra's torrent—still a unit to be considered even while the division's front continued to come under heavy attack.

Over the radio Herbert received word that the 6th ROK's were holding a position on Hill 1010, a height that could be reached by a ridgeline from Hill 628 and about three miles north. Though skeptical that anyone but CCF could be on Hill 1010, his order of the day was to move out along the connecting ridge. Unfortunately Ranger strength was not sufficient to leave a force on Hill 628 as a fall back position. Fortunately Hill 1010 did not turn out to be occupied by CCF, and after an arduous four-hour approach, 8th Company set up an all-around defense ringing its summit, using the foxholes dug by the long gone ROK's.

This was an excellent position, providing long range obervation to the southeast and southwest. In the latter direction there was nothing to be seen of the 6th Division; in the former, Herbert watched plenty of movement, most notably trains of pack horses moving CCF and their supplies toward Seoul. He looked forward to a field day by DIVARTY as his FO began calling in fire missions.

The answer produced astonishing disappointment. DIVARTY came back saying that elements of 6th Division were reported to be still present at the coordinates of Herbert's targets. He retorted that only CCF had pack horses, and that through his binoculars he could distinguish Chinese from ROK's. No joy—the fire missions were denied pending "confirmation".

Confirmation came that night when these same Chinese columns began attacking 24th ID's rear, even knocking out the division switchboard and taking the divsion airfield and rear CP under fire. In the intervening hours Herbert had much to mull about the Rangers situation which was this:

The Chinese were not bothering with his company-size OP on Hill 1010; instead they were following CCF doctrine, bypassing such strongpoints and taking paths of least resistance into the Amercans' rear. Herbert monitored the enemy's success on the radio: the 19th Infantry (in the center of 24th ID's defenses) bent under extreme pressure, causing Gen. Bryant to pull his entire line back about three miles. This displacement put the 21st Infantry adjacent to Hill 628 which the Rangers had just vacated, leaving them now about three miles north of friendly lines.

At dusk on 24 April, Herbert expected visitors as the CCF started to clear their rear, just as 24th ID was trying to do with their own rear. Two Chinese almost reached the summit of Hill 1010 before Rangers killed them. Herbert sent out small patrols to learn where these probes originated. He hoped that next morning DIVARTY would not be deluged with too many other fire missions: he needed howitzers to keep the CCF's heads down.

One of his night patrols revisited Hill 628 and found it in enemy hands. Killing four of the inflitrators, the patrol took one prisoner and learned that the 8th Company on Hill 1010 was diverting the CCF to the east. With this information, Herbert expected more attention from the enemy the next day. Before then the Rangers would need more ammo, food and water. Only an optimist on good terms with God could hope that such help would be coming from the hard pressed 21st Infantry who were in full-scale fire fight all night long.

CPL Tom 'Mad Dog' Uldall was not a chaplain by MOS but on 25 April he began answering Ranger prayers. When the CCF offensive began he was participating in non-religious activities at an R&R hotel in Seoul. Awakened by the far away rumble of constant artillery, Uldall abandoned his temporary companion and started hitchiking north to find his Ranger companions. Around 1900 he reached the embattled positions of the 21st Infantry.

LTC Gines Perez, the 21st's CO at the time, is another prominent candidate for 'Honorary Ranger'. From his slender resources he had already cobbled together a train of KATUSA porters to try to reach 8th Company with much needed supplies. The train was to be led by a volunteer from the 21st Infantry, a corporal who soon lost his life on the gallant mission; his death left Uldall in charge.

For six hours as tense as piano wires, Uldall led his porters through darkness crisscrossed by tracers, up hills and along ridges, dodging CCF patrols most of the way. Fortunately the Chinese were noisey so Uldall could hear them coming. Then his problem was to prevent his porters from bolting or chattering among themselves. As he approached Hill 1010, Uldall knew he might be entering the most dangerous area of the night because hair-triggered Rangers were ringed around the top. Fortunately Perez had radioed ahead to Herbert that the supply mission was coming, so Uldall's column was greeted with only thirsty throats, empty stomachs—and open arms.

Just before dawn, a group of unidentified figures appeared in front of 8th Company's perimeter. They were challenged, did not respond, and were taken under fire. Because of good luck or bad marksmanship none was hit. At first light some stood and identified themselves as ROK's. The Rangers remained suspicious because CCF often impersonated their Korean enemies. But this group appeared genuinely terrified, so they were accepted as ROK's. The only English they could say was "Many, many Chinese!" pointing over their shoulders. That was not new intelligence for Herbert. His questions were how to get through the CCF and in what direction.

The first step toward the answers came when 8th Company was released from its original mission: forget about 6th Division, Herbert was told—"About time", Rangers muttered—and increase interdiction of the CCF follow-on forces. No problem finding them either: they were rolling down the valleys like tumbleweed.*[*FN: In *Devil,* Bob Black begins his chapter about the action that followed with a very appropriate quote from Agis of Sparta (415 BC): "The Spartans do not ask how many the enemy number, but where they are."]

In order to disguise the location and strength of his position as long as possible, Herbert let howitzers do the interdicting. Adjusting fire from behind the enemy presented unique problems for his FO who had to sense overs as unders, up as down, left as right, and *vice versa.* "They ought to teach this at Ft. Sill," he grumbled while he directed much whizzing steel on the flinching Chinese in the valleys.

At 0700 8th Company's mission was further changed by orders to reoccupy Hill 628 as the 21st Infantry (with an attached medium tank company) pulled back farther to a line adjacent to Hill 1168. 8th Company was to come under operational control of the 5th RCT which was the division blocking force. However, this transfer was not accomplished because of no radio contact between the Rangers and 5th RCT, a great concern for Herbert because the RCT was to provide covering fire for the Rangers' withdrawal.

Except for some haze and drifting smoke, visibility was excellent for the movement back to Hill 628. Enroute, 21st Infantry reported that they had probed toward Hill 628 from the west and had met strong resistance. As they descended from Hill 1010, 8th Company locked and loaded for a tough fight on a narrow front. Their expectations were met in every way.

Bob Black was in the thick of the raging combat that developed: *Ranger fire struck among the enemy and was fiercely returned.. from the front, right front, and left flank.*

The narrowness of the ridge and its steep sides prevented establishing the traditional base of fire and maneuver elements. A misstep and a man would tumble from the ridge. Some were shot off. In several locations it was necessary to move single file.

Close to Hill 628 the lst Squad, lst Platoon found space to move into a skirmish line and advance by marching fire. As lst Squad took casualties, 2nd Squad found space to join them on line.

The artillery FO with 8th Company obtained priority of fires as the Rangers were in the hottest action on the division front that morning. Howitzers pummeled Hill 628, but 3rd Platoon also provided a bit of indirect fire support for lst Platoon by rigging 60 mm mortar rounds to grenade launchers on their M1's. The rounds traveled only about 50 meters but were helpful because they could be directly aimed and adjusted by the firers. The only problem was that some rifle stocks cracked from the recoil!

3rd Platoon ran down the ridgeline to assist, shooting from higher elevations in support of lst Platoon's attack.. Automatic weapons fire from the left flank broke out, raking 3rd Platoon. MSG William Cox deployed 3rd Squad to return this fire.

Rangers were still wearing their packs, most with rolled sleeping bags attached. As one man raced down the ridgeline, a bullet cut a set of cords and his sleeping bag trailed out leaving him feeling very conspicuous, like a bride at a macabre wedding.

The din was deafening, a constant crackling roar like a forest fire. Soon the deeper sounds of Chinese mortars came into play.

Because 2nd and 3rd Platoons were descending from higher ground toward the saddle between Hill 1010 and 628, they had plunging fire on the Chinese taking on 1st Platoon. They found themselves knocking out the same CCF machine gun nest every few minutes. As fast as the Chinese replaced a crew, BAR's from 2nd or 3rd Platoon would kill them with long range fire. Finally the gun itself was destroyed.

Rangers Diaz and Cisneros became a designated machine gun killer team. A crack shot, Diaz loaded his M1 with tracers. When he spotted enemy automatic weapons, he marked them for Cisneros' BAR. In this way machine gunner became a Chinese MOS with a very short life expectancy. Yet there were always more, a seemingly unlimited number of Chinamen bringing automatic weapons up the back slopes of Hill 628. Herbert had artillery pound these slopes though the FO could not see the effect.

As Rangers fell wounded, Herbert was glad to have the seventeen or so ROK's that had collected during the previous days. Inspired by the Rangers' dauntles example, they joined the battle in a way that contradicted 6th Division's reputation. Not so for the KATUSA porters. They slipped off into defilade, hiding in ravines where they pillaged the packs of the wounded.

Toward the front, Phil Moore and Paul Chagnon had just killed some Chinese and were running forward to catch up with lst Platoon when Moore was hit in the thigh and his map case shot away. The impact knocked him off the ridgeline, sending him rolling over and over down the slope. Managing to halt his fall, Moore resumed firing while Medic Potter bandaged his wound...

Standing, sitting or kneeling as the ground permitted, fighting from behind boulders and from folds in the earth, Rangers kept up a heavy fire. Tracers played among the rocks, splattering and singing off stone...

Man after man was hit. Emanuel Rish, Harold Hooks and Marko Sparko went down. Ron Smith had a bullet part his hair. Bill Tiemeyer and another Ranger were hit simultaneously and tumbled down the side of the ridge together.

In the hot fight there were some cool heads: A Ranger's BAR fouled. On the open hillside, he removed his fatigue jacket, spread it out, disassembled and cleaned the weapon, put it back together and resumed firing...

All over the ridgeline Rangers were fighting as individuals and groups, attacking here, defending there, engaging at long range in one place, firing point blank at another. Ingram and Chagnon were hit. Harry Trout, a former medic, ran through a hail of fire to help the wounded until he was killed by a machine gun, his jacket burning from the tracer that pierced his body.

Slowly but steadily the Rangers fought their way down the ridgeline onto the slopes of Hill 628.

Herbert was cheered by this success, costly as it was, but realized that taking the hill would by no means result in salvation for 8th Company. It was clear to him that they would still be isolated, beyond the help of the 21st Infantry or 5th RCT. Vainly Herbert had been trying to reach the latter regiment to coordinate passage of lines if the Rangers could ever get down from Hill 628.

Actually 8th Company was acting as a small break water as the CCF tide rolled in to demolish 24th ID's right flank. Already two battalions of the 21st had swung 90 degrees to face west and refuse the flank. At 0830, 5th RCT sent out a two-platoon contact patrol to help what Herbert hoped would be an extraction from Hill 628. The contact patrol, however, came under such heavy fire that they aborted the mission after two hours. The 5th's S-3 asked that this helpful message be relayed by the 21st to Herbert: "Get out the best way possible."

But LTC Perez of the 21st was not going to leave Herbert in the lurch. As his regiment began to displace farther south, he detached a platoon of tanks to link up with the Rangers. Where that would be no one could guess. At some point 8th Company would have to veer west from Hill 628 and head for U.S. lines. That might not be the order from division but forced by necessity.

Herbert did not know that Perez was sending a platoon of tanks (under LT Dave Teich, another 'Honorary Ranger' nominee) to help the extraction, so for the Ranger CO the situation seemed increasingly dire. Already he had seven dead and about triple that number wounded. That was at about 1030.

8th Company needed a miracle but what they got were more losses. First, both the artillery FO and his sergeant were killed by mortar fire they were trying to suppress. Among the last to leave the ridgeline onto Hill 628, LT Strong was stiched across the stomach by machine gun fire. Herbert was shocked to see him climbing the last slope, holding his rilfe in one hand, pressing guts into his abdomen with the other.

Then Herbert—who had been all over the hill, directing fire, leading and moving troops—was torn up himself, a heavy claiber bullet piercing his neck, breaking the collar bone and blowing away the deltoids as it exited. He rapidly lost consciousness and a huge amount of blood. Word was passed to the XO, LT Giacherine, that he was in command.

While bringing up the rear, Giacherine had noted a possible escape route from Hill 628. With the only operable radio remaining in the company, he relayed through the 21st Infantry for artillery fire. With one platoon covering by fire, he moved the other two back and off on the least exposed side of the ridgeline. This quick thinking and prudent action gave up part of Hill 628 but the Chinese were slow to grasp the hornets nest that had stung them so badly.

Giacherine had 8th Company close ranks, wounded in the center of a spearhead that would launch for friendly lines, come hell or high water. Orders were being passed to execute this maneuver when an OP on Hill 628 shouted that some friendly tanks were coming up the nearest valley to the east. Giacherine could not raise them on the radio but the tanks were now a lodestar toward which the Rangers could aim their attack. They boiled over the crest of Hill 628, preceded by very close and heavy artillery fire that broke open a sort of corridor through CCF positions. The last thing the Chinese expeced was a concentrated attack but that's what they got as 8th Company began to break out.

The Chinese reacted with heavy mortar fire that caused Giacherine's spearhead to split up. As Rangers approached, Chinese came out of their holes in banzai attacks. They were mowed down—nothing was going to stop 8th Company unless every man took a bullet.

But the first thrust was not strong enough to push on along the most direct route to the waiting tanks. Giacherine ordered his trailing platoon to divert into a ravine he had noted earlier, one that led away from the tanks, but entered the valley where the platoon could loop around and reach their destination. Chinese were everywhere but seemed disorganized. As they popped up some Ranger would spot them, fire a killing burst or grenade them.

Black: *MSG Ellis and six men moved off on point along the route that Giacherine directed. The wounded went next, assisted and guarded by other men... Movement in any direction required running a gauntlet through enemy positions. For some it was a race down the slopes, firing into enemy foxholes on each side.*

Medic Hardground tried to prepare Herert for the descent. Hardground's counsel was not reassuring to his patient: "Sir, the book says to apply a tourniquet between the wound and the heart. If I do that I'll have to tighten it around your neck." Though he was in shock, Herbert recommended an alternative! It was to thrust his finger into his gushing neck wound and hold onto the litter with his other hand.

The steep descent for other Rangers was faster, none more so than Joe Slavinski who dove off the ridgeline as if he were going to swim a river. He bumped along but picked up speed, too fast a target for the Chinese whose rounds all struck behind him. Slavinski's purpose was to brake as he reached a boulder about a third of the way down, but instead he accelerated right by it. Except for the jarring rocks, the feeling was like sledding when he was a kid. The ride ended in a crash at the bottom, but he got to his feet and discoverd how he had descended so rapidly—he had been sliding on four grenades attached to the front of his harness!

The rear guard had the toughest fight getting off the hill. A mixed platoon of CCF and NKA came over the crest from the east as the Rangers went into a hasty reverse slope defense. Six BAR's fired in unison. The hill crest seemd to jump into the air as a dozen enemy dropped as one. That ended the first assault on the rear guard. The next was held off by a Navy Corsair dropping napalm on the crest. The rear guard hastened to catch up with the main body which battled on under the tight control of LT Giacherine who directed the mass fires of BAR's to overwhelm enemy strong points. On Slavinski's route, elements of two platoons were farther behind but could guide on a dirt road that led to the tanks.

LT Teich, the tank platoon commander, had come up on 8th Company's frequency. He told Giacherine that he had some Rangers in sight (the advance guard) and he would try to hold fire unless threatened by Chinese who were bypassing his tanks on all sides. With still more than a quarter mile to go for the main body, Giacherine replied that, "If we can't get help we'll fight it out here." Teich wasn't going to leave 8th Company stranded like that. He had a debt to pay: during WW II in Europe, Rangers had come to his aid when he was about to be cut off. He'd do the same for the Korean Rangers.

Forty-four years later, Teich provided his recollections of the extraction: *I was informed by my company conmmander that we were to pull back immediately. I informed LT Giacherine and he asked me to wait till the Rangers completed their fight down the mountain. I relayed this request to my company commander who replied, "That's their problem. We are under orders to pull out and that's what we're going to do." I answered that the unit asking for help were Americn infantrymen, and it was up to us tankers to stay and wait for them.*

He still wouldn't acknowledge my request and repeated the order for my platoon to withdraw. I then approached him with, "What if I volunteer to stay behind and bring them out?" He said that was up to me, and then pulled out with the rest of our tank company."

My platoon at that time consisted of four M-46 tanks, painted to look like tigers. I briefed each of my crews and set up a perimeter defense. Hundreds of Chinese bypassed my tanks on both flanks...

*I had to go from tank to tank, reassuring my men that we could rescue your company, and use extreme caution not to fire on the Chinese who were leaving us alone. I was amazed that they did not stop and anihilate our platoon.**[*FN: The author speculates that the Chinese—under rigid orders to advance as far south and as rapidly as possible— may have assumed that the four tanks were already knocked out because they did not swivel their turrets or otherwise indicate signs of life. Nevertheless it took cool presence of mind for the tankers to play dead when at any time they could have been dead indeed.] *We didn't have enough ammo to sustain a fire fight.*

Finally, toward the end of the afternoon, the sound of your small arms kept getting closer. Then out of a small gully on our right flank, a group of Rangers started approaching our tanks. At the same time, we received fire from the Chinese. Your wounded company commander was carried into my perimeter on a litter...

He was placed on my tank's right fender, and we proceeded to load approximately 120 men from your company along with some Korean stragglers you had picked up along the way. There were so many infantrymen on the outside of our tanks you couldn't see anything but the tracks...

Orders were given to move out and for everyone to hold on real tight. If we ran into a CCF road block, your men were to return fire becasue with all of you on the tanks it was impossible for us to fire any of our weapons.

We took off in a swirl of dust and raced toward the rear. At this point I had no idea where our front lines were. About five to seven miles down a dirt road we came upon a large group of American troops who were setting up a block. With them was a bird colonel, some tanks from the 5th RCT, and a battery of 105's.

Your wounded company commander was

Recovering bodies of dead G.I.s, Pork Chop Hill, Korea. November, 22, 1951.

in really bad shape so he was loaded on a small chopper and evacuated. Others of the wounded were placed on tanks because the ride back would be smoother than in a litter jeep. I cautioned everyone to look out for hundreds of Chinese that had been bypasing us all day.

The colonel seemed to have the situation under control so he released us to mount up again and continue to the rear. As we proceeded around a narrow bend at the end of the valley, a tremendous fire fight broke out behind us at the road block we had just left. The Chinese had hit it with everything they had. We continued down the dirt road until we reached the main lines where we stopped to let you off.

I never got to see or hear from any of you people after that, and always wondered what became of your company. The only comment I got from my company commander was that I was a damn fool and lucky to be alive. He was right about that twice over, because the road block where we dropped off your CO got wiped out by the attacking Chinese along with nine tanks from our D Company that went in to rescue them. So I guess we sure were lucky and had the good Lord on our team.

Though there was no lull in the fighting anywhere on the division front, MG Bryan took time to come down and congratulate the Rangers, telling Giacherine—the only officer left—that 8th Company had plugged a vital hole on the right flank and prevented the 21st Infantry from being cut off, endangering the entire division. From Hill 628 enemy casualties had been inflicted the hundreds.

Only 69 Rangers (dozens would soon be promoted) were present for duty to hear that praise. According to the G-l's formula, 8th Company was now so understrength that it had to be declared 'combat ineffective', a designation resented by hard-fighting Rangers. Nonetheless, there weren't many grumbles when their next mission was to secure the division CP.

On 28 April, the company moved out to protect a key bridge on 24th ID's right flank. On the 29th, they went into division reserve as part of a task force to back up the boundary with 25th ID. U.N. forces had finally stopped the first impulse of the CCF spring offensive a few miles north of Seoul on what was named the No Name Line. From here the U.S. divisions began to reverse the momentum.

Black:*The send-off for the 24th Infantry Division was a devastating salute. Division Artillery marked the Communist May Day celebrations on 1 May with a massive time on target salvo that rained on the Chinese parade. From 1-19 May, the division maintained an active defense; on 20 May it launched a series of attacks. Heavy patrol and probe activity throughout the month meant frequent calls for the Rangers.*

As airborne personnel, the Rangers did not receive combat pay. Hazardous duty pay was dependent on making at least one jump every three months. lst Sgt. Charles Craig made arrangements for a company jump. Unfortunately the DZ was in the I Corps sector; they objected and planning resumed. [Meanwhile] *eleven replacements arrived and some of the lightly wounded returned.*

Back to the line with no jump pay, 8th Company screened the division flank and kept contact with 25th ID. A second try for a jump was seheduled for 18 May but would have required a night withdrawal to the marshalling area. There was too much enemy activity for the Rangers to be relieved, even temporarily, so again the jump was scrubbed.

On 19 May, as part of a tank-infantry task force, 8th Company clashed with a CCF battalion hell bent to destroy a lot of U.S. armor. Combined arms teamwork prevented this—on one occasion a Ranger shot a Chinaman shoving a Bangelore torpedo under a tank— inflicting 300 casualties, losing but three tanks and no KIA.

24th ID was now building momentum as part of the corps offensive to regain much lost ground. Still part of the task force, 8th Company moved out at 0330 to seize the village of Chonggang-Ni, five miles away, and high ground that controlled it, Hill 112. The fight to reach these objectives was tougher than their seizure, as the CCF made best use of delay positions on intervening terrain.

Each strong point had to be worked over by the tanks before Rangers moved in to root out the defenders. Some CCF stayed back so sniper fire was a constant harrassment. None of this was as bad as when a squad of 3rd Platoon was taken under friendly fire by a quad-50.

Black: *Suddenly the air seemed filled with a howl that tore at a man's soul, a sound like a thousand freight trains rushing by. Trees on the crest above the men began to topple as a swath of fire marched downhill toward them.*

The half track mounted quad-50 below had opened concentrated fire on the hill.

As one the Rangers dove for cover, a slight depression, just a wrinkle in the earth, held the only chance for survival. There was barely time to press body to earth before the ground began to convulse. All around them were trees chopped to pieces crashing to earth and showering the men with splinters and branches.

To be caught by enemy fire is unpleasant enough, but to be under American firepower is an experience one never forgets.

The quad-50 crew was persuaded by screams and oaths to take more appropriate targets under fire. The task force moved on, assaulting hill after hill despite minefields and drumming salvos from CCF 120mm mortars. The Chinese delay had been well planned, and plenty of ammo moved forward to assure heavy fires from blocking forces. Often Rangers walked behind tanks that set off anti-personnel mines, but then mortars concentrated their bursts to prevent such protection.

The Chinese gave way slowly, then dug in to fight to the death in front of Chonggang-ni. As squad leader George Hall went about checking his men's positions, he was killed by long range fire. Bob Shinavier was also wounded, so badly that he would lose a leg, while three other Rangers were hit. Giacherine recommended to the task force commander that the town and the high ground around it be thoroughly softened up before it was assaulted. This was done, reminding Black of a fire power demonstration at Ft. Benning. Artillery, tanks, quad-50's and massed fire from infantry weapons worked over enemy defenses for more than an hour, shredding CCF positions and taking the fight out of those that survived.

Tank-infantry teams moved in with a coordinated attack. Chinese broke and fled, to be gunned down from the back. They had stayed too long, defying the principle that a delay force should not become decisively engaged. Only remnants of a CCF battalion got away, leaving 400 casualties.

On the fifth day of the TF's deep reconnaissance in force, 8th Company pressed on to exploit the Chinese withdrawal, finding and destroying a major ammo dump. Rangers like to work alone, but against stiff resistance they proved once again that, as in World War II, they can add punch and bad terrain mobility to combined arms operations.

With 8th Company—down to 90 men foxhole strength— on the spearhead, 24th ID's counterattack had regained only about ten miles of rugged terrain against stubborn resistance. Pausing at the road-rail junction of Kapyong, MG Bryan gathered strength to continue the push. The hills ahead were cross-compartmented against his axis of attack. On his right, was the unreliable 6th ROK Division, and overhead the skies opened a deluge.

These factors prevented any major envelopment of retreating communist forces. Nevertheless Eighth Army's offensive slogged ahead and, by the end of May, were once more across the 38th Parallel. His objectives limited by JCS cautions, Van Fleet had to settle for small-scale (but abundantly supported) attacks to straighten out his line on the most defensible terrain. He also hoped to lure the CCF into costly attacks but the fighting of May, in which they had lost over 200,000 men, seeemed to have sapped their ardor. Big enemy surrenders were now much more commonplace.

Few in 8th Company saw the end in sight. Their eyes were still fixed on a basic Ranger function— to jump! When, in early June, 24th ID went into corps reserve, an airborne opportunity was finally at hand. It was the good luck of sixteen replacements to arrive just in time at the marshalling area. On 3 June they went out of C-47's with the company, the first, last and only jump for the Devils during the war. Finally they drew hazardous duty pay after all the shock, shot and shell of previous combat for which they had not received an extra nickel. To add to the irony, 'Mad Dog' Uldall of Hill 1010 fame, suffered a jump injury that hospitalized him. Thus he was knocked out of the war, a feat the CCF could never accomplish.

Other personnel changes were CPT Bob Eikenberry, former CO of 7th Company at Ft. BEnning, taking command from LT Giacherine. LT Strong returned— his stomach stitched like a baseball— from convalescent leave, and a new lieutenant, Roy Fegley, reported in.

The now static front meant static duty for 8th Company. On 5 June they relieved 4th Ranger Company as security for the IX Corps tactical CP. Since the days of Darby, high headquarters have called for Ranger security when nothing more exciting was going on. Ranger security means vigorous patrolling and a tight perimeter but Rangers and REMF's do not mix well, and sometimes a high level CP will find they will be losing more equipment than they would have to enemy infiltrators. What 8th Company never had and always wanted was some vehicular mobility, and as trained night operators they obtained an abundance through 'midnight requisitions'. A favorite opportunity was when REMF drivers attended outdoor movies.

Another Ranger practice was to stage mock firefights (complete with CCF weapons and Chinese war crys) to check the CP's reaction drills—but this was as much a means to pull the REMF's chain. Such pranks and pilferage resulted in 8th Company's leash being shortened. As the Ranger 'motor pool' became too large to conceal, unit representatives came by to reclaim vehicles by checking engine numbers (bumper markings had long since been conterfeited). Fortunately while the Rangers were outliving their welcome at IX Corps CP, a schedule of R&R's came down to divert them. Each of the select few who went to Japan was charged as a Class VI officer to return with enough *sake* to ease the regrets of those who did not go.

On 23 June, 8th Company rejoined 24th ID as it moved back on line. Operating with tanks and elements of two regiments, the Rangers were once again put on the spearhead and became specialists at reducing roadblocks. Forging ahead more than a mile from friendly lines, tank-Ranger teams would slice through resistance until stopped by roadblocks. While these were taken down, line infantry mopped up in the zone between the roadblocks and friendly lines, destroying enemy field fortifications and uncovering caches. This territory was therefore scorched earth if the enemy tried to advance over it in the future.

Elsewhere, U.S. regiments were shortening their lines by withdrawing from indefensible terrain. 8th Company performed as rear guard and covering force for one such withdrawal. These operations, which lasted six days, were credited with the destruction or capture of an enemy howitzer, 33 mortars, plus 135 rifles and automatic weapons—very impressive numbers indicating not so much the intensity of the fighting but U.S. superiority at this point in the war.

Recognizing that superiority, the communists asked to talk about a cease fire. When it came to talking, they proved to be much superior, trying to convince the world that it was the United Nations who was suing for peace. Van Fleet would not let that impression last, and began to gnaw on the CCF again, insisting that fighting would continue until a truce was signed. In the interim, both sides agreed not to engage in major offensives.

Black: *To the individual soldier, July 1951 was a bewildering period. Ordered to halt while the enemy fled, men watched the communists gain breathing space to regroup and fortify. Positions that could have been taken without cost soon bristled with barriers and gun emplacements. As the talks dragged on, attacks were ordered to seize heavily defended terrain. After costly success, the wounded and dead were brought down and the newly won positions often ordered evacuated without enemy pressure.*

The roll of 8th Company (now up to a strength of 108) in this strange war was combat patrolling, initially as part of a battalion-size patrol base forward of U.S. lines as those lines were fortified on a World War I scale. On 3 July the Rangers were advanced guard (actually pathfinders) for a raid by 2/21st Infantry to kick the enemy off Hill 851 and temporarily occupy it. Through fog and drizzle, 8th Company led the way over heavily vegetated razor back ridges broken by chimney rock, with 60 to 70 degreee slopes slick with mud. The Rangers peeled off after guiding the infantry into their attack assembly area, then held open their resupply route.

2/21st's attack was successful, killing 145 CCF, and repulsing a counterattack. 8th Company became their rear guard as they pulled off Hill 851 to return to friendly lines. While occupying blocking positions they had unearthed a number of CCF land mines. These were relocated onto avenues of enemy approach, and several were heard detonated by their previous owners.

On 11 July, once again in support of a 21st Infantry attack, the

Devils were given the sheerest, steepest mountains to cross and Hill 1118 to seize. The approach was so difficult the Chinese had not considered defending it, so the Rangers found their objective unoccupied. From what amounted to a fire base in the enemy rear, 8th Company interdicted fleeing CCF forces.

The U.S. drive continued, pivoting on the Ranger spearhead against Hill 851 where battle raged for three days. Eikenbarry had a 50 cal. manhandled onto high ground to add firepower to the slugfest. A ridgeline between his position and Hill 851 was used for attacks by both sides. Snipers also conducted marksmanship contests with each other. Small night patrols clashed at close range among the boulders; there was much grenading and exchange of indirect fire.

All this kept a CCF battalion on Hill 851 fully occupied by day and night, allowing 21st Infantrymen to bypass the hill and continue to deeper objectives. Their mission accomplished at great loss to the enemy, they circled back to friendly lines, the Rangers left to screen. On 16 July, 8th Company returned with a body count of 80 CCF. The Devils had lost one killed and five slightly wounded.

The new men were no longer cherries, new leaders had won confidence, and the company had shaken down into high efficiency. Coordination with supporting arms and supported units was at a peak, so when G-3 called for Elkenberry to report he expected another mission like the last two for which the company had been highly praised. All the regiments were asking for Rangers in their task forces. What Elkenberry feared was that the Devils would have to split up into platoons.

What he learned was that the company would be split forever, split into atoms by the DA abolition of Ranger companies. As if they had all been taken prisoners of war, the Devils turned in division equipment and prepared for the train ride to Pusan and a sad reunion with their sister companies to furl guidons.

Afterword

Like most Vietnam veterans, the author had normal grumblings about the lack of public appreciation for that long and ghastly war. But in 1982 a feeling of compensation began to stir in us when The Vietnam Wall was dedicated at a most prominent location in Washington, D.C. Then a sense of closure and completeness arose from the Gulf War Victory Parade which marched by The Wall in 1991 as the president said it was our parade too, a long deferred tribute to overlooked sacrifices.

In researching and writing the foregoing accounts of Ranger companies in Korea, I developed a biting realization that for Korean vets their compensation has still never been paid, their parade never assembled, their monument never erected (though one near The Wall is planned for 1995). Korea is in fact the neglected war, its veterans the most unsung heros, its casualties the least memorialized. These casualties include nearly the same number of American dead as Vietnam and World War I, and far more MIA's.

It seems the Korean vets' only tribute has been from their own post-war lives. Their collective lack of bitterness. Their forebearance from bitching about lack of deserved recognition. This may be because they were mobilized but their nation was not. They went to war while their countrymen remained at peace. They fought, they bled, they won. Then they returned. Having given their all, they asked for nothing—and that's just what they got.

Consequently, in no other American war has the importance of duty been so prominent. Duty, in its simplest terms, that meant doing a job because someone had to do it. Ungarnished, unexalted duty.

For the Rangers there was an additional and rare ingredient that distinguished them: the craving to fight. They were the pit bulls in a U.N. kennel of many breeds, the attack dogs of a war fought with canine cruelty oblivious to quarter. To understand the setting in which Rangers battled so ferociously one must contemplate the prevelant attitude, the atmosphere, the environment of that war. Those components, and their contrast to America's home front at that time, are mystically present in Robert T. Fallon's essay which follows.

Fallon was an infantryman, badly wounded in 1950. His essay appeared, among other publications, in *The Saturday Evening Post, Reader's Digest, Coliers,* and *The Washington Post.* It is titled, 'I want to tell you something about a war.'

Fallon juxtaposes the comforts of home with the bleak chill of some nameless piece of frozen ground during a Korean winter; and though the heat of Korea can be hellish and throat parching, the deep cold is probably best remembered by veterans like this solitary soldier, symbolic of so many we have yet to sufficiently honor.

I Want to Tell You Something About a War

I want you to imagine for a moment that it is a very cold, wet wintry evening at about 10 o'clock. You have been sitting comfortably by your fireside reading the evening paper. It's been a pleasant day, dismal outside, but warm and restful by the fire. You decide to step outside on the porch for a breath of air before turning in. I'm sure you've done it often. But on this particular evening, a strange sight greets you.

There's a great hole right in the middle of your front lawn, and the dirt has been thrown up all around it, outlined sharply against the white, even snow. Squatting in the hole is a hunched figure.

Let me tell you something about him.

He's been in this area now for about three weeks, living in a dozen holes just like this one on your front lawn. The most apparent thing about him is that he is cold; that's because out on your lawn it's about 20 degrees colder than where you're standing. Every now and then he'll grab his shovel and dig a little deeper in the hole just to get warm. That's the only way he has, because he'll be seen if he builds a fire, and that might bring mortar fire into your living room.

He's been cold for a long time— cold and wet. He can't feel his feet; that worries him because he's afraid they might be frostbitten. It's going to be a long night, and it's going to get colder.

He's very dirty and itches. The grease from a hundred 'C' rations is frozen to his parka and gloves, coating the two weeks' beard which covers his face. Soot from small fires he dares make during the day is all over his pants and boots.

But he's dirty all the way through. He hasn't changed his underclothes in over a month, and he doesn't intend to for some time to come. It's too cold to go down that far. You can smell him really bad.

He's pretty hungry too. They didn't get his rations up to him until after dark, so he couldn't build a fire to thaw them out. He'll have to wait until morning. That's just another thing he'll have to do without, and right now a cup of hot coffee would sure taste good.

He looks old with that beard and sort of hunched-over posture. But he's only 19—though not like any other 19-year-old you've ever seen.

You may wonder what he's thinking about as he sits there during those long, solitary hours. Well, it's not much. Just how cold it is and again how nice that coffee would be. Maybe now and then he thinks of home but that's a long ways off, while the cold, his feet and his hunger are much more immediate. You'd be surprised how those three things can fill your mind.

You notice that he's cut a hole through your hedge. His rifle is sitting on the pile of dirt, pointing in readiness through the opening. That's another thing he's thinking: When are they coming again? He gets a little scared out there all alone. He'd like to go over and talk to his buddy in a similar hole about two houses up, but it's not a good idea to be crawling around at night. He wishes they weren't so far apart, but it seems they always have a big sector to cover with never enough men to cover it. It doesn't help being so far apart.

What would you like to do for this man? Ask him to your fireside, get him that cup of coffee? Would you like to loan him your razor, let him take a hot shower? Give him a bed to sleep in instead of the frozen dirt of his foxhole? Sure you would! You wouldn't think twice about it. But I'm afraid you can't. Over on that hill there are people who want to get in your front door. His platoon leader told this soldier to stop them—his orders are to protect your house and the one next door. So he can't come in. Neither can you reach him. He's actually very far away.

But you come out in the morning and he's still there, huddled over a little fire, thawing out his hands and his rations, trying to get the feeling back into his feet. By this time his hole is pretty deep from all the digging he's done during the night. He's also cut down a little more of your hedge to improve his field of fire.

During the day icy winds build up, but he's still there when you come home from work. While you are greeted by a comfortable fire in the living room, he's hunkering down for another even colder night.

I have told you about this soldier, and placed him on your front lawn, because I want you to realize that every desolate hill this soldier defends in a far-off land is in reality your front door. There are enemies who want to get in to do you harm; it's his job to keep them away. He's going to do his job and he's not going to ask you to do it for him. If he has to crouch on your front lawn in the cold, that's just the way things go. He's not going to begrudge you the comforts of your fireside and dinner table, but he'll be mighty bitter if he finds out that you're not doing your part of the job.

What is "your job"? Well, it's not really important what I happen to think "your job" may be. But I think it is important that you find out soon what it is and start doing it; for whatever it is, it's a vital part of our country's existence. Perhaps it would be more clear to you what your job is if you returned home tonight to find that hunched, shivering figure sitting in a hole on your lawn covered with snow.

First Induction

July, 1993

COMMAND SERGEANT MAJOR HENRY CARO is inducted into the Ranger Hall of Fame for extraordinary valor, exceptional gallantry, and heroic service to the nation and the U.S. Army as a Ranger Leader. Culminating three tours in Vietnam, two as a Vietnamese Ranger Battalion advisor, and one in direct action with Company N, 75th Infantry, 173rd Airborne Brigade, CSM Caro was awarded the Bronze Star with Valor five times, and the Vietnam Service Medal with nine Battle Stars. Command Sergeant Major Caro earned the reputation as a man who stood tallest among men. While the smoke of the battle field still hung in the air, and in action in which then SFC Caro had distinguished himself, an admirer of his, a Colonel, remarked to his company commander, "There's a real man- one that can be counted on when all the chips are down. There is not a phony cell in him!" As Command Sergeant Major Caro demonstrated throughout his 18 years of distinguished and devoted service, he was a true warrior. A Ranger that led from the front, his dedication to duty and superb leadership are a hallmark that continue to set the example for Rangers past and present.

STAFF SERGEANT DAVID C. DOLBY, is inducted into the Ranger Hall of Fame for extraordinary heroism and valor while assigned to Company B, 1st Battalion (Airborne), 8th Calvary, 1st Cavalry Division (Airmobile). SSG Dolby was awarded the Congressional Medal of Honor for gallant actions on May 21, 1966. While on patrol in the Republic of Vietnam, SSG Dolby's platoon came under intense fire. Six members of the platoon were killed instantly and a number were wounded, including the platoon leader. Aware that the platoon was in a precarious situation, SSG Dolby moved the wounded men to safety and deployed the remainder of the platoon to engage the enemy. Under orders from his dying platoon leader to withdraw, SSG Dolby, with utter disregard for his own safety, positioned able-bodied men to cover the withdraw of the forward elements, assisted the wounded to the new position, and he alone, attacked the enemy positions until his ammunition was expended. Replenishing his ammunition, Dolby single-handedly killed three enemy machine gunners and neutralized the enemy fire. Ranger Dolby's actions of unsurpassed valor during four hours of intense combat were a source of inspiration to his entire company, and contributed to the overall success of the assault on the enemy position.

COMMAND SERGEANT MAJOR NEAL R. GENTRY is inducted into the Ranger Hall of Fame for extraordinary valor and heroic service to the nation and the U.S. Army as a Ranger Leader. His service in the Ranger community began with a volunteer tour in combat with Company L, 75th Infantry, 101st Airborne Division (Air Assault), where he was recognized numerous times for acts of valor and service. CSM Gentry was hand-picked by then LTC K.C. Leuer to serve as the first modern era Command Sergeant Major for the 1st Ranger Battalion. Author of "The Ranger Creed," he established a code of ethical conduct in which the Ranger of today sets his beliefs. CSM Gentry was responsible for establishing the cadre of noncommissioned officers that would build upon and set the standards and successful activation of the 1st Ranger Battalion. He was instrumental in the development of the Coat of Arms and Distinguished Unit Insignia (DUI) for the Regiment. The DUI is still worn as an accouterment on the Ranger Dress uniform and black beret. The standards established by Command Sergeant Major Gentry were demonstrated in every combat operation that Rangers have participated in since activation in 1974. CSM Gentry's dedication to duty and superb leadership continue to lead the way for Rangers past and present.

NATHAN HALE is inducted into the Ranger Hall of Fame for his outstanding service to the infant United States of America as a Ranger of the Continental Army. During the late summer of 1776 Captain Hale was selected to command a company in Thomas Knowlton's Battalion of Connecticut Rangers. A short while later, Gen. George Washington asked for a Ranger officer to gather intel-ligence behind British Army lines in New York. Hale quickly volunteered despite a friend's attempt to discourage him. He penetrated British lines, disguised as a school-master. After gathering information on the dispositions of British regiments for about a week, he was com-promised and arrested. He proudly admitted that he was an American soldier and that he was spying on his captors. The British commander ordered him to be executed, without trial. The next day, September 22, 1776, as he was being readied for hanging, he declared to his British captors, "I only regret that I have but one life to lose for my country." He was buried in an unmarked, Manhattan, New York grave. He knew that his special mission was extremely dangerous, but he did not hesitate to serve his country with his very life. During the last two hundred years Captain Hale's dying words have become the classic expression of American patriotism.

COLONEL CHARLES N. HUNTER is inducted into the Ranger Hall of Fame for extraordinary courage and gallantry as a combat leader while serving with the 5307th Composite Unit (Provisional) known as Merrill's Marauders. In the fall of 1943, Colonel Hunter, the senior officer among the volunteers, was appointed commander of the battalion and he prepared the men for combat, and would serve as second in command under Major General Frank D. Merrill. The Marauders would become legendary for their trek through the jungles of China, Burma, and India, conducting operations behind enemy lines in the steaming jungles to sever Japanese communications and supply lines. Colonel Hunter's foresight, keen analytical mind, and thorough planning abilities had a direct and overwhelming impact on the success of the Marauders against the Japanese. Colonel Hunter led the 1st Battalion and the Chinese 150th Regiment known as "H Force" to capture the all weather airstrip a Myitkyina. This was to be the final victory of the 5307th Composite Unit. Following the capture of the airstrip the Marauders participated in a campaign to capture and secure an objective of great tactical importance, the town of Myikyina. In five major and 30 minor engagements they met and defeated the veteran soldiers of the Japanese 18th Division. Colonel Hunter's dedication and proven combat abilities in service of our nation are in keeping with the highest traditions of selfless service.

ABRAHAM LINCOLN is inducted into the Ranger Hall of Fame for his outstanding service to the United States of America, culminating with his service as President (1861-65). Early in 1832 he was elected captain of a company of the 4th Illinois Regiment, which served during part of the Black Hawk War (1832). When his company was discharged, Lincoln volunteered as a private in a company of the Illinois frontier guard, whose soldiers were known as Rangers, scouts, and spies. For two and one-half weeks during the late spring of 1832 he patrolled the northwestern frontier of Illinois with his company, on the lookout for Indian war parties. When the company's Rangers were discharged, Lincoln immediately reenlisted in another company of the frontier guard. He and the company served a three week tour of duty during the early summer by scouting in advance of the army as it moved northward into Wisconsin. After his military service Lincoln became a prominent Illinois attorney, a popular political leader, and President of the United States. His strong, successful leadership as the Union's President during the War Between the States (1861-65) made him a great American folk hero. He is often considered the best example of the greatness that can be produced by a democracy.

COMMAND SERGEANT MAJOR GARY L. LITTRELL is inducted into the Hall of Fame for conspicuous gallantry in action as a Ranger leader. CSM Littrell was awarded the Medal of Honor for his actions on April 4-8, 1970, when serving as and Infantry advisor with the 23rd Battalion, 2nd Ranger Group near Dak Seang in Kontum Province, Republic of Vietnam. After establishing a defensive perimeter on April 4, the battalion was subjected to an intense mortar attack which killed the Vietnamese commander, one advisor, and seriously wounded all the advisors except SFC Littrell. Sergeant

First Class Littrell single-handedly bolstered the besieged battalion, directed artillery and air support by day and marked the unit's position by night, despite the heavy concentrated enemy fire. Assault after assault was repulsed as the battalion responded to the extraordinary leadership and personal example exhibited by SFC Littrell. He continuously moved to those points most seriously threatened by the enemy, redistributed ammunition, strengthened faltering defenses, cared for the wounded and shouted encouragement to the Vietnamese in their own language. When finally ordered to withdraw, numerous ambushes were encountered. SFC Littrell repeatedly prevented widespread disorder, by directing airstrikes within 50 meters of their own position. Ranger Littrell's extraordinary courage and selflessness displayed over an extended period of time clearly demonstrates that "Rangers Lead The Way!"

LTC ANDRE C. LUCAS' career in the U. S. Army, and the ultimate sacrifice of his life truly epitomize the Ranger Creed and the honor of the U.S. Army. He began his career in the Army as an enlisted soldier and obtained the rank of Corporal. He competed for and received a Presidential appointment to the United States Military Academy at West Point. He graduated and was commissioned on June 7, 1954 as an officer in the U.S. Army. From his attendance at the Infantry Officer courses, Airborne and Ranger schools his numerous duty positions from platoon leader to battalion commander, including instructor, Florida Ranger Camp, LTC Lucas exceptionally performed all of his duties a positively outstanding manner. While serving as commander, 2nd Battalion, 506th Infantry, 101st Airborne Division (Airmobile), LTC Lucas distinguished himself by extraordinary heroism during the period from 1-23 July, 1970 at Fire Support Base Ripcord in the Republic of Vietnam. His actions during this extended period inspired his men to heroic efforts and was instrumental in saving the lives of many of his fellow soldiers. For his conspicuous gallantry and intrepidity of action, at the cost of his own life, LTC Lucas was awarded the Congressional Medal of Honor. His actions were in keeping with the highest traditions of military service and reflect great credit on himself, his unit, the Ranger Creed and the U.S. Army. LTC Lucas provides all soldiers and especially Rangers an exceptional leader to emulate and a level of performance to achieve.

PRIVATE FIRST CLASS ROBERT L. MASTIN is inducted into the Ranger Hall of Fame for extraordinary heroism in military operations as a Ranger trained soldier. Private Mastin trained as a Ranger with the first 9th Ranger Infantry Company (Airborne) at Fort Benning during the period February through April, 1951. Due to numerous casualties in the six Airborne Ranger companies then in Korea, the first 9th Company was disbanded and many of its members were sent to Korea as Ranger replacements. Private Mastin joined the 1st Airborne Ranger Company just before an extremely heavy action known as the "May Massacre" in which 1st Company was almost annihilated. In this major action near Kunmul-Gol, Korea, May 17-18, 1951, Private Mastin had been assigned as a platoon aidman. The company, committed to secure commanding ground on the left flank of a 2nd Division battalion, had moved into position under withering automatic weapons and small arms fire. Private Mastin constantly braved intense hostile fire as he ministered to the wounded and assisted in their evacuation. Although wounded himself in the initial action, he continued to perform his duties with determination. The enemy launched repeated assaults, showering friendly positions with grenades, inflicting further casualties. Disregarding his own personal safety, Private Mastin moved freely among the wounded, continued to render aid, and assisted his fellow Rangers to places of safety. Despite additional wounds received during the continuing action, he unhesitatingly made his way to the assistance of a wounded comrade through heavy enemy fire. While continuing his heroic performance of duty and directing others in caring for the wounded, he lost his life. Private Mastin was recommended for the Congressional Medal of Honor and was awarded the Distinguished Service Cross posthumously. He has set an example of dedication to his Country and to his fellow Rangers that will stand forever in the Ranger Hall of Fame.

MASTER SERGEANT ROY H. MATSUMOTO is inducted into the Ranger Hall of Fame for extraordinary courage and service with the 5307th Composite Unit (Provisional) known as Merrill's Marauders. In the fall of 1942, Matsumoto and many other young Japanese Americans incarcerated in internment camps, volunteered for service in the United States military. His unit would trek the jungles of Burma, conducting operations behind enemy lines in the steaming jungles to sever Japanese communications and supply lines. Master Sergeant Matsumoto accomplished many dangerous missions by infiltrating Japanese lines every night to listen and snipe. During the siege at Nhpum Ga on an intelligence gathering mission, he returned one night with news of an impending attack. The Marauders met the enemy with devastating fire. When the attack failed, Matsumoto stood and yelled in Japanese for the attack to continue. The next wave met the same fate as the first. The siege was finally lifted after ten days by an attack of the remaining Marauders. The Marauders continued to march and capture the all weather airstrip at Myitkyina where Matsumoto was in the last group of 17 men to be evacuated after the mission. He then joined the 475th Infantry "Mars Task Force" where he was attached to the Chinese Nationalist Army guerrilla forces behind enemy lines to serve as an Intelligence NCO near the French Indochina border. Master Sergeant Matsumoto's dedication and proven combat abilities in service to our nation are in keeping with the highest traditions of selfless service.

FIRST SERGEANT HARVEY L. MOORE, JR. is inducted into the Ranger Hall of Fame for his dedication to duty and contributions to the U.S. Army as a Ranger qualified leader. First Sergeant Moore began his career of military service as a Rifleman with the 4th Battalion, 54th Infantry in 1978. He was assigned to 1st Battalion (Ranger), 75th Infantry Regiment in 1980 and served as a fire team leader and squad leader. During this period First Sergeant Moore consistently displayed unparalleled leadership and an innate ability to teach and coach Rangers. In 1984, First Sergeant Moore was selected to serve as a Reconnaissance Specialist for the 75th Ranger Regiment. From 1986 to 1989, First Sergeant Moore served as a Ranger Instructor and Senior Ranger Instructor for 3rd Ranger Company, The School Brigade and 4th Ranger Training Battalion at Fort Benning, Georgia. First Sergeant Moore returned to 1st Battalion, 75th Ranger Regiment in 1989 to serve as Platoon Sergeant, 3rd Platoon, Company C. First Sergeant Moore's awards and decorations include the Meritorious Service Medal with Oak Leaf Cluster, Army Commendation Medal with 3rd Oak Leaf Cluster, Army Achievement Medal with 2nd Oak Leaf Cluster, Noncommissioned Officers Professional Development Ribbon (3rd Award), Armed Forces Expeditionary Medal (Grenada), Armed Forces Expeditionary Medal with Bronze Arrowhead (Panama), National Defense Service Medal, Southwest Asia Service Medal with Bronze Service Star, Expert Infantryman's Badge, Combat Infantryman's Badge, Master Parachutist Badge, Pathfinder Badge, Diver's Badge, British Army Parachutist Badge, and the coveted Ranger Tab. First Sergeant Moore's dedication to duty throughout his 14 year career brings immeasurable credit to himself, his country, and the U.S. Army Rangers.

SERGEANT MAJOR OF THE ARMY GLEN EDWARD MORRELL is inducted into the Ranger Hall of Fame for outstanding service to the nation and to the U.S. Army as a Ranger qualified non-commissioned officer leader. SMA Morrell began his career of military service when he enlisted into the Army in November, 1954, from Wick, West Virginia. As SMA Morrell rose through the enlisted ranks he held virtually every possible leadership position for a noncommissioned officer, culminating by his being sworn in as the seventh Sergeant Major of the Army on July 1, 1983. He was the first Ranger qualified soldier to hold that position. During his distinguished 32 year career, SMA Morrell has served in numerous assignments stateside and many overseas, to include Europe, Panama and three tours of combat duty in the Republic of Vietnam, He has earned the Combat Infantryman's Badge, Master Parachutist Badge, U.S. Army Recruiter Identification Badge, the Special Forces Tab, and the coveted U.S. Army Ranger Tab, which he earned at age 41. SMA Morrell is a graduate of the Sergeants Major Academy, Class 2 and has numerous other campaign and service medals. SMA Morrell's dedication to duty throughout his career brings great credit upon himself, the United States Army and the Rangers past, present, and future.

COLONEL ROY A. MURRAY is inducted into the Ranger Hall of Fame for extraordinary valor and heroic service to the nation as a Ranger leader. Colonel Darby chose then Captain Murray to command Fox Company, 1st Ranger Battalion. As Fox Company Commander, he led the Ranger element on August 18, 1942 Dieppe Raid, landed in North Africa at Arzew, and fought in the raid on Sened Station. Murray was named Commander,

4th Ranger Battalion as a Captain. He led the battalion in the invasion of Sicily at Gela, and was promoted to Major. The battalion then fought in the amphibious assault at Maori-Salerno and Murray earned a battlefield promotion to Lieutenant Colonel. He then led his battalion behind enemy lines to cut the road at Ceppagna, Italy. Murray also led his Rangers on the Anzio invasion. During this battle, German forces surrounded 1st and 3rd Ranger Battalions at Cisterna, LTC Murray led the 4th Ranger Battalion in several gallant, though unsuccessful attacks to rescue their comrades. In 1953, Colonel Murray commanded the 32nd Infantry Regiment in Korea and then served as I Corps Deputy Chief of Staff. Other notable positions include: Infantry School Director of Instruction, Berlin Brigade Chief of Staff, and Combat Surveillance School Commandant. His awards include the Silver Star, Purple Heart, Bronze Star with Oak Leaf Cluster, Legion of Merit, and the Chief Emperor Minilik II Medal. Colonel Murray's selfless dedication to duty and valorous leadership exemplify that "Rangers Lead The Way!"

COLONEL GEORGE A. PACCERELLI devoted twenty-six years of his life to leading and serving American soldiers. As an enlisted soldier, he served in artillery, infantry, airborne, and special forces units around the world, to include a combat tour as a Special Forces adviser in Laos in 1961-1962. As an Infantry officer, he continued to serve in troop and special forces units. in 1966, while serving as a Special Forces A Team commander in the Republic of Vietnam, then CPT Paccerelli was wounded in action, but courageously returned to complete a tour of duty with the Special Operations Group, Military Assistance Command Vietnam, In 1968-1969, while assigned to the 1st Cavalry Division in Vietnam, CPT Paccerelli commanded Company E, 52nd Infantry (Long Range Reconnaissance Patrol), later redesigned as Company H (Ranger), 75th Infantry (Airborne). He retrained, reorganized, and transformed the LRRP/Rangers into a self-confident, skilled, and professional military force that dominated the III Corps Tactical Zone and carried the battle to the North Vietnamese Army. In later advisory, staff, and battalion command assignments, COL Paccerelli made many other significant contributions to the U.S. Army and the nation. Through his dedication to duty, his manifest love and respect for soldiers, and his numerous examples of selfless courage COL George A. Pacerelli led the way for all Rangers, past, present, and future.

SERGEANT DONN F. PORTER is inducted into the Ranger Hall of Fame for extraordinary courage and gallantry in action as a Ranger qualified leader. Sergeant Porter was the only Airborne Ranger to be awarded the Congres-sional Medal of Honor during the Korean War. He trained with the second 9th Ranger Infantry Company (Airborne) from May through September, 1951 and was awarded the Ranger Tab on graduation. When the Airborne Ranger com-panies were disbanded as cease fire negotiations started in Korea, Sergeant Porter and 24 other 9th Company Rangers vol-unteered for duty in Korea and were assigned to the 187th Airborne Regimental Combat Team (ARCT). At that time, the 187th was in strategic reserve on Kyushu Island, in southern Japan, so Sergeant Porter volunteered for combat duty on line in Korea and was assigned to Company G, 14th Infantry Regiment, 25th Infantry Division. He was soon promoted to Sergeant and placed in charge of a combat outpost in the vicinity of Mundung-ni and "Heartbreak Ridge" at a key approach into the main line. An enemy force estimated at 400 men probed the position with artillery and mortar for several hours. In the first main assault, Sergeant Porter's group of four men expended 4-6 cases of grenades that were on position, killed 15 enemy soldiers and wounded many more. Two of Porter's men were killed. In the second assault, Sergeant Porter jumped from his position with bayonet fixed and meeting the onslaught in close combat, killed six hostile soldiers and routed the attack. When his body was found, he still clutched his rifle with the fixed bayonet rammed into the body of a North Korean soldier. His supreme sacrifice has set the example for all who wear the Ranger Tab on into the future. The Porter Range at Fort Benning has been named in his honor.

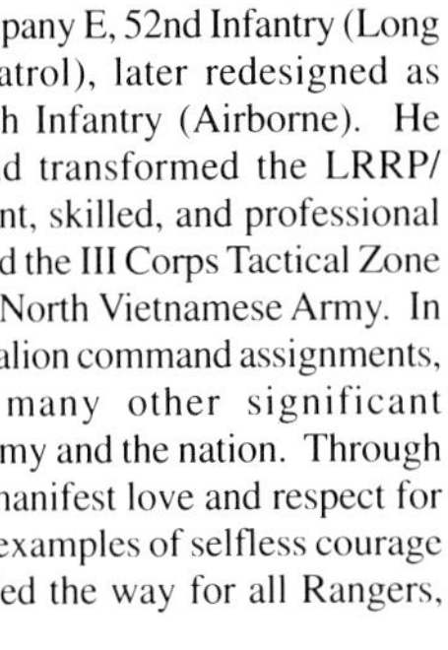

COLONEL ARTHUR C. STANG, III's service to the U.S. Army and the nation was exceptionally steadfast, distinguished, and loyal. While serving in various command and staff positions in peacetime and war, his performance was magnificent in every respect. His diligent efforts as Brigade S2 saved countless lives and contributed significantly to the combat effectiveness of the 173rd Airborne Brigade in the Republic of Vietnam. He was highly regarded by the Joint Chiefs of Staff and the Navy Staff while serving as a member of the Army Staff. In this capacity, he far exceeded the normal outstanding performance of duty expected of Office of the Deputy of Chief of Staff for Operations (ODCSOPS) staff officers. His professional competence, creativity, and diligence were instrumental in the production of the Land Forces Book, Volume II, Joint Standing Operating Procedures. By virtue of his dynamic personality, professional qualifications and exacting standards as commander of 1st Battalion, 17th Infantry, 2nd Infantry Division, he electrified and energized the battalion to the point where it enjoyed the well deserved reputation "Second to None." Because of his superb performance as a battalion commander, he was selected to command the 1st Battalion (Ranger), 75th Infantry. As a Ranger battalion commander, he welded his men into a highly motivated, elite, skilled unit with everyone working as a team. Through his leadership, the battalion reached a state of motivation unmatched by his peers. Through his unit, Colonel Stang clearly exemplified the motto "Rangers Lead The Way." During his tour of duty as the Commander of 3rd Brigade, 82nd Airborne Division, culmination as the Chief of Staff, 82nd Airborne Division, Colonel Stang demonstrated through dynamic, caring leadership and recognition, the theory that the individual soldier would be more loyal, dedicated, and professional if subject to these virtues. Colonel Stang's service which was characterized by selfless dedication to the defense of the country has been distinguished in every regard.

LIEUTENANT GENERAL SAMUEL V. WILSON is inducted into the Ranger Hall of Fame for heroism, extraordinary achievement, and continued service to his country and the special operations community. General Wilson began his special operations service with the Office of Strategic Services and subsequently with the 5307th Composite Unit (Provisional), popularly known as Merrill's Marauder, in the China-Burma-India Theater during World War II. As a highly decorated combat veteran, General Wilson returned to the United States where he entered the Army's Foreign Area Specialist Training Program. Upon completion of the program, he was assigned to various tours involving the Iron curtain countries and the Soviet Union. His expertise was recognized quickly and he was assigned to several high level positions within the Department of Defense and Department of State. As a general officer, some of his assignments included: Assistant Division Commander (Operations), 82nd Airborne Division; United States Defense Attaché to the Soviet Union; Deputy to the Director of Central Intelligence; and Director of the Defense Intelligence Agency. General Wilson's expertise in the area of special operations has been recognized and he remains a consultant to the Secretary of Defense and the Congress for matters concerning both special operations and intelligence.

Second Annual Induction August 1994

SERGEANT MAJOR FRED E. DAVIS distinguished himself throughout his 33-year U.S. Army career while serving in combat during three wars in Ranger and Special Operations assignments. Always a volunteer, Sergeant Major Davis served with distinction during World War II as a member of the 3rd Ranger Battalion. While engaged in heavy fighting on the Anzio Beachhead in Italy, he was wounded and subsequently captured by enemy forces. After numerous attempts, Sergeant Major Davis successfully escaped from his German Captors and made his way to Allied Lines. In 1950 during the Korean War, a call for Ranger Volunteers was again issued and Sergeant Major Davis rose to the challenge and joined the 3rd Airborne Ranger Company. After heavy fighting at the battle for "Bloody Nose Ridge," he was recognized for his superior leadership ability and awarded a battlefield commission. Following deactivation of the Airborne Ranger Companies in Korea at the start of Armistice Talks, he served with the 187th Airborne Regimental Combat Team and then volunteered to return to Korea and participate in extensive Ranger-related Unconventional Warfare activities. These operations served as forerunners for the establishment of Special Warfare Schools and later Special Forces units. After reverting to enlisted rank Sergeant Major Davis served several tours in Vietnam as both A and B Team leaders. Sergeant Major Davis' long and distinguished career reflects great credit on himself, past and present Ranger units, and the military service.

CAPTAIN GREGORY ELDON GARDNER is inducted into the 1994 Ranger Hall of Fame for his actions above and beyond the call of duty in saving fellow Rangers from certain death following the fiery crash of their C-130 Transport Aircraft. On the night of September 21, 1981, Captain Gardner was serving as the 2nd Ranger Battalion surgeon and participating in a night assault landing exercise at an airstrip near

Las Vegas. The cargo plane crashed short of the runway and exploded into flames killing several soldiers including the battalion commander. Having survived the initial crash and escaping from the burning aircraft, Captain Gardner quickly established a first aid post for the injured Rangers. Fearing more Rangers were trapped in the burning plane, Captain Gardner continued to risk his life to help his fellow Rangers by making several trips into the aircraft to bring out a number of injured soldiers. It was during the final trip into the aircraft that it exploded, killing Captain Gardner. For his courage, self sacrifice, and refusal to leave a fallen comrade Captain Gardner was posthumously awarded the Soldiers Medal. His actions are in keeping with the highest traditions of the Rangers and the United States Army and sets the example for others to follow.

LIEUTENANT COLONEL WILLIAM E. GORE is inducted into the Ranger Hall of Fame for his contribution to the United States of America as a courageous and successful Ranger. He commanded a company of South Korean Marine Rangers during the Korean War. He had previously demonstrated his unusually high degree of battlefield courage and leadership ability when as a junior airborne officer he led a battalion command group in the successful defense against a superior attacking enemy force. Later, he organized, trained, an led a company of Korean Marine Rangers. Known as Task Force Red Wing and placed under the operational control of the 8086th Army Unit. Lieutenant Colonel Gore and his Korean Rangers conducted a series of successful reconnaissance, raid, and ambush operations on the west coast of Korea. Task Force Red Wing remained on the move, conducting their operations behind enemy lines for five months while successfully avoiding a pursuing enemy force. He lost only three men during his time in action behind enemy lines and caused the North Korean and Chinese Armies to commit large numbers of troops to his pursuit that could have been used to fight against the United Nations forces. His actions have been chronicled in books and articles. He has been cited as an example of the value of courage. Lieutenant Colonel Gore was an outstanding combat Ranger and his actions and deeds set the example for all future Rangers to follow.

LIEUTENANT COLONEL WILLIAM C. GRISSOM is inducted into the Ranger Hall of Fame for his actions behind Japanese lines as a member of Merrill's Marauders during World War II. Then, First Lieutenant Grissom commanded the Intelligence and Reconnaissance Platoon of the 2nd Battalion 5307 Combat Unit Provisional (Merrill's Marauders). As the Marauders always operated deep behind the enemy lines, in Burma Lieutenant Grissom constantly, at great risk, ranged far out in front of the two combat teams of the second battalion. He collected intelligence, scouted the enemy positions and strength, and picked the safest routes for the Marauders to travel. Lieutenant Grissom was involved in numerous skirmishes with the enemy and participated in the thirty-five battles that the Marauders were involved in during the campaign. Lieutenant Grissom fought with the Marauders until the main objective, the capture of Myikyina, the only all weather air strip in Burma, and the opening of the road into China from Burma and India was completed. Following his release from active duty Lieutenant Colonel Grissom continued to serve the military as a training officer with the Veterans Administration. His courage, strength, and willingness to face constant hardships and danger are in the finest tradition of the Rangers and the United States Army.

FIRST LIEUTENANT MAYO S. HEATH is inducted into the Ranger Hall of Fame for distinguishing himself in action while serving with the United States Army Rangers during the Korean War. First Lieutenant Heath enlisted in the U.S. Army upon graduation form Carnegie Mellon University. He participated in the Parachute Assault into France and Holland. He was awarded the Bronze Star for his actions during these campaigns. When the Korean War broke out First Lieutenant Heath again volunteered for service. He served with the 1st Ranger Infantry Company (Airborne). In Korea, he led numerous patrols behind enemy lines. Once for four days, he raided 20 miles behind enemy area of operations. First Lieutenant Heath was awarded his second Silver Star for this action. In mid February, 1951, First Lieutenant Heath's Ranger Unit was assigned to the 23rd Infantry Regimental Combat Team, then completely surrounded at Chip Youngni. Day and night the enemy launched continuous attacks from every quarter. This became a "stand or die" situation. When the Chinese broke the Combat Team's perimeter, the 1st Ranger Company was in reserve. The Combat Team's Commander ordered a counter attack to reestablish the perimeter. First Lieutenant Heath was killed leading the attack. Inspired by Lieutenant Heath's bravery, his men continued the attack driving the enemy from his newly won positions. First Lieutenant Heath's actions followed the tradition of the Rangers, in leading the way and completing the mission.

STAFF SERGEANT JACK E. KUHN is inducted into the Ranger Hall of Fame for his extraordinary feats of courage, while serving with the 2nd Platoon, Company D, 2nd Ranger Battalion at Pointe du Hoc, Normandy, France. On D-Day June 6, 1944, he was an acting Platoon Sergeant. The actual Platoon Sergeant had been transferred to acting First Sergeant just prior to D-Day. Staff Sergeant Kuhn and his platoon fought their way up the 100 foot cliffs of Pointe du Hoc through the German defenses and continued fighting out to the coast road, about one mile inland, and helped establish the D Company road block. He then went on a two-man reconnaissance patrol and helped his acting platoon leader First Sergeant Leonard G. Lomell find and destroy the five large mobile coastal guns that were missing from the Pointe. He returned to his outpost duty, setting the example for the rest of D Company. When engaged in helping repel three determined German counterattacks by a numerically superior enemy force. Staff Sergeant Kuhn refused to surrender the road block position until properly relieved, which occurred on D-Day plus two. The gallantry and courage displayed by Staff Sergeant Kuhn reflects great credit on his leadership, courage, and determination. Staff Sergeant Kuhn was later promoted to First Sergeant of D Company and later served in the Korean War. Staff Sergeant Kuhn actions were in the highest traditions of the United States Army Rangers and the military service.

SERGEANT PETER C. LEMON is inducted into the Ranger Hall of Fame for conspicuous gallantry and intrepidity in action at the risk of his life above and beyond the call of duty. On April 1, 1970, Sergeant Lemon distinguished himself while serving as an assistant machine gunner during the defense of Fire Support Base Illingworth in Tay Ninh Province, Republic of Vietnam. Sergeant Lemon engaged a numerically superior enemy with machine gun and rifle fire from his defensive position until both weapons malfunctioned. He then used hand grenades to fend off the intensive enemy attack launched in his direction. After eliminating all but one of the enemy soldiers in the immediate vicinity, he pursued and disposed of the remaining soldier in hand-to-hand combat. Despite fragment wounds from an exploding grenade, Sergeant Lemon regained his position, carried a more seriously wounded comrade to an aid station, and as he returned, was wounded again by enemy fire. Disregarding his personal injuries, he moved to his position through a hale of small arms and grenade fire. Sergeant Lemon immediately realized that the defensive sector was in danger of being overrun by the enemy and unhesitatingly assaulted the enemy soldiers by throwing hand grenades and engaging in hand-to-hand combat. Fully exposing himself to enemy fire Sergeant Lemon engaged the enemy until he collapsed from his wounds. After regaining consciousness he refused medical evacuation until those more seriously wounded had been evacuated. Sergeant Lemon's actions epitomized the Ranger Creed. "Surrender is not a Ranger Word."

STAFF SERGEANT GARY L. LEMONDS is inducted into the Ranger Hall of Fame for extraordinary heroism in connection with military operations involving conflict with an armed hostile force in the Republic of Vietnam. Staff Sergeant Lemonds distinguished himself by exceptionally valorous actions on the 9th, and 10th of April, 1969 while serving as team leader during an ambush patrol maneuver. When an enemy force launched a massive attack Sergeant Lemonds led his men in an assault on the enemy emplacements. He single-handedly charged a bunker and destroyed it with grenades. Sighting another fortification, he crawled through the fierce enemy fire and silenced it with his rifle. After eliminating the hostile entrenchment system, he continued his patrol mission. On the 10th of April his patrol was attacked a second time. From his patrol position in a bomb crater, he called in artillery and air

support on the large enemy element. When withdrawing his troops to a nearby landing zone for evacuation, he came under sniper fire from two North Vietnamese soldiers, whom he quickly eliminated with a well-thrown grenade. Intense enemy fire thwarted the extraction aircraft's first landing attempt. He then directed a gun ship on the enemy location and effected a successful extraction. Staff Sergeant Lemonds' extraordinary heroism and devotion to duty are in keeping with the highest traditions of the Rangers, and reflect great credit upon himself, his unit and the United States Army.

GENERAL JAMES J. LINDSAY is inducted into the Ranger Hall of Fame as one of the foremost soldiers of his time, through leadership and innovation, leaving a lasting imprint on the United States Army, Special Operations, and Ranger communities. Throughout his career, he demonstrated a deep understanding of what motivates the American fighting man, and used this insight to create superb combat units which played key roles in the security of the United States. His innovations in training were critical in the revival of light infantry and special operations forces in wake of the Vietnam War. As a general officer his command assignments include: 82nd Airborne Division, United States Infantry Center, XVIII Airborne Corps, United States Readiness Command, and the United States Special Operations Command. As the first Commander-in-Chief of the United States Special Operations Command, General Lindsay engineered the form and substance of the new unified command, and established himself as the nation's foremost leader and authority on special operations. Special Operations, Psychological Operations, and Civil Affairs Forces conducted over 500 operations in more than 60 countries, to include spearheading the highly successful Operation JUST CAUSE into Panama. His selfless dedication to the service of this nation, the United States Army, and the special operations community, to include the Rangers reflects the highest credit upon him and truly set an example for all Rangers and soldiers to follow.

FIRST SERGEANT LEONARD G. LOMELL is inducted into the Ranger Hall of Fame for his outstanding acts of heroism during the D-Day invasion of Normandy on June 6, 1944. While serving with Company I, 2nd Ranger Battalion, First Sergeant Lomell, though wounded, climbed the 100 foot cliffs of Pointe du Hoc, under fire from the enemy. As an acting platoon leader he led his 2nd platoon through the heaviest kind of automatic weapons fire to destroy an enemy machine gun position. He continued his courageous assault on the enemy while under devastating artillery bombardment, through two enemy lines of resistance to the rear of their position. Surrounded by the Germans, he, together with SSG Jack E. Kuhn, found the five big mobile coastal guns of Pointe du Hoc that were missing from the Pointe. They were hidden in an apple orchard over a mile inland from the Pointe. The guns were momentarily unguarded. First Sergeant Lomell alone seized the opportunity to slip into the position and silently, with the use of thermite incendiary grenades melted together the moving parts of the traversing and elevation mechanisms and the breech blocks of the guns, thus rendering them inoperable by 0830 hours D-Day. Many thousands of lives of soldiers on the landing beaches of Normandy were spared by First Sergeant Lomell's bold, brave, and outstanding leadership in the face of such superior numbers. First Sergeant Lomell's successful mission was in keeping with the highest traditions of the Rangers and the military service.

BRIGADIER GENERAL FRANCIS MARION is inducted into the Ranger Hall of Fame for his distinguished service as a Ranger leader during both the French and Indian War and the American Revolution. During the French and Indian War, he fought against hostile Cherokee Indians who were making raids on frontier settlements in South Carolina. He and his men adopted many tactics used by the Indians and used them to their advantage. In the American Revolution, Brigadier General Marion directed Ranger-type operations against the British throughout the southern colonies. Marion's men conducted raids on British encampments and ambushes on their supply lines. The South Carolinians once captured a British prisoner of war camp freeing 150 Colonial survivors of the Battle of Camden. It was during this time that Brigadier General Marion earned the nickname the "Swamp Fox" because of his ability to use the cover and concealment of the swamps to strike British and Loyalist forces at will. The operations conducted by Francis Marion were key in preventing the British from securing the southern colonies. Brigadier General Marion brought great credit upon himself, the state of South Carolina and the United States Army. His actions and outstanding service to the United States of America truly exemplify what it means to be a Ranger.

STAFF SERGEANT ANDREW B. PUNG is inducted into the Ranger Hall of Fame for actions behind Japanese lines while serving with the 5307 Combat Unit Provisional (Merrill's Marauders) during World War II. Staff Sergeant Pung distinguished himself in action at Walawbum, Burma. As a member of the Orange Combat Team his battalion swept around the Japanese right flank in a pincers movement that overran the enemy position. This maneuver killed over 800 enemy soldiers who were attempting to escape from the trap. Staff Sergeant Pung then climbed a tree and perched 50 feet off the ground. He kept his unit posted on the Japanese activity via a walkie-talkie radio. Upon seeing the Japanese advance he alerted his unit, then waited until the enemy was 50 yards away before giving the order to open fire. The Japanese soldiers running into a solid sheet of gunfire, broke and ran leaving another 200 dead behind. During this battle Staff Sergeant Pung's canteen was shot off his hip and his radio knocked out of his hand by a bursting artillery shell. Shakily he managed to climb down from his tree and rejoin his unit. He served with the Marauders until the completion of their mission, the capture of Myitkyina, the only all weather air strip in Burma, and the opening of the China road. Staff Sergeant Pung continued to serve with the United States Army in Korea and Germany. Staff Sergeant Pung retired in July, 1967, after completing 30 years of dedicated military service, upholding the Ranger traditions and leading the way for future generations of Rangers to follow.

MAJOR JAMES C. QUEEN is inducted into the Ranger Hall of Fame for his valiant actions while serving as a Ranger leader in the Korean War. While serving with the 4th Ranger Infantry Company (Airborne) (redesigned the 2nd Ranger Infantry Company (Airborne)) Major Queen led numerous patrols and controlled the company heavy weapons. He made the combat jump with the company at Munsan-ni and saw action at Tangerine Pass. In May of 1951 Major Queen led the company on a mission to take Hill 581. During this operation he acted as the Company Commander for the 2nd Ranger Company. They moved as the lead element for the 7th Infantry Division and with only 65 men came into contact with a battalion of Chinese regulars. Major Queen immediately called for artillery fire and led his men into battle. They fought the enemy for three hours but after taking 50% casualties the company fell back to the 7th Division lines. At dawn the following day, Major Queen led the remainder of his company in a counterattack on the Hill. After nearly one hour of intense fighting, the Rangers pushed the Chinese back and retook the Hill. Major Queen's leadership ability and courageous actions are in the finest traditions of the Rangers and the United States Army.

SERGEANT FIRST CLASS RANDALL D. SHUGHART is inducted into the Ranger Hall of Fame for extraordinary heroism in connection with military operations involving conflict with an armed hostile force in Mogadishu, Somalia. In August, 1993, while assigned to Task Force Ranger, Sergeant First Class Shughart's mission contingency plans included rescue of any downed aircraft crew. When the first helicopter crashed into the Magadishu streets, he provided protective covering fire allowing the Ranger rescue team to secure the area around the helicopter. When the second helicopter crashed, Sergeant First Class Shughart could see movement inside. Realizing friendly ground troops could not reach the downed helicopter in time, he requested permission to land. Making their way to the crash site he and another Special Operations soldier removed the seriously wounded helicopter crew from the aircraft. Armed with only a rifle and pistol he moved around the crash site attempting to stop the rush of an angry Somalian crowd. Checking the downed helicopter for weapons and ammunition, he gave the sole survivor, CWO Michael Durant a rifle and five rounds of ammunition. Armed with only a pistol he continued to safeguard the wounded pilot until he, too, received a fatal wound. When CWO Durant was released from captivity 11 days later, he credited his survival to Sergeant First Class Shughart's unselfish actions. His refusal to leave a fallen comrade, even at the expense of his own life is in keeping with the finest traditions to the Rangers and the United States Army.

SENATOR STROM THURMOND predated the advent of the United States Army Ranger Course. On

December 11, 1941, the day the United States declared war on Germany, Senator Thurmond took a leave of absence from the Judicial Bench to join the United States Army. Having been in the Reserves since 1924, he was com-missioned a Lieutenant with the First Army's Eighty-Second Airborne Division. On D-Day in 1944, his troop glider crash-landed behind enemy lines in France. Sustaining minor injuries, he and the other men fought for two days in isolation before they succeeded in linking up with American Forces advancing from their beachhead in Normandy. This type of mission demonstrates a classic Ranger-style operation which included infiltration of an enemy held area, and disruption of enemy lines of communication and his order of battle. Senator Thurmond later served in the Pacific. By the time he was discharged, he held the rank of Lieutenant Colonel. He retired from the United States Army Reserve as a Major General in 1960. In his later public service, Sen. Thurmond advocated a strong Nat'l Defense including the full participation of the United States Army Reserve. He strongly supported the establishment and maintenance of an extraordinary fighting force, one that is capable of operating in all types of terrain, weather conditions, and without regard for the potential enemy hampering of the operation. This idea would later be embodied as the United States Army Rangers.

GENERAL LUCIAN KING TRUSCOTT, JR. is inducted into the Ranger Hall of Fame for his outstanding service to the United States Army and his demonstrated commitment to Ranger operations. General Truscott holds the sole claim of rising in rank from Colonel to Lieutenant General through successive battle commands of Regiment, Division, Corps, and Field Army in World War II. In 1942, General George C. Marshal assigned General Truscott to Admiral Louis Mountbatten's Combined Operations Headquarters in London and gave him the responsibility of organizing an American counterpart to the British Commandos. Truscott visited every Commando unit training in England and Scotland and, as a result, created an American Ranger Unit, the first of its kind sense the days of the American Revolution. He derived the name from Rogers, Rangers who fought in the French and Indian War. Truscott recommended William O. Darby for the command of the new unit. General Truscott accompanied the Rangers on the Dieppe Raid and to North Africa. He received the Distinguished Service Medal in part for his organization of the new Ranger Unit. General Truscott's military service and dedication to the Ranges are of the highest degree. They reflect great honor upon himself, the Rangers and the United States Army.

Third Annual Induction

July 1995

FIRST SERGEANT ANDERS K. ARNBAL was drafted into the U.S. Army in 1941 and volunteered for service with the Rangers shortly thereafter because he was impressed with the small unit tactics of the British Commandos and longed for similar training and a challenging environment. He trained at Carrickfergus as a member of Darby's original Rangers and was assigned to Company B, 1st Ranger Battalion. He participated in the Torch Landings at Arzen, Algeria as a machine gun corporal and fought valiantly in Tunisia and at El Guettar. Later, he was part of the original cadre of 3rd Ranger Battalion and was assigned as a platoon sergeant in Company D. Ranger Arnbal made the landings in Sicily and Italy and was later promoted to First Sergeant of Company D. Within hours of LD time for the Cisterna attack, he and three other members of 3rd Ranger Battalion- all original Rangers, received orders for return to the United States. Both Colonel Darby and Major Miller, then commander of 3rd Ranger Battalion, gave Ranger Arnbal and his three comrades orders not to participate in the attack as planned. Thus, he remained on the beachhead the next morning when news of the disaster and casualties started coming in. Ranger Arnbal took charge of the rear detachment Rangers to evacuate, treat, and account for the survivors. He attempted to launch a relief effort, but was overruled. First Sergeant Arnbal spent two grueling years overseas, fighting ferociously during four invasion landings and six major campaigns- all with the Rangers! RANGERS LEAD THE WAY!

MASTER SERGEANT AUBREY M. BATTS was born in Washington, DC on 4 March 1939. He enlisted in the military service through the Reserves at Fort Jackson, SC, on 11 April 1956. Following six months in basic training and advance training, MSG Batts went to civilian life for one year. He reentered the Army on 1 August 1958. In March of 1964, he attended Ranger School (44th Company (Ranger), 4th Student Battalion). Upon completion of the Ranger Course in May of 1964 he was reassigned to A Company, 2d Battalion, 405th Infantry, 82nd Airborne Division until December 1964. Following this assignment, he was reassigned to Germany with Company B, 1st Battalion 509th Infantry Division as squad leader and platoon sergeant. He served two tours in Vietnam being assigned to MACV as a light infantry instructor for the Vietnamese Army and C Company, 327th Airborne Infantry, 101st Airborne Division. From June 1969 to September 1970, MSG Batts was assigned to 3d Ranger Company, Ranger Department. His final duty assignment was with the Ranger Department (3d Ranger Company, City Phase) as a Chief Instructor from January of 1972 until retirement in July of 1978. During his twenty year military career, MSG Batts attended several schools which include Airborne School, Ranger School, Jumpmaster School, Pathfinder School, NCO Academy (Europe), and Recondo School. His awards and decorations include the coveted Ranger Tab, Master Parachutist badge, CIB, GCMDL (4th Awd), Army Commendation Medal with OLC, the Crewman's Badge, the Meritorious Service Medal, and Overseas Bars. His Great service brings great credit to himself and the United States Army.

COLONEL ROBERT W. BLACK distinguished himself throughout his military career rising through the ranks of Recruit to Colonel and serving in ground combat during the Korean and Vietnam Wars. Col. Black saw combat action in eight campaigns. Twice awarded the Combat Infantryman Badge, he holds the Silver Star, three Bronze Stars (two for valor), the Legion of Merit, the Air Medal, the Joint Service Commendation Medal, two Army Commendation Medals, the Vietnam Cross of Gallantry and the Vietnam Cross of Honor. Col. Black is the foremost historian of the American Ranger. His research and writings have proven ranger accomplishments from the 1600s to the 20th century. He is the author of "Rangers in Korea", and "Rangers in World War II". These and other publications informed the public of Ranger contributions to our national development and security. Thirty years after the Korean War, Col. Black led the successful effort by Rangers Associations to retrieve Ranger lineage and battle honors that were given to Special Forces. He secured official recognition of the 8th Army Ranger Company and the award of the Ranger Tab to the men who served in that unit in Korea. His research resulted in Rangers and their units receiving decorations and awards earned, but not received. He developed programs that resulted in Ranger monuments, the Ranger Regimental Bowl and grave markers. He was the founder of the Ranger Research Collection at the United States Military Institute which includes the largest collection of Ranger Photographs in existence. Col. Black was the founding President of the Association of Ranger Infantry Companies (Airborne) in the Korean war. Col. Black's distinguished service reflects great credit upon himself, the Rangers, and the military service.

COLONEL CHARLES E. BEACH II was born 5 May 1908 in Cincinnati, OH. He enlisted in the National Guard in 1926 and was assigned to the 147th Infantry, 37th Division. He was commissioned as a 2nd Lieutenant on 4 January 1930. Col. Beach was called into active duty on 15 October 1940 and was promoted to Lieutenant Colonel while he was on rest duty on the island of Samoa in late 1943. Shortly thereafter he volunteered to serve with "Merrill's Marauders" under Frank Merrill and Joe Stilwell. He was given command of 3rd BN 5307 Composite Unit Provisional. He volunteered to lead the 3rd Battalion Marauders and two combat teams at the outset of the Burma campaign and continued on to the capture of the airfield at Mytkiana, the city and the rail head. His schools and awards include being a graduate of the Infantry Schools Company Officers Course, the Infantry Officers Advanced Course, the Command and General Staff College. He received the Silver Star with OLC, Presidential Unit Citation, Combat Infantry Badge and Bronze Star. He was promoted to Colonel in 1946 And was very honorably discharged from active duty 1 February 1946. His service brings great credit to himself and the United States Army.

MAJOR GERALD M. DEVLIN is inducted into the Ranger Hall of Fame for his contribution to the United States Army as a senior advisor to the South

Vietnamese Biet Dong Quan, or Rangers. Devlin accompanied the 44th Vietnamese Ranger Battalion on numerous combat missions in the Mekong Delta. By late 1965 the 44th was one of the most successful South Vietnamese fighting units, due in large part to Devlin's leadership by example, his tactical advice, and his fire support coordination skills. He was awarded the Bronze Star for Valor, the Purple Heart, the Distinguished Service Cross, and multiple awards of the Vietnamese Cross of Gallantry, earned during combat assaults with the 44th. During his last action as a Ranger advisor the 44th was counterattacked at night by a brigade strength Viet Cong unit. The battalion commander was killed and Devlin was severely wounded. When the 44th began to give way Devlin took command. Unable to walk and working in darkness with Rangers who understood only the Vietnamese language, he used a radio and crawled from position to position directing fire, adjusting fighting positions, directing aviation fire support, securing helicopter evacuation for wounded Rangers, and assuring the Rangers that they could hold. The 44th held and before dawn the Viet Cong withdrew. Devlin's example of Ranger leadership during his tour as a Ranger Advisor with the 44th Vietnamese Ranger Battalion reflects great credit on himself and all Rangers, past and present.

LIEUTENANT ROBERT T. EDLIN joined the Indiana National Guard in 1939 and was assigned to B Company, 152nd Infantry, 38th Division. In April, 1941, he was promoted to Staff Sergeant and made a Platoon Leader, and his unit participated in Louisiana Maneuvers. In 1942, he entered Officers Candidate School at Fort Benning, GA, where he was promoted to 2nd Lieutenant in September of 1942, and assigned to Company K, 112th Infantry, 28th Infantry Division, at Camp Livingston, Louisiana. In May of 1943, he was promoted to 1st Lieutenant and given further training at Camp Carrabelle, Florida, where he became an expert on the Browning Automatic Weapon (BAR), light machine gun, and M-1 rifle. Later he was assigned as the Division Bayonet Instructor at Camp Pickett, Virginia. In November of 1943, Lieutenant Edlin was shipped to England with the 28th Infantry Division where he attended a British Commando and Battle School. In December of 1943, he volunteered for the Rangers and received Ranger Training at Braunton, England, and was assigned as a Platoon Leader, Company A, 2nd Ranger Battalion at Bude, Cornewall, England. On June 6, 1944- D-Day, Lieutenant Edlin led his platoon across Omaha Beach and received two wounds in his leg, incapacitating him. The following day, he was evacuated to the 97th General Hospital, in England, where he remained until rejoining his platoon at St. Jean de Daye, France, on July 15, 1944. It was during the Brest Campaign on September 9, 1944, that Lieutenant Edlin led one of the most amazing assaults of the war by sneaking a small patrol into the most powerful fortress on the Brest Peninsula. At great personal risk, he forced a captured officer at gunpoint to lead him to the fortress commandant whom he threatened with a live grenade to surrender the fortress; this ultimately resulted in the capture of over 800 German soldiers and caused a cascading of surrenders throughout the area. For this action he was awarded the Distinguished Service Cross. In November of 1944, while leading a patrol in the Hurgent Forest, a mine exploded wounding Lieutenant Edlin in the hand and face. Subsequently, he was granted leave, and returned to the 2nd Ranger Battalion and participated in the fighting across Germany. Lieutenant Edlin eventually returned home and married his High School sweetheart on December 7, 1945. He was released from active duty in March of 1946. In 1948, Mr. Edlin joined the Indiana National Guard as a Captain of the 152nd Infantry Regiment in Company K. He resigned his commission in 1954. His valiant service brings great credit upon himself and the United States Army.

GENERAL JOHN W. FOSS distinguished himself during a career spanning over 35 years as a tireless advocate of the Ranger and Light Infantry Communities. He served with and commanded infantry units from platoon to corps level, participating in the 1958 U.S. intervention in Lebanon, two combat tours in Vietnam, and commanding the 82nd Airborne Division, the U.S. Joint Task Force in Grenada, and XVIII Airborne Corps. Gen. Foss also played a significant mentoring role: he was a Ranger Instructor and later Commandant at the Infantry School, instructed tactics at the U.S. Military Academy, served as an infantry instructor at the Royal Military Academy in Sandhurst, England, and culminated his career as Commander of U.S. Army Training and Doctrine Command. Gen. Foss' illustrious career brings great credit to him, the Ranger and Light Infantry Communities and the military service.

MASTER SERGEANT GARY I. GORDON is inducted into the Ranger Hall of Fame for extraordinary heroism in connection with military operations involving conflict with an armed hostile force in Mogadishu, Somalia, while attached to Task Force Ranger. On October 3, 1993, after providing protective fires during an assault, he and his team member were directed to the first helicopter crash site where they provided protective covering fires that allowed the Ranger Rescue Team to secure the area. When a second helicopter crashed, Master Sergeant Gordon unhesitatingly volunteered to be inserted to protect the critically wounded crew. When his third request to be inserted was granted, he and his fellow sniper fought their way through enemy fire and a dense maze of shanties and shacks to reach the critically injured crew. Master Sergeant Gordon pulled the pilot and other crew members to safety. He and his fellow sniper then secured the perimeter of the crash site, placing themselves in the most vulnerable positions. He then used his side arm and long range rifle to kill and undetermined number of attackers. Master Sergeant Gordon then went back into the wreckage and recovered some of the crew's weapons and ammunition. Despite the fact that he was critically low on ammunition, he provided some to the dazed pilot and continued to radio for help. With his team member now fatally wounded, Master Sergeant Gordon continued to travel the perimeter and provide protective fire. He gave his team member's rifle along with the last five rounds of ammunition to the downed pilot with the words "good luck." Then, armed only with his pistol, he continued to fight until he was fatally wounded. His actions saved the pilot's life. Master Sergeant Gordon's extraordinary heroism and devotion to duty were in keeping with the finest traditions of the Rangers, and reflect great credit upon himself and the United States Army.

FIRST SERGEANT RANDALL HARRIS is inducted into the Ranger Hall of Fame for uncommon valor and selfless gallantry as he Led the Way. Among the first to volunteer for 1st Ranger Battalion, activated in Carrickfergus in Northern Ireland, June 21, 1942, he was assigned to F Company as a mortar squad sergeant. After surviving the grueling and realistic Commando Training at Achnacarry, Scotland, he took part in the North African Invasion at Arzew and the Tunnisian Campaign where his performance earned him the rank of 1st Sergeant when the 4th Ranger Battalion was activated in May of 1943. He was awarded the Distinguished Service Cross and earned a battlefield commission for his extraordinary performance during the invasion of Sicily. His platoon was caught in a mine field that killed the company commander, wounded the remaining officers and also wounded Harris. There was a gash in his abdomen through which his intestines started to extrude; whereupon he pushed his innards back in place and covered the gaping hole with his web belt. Then he led the remnants of his company out of the mine field. Still in the battle, he attacked and silenced enemy pillboxes that threatened the landing area with machine gun fire. Later when the fight was over he refused medical help until his men, less wounded than he, were treated. He recovered from his injuries and returned to lead F Company through battles at Ceppagna and Venafro. He also participated in the Anzio landings where he led his company on the deepest inland penetration and successfully held off enemy counter attacks.

SENATOR J. ROBERT KERREY is inducted into the Ranger Hall of Fame for conspicuous gallantry and intrepidity at the risk of his life while serving as a Ranger-qualified SEAL team leader during action against the Viet Cong forces during the Vietnam War. Lieutenant Junior Grade Kerrey led his SEALs on a mission to capture important members of the enemy's area political cadre known to be located on an island in the bay of Nha Trang. While leading his men in a treacherous downward descent under intense enemy fire, Lieutenant Kerrey received massive injuries from a grenade. Although bleeding profusely and suffering great pain, he displayed outstanding courage and presence of mind and immediately directed his element's fire into the heart of the enemy camp. After successfully suppressing the enemy's fire, and although immobilized by his multiple wounds, he continued to maintain calm, superlative control as he ordered his team to secure and defend an extraction site. Lieutenant Kerrey resolutely directed his men, despite near-unconscious state, until he was eventually evacuated by helicopter. The havoc brought to the enemy by this very successful mission cannot be overestimated. The

enemy soldiers who were captured provided critical intelligence to the allied effort. For this action he was awarded the Medal of Honor. Lieutenant Kerrey's courageous and inspiring leadership, valiant fighting spirit, and tenacious devotion to duty in the face of almost overwhelming opposition sustain and enhance the finest traditions of the Rangers and truly set an example for all to follow.

GENERAL FRED K. MAHAFFEY is inducted into the Ranger Hall of Fame as an officer whose distinguished 31-year military career left a lasting imprint on the Ranger community. In combat during the Vietnam War, General Mahaffey exemplified the Ranger Creed, and was always found "leading the way" whenever his unit engaged the enemy. His selfless and heroic leadership was recognized with seven Valor awards. He encouraged his men to attend Ranger School as a way to "make a good soldier better." General Mahaffey's unparalleled leadership, knowledge of training and team-building, and technical and tactical proficiency ensured that his units always possessed the highest levels of combat readiness, cohesiveness and Ranger-like proficiency. As the Assistant Commandant of the United States Army Infantry School, he was a proponent of Ranger training for the entire Army. While the Deputy Chief of Staff of operations for the U.S. Army, General Mahaffey was instrumental in the reactivation of the 3rd Ranger Battalion and the 75th Ranger Regiment. On May 16, 1989, the 3rd Infantry Division dedicated its Headquarters building in his honor. At that ceremony, he was remembered as a Commanding General who fostered the environment where all soldiers could develop their potential to the fullest, and as a soldier whose uncompromising loyalty and devotion to duty served the U.S. Army superbly for over a quarter of a century. General Mahaffey's illustrious career truly epitomizes the Ranger Credo of "Rangers Lead The Way."

COMMAND SERGEANT MAJOR MICHAEL N. MARTIN is inducted into the Ranger Hall of Fame for his contribution to the United States Army as courageous and successful Ranger for a period of thirty years, twenty one of which were volunteered overseas service in Ranger units, Airborne units and V Corps Long Range Reconnaissance Company. While in Vietnam, he advised the 32nd Vietnamese Ranger Battalion and later volunteered to join the 44th Vietnamese Ranger Battalion, when they sustained heavy advisor casualties. Based on his valor and Ranger skills, he was selected to advise a Provincial Reconnaissance Unit (PRU) in Chau Duc Province, covertly organized and secretly operated as part of the Central Intelligence Agency program to infiltrate into Viet Cong controlled territory and to interdict the Viet Cong infrastructure. His unit was successful in the Mekong Delta and Martin was decorated by the South Vietnamese government. Returning to CONUS, selection as one of four original First Sergeants of the 1st Ranger Battalion in 1974. Command Sergeant Major Martin's long and distinguished career is the embodiment of the Ranger creed and reflects great credit on himself, past and present Ranger units, and the military service.

BRIGADIER GENERAL JOHN HUGH MCGEE was a professional soldier, West Point graduate, guerrilla leader, husband, and father. During World War II, he was a prisoner of war who escaped by jumping overboard at night from a POW ship and swimming through a hail of bullets and joined the Philippine guerrillas. He was selected by 8th Army Headquarters to train men for Ranger type missions during the Korean War. He organized and trained the 8th Army Ranger Company and a Company of Korean Rangers. He also had other small groups of American and United Nations men doing Ranger Missions. The combat effectiveness of these units is a testimonial to General McGee's ability to train and lead Rangers or Guerrilla Warriors. A soft spoken man whose integrity was beyond question, Ranger McGee was well liked and respected by both subordinates and superiors, he was a true friend and comrade to all Rangers. He was awarded the Silver Star, Legion of Merit with two Oak Leaf Clusters, the Bronze Star, and numerous campaign medals.

LIEUTENANT PHILIP B. PIAZZA enlisted at Fort Wheeler, GA, where he received basic training and was expeditiously promoted to the grade of First Sergeant in a period of eight months. Turning down a commission in the medical field, he was sent to the Infantry Officers Candidate School. He was commissioned in 1942, and was selected to ship out to combat duty in Africa, but a change in orders sent him to the Dutch West Indies then to Trinidad. Here he was assigned to the 33rd Infantry where he served as Executive Officer of Company E. In September, 1943, he volunteered for a dangerous and hazardous mission somewhere, and later moved with elements of the 33rd back to the ZI only to ship out from Camp Stoneman on the Lureline to Bombay, India. The now First Lieutenant was placed in command of the heavy weapons of one platoon of the six combat teams composing the famous 5307 Provisional Unit known as Merrill's Marauders. The Combat mission behind enemy lines began with a 125 mile conditioning foot march and terminating about 750 miles behind the Japanese units that were confronting friendly Chinese troops in the Mogaung/Hukong valleys of north and central Burma. The unit got the orders to establish and hold a road block for 24 hours on the enemy M.S.R. They held for 38 hours. Finally being able to extract from combat due exclusively to the heavy weapons supporting fires of Lieutenant Piazza with disregard for his personal safety observing from an advance position for his mortar and machine guns. The success of this mission denied General Tanaka and the Japanese 18th Imperial Division to retake approximately 35 miles of the Hukong valley which had just been liberated by our aggressive position. At the south end of the Mogaung valley, the unit was hit by a series of banzai attacks. Once again, while observing the machine gun and mortar fire at his position from an advance location Lieutenant Piazza was hit in the head by two rounds of enemy small arms fire and in the leg by a mortar shell burst. Once again, the effectiveness of heavy weapons enabled the battalion to withdraw to safety. Lieutenant Piazza was litter borne about eight miles to an evacuation strip from which, seven days later, he was flown to the 20th General Hospital in India. He spent the next year convalescing before being discharged from active duty. Military awards include the Bronze Star, Combat Infantryman Badge, Purple Heart with Oak Leaf Cluster, Presidential Unit Citation, and several campaign medals. Also awarded by the National Chines Army is the Chinese War Memorial Medal.

SERGEANT MAJOR FRED M. SHEPHERD retired, a three time recipient of the Combat Infantryman Badge served in Europe as a member of the 101st Airborne Division during World War II. His service included two combat Parachute assaults at Normandy and Holland and fighting in Belgium when the 101st Airborne Division that withstood assaults from four German Divisions during the Battle of the Bulge. During the Korean War Sergeant Major Shepherd was awarded the Silver Star for Gallantry in Action when, after a total disregard for personal safety, he moved forward under heavy enemy fires three times to assist wounded Patrol members to withdraw to safety; after deploying his patrol and directing fires on the enemy. He provided covering fires for withdrawing members of his unit, withdrawing only after being ordered to do so. During 1959, SGM Shepherd graduated from the Army Ranger School after which he served as a Sergeant Major of a Battalion for the 173rd Airborne Brigade and retired after thirty-two years from the 3rd Ranger Company, Fort Benning, Georgia. As a Police Captain of Security at Columbus, Georgia Airport, he continued his service to the Nation apprehending several armed individuals who attempted to board Commercial Aircraft. His service under the FAA in coordination with the Federal Bureau of Investigation and the Secret Service provided security for President Carter, his family, and other Military and civilian dignitaries.

BRIGADIER GENERAL JOSEPH S. STRINGHAM, a Ranger from the beginning of his career in 1961, 2nd Lieutenant Stringham was appointed the Executive Officer of the 7th Infantry Division Long Range Patrol Detachment in the Republic of Korea on his first assignment after being commissioned from the USMA. 1LT Stringham served in the Republic of Vietnam in 1963 through 1965 with Special Forces and in 1967-68 with the 31st Infantry as a Company Commander and Battalion S-3. In January, 1965, Captain Stringer assumed command of a Special Forces A Detachment Camp at Ben Cat, then under heavy pressure from a large Viet Cong unit. Here Captain Stringham gained his reputation as a combat leader who could and would gain and maintain contact with the enemy. Also during the 1965 offensive, CPT Stringham was directed to gather what remained of his camp's indigenous forces after the operation was neutralized by a major communist attack and form a Chinese mercenary battalion, code name "MIKE FORCE," the first of its kind and the subject of the John Wayne movie, "The Green Berets." In his third tour, CPT Stringham again served as a Company Commander and Battalion S-3. During the first moments of the Tet Offensive in 1968, C 4/31, under his command,

threw back three all out assaults from a North Vietnamese Regiment. His personal example of calm resolve in the face of danger held the unit together, CPT Stringham personally manned one of his company's machine guns while his company successfully held off the assault from the enemy. From 1975-1979, Major, and then LTC, Stringham commanded three Infantry Battalions, one as a Major, culminating in his assumption of command as the fourth commander of the 1st Ranger Battalion. While he commanded the 1st Ranger Battalion, he gained the nickname by his troops of "Smokin' Joe." His tenacity, integrity, lead-from-the-front leadership style and demanding of the highest standards of performance remain a part of the battalion's fabric today. In the decade of the Eighties, COL Stringham would be assigned to yet another combat assignment as Commander of the U.S. Military Group in El Salvador during the critical phase of the insurgery. Under his direction and hands-on style, the Military Group successfully trained and developed the Salvadoran Armed Forces to the point where they created the security climate in which free elections were successfully held. For his service, COL Stringham received El Salvador's highest military decoration for bravery and meritorious service, "The Medal of Gold." After a successful experience in El Salvador, COL Strigham took command as the fourth Colonel of the newly formed 75th Ranger Regiment. Under his guidance the regiment became the mature and recognized organization it is today. COL Stringham remained in the SOF community upon leaving the 75th Ranger Regiment as the DCG of the USASOC, then 1st SOCOM. BG Stringham retired from active duty on November 30, 1992. He holds two DSMs, and four LOMs, seven combat decorations for valor to include the Silver Star and six foreign decorations for meritorious and combat service. He was awarded several foreign parachutist wings from the U.K., RVN, Israel, Canada, Brazil, Mexico, and Panama. His military service brings great credit upon himself, the Special Operations community, and the United States Army.

COMMAND SERGEANT MAJOR PATRICK TADINA's service during the Vietnam era was one of extreme valor and service as a member of the 173rd Airborne Brigade Long Range Reconnaissance Patrol (LRRP), 74th Infantry Detachment Long Range Patrol (LRP) and Company N (Ranger), 75th Infantry. His service during the Vietnam War for over five straight years in the above mentioned units make him the longest continuously serving Ranger during the Vietnam War. While serving as a team leader for five years, he never had one of his team killed. This was accomplished at the risk of his own life during the many times he was cited for valor (2 Silver Stars), 10 Bronze Stars (seven with "V" Device), four Army Commendation Medals (two with "V" Device), and three Purple Hearts. His service before the Vietnam War in the Dominican Republic and during Desert Storm reflect great credit upon himself, the units he served in and the United States Army.

Honorary Inductee, 1993

HONORABLE JOHN O. MARSH, JR. is inducted into the Ranger Hall of Fame for his dedication and service to the nation, the United States Army and the Rangers. Mr. Marsh enlisted in the United States Army and was commissioned from Officer Candidate School in 1945 at age 19. He served with the occupation forces in Germany and was assigned to the 116th Infantry Regiment, the 29th Infantry Division and the State Headquarters Detachment to the Virginia National Guard from 1951 to 1976, retiring with the rank of Lieutenant Colonel. Mr. Marsh was a four-term Congressman from Virginia's 7th District and has served in such positions as Assistant Secretary of Defense for Legislative Affairs, Vice President Ford's assistant for National Security Affairs, then Counsellor, with Cabinet Rank, to President Ford. Mr. Marsh returned to private law practice in January, 1977 until he was nominated and then sworn in as Secretary of the Army to President Ronald R. Reagan on January 30, 1981. He served in that position until August 14, 1989, making him the longest serving Secretary of the Army or Secretary of War in the history of the nation. Among Mr. Marsh's most notable contributions to our national security was his revitalization of the special operations community while serving as Secretary of the Army. He was a graduate of the U.S. Army Airborne School, Jumpmaster School and was awarded the Senior Parachutist Badge. His selfless dedication to the service of this nation, the U.S. Army and the Rangers has truly set an example for all Americans to follow.

SERGEANT WILLIAM C. CLARK, JR. The Second Ranger Bn. was divided on D-Day, Sgt. Clark was assigned to be in charge of the Aid Station and Company "medics" assigned to A and B Companies. Upon landing at Omaha Dog Green Beach, both companies suffered heavy losses; Sgt. Clark, without hesitation, calmly and with great presence of mind, treated the wounded, one with an open pneumothorax (sucking chest wound), wherever they had fallen. Most of the other medics already were either WIA or KIA. Later, while he was B Company's aid man, the Company, ordered to protect the left flank of the 4th Div., moved into an enemy mine field at night, directly in front of three enemy strong points. Three mines were detonated, killing two and wounding six, two seriously. Immediately, he moved in, groping in the darkness to tend to the wounded and, eventually, remained all night with the seriously wounded in order to car for, and guard them. Bill Clark display such courage and devotion to his duty and to fellow Ranger repeatedly during all the engagements of the Battalion.

Upon his return home, Bill Clark became a police officer in Washington, D.C. After serving on both the vice and auto squads for some years, he retired as Detective Sergeant. He died Feb. 16, 1992 and is buried at Arlington National Cemetery.

The awards of the Silver Star, Purple Heart with cluster and Bronze Star with cluster, attest to the actions of William C. Clark, Jr. and his exemplification of a Ranger Medic.

Rangers Roster

ABBOTT, RALPH P.
ABLE, LEE R.
ACOSTA, JOSEPH M.
ADAMAITIS, WILLIAM L.
ADAMS, CHRIS
ADAMS, EDWARD D.
ADAMS, JAMES Y.
ADAMS, LOUIS M.
ADAMS, WILBUR L.
ADKINS, JACK D.
ADKINS, KIRK P.
ADKINSON, JACK
AIKENS, LEGREE
AKERS, ISHMAEL W.
ALARID, MAXIMINO
ALBARO, ERNEST V.
ALDRIDGE, MARGARET
ALLEN, DONALD
ALLEN, FARRELL J.
ALLEN, MARY P.
ALMEIDA, JOE
ALSTON, MARION A.
ALVA, JOSE
ALVAREZ, ARTHUR J.
AMATO, SAM L.
AMBURN, THOMAS S.
AMES, WARREN H.
AMMERMAN, JAY F.
ANDERSON, FRANCISCO
ANDERSON, JESSE
ANDERSON, JESSE E.
ANDERSON, RICHARD E.
ANDERSON, WILLIAM E.
ANDRADE, MARVA
ANDRES, TEEDIES
ANDREWS, WILLIAM , .
ANDRUS, HERBERT B.
ANDRZEJCZYK, FRANK J.
ANGELL, JOHN L.
ANGLIN, JOHN P.
ANGLIN, ROBERT
ANTROBUS, CHARLES M.
ANWEILER, ROBERT
ANZALDUA, MOISES
APPUGLIESE, JAMES V.
ARENS, JOHN W.
ARKELL, LYLE J.
ARNAUD, HARRY
ARNER, ROBERT B.
ARNOLD, GERALD A.
ATKINS, ALBERT E.
ATKINS, CLEM
ATWOOD, WILSON R.
AUSBORN, TERESA
AVERY, FAYE
AYERS, ROBERT W.
BACH, MILISSA
BAETEN, RICHARD N.
BAGBY, EDWIN C.
BAILEY, ROY L.
BAILEY, WILLIAM N.
BAKER, EARL RAY
BAKER, JACKIE L.
BAKER, SYLVIA
BALDRIDGE, ORAL L.
BALDWIN, DEREK , .
BALL, ALFRED L.
BALL, JESSE N.
BALLARD, ALFRED J.
BALLOU, ROLAND N.
BALLOU, SELAH D.
BALOMENOS, ARTHUR C.
BALSLEY, ELMER E.
BANBURY, JACK L.
BANKS, NELSON C.
BARBER, HAROLD L.
BARBER, WALLACE R.
BARBOE, HOWARD L.
BARCROFT, CURTIS L.
BARLEY, MARIE
BARNETT, PAUL
BARNETTE, JAMES L.,
BARR, JUDITH
BARRETT, ROY C.
BARRIO, EDWARD G.
BARTHOLOMEW, RALPH J.
BARTLETT, ANDREW V.
BARTO, JOSEPH W.
BASS, DAVID P.
BASSETT, GEORGE E.
BASTIAN, RANDOLPH
BATCHA, GEORGE M.
BATCHELDER, HARRY J.
BATES, HUGH M.
BATSON, RICHARD L.
BAUER, JAMES A.
BAXLEY, MARCUS L.
BAXTER, RALPH E.
BAYNE, ROBERT D.
BEATTY, JAMES D.
BEATTY, ROY
BEAUCHAMP, RONALD G.
BEAUREGARD, JOSEPH A.
BEBEAU, HOWARD C.
BECHER, THOMAS A.
BECK, EARL P
BEEBE, DONALD E.
BEERS, HARRY D.
BEERS, WARREN E.
BEERY, ROBERT C.
BEINER, ALLEN
BELFORD, WARREN
BELKNAP, VIRGIE
BELL, DONALD K.
BELL, ROBERT H.
BELL, TOMMY L.
BELLOUMINI, RUDOLPH S.
BELTER, RONALD D.
BELTON, HENRY , .
BENJAMIN, DONALD G.
BENSON, ALVIN J.
BENT, CHARLES K.
BERG, ROBERT W.
BERGERON, HAROLD G.
BERRY, CHARLES J.
BESS, HAROLD M.
BEST, EUGENE E.
BETIT, CHARLES W.
BIGART, ROBERT H.
BIGELOW, GENE H.
BIRD, JAY G.
BIRTCHER, RICHARD L.
BISHOP, DANIEL
BISHOP, JOHN H.
BISSONNETTE, AIME J.
BITTORIE, ERICA
BIVENS, GEORGE E.
BLACK, ROBERT W.
BLACK JR, CHARLES J.
BLACKBURN, CLYDE B.
BLACKBURN, KEN
BLACKKETTER, CLAUDE P.
BLACKWELL, BETTY
BLIXRUD, ALDEN M.
BLODGETT, JOSEPH D.
BOATMAN, ROY
BOATNER, CARY F.
BOATWRIGHT, DANIEL
BOCAN, SAMUEL M.
BOCK, KARL M.
BODROGHY, ROBERT F.
BOELTER, RICHARD W.
BOGARDUS, CHARLES O.
BOHIN, JR, HARRY
BOHLANDER, MARILYN
BOLDT, HERMANN
BOLGRIN, FLOYD E.
BOLIN, GEORGE E.
BOLK, PHILLIP J.
BOLSTER, GERALD P.
BOOKER JR, DAVID L.
BOOTH, JERRY
BORDELON, LAURY J.
BORDON, EMIL
BORELLA, WILLIAM W.
BORK, WALTER E.
BORRIELLO, RALPH
BORUSIEWICZ, STANLEY
BOULAPHA, BOUN MAY
BOWDEN, FRANCES
BOWEN III, CHIYO
BOWER, KENNETH J.
BOWERS, MARVIN A.
BOWERS, ROMAN A.
BOWLES, DELBERT D.
BOWLING, GEORGE L.
BOYD, HENRY A.
BOYKIN, SAM M.
BOYNTON, ALLEN
BRACE, ROBERT D.
BRADLEY, EDWARD J.
BRADY, JOHN , .
BRADY, PATRICK J.
BRAMONTE, SEBASTIANO
BRANDENBURG, JAMES
BRANDER, FRED C.
BRANDT, JAN.
BRANHAM, RICHARD D.
BRANNON, GERALD N.
BRASSEUR, RONALD A.
BRAYMAN, WILLIAM
BREEDLOVE, BARNELL
BRENNAN, ARTHUR
BRENNAN, THOMAS S.
BRENT, ROBERT H.
BRENTLEY, HAROLD R.
BREWER, LUCIEN , .
BREXEL, CHARLES
BRIDGES, JAMES E.
BRIGGS, EDWIN
BRIGGS, JACK H.
BRIGHTWELL, PATRICIA
BRILLA, MARILYN
BRILLA, TOM J.
BRISCOE, RICHARD
BRITT, LILLIAN
BRITTAIN, FRANK W.
BRITTON SR, JOSEPH D.
BROOKS, CARL G.
BROOKS, DONALD E.
BROTHERS, ISRAEL J.
BROWN, CATHY
BROWN, JOHN W.
BROWN, LEE E.
BROWN, MILDRED
BROWN, ROBERT L.
BROWNLEE, LOUIS C.
BRUCE, NELSON E.
BRUTSCHE, THOMAS
BRYANT, CARL
BRYCE, WALTER J.
BRYER, EDWARD G.
BUCEY, PAUL E.
BUCHANAN, ROBERT G.
BUCK, RICHARD R.
BUCKALEW, ALVIN H.
BUCKALEW, BERNARD B.
BUCKLES, HOMER L.
BUDYNKIEWICZ, PHYLLIS
BUKATY, ALFRED M.
BUNDY JR, VICTOR S.
BUNGARZ JR, FRANK
BUNN, CHARLES N.
BURAS, FRANK , .
BURBIE, PATRICK J.
BUREL, THOMAS F.
BURGESS, NELDA
BURNHAM, ERNEST D.
BURNHAM JR, ELMER
BURNS, JOHN F.
BURNS, PAUL F.
BURT, JOHNNY B.
BUSH, HOMER
BUTCHKO, JOSEPH
BUTLER, WILLIAM E.
BUZA, WILLARD J.
CADENA, MARCELLO S.
CAIN, JANE
CAIN, WILLENE
CALDERHEAD, JAMES R.
CALVIN, ARLEN D.
CAMACHO, RAY
CAMBRA, JACK
CAMERON, RUSS
CAMPBELL, GRADY B.
CAMPBELL, RICHARD C.
CAMPBELL, VINCENT
CAMPBELL III, GEORGE W.
CAMPOS, VICTOR
CANON, EARL
CANTWELL, BILL
CAPE, BUSTER
CAPONE, RICHARD F.
CAPPELETTI JR, VICTOR
CARAWAN, ANN
CARBONEL, JAMES J.
CARGAN, JOHN J.
CARLL, FREDRICK J.
CARLSON, DOUGLAS W.
CARMICHAEL, HENRY A.
CARNEY, JONATHON D.
CAROZZA, JOHN
CARPENTER, JO ANNE
CARR, HARRY B.
CARSTENS, EARL M.
CARTLEDGE, JAMES C.
CARUTH, CHARLES W.
CARVELL, RICHARD F.
CASELLA, EMIL
CASERES, JUAN
CASHMAN, RICHARD J.
CASNER, FRAN
CASSATT, ELMER H.
CASSIDY, HARRY E.
CASTELLO, ALFONSO G.
CASTER, HAROLD W.
CASTO, CLIFFORD E.
CATES, BEN A.
CATINEAULT, ROGER G.
CATLETT, BILLY F.
CAVIN, ARLEN D.
CENTOFANI, VINCENZO
CHACOS, JAMES M.
CHADA, BILL
CHAGNON, AUDREY
CHALMERS, JOHN
CHAMBERLAIN, COLIN
CHAMPAGNE, ROBERT G.
CHAMPION, ROGER R.
CHANDLER, JOHN C.
CHANEY, DON
CHANEY, JOE
CHANNON, ROBERT I.
CHAPMAN, JAMES D.
CHARPENTIER, MELVIN R.
CHARTIER, RAYMOND A.
CHAVIS, FRAN
CHEEK, RALPH L.
CHESTER, JOHN A.
CHICKAUL, JOHN
CHIKI, RUDOLPH
CHILDERS, DONALD D.
CHILLION, LOIS
CHRESTMAN, CHARLES
CHRISTENSON, JACK
CHRISTMAS, MERRY L.
CICUZZA, CHARLOTTE
CINICOLA, FRAN
CISNEROS, ARTHUR
CISNEROS, PILAR
CLARK, DAVID
CLARK, ROBERT P.
CLARK, VERNON W.
CLARK, WILLIAM L.
CLARKE, DAVID A.
CLEAR, JOHN W.
CLEMENT, ALBERT V.
CLEMENT SR, JAMES L.
CLEVELAND, CLINTON
CLEVELAND, GEORGE
CLEVENGER, WALTER
CLEVENGER, WALTER R.
CLIETTE, ALBERT
CLIFTON, JIM H.
CLINE, OSCAR R.
CLINGER, WILLIAM S.
COCKAYNE, JOHN D.
COFFEY, THERESA
COFFMAN, CARL C.
COHEN, ALLEN
COKER, HORTON S.
COLBERT, CLEATUS L.
COLE, FRANZ
COLE, GAIL
COLE, ROBERT L.
COLE SR, HAROLD D.
COLEMAN, CLINTON (JIM)
COLEMAN, LONNIE
COLEMAN, ROBERT
COLLINS, DONALD
COLLINS, WILLIAM
CONDON, JOHN J.
CONKLIN, THEODORE L.
CONNALLY, WILLIAM J.
CONNELLY, ANDREW
CONWAY, EDWARD J.
COOK, HERSCHEL H.
COOK, JAMES
COOK, NOEL
COOK, ROBERT A III
COOMBS, ROBERT , .
COOPER, BEN
COOPER, DOROTHY H.
COOPER, ERNEST J.
COOTS, BILLY G.
COPELAND, AUBEY
COPLEY, LEWIS D.
COPPING, CHARLES
CORBIN, CLARENCE R.
CORREA, EDWARD
CORTEZ, JOSE E.
CORYELL, DAVID C.
COSMAN, EARL
COTE, ROGER J.
COTTER, LOUIS
COULTER, JACK
COURNOYER, JOSEPH R.
COURTNEY, ROBERT
COURTS, CURTIS E.
COVELLO, HENRY
COX, NORMAN , .
CRABTREE, CHARLES E.
CRAIG, CLARENCE W.
CRAIG, RAYMOND F.
CRAIG JR., CHARLES R.
CRAMER, FLORENCE
CRAN, JAMES A.
CREAGER, ROBERT E.
CRESON, ROBERT F.
CREW, MORRIS
CRISPINO, FRED
CRITCHLEY, JOHN J.
CROMER, RALPH
CRONIN, ETHEL
CROOK, WILBURN
CROSBY, ROY
CROW, WILLIAM
CROWDER, KENNETH
CRUZ, TRINO
CULPEPPER, JOHN H. JR
CULVER, STANLEY H.
CUMMINGS, CHARLES , .
CUNNINGHAM, MAURICE
CURTIS, CHARLES H.
CUSTER, HAROLD W.
CZERNIEWSKI, STANLEY
DA SILVA, JACK
DAHL, GLENN L.
DAHLHEIMER, GENE D.
DALLMAN, DONALD P.
DALY, JOSEPH
DANIELS, SHERMAN
DANZI, DOMINIC P.
DAUGHERTY, DARREL W.
DAVIDSON, ANDREW D.
DAVIDSON, HOMER
DAVIS, ALBERT H.
DAVIS, FLOYD H. (MULE)
DAVIS, FRANK J.
DAVIS, FRED E.
DAVIS, HOWARD K.
DAVIS, MARJORIE
DAVIS, NOAH
DAVIS, WILLIAM
DAVIS, FRED E.
DAWSON, FRANCIS W.
DAWSON, MAX
DAY, LESLIE H.
DE ROUCHEY, DELOS D.
DEAN, MURRIEL E.
DEAVER, ALLEN
DECKER, RON
DECRISTOFERO, NANCY
DEERY, JAMES W.
DEHART, JEAN
DEL TORO SR, MARIO
DELANEY, RICHARD
DELFINE, KENNETH A.
DELGADO, JANE
DELMASTRO, MICHAEL J.
DELUCA, JOHN
DELUKEY, ANN
DELVALLE, CLEMENT
DEMENT, ROLAND F.
DEMLOW JR, GEORGE W.
DENNIS, VINCENT , .
DERRYBERRY, LARRY
DERWIN, JAMES M.
DESMORE SR, ERNEST J.
DESMUKES, RAY L.
DEVERY, JAMES J.
DEWBERRY, OSCAR H.
DEWITT JR, HOWARD H.
DEXTER, LEWIS F.
DIACK, DONALD
DIAS, HERCULANO G.
DIAZ, MANUEL M.
DIAZ, ROBERT A.
DICKENS, TROY L.
DIERKS, NANCY
DILIBERTO, JOHN K.
DIXON, EDGAR
DOBSON, MARLENE
DODD, GERALD R.
DODGE, RUSSELL E.
DOERR, WAYNE C.
DOFFING, SYLVESTER (CY) F.
DOLIN, LOUIS F.
DONALDSON, CHARLES N.
DONNELLY, HARRISON C.
DONNELLY, JAMES F.
DOTY, PAUL L.
DOUGLAS, LAURA
DOWER JR, ALBERT M.
DOWLING, ROY M.
DRAHEIM, JOHN J.
DROST, CARL W.
DRUM, CONNIE M.
DUBE, LEONARD
DUBREUIL, EDMUND J.
DUCKWORTH, ROBERT
DUDA, CECILIA
DUFFY, GERALD J.
DUFFY, PETER
DUGAN, JAMES O.
DULANEY, BETTY
DUNCAN, LAWRENCE
DUNHAM, FRANCIS W.
DUNN, EDWIN L.
DUNN, JOHNNIE W.
DUNN, RUBY JEAN
DUPONT, CLAUDE
DUQUEMIN, GORDON J.
DURHAM, CHARLIE R.
DUTKO, VELMA
DVORAK, JAMES A.
DYE, RONALD E.
DYE SR, JOSEPH
DYER, FREDDIE A.
DYKE, DONALD M.
EALY JR, HAROLD O.
EATON, RICHARD E.
EBEL, MARY LOU
ECKHARDT, WALTER
EDDY, WALTER E
EDGE, ZANE
EDWARDS, CLIFTON R.
EDWARDS, EVERETT L.
EDWARDS, THEODORE M.
EDWARDS, WILLIAM F.
EDWARDS, JERRY D.
EGELER, WARREN
EIDELL, KURT
EISMAN, JOHN V.
ELLIS, GAYAL H.
ELLIS, GEORGE B.
ELLMERS, RICHARD D.
EMERSON, JOHN (BO) M.
ENGSTROM, JOHN W.
ERB, ELLWOOD P.
ERB, NORMAN F.
ESCALERA, JOSE A.
ESEPPI, PAUL
ESPINOZA, TORIBIO
ESTELL, LAWRENCE L.
ESTERLINE, TOM
ESTRADA, FERNANDO
ESTRADA, RICHARD
EUSTICE, RAYMOND (JOE) J.
EVANS, ALBERT G.
EVANS, BEN M.
EVANS, JAMES C.
EXLEY, ROBERT L.
FABRE, MICHAEL D.
FAIOLA, HENRY
FAISTENHAMMER JR, LUDWIG
FALVO, JOSEPH N.
FANNING, JOHN V.
FARR, BOBBY D.
FATLAND, EVAN
FAVORS, JOSEPH H.
FAZENBAKER, ELVON B.
FEDORCHAK, WILLIAM JR
FEHSER, RICHARD M.
FELDER, DONALD
FENLEY, SHARON RAE
FEREBEE, ROBERT L.
FERGUSON, BERNARD
FERNANDEZ, PEDRO P.
FERRALL, RICHARD R.
FERRIS, JOHN R.
FICK, LEROY G.
FIELDS JR, JAMES H.
FIGUREL, JOHN
FIKE JR, EMMETT E.
FINKE, EVELYN
FIORE, ANTHONY S.
FIORI, CHARLES
FIORITO, JOHN J.
FISH, DEAN S.
FISH SR, VINCENT
FITCH, FAYE
FITCH, LESTER L.
FIZUR, ROBERT F.
FLAHERTY, RAY
FLEISHMAN, JERRY
FLEMING, RALPH L.
FLETCHER, CHARLES E.
FLETCHER, HENRY
FLOYD, JAMES
FOLEY, GERALD F.
FOLKS JR, S. JOHN
FONSECA, RAUL
FOOTE, OSCAR H.
FORBES, FRANK M.
FORCIER, EDGAR M.
FORD, CHARLES R.
FORD, VIVIAN
FORREST, JOSEPH R.
FORSETH, RICHARD H.
FORSMARK, ARTHUR
FOSHEE, RUDOLPH P.
FOSTER, ALBERT
FOX, GEORGE P JR
FRANKO, JOSEPH J.
FRAZIER, ANNELIESE
FRAZIER, ARVIE
FREEMAN, JAMES E.

FRENCH, THOMAS O.
FRIEND, HOWARD H.
FROST, HUBERT H.
FRYE, MARVIN H.
FRYSON, RAYMOND
FULLER, ROBERT N.
FULWILER, HAROLD K.
FURGUILE, JOHN
FURRY, LOYAL
GABOSSI, TONY
GAGNON, JOSEPH R.
GAINER, RICHARD (DICK)
GAINOR, CHARLIE W.
GAITHER, CHARLES R.
GALA, DEE
GALINGTON, OLIVER W.
GALLAGHER, DANIEL
GALLAGHER, PATRICK J.
GALLAGHER, ROBERT , .
GALLOWAY, DAVID B.
GALLOWAY, RICHARD W.
GALT, CHADBOURNE W.
GAMBLE, CHARLES B.
GANTT, MILDRED
GARCIA, THOMAS
GARDNER, GEORGE M.
GARDNER, RICHARD
GARRETT, CARLOS W.
GARSIDE, LORRIN H.
GATES, EMERSON A.
GAUNT, ELDON S.
GAUNT, ROBERT W.
GAUT, DON P.
GAVIN, JOHN R
GEER, ROBERT
GEGICH, BRANKO
GELINAS, WILLIAM P.
GENTRY, LAVERNE D.
GEORGE, BERTIE
GEORGIA, WILLIAM A.
GEORGIOU, CONSTANTINO C.
GERMAIN, GERARD
GIBLIN, THOMAS F.
GIBSON, LORI
GIBSON, ROBERT E.
GIBSON, WILLIAM D.
GIBSON JR, WILLIAM V.
GILL, BARNEY
GILLAND, RICHARD B.
GILLILLAND, WILLIAM S.
GILMORE, JOHN
GILMORE, ROBERT W.
GILSDORF, RONALD B.
GIROLMO, JOHN D.
GLENDENNING, CHARLES F.
GLENN, JOHN
GODDARD, NORMAN G.
GODWIN, PATRICIA
GOETTE, JAMES A.
GOLDWORTHY, JAMES H.
GONZALES, JESSE
GONZALES, ROBERT
GOODE, MONTE H.
GOODPASTURE, JOHN A.
GOOLSBY, BETTY
GORDON, ANDREW
GORDON JR, MORGAN S.
GORE, WILLIAM E.
GORMAN, DAVID F.
GOTNEY, VERA
GOULD, JOHN W.
GOWER, GENE
GOZA, ROBERT
GRACE, JAMES E.
GRAHAM, EMERY B.
GRAHAM, JOHN P.
GRANT, W.E.
GRASTY JR, ISAAC
GRAY, LLOYD M.
GRAY, ROBERT W.
GRAY JR, THOMAS J
GREEN, JAMES
GREEN, KIYOKO
GREENFIELD, JAY L.
GREENMUN, FRANCIS E.
GREENWAY, RICHARD D.
GREGORY, GILL M.
GREGORY, WILLIAM
GREYSHOCK, DONALD G.
GRIFFIN, JOE
GRIFFIN, MARVIN E.
GRIFFITH, KENNETH S.
GRISWOLD, DONALD E.
GROOMS, ANGIE
GROSSMAN, ALBERT H.
GROVE, ERNEST R.
GROVES, RAYMOND P.
GRUETZMACHER, WILLIAM F.
GUDE, JAMES H.
GUNDERSON, JUNIOR B.
GURIA, JOHN E.
GURZYNSKI, JOHN G.
GUSTAFSON, BENJAMIN P.
GUSTAFSON, ROLF L.
GUSTAFSON, THOMAS
GUYOT, ROBERT
HAAS, JACK D.
HACKLER III, JOHN P.
HAGGARTY, ROBERT J.
HAGUE, THELMA
HALE, JOHN H.
HALL, ROY
HALL, BERNARD
HALLIGAN, JOHN P.
HAMANN, EDWARD J.
HAMER, JOHN
HAMILTON, PETER
HAMILTON, WELDON W.
HAMLIN, FRED
HAMPEL, JEROME A.
HAMRICK, PAUL R.
HAND, DONALD A.
HANDLEY, MERLIN J.
HANES, PHILIP D.
HANLON, JIM
HANNON, CATHERINE
HANSCOM, WILLIAM E.
HANSEN, EDGAR
HANSON JR, WILLIAM G.
HARDWAY, ROBERT G.
HARDY, JIM
HARDY, RICHARD I.
HARGETT, RICHARD A.
HARL, GEORGE D.
HARLOW, GILES D.
HARMON, WILMA
HARRINGTON, JAMES R.
HARRIS, CLAUDE
HARRIS, DAVE A.
HARRIS, MERLE E.
HARRIS, WILLIAM C.
HARRIS II, FRANK B.
HART, BEDFORD L.
HARTNETT, JAMES F.
HARVEY, ALTON H
HARVEY, JACK H.
HARVEY, JOHN H.
HARWOOD, VICTOR K.
HASKEW, WM "DOUGLAS"
HATHAWAY, ROBERT C.
HATHCOAT, HOWARD C.
HAWES JR, EGBERT D.
HAY, GILBERT E.
HAYES, ROBERT J.
HAYES JR, ERNEST , .
HAYNES, ROBERT
HAYNES JR, JAMES D.
HAYWOOD, ROBERT S
HEALY, MICHAEL D
HEASLEY, ALFRED C.
HEATH, ARLINE
HEATH, JAMES
HEEDT, FRED
HEFFERNAN, JAMES M.
HEIL, JOSEPH
HELIN, RUTH
HEMMING, NORBERT G.
HENDRIX, RALPH W.
HENDZEL, FRANK R.
HENNESSY, PAUL K.
HENRICKSON, HILBERT J.
HENRY, RONALD B.
HENSLEY, JACK B.
HERBERT, JAMES "JIM"
HERRING, THOMAS E.
HESS, GLEN F.
HESTER, ERNEST W.
HIBDON, JAMES E.
HICKEY, EDWARD J.
HICKMAN, MARION L.
HICKS, IRVIN L.
HICKS JR, MELVIN L.
HIGGINBOTHAM, BETTY
HIGGINS, DAVID D.
HIGGINS, EARL A.
HILBERT, ELLSWORTH F.J.
HILBORN, VIRGIL B.
HILL, ALFRED
HILL, ARNOLD
HILL JR, FELIX J.
HILL JR, ROLAND J.
HILLER, JOSEPH G.
HILPERT JR, EDWARD , .
HILTON, JIM M.
HILTON, MEDFORD
HINKLEY, JOHN
HINTZ, HENRY J.
HOBBS, ROBERT N.
HODAK, EVELYN
HODGKINSON JR, HAROLD J.
HODNETT, JOHLIAM J.
HOFFMAN, MARVIN H.
HOFFMAN, RANTER L.
HOFFMASTER, GENE
HOFFMASTER, JACK
HOGE, CAROL
HOGUE, FREDERICK
HOLAPPA, FRED C.
HOLBERT, CHESTER L.
HOLLEY, HORACE E.
HOLLIS, LOLA
HOLLOWAY JR, JOE
HOLMAN, HURSHEL O.
HOLMES, THEODORE S.
HOMER, W. DALE
HONEYCUTT, HAROLD F.
HOOPER, FRED
HORINE, JAMES R.
HORNBEAK, JOE D.
HORNER, BILLY J.
HORST, NORMAN
HOSCH, JOSEPH E.
HOWARD, LEE
HOWE, THOMAS E.
HOY, FRED A.
HUBBARD, EZRA
HUCAL, MICHAEL
HUDSON, ED
HUDSON, HOWARD
HUDSON, REMUS A.
HUFFAR, WALTER E.
HUFFINE, JONATHON F.
HUGHES, ARCHIE
HUGHES, LEE
HUMPHREY, FRANK N.
HUNT, JOSEPH E.
HUNYADY, LESLIE C.
HURD, BENNY E.
HURLEY JR, ROBERT J.
HURST, KENNETH J.
HUTCHESON, WILLIAM M.
HUTTO, SADIE
HYATT, RICHARD D.
IAEA, SHERMIAH K.
IDELL, ROBERT F.
ILLUZZI, JOHN
IMPERIALE, JEAN
INGRAM, BILLY H.
INGRAM, ROBERT C.
INMAN, LOUIS F.
IVEY, GEORGE , .
JABBOUR, ALBERT G.
JACKSON, ASHBY H.
JACKSON, GEORGE W.
JACKSON, HERMAN
JACKSON, KENNETH R.
JACKVONY, THEODORE C.
JACOBY, PHILIP A.
JAROSKY, CHESTER M.
JARRETT, THOMAS H.
JAYNES, BILLY JOE
JENKINS, FRANK J.
JENKINS, JOHN B.
JENNINGS, CURTIS M.
JEWELL, ROBERT M.
JEWELL, THEODORE P.
JOHNSON, BOBBY G.
JOHNSON, BRUCE A.
JOHNSON, CARL W.
JOHNSON, CHARLES R.
JOHNSON, CLARENCE G.
JOHNSON, CURLEE
JOHNSON, ELWOOD F.
JOHNSON, HARLAN
JOHNSON, HAROLD A.
JOHNSON, JULIUS D.
JOHNSON, LEONARD F.
JOHNSON, TED C.
JOHNSON, THOMAS W.
JOHNSON, TWILA
JOHNSON, VALENTINO
JOHNSON JR, GRAYDON W.
JONES, BOBBY L.
JONES, EDNA
JONES, EDWARD E.
JONES, FRED M.
JONES, HAZEL
JONES, NEIL
JONES, RAYMOND A.
JONES, WILLIAM R.
JORDAN, BILL E.
JUDGE, CHESTER F.
JUDY, WILLIAM L.
JULIAN, BETTY
JUNG, GORDON C.
JUNT, JOSEPH
KADER, JAC
KALAKAUSKIS, EDWARD A.
KALISH, MIKE
KAMM, WILLIAM F.
KAPRAL, DANIEL M.
KARETSKY, RICHARD M.
KASKUSKIE, STANLEY L
KATSIKARIS, ARISTOTELOS
KAUFFMAN, ALBERT L.
KAVOLIUS, VYT J.
KAYLOR, JOHN
KAYS, HELEN
KEEGAN, FRANCIS V.
KEEGAN, JOHN
KEELER, CHARLES I.
KEIR, ROGER L.
KELCH, HOWARD
KELLBACH, CLARENCE A.
KELLEHER, JAMES E.
KELLEY, FLOYD E.
KELLY, ALFRED F.
KELLY, FREDERICK J.
KELLY, JAMES J.
KELSO, MINOR
KELSO, MYRON M.
KELTNER, JAMES L.
KEMMER, WILLIAM R.
KEMMERER, RUS
KEMP, JOHN , .
KENDALL, CLAUD A.
KENNEDY, DONLEY C.
KENNEDY, JAMES N.
KENNEDY, PAUL
KENT, J.W.
KENT, WILLIAM A.
KENYON JR, KENNETH E.
KIEFER, BYRON D.
KILDAY, JAMES F.
KILGORE, RILEY
KILKUSKIE, THEODORE F.
KILPATRICK, JOHN C.
KIMBALL, IRENE
KIMMEL, WALTER E.
KING, EUGENE W.
KING, FRANCIS W.
KING, FRANK L
KINNEY, WILLIAM J.
KIRKLAND, CHARLES W.
KLEIN, A. DANIEL
KLEIN, LAVERNE H.
KLEINDORF, SAM
KLITZKE, MERLE E.
KLOBCAR, WILLIAM J.
KLOKE, HARLYNN
KLUEGAL, THOMAS
KNIGHT, RAY A.
KNOEBEL, BENNY L.
KOBILIS, ROBERT "SMOKEY"
KOCH, VERNON S.
KOLBY, ED
KOMISAREK, JOHN R.
KONOPKA, JOHN F.
KOONCE, JAMES L.
KOOP, ALFRED E.
KOOP, LEONARD V.
KOSINSKI, ANTHONY L.
KOSKI, EDWIN S.
KOSS, MICHAEL S.
KOVALCIK, JOHN N.
KRICK, ARLENE
KRIEGER, HENRY , .
KRISKO, CLAUDETTE
KRUPNICK, ROBERT D.
KUCK, DANIEL F.
KUHN, CHARLES E.
KUNAU, ROBERT L.
KURTZ, DOROTHY
KURTZ, RICHARD W.
KUTA, EUGENE A.
LA RUE, DOUGLAS J.
LABROZZI, ANTHONY
LACATENA, MARCO
LACATENA, MARCO
LADNIER, HERMES M.
LADOW, ROBERT P.
LAFONTAINE, JOSEPH "LEFTY" R.
LAIRD, REGIS "RED" C.
LAKEY, DONALD L.
LAMBERT, LUCIEN
LAMBRIES, EUGENE "BUD"
LAMICA, STANLEY B.
LAND, ELMORE
LANDERS, BILLY S.
LANDIS, VANDA J.
LANEY, LUTHER M.
LANG, ALAN
LANGAN, GERALD "JERRY" J.
LANGHAM, CLARENCE
LANGHAM, RAY
LANGNER, STANLEY F.
LANGSTON, JAMES W.
LANIER, JAMES C.
LANNOM, R.L.
LANT, ROBERT G.
LANZILLO, JOSEPH G.
LAPORTE, JOSEPH
LARKIN, HOWARD E.
LARKIN, LEROY E.
LAROCHE, ALDEN
LARSCHEID, ROBERT A.
LAST NAME, FIRST NAME
LAURSEN, ROBERT L.
LAVERTY, DONALD C.
LAW, MONTE K.
LAWRENCE, FRED M.
LAWRIE MD, WILLIAM L.
LAWTON, ELWYN , .
LAYNE, PHILLIP E.
LEATHERS, GENE W.
LEE, JOHN
LEE, PAULA
LENZ, CHARLES N.
LEONARD, EDWARD B.
LEONARD, EUGENE F.
LESLIE, BEN H.
LESURE, DAVID
LEVACK, LEN
LEVINE, CLARENCE
LEWALLEN, WILLARD S.
LEWIS, EDWARD L.
LEWIS, JOSEPH L.
LEWIS, SAMUEL C.
LEWIS, STANLEY
LEWIS, WILLIAM
LEWIS, WILLIAM (BILL)
LEYBA, RICHARD
LILLARD, PAUL , .
LINARDAKIS, ANASTASIO
LIND, OTTO J.
LINNEAR, LEROY E.
LINZELL, CLINTON F.
LISH, VERL
LISI, JOSEPH P.
LIVINGSTON, HARRY J.
LOCKIE, CHARLES
LOEWEN MD, CURTIS
LOHSE, WILLIAM F.
LOLLAR, CLIFTON D.
LONG, EDWIN B.
LONGO, JOHN R.
LORBECK, EUGENE D.
LOUDNER, CLIFFORD E.
LOUGEE, DONALD E.
LOVEGROVE III, JOHN E.
LOVETTO, VINCENT V.
LOVING, JOSEPH S.
LOWENSTEIN, LEE ROY S.
LOZA, EDWARD C.
LUBINSKI, JOSEPH
LUHMANN, GEORGE H.
LUKASIK, ANTHONY
LUNA, MANUEL
LUNDBERG, RICHARD
LUNSFORD, JOHN E.
LUPO, JOHN M.
LUTTRELL, BOB G.
LUTTRELL, JOHN J.
LUVARRA, JOHN J.
LUX, HARRY E.
LYKKEN, DONALD L.
LYLES, PAUL , .
LYNCH, JOSEPH F.
MAACK, ELROY V.
MACAULAY, WILLIAM A.
MACDONALD, KIRBY P.
MACIEL, DAVE
MACK, J.B.
MACKEY, PATRICIA
MACRI, ANGELO C.
MADDRY, TILLIE
MADISON, EUGENE H.
MADURA, JOHN C.
MAHER, TRUDY
MAHLER, WILLIAM F.
MAILHOT, MAURICE J.
MAINES, PHILLIP D.
MALLORY, WILLIAM
MALONE, CECIL
MANDRELL, TIMMIE
MANGES, LEWIS B.
MANGRUM, EPHRIAM "FRANK"
MANLEY, CHARLES L.
MANNING, CLYDE L.
MANNION, THOMAS F.
MANRIQUEZ, EUGENE
MANTHEY, JOHN H.
MAPP, JAMES H.
MARCISZEWSKI, STANLEY M.
MARCOTTE, WILLIAM J.
MARIK, RAYMOND L.
MARINO, RALPH
MARKS, WALTER S.
MARR, LEONARD V.
MARSHALL, OSCAR E.
MARSHALL, SETH
MARSHALL, TED
MARSHALL, DOUGLAS G.
MARTIN, LEVI R.
MARTIN, MARIA J.
MARTIN, THOMAS
MARTINEZ, GEORGE G.
MARTINEZ, RUTH
MARTINOVICH, THOMAS G.
MARX, FRANK C.
MASON, JACOB
MASOVCEVICH, ROBERT R.
MASTERS, REX G.
MASTIN, LORETTA
MASTIN, ROY
MATHIS, BILL
MATHIS, ROBERT E.
MATTEO, CHARLIE
MATTEO, GREGORY A.
MATTESON, MARTIN "SNAKE" P.
MATTHEWS, ROY A.
MAX, WILLIAM E.
MAYFIELD, MURRELL D.
MAYNE, ROBERT L.
MAYS, GRACE
MAZUR, BRENDON E.
MAZUR, BRIGETTE
MCABEE, GARLAND G.
MCALLISTER, MAURICE E.
MCANDREWS, CAROLYN
MCCANDLESS, ARCHIE G.
MCCAULLEY, GERALD E.
MCCAULLEY, KENNETH H.
MCCAULLEY, KENNETH H.
MCCLAIN, HENRY K.
MCCLAIN, JOSEPH R.
MCCLAIN, LOUIS W.
MCCLELLAN, HENRY G.
MCCLOSKEY, GERALD I.
MCCLOSKEY, WILLIAM J.
MCCOLLUM, MARGARET
MCCOY, JAMES F.
MCCUNE, HERBERT F.
MCDERMOTT, JOHN "JACK" A.
MCDONALD, FRED S
MCDONALD, WILLIAM J.
MCDONOUGH, EDWARD , .
MCDOWELL, CLYDE
MCENRUE, JAMES P.
MCENTEE, JOHN P.
MCGEE, ANNE
MCGILL, ROY , .
MCGLOTHLIN, BILLY
MCGOVERN, NEIL P.
MCGUINNESS, TOM P.
MCGUIRE, WILLIAM G.
MCINTOSH, LOUIS
MCKELVEY, BILLY J.
MCKINNEY, SAMUEL L.
MCLESKEY, BROADUS L.
MCLOUTH, ANN
MCMAHON, JOSEPH P.
MCMAHON, ROBERT L.
MCMILLAN, WILLIAM N.
MCNAMARA, LAWRENCE F.
MCNAMARA, ROBERT J.
MCNEVIN, GLORIA
MCPARLAND, ALPHIE
MCPHERSON, GEORGE C.
MCPHERSON, WILLIAM J.
MCQUIEN, RICHARD W.
MEDDERS, TRUMAN L.
MEDIEROS, FELIX L.
MEDRANO, RALPH
MEEHAN, MICKEY J.
MEEKS, CHARLES A.
MEEKS, ROY E.
MEGINNESS, KENNETH E.
MEHRLE, JOHN R.
MEISENHEIMER, ROBERT C.
MENA, JESUS
METCALF, GLENN W.
METHENY, DELBERT J.
METZLER, ROBERT
MEYERS, EDDIE M.
MICCICHE, JOE
MICHAELSON, MIKE
MICHOLA, IRENE
MIDDIONE, FARRAN
MIDKIFF, LON
MIKEZ, EDINA
MILARDO, VINCENT J.
MILES, BOBBY C.
MILES, RALPH G.
MILES, RICHARD
MILLER, FRANCES
MILLER, FRANK
MILLER, JACK L.
MILLER, JAMES
MILLER, MAMIE
MILLER, ROBERT R.
MILLER, SAMUEL C.
MILLER, WILLIAM G
MILLS, WILLIAM E.
MINKA, WALTER
MIOTKE, THOMAS F.
MITCHELL, JEROME L.
MITCHELL, JOSEPH M.
MITCHELL, MAX H.
MIZER JR, GLEN L.
MOCK, JAMES W.
MOLOCK, JOHN R.

MONJE, HECTOR
MONSARRAT, GEORGE F.
MONTE, JAMES L.
MONTGOMERY, KENNETH
MONTOYA, LOUIS C.
MOODY, ALFRED L.
MOON, MORRIS P.
MOONEY, GARY H.
MOORE, GENEVIEVE
MOORE, GORDON L.
MOORE, HARRY B.
MOORE, HILTON E.
MOORE, JAMES F.
MOORE, JOHN C.
MOORE, PHILLIP D.
MOORE, WILLIAM M.
MORDENTE, JOSEPH "JOE"D.
MORENO, LOUIS J.
MORGAN, GRADY
MORGAN, ROBERT W.
MORI, RIKIO
MORIARITY, MICHAEL G.
MORLEY, ADRIAN
MORRISON, ARTHUR M.
MORRISON, WILLIAM F.
MORRISSEY, MELVA
MORROW JR, JESSE L.
MORSE, BOBBIE
MORSE, CALVIN A.
MORTON, DOLORES
MOSELEY, CECIL R.
MOSES, RICHARD R.
MOSS, KENNETH N.
MOTT, RONALD H.
MOUNCE, MAC D.
MOXLEY, LUCIUS C.
MUIR, WILLIAM H.
MULAC, ANTHONY M.
MULLIN, JERRY F.
MULLINS, EMMETT
MUNK, WAYNE M.
MUNOZ, JOHN J.
MURDOCK, BERNARD E.
MURPHY, CHARLES W.
MURPHY, FREDERICK J.
MURPHY, JAMES M.
MURPHY, KENNETH J.
MURRAY, DONALD B.
MYERS, CHARLES E.
MYERS JR, LOYS F.
MYERS JR, PAUL F.
NAGLE, DENNIS F.
NAKAJO, MAS M.
NAYLOR JR, WILLIAM H.
NEAL, ALBERT S.
NEAL, SHIRLEY
NEAL, THOMAS L.
NEAULT, GERRARD A.
NEEL, MARTHA
NELSON, HAROLD M.
NELSON, ROBERT B.
NELSON, ROBERT W.
NELSON, WILLIAM S.
NETHERLAND, GEORGE E. JR
NETTLES, JOYCE
NETTLES, ROBERT A.
NEVSETA, RONALD G.
NEWMAN, RICHARD L.
NICE, JOSEPH W.
NICHOLS, BOBBY C.
NICHOLS, EARL L.
NICHOLS, EUGENE
NICHOLS JR, ALBERT E.
NICHOLSON, DARL
NICHOLSON, THOMAS W.
NICKELS, FRANK C.
NIEMI, KENNETH A.
NIGH, GLENN J.
NIMS, DONALD F.
NOLLAU SR, STEVAN
NORBURY, MICHAEL A.
NORRIS, GLORIA
NORRIS, JAMES E.
NOWICKI, EUGENE F.
NUNLEY, JOHN E.
NYKIEL, EDWIN C.
O'BRIEN, EDWARD
O'BRIEN, EDWARD
O'CONNOR, MICHAEL J.
O'DONNELL, PAT
O'KAIN, ROBERT P.
O'NEILL, HARRY B.
OATLEY, JOHN F.
OGBURN, HOWARD
OGLE JR, WILLIAM
OHS, KENNETH L.
OKON, MARILYN
OLINGER, WILLIAM E.
OLIVER, HOWARD C.
OLIVER, JOE
OLSON, DORAL D.
OLSON, HARVEY S.
OLSON, ROLLAND E.
OPP, JOANN
ORYSIEK, HENRY E.
OSBORN, SHERRY
OSBORNE, BETTY
OSTEEN JR, JOHN L.
OUIMETTE, CHARLES E.
OWEN, HAROLD D.
OZGA, EVELYN
PACCIO, DONALD W.
PAGANO, FRANK
PAHL, IRVING C.
PALMER, MICHAEL B.
PALMER, ROBERT A.
PALMER, ROBERT C.
PALMER JR, A. AUGUST
PALOMBI, ROCCO A.
PANEK, RICHARD M.
PAPPAS, JOHN
PARISH, RON
PARKER, ARTHUR C.
PARKER, EARL B.
PARKER, ROBERT L.
PARKER, ROBERT M.
PARKER, THOMAS P.
PARKS, JATTIE
PARNELL, GEORGE A
PARRISH, CHARLES W.
PARSONS, EARL
PASSINI, MILDRED
PATERSON SR, HORACE E.
PATTERSON, ALTON H.
PATTERSON, LAURENCE
PATTERSON, LLOYD I.
PATTERSON, SALDO A.
PAULDING, CRAIG
PAYNE, JOE A.
PAYNE JR, SAMUEL L.
PAZANOWSKI, BERNARD
PEASE, LAWRENCE H.
PEDLEY, RICHARD A.
PEERS, MARGARET
PEKAY, STANLEY F.
PELLON, FREDERICK
PENA, M.E.
PENNELL, SHIRLEY
PEPIN, JOHN L.
PERCIVAL, CLEVA (DAVIS)
PEREA, GILBERT
PERRY, ALVIS
PERRY, ROBERT M.
PETERSON, AXEL B.
PETERSON, CARMEN R.
PETERSON, LEE E.
PETROGEORGE, MICHAEL C.
PFIELSTICKER, RICHARD M.
PHILLIPS, ALBERT L.
PHILLIPS, BETTY JO
PHILLIPS, JOSEPH E.
PIEMONTESE, JOSEPH
PIERCE, PATRICIA
PIERSALL, MITSUKO
PIERSZA, ROBERT
PIETROWSKI, EDWARD F.
PIKE, EUDORSEY
PINCKNEY JR, MILES M.
PINEDO, AUGUSTINE R.
PIRES, GILBERT
PITTENGER, RICHARD A.
PITTMAN, BILLY D.
PITTMAN, RUFUS
PITTS, CHARLES L.
PLACKIS, JAMES J.
PLATER, JAMES M.
POGREBA, LAWRENCE J.
POLICICHIO, ANTHONY
POOLE, ELVERT E.
POOLE, FRANCIS K.
POPE, JAMES A.
PORTER, ROBERT E.
PORTERFIELD, WILLIAM A.
PORTILLA, JEAN
POSEY, EDWARD L.
POTEMPA, EDWARD J.
POTTER, HAROLD H.
POWELL, CHESTER F.
POWELL, JEANNE
POWELL, WILLIS G.
POWELL, THOMAS W.
PRATT, CARL R.
PRESTON, RICHARD B.
PREWITT, HAROLD C.
PRICE, WELDON R.
PRITTS, ROBERT K.
PRON, LOUIS
PRONIER, ROBERT A.
PROVOST, NORMAN J.
PRUITT, JAMES B.
PUCKETT, RALPH
PULLINS??, BETTE
PULLON, MELVIN F.
PURCELL, PATRICK J.
PUTNAM, SUSAN
PYBURN, ALLAN R.
QUEEN, JAMES C
QUIGLEY, ROBERT J.
RAEIN, WILLIAM N.
RAMEY, TRACY A.
RAMOS, VICTOR E.
RAMOS JR, GUADALUPE
RAMSEY, WILLIE D.
RANF, CARL V.
RANKINS, GEORGE
RASENS, ROBERT J.
RASH, WESLEY
RASMUSSEN, WAYNE
RASTOFER, LAWRENCE E.
RAVAN, DONALD W.
RAWLS, JOANNA
RAY, CHARLES A.
RAY, ERNEST A.
RAY, LILLIE
RAY, RITA
RAYMOND, RAWLIN R.
RAYMOND, REUBEN E.
RAZE, DALTON D.
REAMER, GLENN E.
REAMES, WILLIAM A.
REARDON, WILLIAM D.
REAVES, JACK
RECLA, JACK L.
REDDIES, HENRY W.
REDGATE, PAUL
REEDER, EDWARD M.
REEDY, JOAN
REESE, EARNEST
REESE, METTIE
REESE, RICHARD K.
REEVES, GROVER E.
REEVES, PAUL W.
REEVES, ROBERT R.
REGAN, DONALD B.
REGAN, MARY LOU
REGAN, THOMAS C.
REICHMAN, WALDON L.
REINHART, DONALD J.
REISCH, JERRY R.
REISS, HOWARD C.
REMER, FORREST H. (BO)
REMILLONG, ROLAND
RENZ, LEONARD W.
REPACK, JOSEPH
REVELS, GEINON , .
REYNOLDS, JACK
REYNOLDS, JAMES R.
REYNOLDS, JIMMY A.
RHATIGAN, WILLIAM "RIP" C.
RHOADES, RICHARD R.
RHODES, CLAIR L.
RHODES, HORACE C.
RHONE SR, RAY
RICHARDS, HAROLD A.
RICHARDS, LARRY K.
RICHESON, RAYMOND L.
RIDDLE, JACK
RIDENHOUR, CHARLES L.
RINARD, STEPHEN
RISH, EMANUEL P.
RITZ, FRANCES
RIVERA, DANIEL C.
RIVERA, E. CONRAD
ROBERTS, EDGAR , .
ROBERTS, PEGGY
ROBERTS JR, MARION A.
ROBERTSON, SNEAD
ROBINSON, DOUGLAS J.
ROBINSON, KENNETH A.
ROBLES, LUGARDO S.
ROCCO, RICHARD
RODARTE, GILBERT V.
RODRIGUES, JOSEPH A.
RODRIGUEZ, MANUEL A.
ROGERS, EUGENE
ROGERS, SARA
ROGERS, MONTELL M.
ROMANIO, MICHAEL M.
ROSE, JAMES JR
ROSEGRANT, EDWARD J.
ROSEN, GRACE
ROSS, CHARLES G.
ROST, ELMER L.
ROY, JOHN W.
RUSSELL, FRANKLIN , .
RUSSLER, LEE W.
RUSSO, JOSEPH
RYAN, BILL
RYAN, JOHN P.
SAFRAN, ROBERT J.
SAITES, LOUIS N.
SAJER, GERALD , .
SALAS, MARY LOU
SALO, JAMES R.
SALZMAN, ROBERT P.
SANCHEZ, MIQUEL J.
SANDERS, ROBERT C.
SANDERSON, ROY D.
SANSALONE, GOLFREDO D.
SARAMA, ROBERT
SASSER, MARGARET
SAVAGE, ROY C.
SAVIO, AUGUST.
SAVOIE JR, ROY J.
SAXTON, CHARLES R.
SCAGNELLI, JOHN C.
SCARBORO, DOROTHY
SCEPANIAK, JAMES G.
SCHLOSS, IRWIN
SCHLOSSER, TED
SCHNELL, MERLYN W.
SCHOCK, GEORGE R.
SCHRAGE, URBAN A.
SCHROEDER, EDWIN J.
SCHROEDER, GEORGE E.
SCHROEDER, HARRY J.
SCHULTZ, ANSEL C.
SCHULTZ, THEODORE M.
SCHUSTEFF, MARVIN
SCHUSTEFF,, ROBERT
SCOTT, ADRIAN V.
SCOTT, JACK I.
SCOTT, REV BOB D.
SCULLY, ROBERT
SEAMAN, WILLIAM E.
SEAVEY, RICHARD R.
SEDGEWICK, VERA
SEIBERT, ELMER O.
SETTLES, RICHARD E.
SHADE, WILLIAM
SHAFER, JACK
SHAFFER, GERALD H.
SHAHAN, VERNON
SHARE, PETER O.
SHARP, JAMES H.
SHARP, WAYNE E.
SHAW, CLYDIE
SHAW, EUGENE D.
SHAW, WALLACE
SHAW, WARREN E.
SHEA, GAIL
SHELLEY, ROBERT A.
SHELLEY, ROBERT A.
SHELTON, BILLY W.
SHEMITZ, JOAN
SHERRY, EDWARD G.
SHEVLIN, FRANCIS A.
SHIGEMOTO, FUYIKO
SHINE, PAUL
SHINPAUGH, GRADY C.
SHIPMAN, CLARENCE E.
SHIRES, JAMES C.
SHIRLEY, JERRY L.
SHOEMAKER, CLYDE B.
SHORE, NED J.
SHORT, RAYMOND
SHORTS JR, CLIFFORD J.
SHUGRUE, DANIEL F.
SHULER, ELLEN
SHULER, HAROLD
SHUSTEFF, HENRY
SHUSTER, MANUEL A.
SHUTES, MALCOM D.
SHUTT, LADDIE D.
SILAS JR, AUSTIN
SILVERIO, ANTHONY J.
SILVERS, MARJORIE
SIMMONS, DONALD C.
SIMMONS, RALPH E.
SIMMS, BENJAMIN
SIMMS, , . JOSEPH
SIMON, STEVE R.
SIMPSON, MERLE W.
SIMPSON, MILBURN D.
SIMS, , .J.
SINGLAUB, JOHN K.
SINOR, BILLY G.
SISK, PATRICK H.
SISK JR, MARTIN J.
SKELTON, JOHNNY
SKERKOWSKI, WILLIAM L.
SKIVER, ERLINDA
SKOIEN, CLARENCE
SLAUGHTER, DELBERT W.
SLAUGHTER, JERRY D.
SLAVINSKI, JOSEPH P.
SLIGHTER, VERNEA.
SLOAN, GEORGE R.
SLOAN, WILLIAM C.
SLOTA, PATRICIA A.
SLOTTERBACH, WARREN
SMALL JR, WHEELER S.
SMART, JOHN R.
SMATHERS, JOHN M.
SMITH, BERYLE
SMITH, BIRGIT
SMITH, CHARLES E.
SMITH, DAVID
SMITH, DONALD V.
SMITH, JACQUELINE
SMITH, JAMES J.
SMITH, LEO B.
SMITH, MIKIKO
SMITH, ROBERT L.
SMITH, RONALD F.
SMITH, SABATINA
SMITH, EDWARD D.
SMITH JR, LESLIE A.
SMITH SR, EDWARD M.
SMYTH, ROY E.
SNEVA, ROBERT L.
SNIPES, JOHN M.
SNODGRASS, RUBY
SNYDER, DON
SNYDER, ELEANOR
SNYDER, JAMES E.
SODENKAMP, CAROL
SONNIER, WHITNEY J.
SOTUYO, GILBERT
SOUNIA, DONALD L.
SPALDING, ROBERT L.
SPARKO, MARKO R.
SPARKS, JAMES C.
SPEARS, GEORGE
SPEGAL, EDDIE W.
SPENCER, JOE , .
SPENCER, ODIS H
SPEYERER, EDWIN M.
SPICER, MARION A.
SPIESS, JAMES C.
SPRAGINS, CHARLES "PETE" E.
SPRATLIN, ROBERT J.
ST JOHN JR, JAMES
ST MARTIN, JUDE P.
STAMPER, JAMES M.
STANEK, BEVERLEY
STANFORD, ALTON
STANKICH, MICHAEL M.
STANTURF, JACK A.
STARR, MOSES JR
STAUDE, RICHARD
STEELE, CANNON R.
STEELE, DEAN L.
STEELE, PAT
STELLATO, TONY P.
STEPHENS, JERRY R.
STEVENS, CARL E.
STEVENS, THOMAS N.
STEVENSON, WALLACE L.
STEWART, JAMES L.
STEWART, SAMTHIA
STEWART JR, RICHARD "DICK" D.
STILES, DICK M.
STINNARD, ELISE
STIVER, ROBERT A.
STOKES, JOSEPH J.
STONEBRAKER, JOSEPH D.
STOUT JR, EDWARD , .
STRAND, RICHARD S.
STREET, JEN
STRICKER, JAMES F.
STRICKLAND, SAM O.
STRICKLER, HUGH
STROCK, STANLEY
STROM, ANTHONY P.
STROMAN, FRED L.
STRONG, BERKLEY J.
STUART, ROBERT M.
SULLENGER, OVERTON G.
SULLIVAN, JOHN W.
SULLIVAN, VIRGIL
SULLIVAN, ROBERT J.
SUMMERS, JOHN A.
SURFACE, MARTIN L.
SUTHERLAND, LESTER A.
SUTHERLAND, WILLIAM Y.
SUTHERLAND JR, JAMES M.
SWAILS, PALMER L.
SWANK, MARCEL G.
SWEIGART, DONALD G.
SWINGLE, RONALD N.
SWINNEY, ROBERT R.
SYLVA, EDMUND K.
SZCZEPANIAK, KAZIMIERZ R.
TABATA, SADAMU
TAKACH, FRANK A
TANONA, CHESTER P.
TARR, SHERMAN E.
TASSEY, GEORGE
TATE, BILLY
TATE, LAWRENCE J.
TAUFER, DONALD P.
TAUFER, RONALD L.
TAYLOR, DAVID , .
TAYLOR, EVERETT R.
TAYLOR, JAMES
TAYLOR, JOHN B.
TAYLOR, ROBERT D.
TAYLOR, ROYCE E.
TAYLOR JR, NESBY
TEAGUE, BURRELL L.
TEAGUE, JAMES H.
TEBO, CLAUDE
TENER, RICHARD L.
TERANISH, YOSHIO
TERRY, O'NEILL
TEUBER, RONALD W.
THATCHER, IRWIN "CURLY" F.
THERENS, EDWARD J.
THIBODEAUX, IMOGENE
THOMAS, GEORGE A.
THOMAS, HARLEY"DUKE" J.
THOMAS, ROBERT R.
THOMAS, WILLIAM A.
THOMPSON, JACKIE E.
THOMPSON, MARTIN W.
THOMPSON, ROBERT J.
THOMPSON, STANLEY
THOMPSON JR, RALPH A.
THRASHER, BILLY J.
TIDWELL, JESSE C.
TIEMEYER, WILLIAM P.
TIMMONS, JOHN M.
TINGLER, HENRY
TISAK, NICK
TOBIN, JOHN L.
TOCK, ROBERT H.
TODD, BOBBY G.
TOMEI, ALBERT C.
TOMKIEWICZ, LEO
TOMLINSON, JOHN H.
TONER, THOMAS E.
TONNIGES, TONY
TORRES, PETER F.
TOTH, JOHN A.
TOVES, ALFRED B.
TRACY, EARL A.
TRAINOR, DONALD E.
TRAVERS, CHARLES E.
TRAXLER, CLIFF
TRAYNOR, DONALD L.
TREMAINE, DUNCAN R.
TRIBBLE, RICHARD G.
TROCHE JR, VINCENT
TSHAKA, DAVID , .
TUCKER, ANDREW
TUCKER, NORMA
TUCKER, ORRIE
TUCKER, ROBERT A.
TUEPKER, NORMAN O.
TURLEY, ALONZO E.
TURNER, ALLAN P.
TURNER, ARMAND K.
TURNER, BOBBY J.
TURNER, MARION P.
TWIGG, DAVID E.
TWILLEY, DR ROBERT F.
TWINING, JOHN P.
ULATOSKI, JOSEPH "JOE" R.
ULIASZ, JOHN , .
ULRICH, GEORGE P.
UNDERWOOD, RICHARD E.
URBANO, JOSEPH A.
VAILS, ROBERT A.
VALDERAMA, ALFRED C.
VALERY, CLEVELAND
VALVERI, JOHN M.
VAN ALSTINE, KENNETH B.
VAN METER, DOTTIE
VANAMAN, ROBERT "JACK" J.
VAPORIS, PIERRE R.
VARELA, HENRY
VARNELL, WILLIAM E.
VASQUEZ, DOMINICK J.
VAUGHAN, BILLY D.
VECCHIO, DOMINIC , .
VERHAGEN, GERALD L.
VERONESI, NORMAN R.
VESY, RICHARD J.
VICK, WALTER
VICTOR, JULIUS
VIDRINE, GEORGE A.
VILLA, LEWIS "LEW" J.
VOLTON, FRANK M.
VON WINDEGUTH, DONALD L.
VOSS, GORDON M.
VOYTKO, THI-KIM
WADSWORTH, CHARLES G.
WAGAMAN, CHARLES R.
WAGNER, LARRY
WAINWRIGHT, LEONARD L.
WAKEMAN, ROBERT B.
WALDECKER, JOHN F.
WALDROP, WINFRED
WALKER, BILLY D.
WALKER, CARROLL K.
WALKER, WINETTA
WALKER, SHARON
WALL, LEO W.
WALLS, BILLY G.
WALSH, JOHN F.
WALSH, LEWIS G.

WALSH, PATRICK J.
WALSTON, DAVE
WALTERS, GENE A.
WARD, ALLEN P.
WARD, NEAL F.
WARD, THOMAS J.
WARD, TOM A.
WARREN, JAMES E.
WARREN, ROBERT F.
WATERS, CHARLES "CHUCK"
WATERS JR, ALLEN
WATSON JR, WILLIAM
WATTS JR, JOSEPH C.
WAY, JAMES H.
WEATHERSBEE, WILLIAM
WEAVER JR, LAURY K.
WEBB, CARL O.
WEBER, ELDRED E.
WEBSTER JR, STANLEY L.
WEGENER, CARL J.
WEHAGE, VERNON C.
WEINBERG, THOMAS H.
WELCH, CHARLES E.
WELLS, MILDRED
WELLS JR, CHARLES L.
WELSER, LAWRENCE M.
WELSH, TEDD A.
WELTON, WALTER B.
WELTON, WALTER B.
WEST, RAMON
WESTENHEFFER, ROBERT L.
WESTMORELAND, LONER B.
WETZEL, PERRY H.
WETZEL, RICHARD S.
WHEAT, CHARLES "ZEB" B.
WHEELER, JANETTE
WHEELER, JOHN A.
WHITE, MARGARET
WHITE JR, WALTER
WHITEHEAD, CHARLIE
WHITEHEAD, WILLIAM
WHITFIELD, ANNIE
WHYTE, ROBERT M.
WIGGINS, EARL
WIGGINS, LEONARD W.
WILBURN, VINCENT
WILCOXSON, LORNA J.
WILHELMSON, LARRY A.
WILKINS, JOSEPH E.
WILKINSON, CALVIN M.
WILLARD, ORIN L.
WILLEY, ANDRE
WILLEY, ANDRE
WILLIAMS, CHRISTOPHER R.
WILLIAMS, FINA P.
WILLIAMS, GERALD E.
WILLIAMS, IRVING L.
WILLIAMS, JACK C.
WILLIAMS, LARRY M.
WILLIAMS, ROBERT H.
WILLIAMS, ROBERT P.
WILLIAMS, WILLIAM
WILLIAMS, WILLIAM L.
WILLIAMS, JOHN A.
WILLIS, OLAN F.
WILSON, DEVERN E.
WILSON, DONALD N.
WILSON, HENRY
WILSON, KENNETH R.
WILSON, LEON A.
WILSON, WILLIAM R.
WINDER, CHARLES R.
WINTERS, HARRY M.
WISE, DONALD K.
WISNIESKI, WILLIAM
WISNIEWSKI, RICHARD W.
WISPERT, WILLIAM J.
WITCZAK, HENRY
WITT, JAMES H.
WOELFUL, LOUIS H.
WOFFORD, E.G.
WOLFE JR, CHESTER A.
WOMACK, GEORGIA E.
WOMACK, LANCE
WOOD, CALVIN E.
WOOD, KENNARD G.
WOOD JR, MANUAL A.
WOODS, HARVEY J.
WOODWORTH, ROY
WOOLERY, LLOYD
WOOLRIDGE, JIMMY
WORDEN, LEONARD R.
WORMAN, BILL E.
WRIGHT, HAROLD B.
WRIGHT, J. WARD
WYATT, NEAL H.
WYNNE, DAVID L.
WYSINGLE, PRESTON
YBANEZ, LOREN D.
YBARRA, DANIEL E.
YBARRA, RAYMOND JR
YBARRA, TONY S.
YEAGER, PETE
YESTER, JOSEPH A.
YORK, CURTIS W.
YOUNCE, HAROLD D.
YOUNG, HAROLD E.
YOUNG, VIRGINIA
YOVICH JR, STANKO
YOWELL, JOANNE
ZALETA, ARTHUR
ZAMBOS, THEODORE J.
ZEIGLER, ARTHUR
ZENISKY, LEON
ZIMMERMAN, BOBBY L.
ZOOK, HARVEY J.
ZUCHNIK, EDWARD L.
ZWICKL, JOSEPH E.

Rangers Index

Places Index

Ranger patch authorized for wear during World War II. Rangers refused to wear it.

1st Ranger Inf. Co. attached to the 2nd Inf. Div.

4th Ranger Inf. Co. attached to the 1st Cav. Div.

2nd Ranger Inf. Co. attached to the 7th Inf. Div.

5th Ranger Inf. Co. attached to the 25th Inf. Div.

3rd Ranger Inf. Co. attached to the 3rd Inf. Div.

6th Ranger Inf. Co. attached to the 1st Inf. Div.